ELEMENTARY NUMERICAL ANALYSIS

ELEMENTARY NUMERICAL ANALYSIS

Third Edition

KENDALL ATKINSON
WEIMIN HAN
University of Iowa

John Wiley & Sons, Inc.

ASSOCIATE PUBLISHER	Laurie Rosatone
SENIOR MARKETING MANAGER	Julie Z. Lindstrom
ASSISTANT EDITOR	Jennifer Battista
PROGRAM ASSISTANT	Stacy French
SENIOR PRODUCTION EDITOR	Ken Santor
SENIOR DESIGNER	Dawn Stanley
ILLUSTRATION EDITOR	Sandra Rigby

This book was set in LaTeX by Techsetters Inc. and printed and bound by R. R. Donnelley Crawfordsville. The cover was printed by Phoenix Color Corporation.

This book is printed on acid free paper. ∞

Library of Congress Cataloging-in-Publication Data

Atkinson, Kendall E.
Elementary numerical analysis / Kendall Atkinson and Weimin Han.–3rd ed.
p. cm.
Includes bibliographical references and index.
ISBN 0-471-43337-3
1. Numerical analysis. I. Han, Weimin. II. Title.

QA297.A83 2004
519.4–dc21

2003053836

To our children
Elizabeth and Kathryn
Elizabeth and Michael

PREFACE

This book provides an introduction to numerical analysis and is intended to be used by undergraduates in the sciences, mathematics, and engineering. The main prerequisite is a one-year course in the calculus of functions of one variable, but some familiarity with computers is also needed. With this background, the book can be used for a sophomore-level course in numerical analysis. The last four chapters of the book present numerical methods for linear algebra, ordinary differential equations, and partial differential equations. A background in these subjects would be helpful, but these chapters include the necessary introduction to the theory of these subjects.

Students taking a course in numerical analysis do so for a variety of reasons. Some will need it in studying other subjects, in research, or in their careers. Others will be taking it to broaden their knowledge of computing. When we teach this course, we have several objectives for the students. First, they should obtain an intuitive and working understanding of some numerical methods for the basic problems of numerical analysis (as specified by the chapter headings). Second, they should gain some appreciation of the concept of error and of the need to analyze and predict it. And third, they should develop some experience in the implementation of numerical methods by using a computer. This should include an appreciation of computer arithmetic and its effects.

The book covers most of the standard topics in a numerical analysis course, and it also explores some of the main underlying themes of the subject. Among these are the approximation of problems by simpler problems, the construction of algorithms, iteration methods, error analysis, stability, asymptotic error formulas, and the effects of machine arithmetic. Because of the level of the course, emphasis has been placed on obtaining an intuitive understanding of both the problem at hand and the numerical methods being used to solve it. The examples have been carefully chosen to develop this understanding, not just to illustrate an algorithm. Proofs are included only where they are sufficiently simple and where they add to an intuitive understanding of the result.

For the computer programming, the preferred language for introductory courses in numerical analysis is MATLAB. A short introduction is given in Appendix D; and the programs in the text serve as further examples. The students are encouraged to modify these programs and to use them as models for writing their own MATLAB programs. When the authors teach this course, we also provide links to other sites that have online MATLAB tutorials.

The MATLAB programs are included for several reasons. First, they illustrate the construction of algorithms. Second, they save the students time by avoiding the need to write many programs, allowing them to spend more time on experimentally learning about a numerical method. After all, the main focus of the course should be numerical analysis, not learning how to program. Third, the programs provide examples of the language MATLAB and of reasonably good programming practices when using it. Of course, the students should write some programs of their own. Some of these can be simple modifications of the included programs; for example, modifying the trapezoidal rule integration code to obtain one for the midpoint rule. Other programs should be more substantial and original. All of the codes in the book, and others, are available from the book's Website at John Wiley, at

http://www.wiley.com/college/atkinson

The authors also maintain a site for these codes and other course materials, at

http://www.math.uiowa.edu/ftp/atkinson/ENA_Materials

In addition to the MATLAB programs in the text, the authors are experimenting with graphical user interfaces (GUIs) to help students explore various topics using only menus, query windows, and pushbuttons. Several of these have been written, including some to explore the creation and analysis of Taylor polynomial approximations, rootfinding, polynomial interpolation with both uniformly spaced nodes and Chebyshev nodes, and numerical integration. The GUIs are written using the MATLAB GUI development environment, and they must be run from within MATLAB. These GUIs are available from the authors' Website given above, and the authors are interested in feedback from instructors, students, and other people using the GUIs.

There are exercises at the end of each section in the book. These are of several types. Some exercises provide additional illustrations of the theoretical results given in the section, and many of these exercises can be done with either a hand calculator or with a simple computer program. Other exercises are for the purpose of further exploring the theoretical material of the section, perhaps to develop some additional theoretical results. In some sections, exercises are given that require more substantial programs; many of these exercises can be done in conjunction with package programs like those discussed in Appendix C.

The third edition of this book contains a new chapter and two new sections, and the book has been reorganized when compared to the second edition. The section on computer arithmetic has been rewritten and it now concentrates on the IEEE floating-point format for representing numbers in computers; the section on binary arithmetic has been moved to the new Appendix E. The new sections are Section 4.7 on the least squares approximation of functions (including an introduction to Legendre polynomials), and Section 8.8 on the two-point boundary value problem. The new Chapter 9 is on numerical methods for the classic second order linear partial differential equations in two variables. In addition, a number of other parts of the text have been rewritten, and examples and many new problems have been added.

In teaching a one-semester course from this textbook, the authors usually cover much of Chapters 1-6 and 8. The linear algebra material of Chapter 6 can be introduced at any point, although the authors generally leave it to the second half of the course. The material on polynomial interpolation in Chapter 4 will be needed before covering Chapters 5 and 8. The textbook contains more than enough material for a one-semester course, and an instructor has considerable leeway in choosing what to omit.

We thank our colleagues at the University of Iowa for their comments on the text. We also thank the reviewers of the manuscript for their many suggestions, which were very helpful in preparing the third edition of the book. We thank Cymie Wehr for having done an excellent job of creating a LaTeX version of the second edition. It was used in preparing this third edition and it saved us a great deal of time and effort. We thank Brian Treadway for his invaluable assistance in helping us to navigate LaTeX. The staff of John Wiley have been supportive and helpful in this project, and we are grateful to them.

Kendall E. Atkinson
Weimin Han
Iowa City, Iowa
May 2003

CONTENTS

CHAPTER 1

TAYLOR POLYNOMIALS

Numerical analysis uses results and methods from many areas of mathematics, particularly those of calculus and linear algebra. In this chapter we consider a very useful tool from calculus, Taylor's theorem. This will be needed for both the development and understanding of many of the numerical methods discussed in this book.

The first section introduces Taylor polynomials as a way to evaluate other functions approximately; and the second section gives a precise formula, Taylor's theorem, for the error in these polynomial approximations. In the final section, we first discuss how to evaluate polynomials, and then we derive and analyze a computable polynomial approximation to a special function.

Other material from algebra and calculus is given in the appendices. Appendix A contains a review of mean-value theorems, and Appendix B reviews other results from calculus, algebra, geometry, and trigonometry.

There are several computer languages that are used to write programs to implement the numerical methods we study in this text. Among the most important basic-level computer languages are *C, C++, Java,* and *Fortran.* In this text we use a higher-level language that allows us to manipulate more easily the mathematical structures we need in implementing numerical analysis procedures for solving mathematical problems. The language is MATLAB, and it is widely available on all lines of computers. We will provide examples of the use of MATLAB throughout this text, and we encourage students to work with these programs. Experiment using them and modify them for solving related tasks. A very brief introduction to MATLAB is given in Appendix D, and we also reference more extensive introductions given elsewhere.

1.1. THE TAYLOR POLYNOMIAL

Most functions $f(x)$ that occur in mathematics cannot be evaluated exactly in any simple way. For example, consider evaluating $f(x) = \cos x$, e^x, or $\sqrt{x}$, without using a calculator or computer. To evaluate such expressions, we use functions $\hat{f}(x)$ that are almost equal to $f(x)$ and are easier to evaluate. The most common classes of approximating functions $\hat{f}(x)$ are the polynomials. They are easy to work with and they are usually an efficient means of approximating $f(x)$. A related form of function is the piecewise polynomial function, and it also is widely used in applications; it is studied in Section 4.3 of Chapter 4.

Among polynomials, the most widely used is the Taylor polynomial. There are other more efficient approximating polynomials in the context of particular applications, and we study some of them in Chapter 4. The Taylor polynomial is comparatively easy to construct, and it is often a first step in obtaining more efficient approximations. The Taylor polynomial is also important in several other areas of mathematics.

Let $f(x)$ denote a given function, for example, e^x, $\sin x$, or $\log(x)$. [In this book, $\log(x)$ always means the natural logarithm of x, *not* the logarithm of x to the base 10.] The Taylor polynomial is constructed to mimic the behavior of $f(x)$ at some point $x = a$. As a result, it will be nearly equal to $f(x)$ at points x near a.

To be more specific, as an example, find a linear polynomial $p_1(x)$ for which

$$\begin{aligned} p_1(a) &= f(a) \\ p_1'(a) &= f'(a) \end{aligned} \tag{1.1}$$

Then, it is easy to verify that the polynomial is uniquely given by

$$p_1(x) = f(a) + (x - a)f'(a) \tag{1.2}$$

The graph of $y = p_1(x)$ is tangent to that of $y = f(x)$ at $x = a$.

Example 1.1.1 Let $f(x) = e^x$ and $a = 0$. Then

$$p_1(x) = 1 + x$$

The graphs of f and p_1 are given in Figure 1.1. Note that $p_1(x)$ is approximately e^x when x is near 0. ■

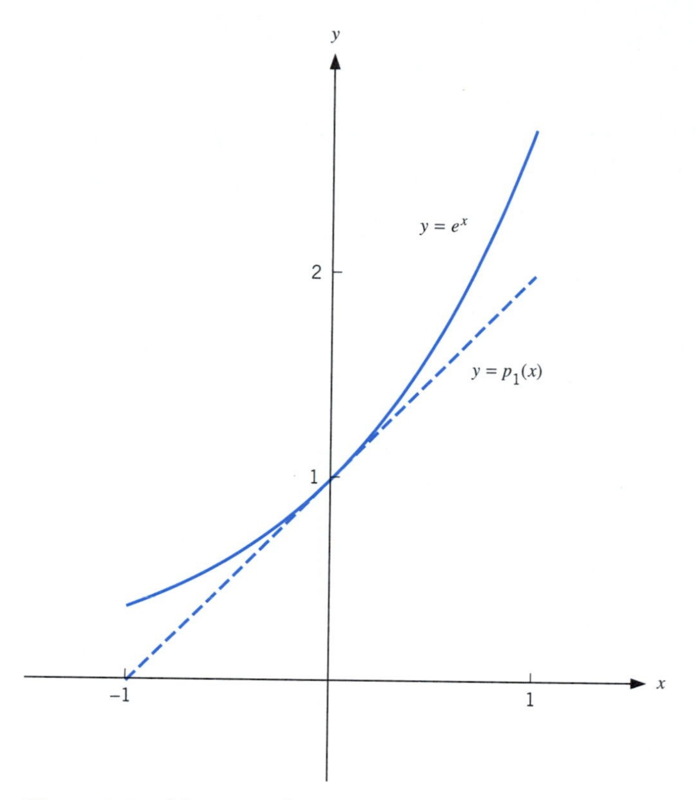

Figure 1.1. Linear Taylor approximation

To continue the construction process, consider finding a quadratic polynomial $p_2(x)$ that approximates $f(x)$ near $x = a$. Since there are three coefficients in the formula of a quadratic polynomial, say,

$$p_2(x) = b_0 + b_1 x + b_2 x^2$$

it is natural to impose three conditions on $p_2(x)$ to determine them. To better mimic the behavior of $f(x)$ at $x = a$, we require

$$\begin{aligned} p_2(a) &= f(a) \\ p_2'(a) &= f'(a) \\ p_2''(a) &= f''(a) \end{aligned} \tag{1.3}$$

It can be checked that these are satisfied by the formula

$$p_2(x) = f(a) + (x - a)f'(a) + \tfrac{1}{2}(x - a)^2 f''(a) \tag{1.4}$$

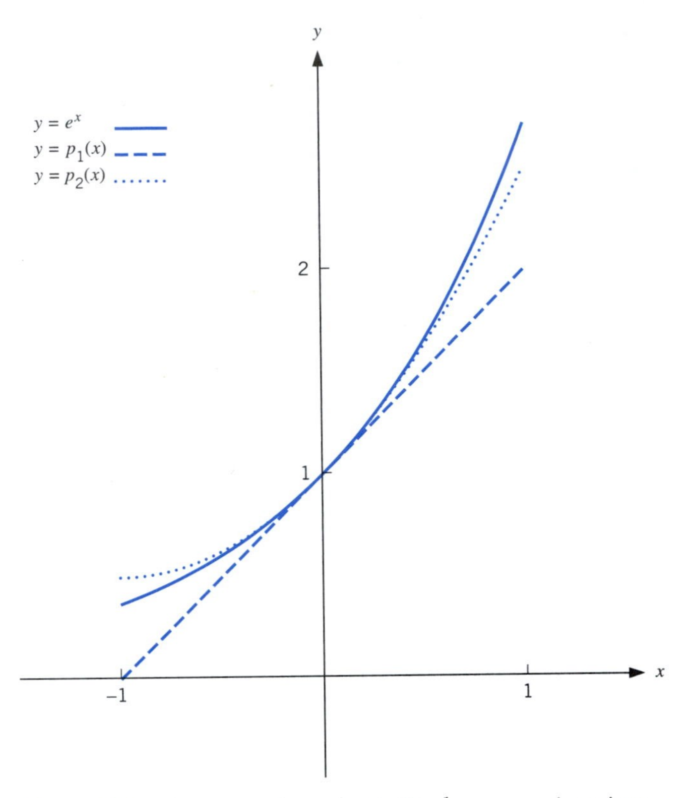

Figure 1.2. Linear and quadratic Taylor approximations

Example 1.1.2 Continuing the previous example of $f(x) = e^x$ and $a = 0$, we have

$$p_2(x) = 1 + x + \tfrac{1}{2}x^2$$

See Figure 1.2 for a comparison with e^x and $p_1(x)$. ■

We can continue this process of mimicking the behavior of $f(x)$ at $x = a$. Let $p_n(x)$ be a polynomial of degree n, and require it to satisfy

$$p_n^{(j)}(a) = f^{(j)}(a), \qquad j = 0, 1, \ldots, n \tag{1.5}$$

where $f^{(j)}(x)$ is the order j derivative of $f(x)$. Then

$$\begin{aligned} p_n(x) &= f(a) + (x-a)f'(a) + \frac{(x-a)^2}{2!}f''(a) + \cdots + \frac{(x-a)^n}{n!}f^{(n)}(a) \\ &= \sum_{j=0}^{n} \frac{(x-a)^j}{j!}f^{(j)}(a) \end{aligned} \tag{1.6}$$

Table 1.1. Taylor approximations to e^x

x	$p_1(x)$	$p_2(x)$	$p_3(x)$	e^x
−1.0	0	0.500	0.33333	0.36788
−0.5	0.5	0.625	0.60417	0.60653
−0.1	0.9	0.905	0.90483	0.90484
0	1.0	1.000	1.00000	1.00000
0.1	1.1	1.105	1.10517	1.10517
0.5	1.5	1.625	1.64583	1.64872
1.0	2.0	2.500	2.66667	2.71828

In the formula, $f^{(0)}(a) = f(a)$, and

$$j! = \begin{cases} 1, & j = 0 \\ j(j-1)\cdots(2)(1), & j = 1, 2, 3, 4, \ldots \end{cases}$$

which is called "j factorial." If in (1.6) we need to reference explicitly the dependence on the point a, we write $p_n(x; a)$. The polynomial $p_n(x)$ in (1.6) is called the *Taylor polynomial of degree n* for the function $f(x)$ and the point of approximation a. [Note, however, that the polynomial $p_n(x)$ has actual degree less than n if $f^{(n)}(a) = 0$.]

Example 1.1.3 Again let $f(x) = e^x$ and $a = 0$. Then

$$f^{(j)}(x) = e^x, \qquad f^{(j)}(0) = 1, \qquad \text{for all } j \geq 0$$

Thus

$$p_n(x) = 1 + x + \frac{1}{2!}x^2 + \cdots + \frac{1}{n!}x^n = \sum_{j=0}^{n} \frac{x^j}{j!} \tag{1.7}$$

Table 1.1 contains values of $p_1(x)$, $p_2(x)$, $p_3(x)$, and e^x at various values of x in $[-1, 1]$. For a fixed x, the accuracy improves as the degree n increases. And for a polynomial of fixed degree, the accuracy decreases as x moves away from $a = 0$. ■

Example 1.1.4 Let $f(x) = e^x$ and let the point a be an arbitrary point, not necessarily zero. Then

$$f^{(j)}(x) = e^x, \qquad f^{(j)}(a) = e^a, \qquad \text{for all } j \geq 0$$

We obtain the formula

$$p_n(x; a) = e^a \left[1 + (x-a) + \frac{1}{2!}(x-a)^2 + \cdots + \frac{1}{n!}(x-a)^n \right] = e^a \sum_{j=0}^{n} \frac{(x-a)^j}{j!}$$

For instance,

$$p_n(x;1) = e^1 \sum_{j=0}^{n} \frac{(x-1)^j}{j!} \tag{1.8}$$

The polynomial $p_n(x;1)$ is most accurate for $x \approx 1$, and the polynomial $p_n(x;0)$ [given in (1.7)] is most accurate for $x \approx 0$. As a problem, compare the accuracy of $p_n(x;0)$ and $p_n(x;1)$ on the interval $[-1, 2]$ for various values of n (cf. Problem 8). ■

Notation In this text, we use two symbols that mean "approximately equals." The symbol "$\approx$" is generally used with symbols. For example,

$$x \approx 5$$

means that x is around 5; and

$$e^x \approx 1 + x, \qquad x \approx 0$$

means e^x is approximately $1 + x$ when x is around zero. The symbol "$\doteq$" is generally used with numbers, as in

$$2\pi \doteq 6.2832$$
$$\sqrt{168} \doteq 12.961$$

The symbol "$\doteq$" is usually used with actual calculational error. We attempt to be consistent in this usage, but sometimes it is not clear which symbol should be used. ■

Example 1.1.5 Let $f(x) = \log(x)$ and $a = 1$. Then $f(1) = \log(1) = 0$. By induction, for $j \geq 1$,

$$f^{(j)}(x) = (-1)^{j-1}(j-1)!\frac{1}{x^j}$$
$$f^{(j)}(1) = (-1)^{j-1}(j-1)!$$

If this is used in (1.6), the Taylor polynomial is given by

$$\begin{aligned} p_n(x) &= (x-1) - \frac{1}{2}(x-1)^2 + \frac{1}{3}(x-1)^3 - \cdots + (-1)^{n-1}\frac{1}{n}(x-1)^n \\ &= \sum_{j=1}^{n} \frac{(-1)^{j-1}}{j}(x-1)^j \end{aligned} \tag{1.9}$$

See Figure 1.3 for graphs of $\log(x)$, $p_1(x)$, $p_2(x)$, and $p_3(x)$. ■

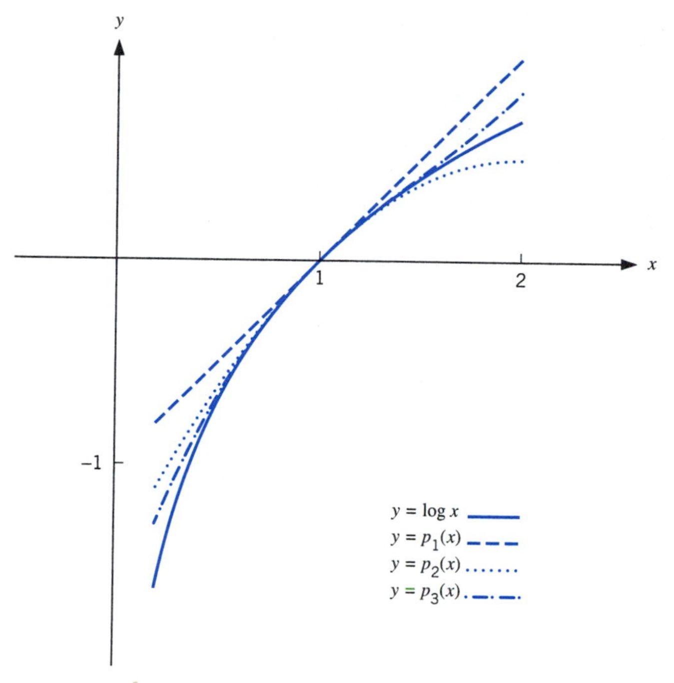

Figure 1.3. Taylor approximations of $\log(x)$ about $x = 1$

Throughout this text we will state a few general observations or rules to use when considering the numerical analysis of mathematical problems.

> GENERAL OBSERVATION:
> When considering the solution of a mathematical problem for which no direct method of solution is known, replace it with a " nearby problem" for which a solution can be computed. (1.10)

In the current situation, we are replacing the evaluation of a function such as e^x with the evaluation of a polynomial.

MATLAB PROGRAM: *Evaluating a Taylor polynomial.* Following is a MATLAB program that will calculate several Taylor polynomial approximations to e^x on an interval $[-b, b]$, with the value of b to be input into the program. The program will evaluate the Taylor polynomials of degrees 1, 2, 3, and 4 at selected points x in the interval $[-b, b]$, printing the errors in them in tabular form. The output will be directed to both the computer screen of the user and to a file `exp_taylor` for later printing.

```
% TITLE: Evaluate Taylor polynomials for exp(x) about x = 0
%
% This evaluates several Taylor polynomials and their errors
% for increasing degrees.  The particular function being
% approximated is exp(x) on [-b,b].

% Initialize
b = input('Give the number b defining the interval [-b,b] ');
h = b/10;
x = -b:h:b;
max_deg = 4;

% Produce the Taylor coefficients for the function exp(x) when
% expanded about the point a = 0.  The coefficients are stored
% in the array c, which will have length max_deg+1.

c = ones(max_deg+1,1);
fact = 1;
for i = 1:max_deg
  fact = i*fact;
  c(i+1) = 1/fact;
end

% Calculate the Taylor polynomials
p1 = polyeval(x,0,c,1);
p2 = polyeval(x,0,c,2);
p3 = polyeval(x,0,c,3);
p4 = polyeval(x,0,c,4);

% Calculate the errors in the Taylor polynomials
true = exp(x);
err1 = true-p1;
err2 = true-p2;
err3 = true-p3;
err4 = true-p4;

% Print the errors in tabular format
diary exp_taylor
disp('    x        exp(x)       err1       err2        err3       err4')
for i = 1:length(x)
  fprintf('%7.3f%10.3f%14.3e%14.3e%14.3e%14.3e\n',...
           x(i),true(i),err1(i),err2(i),err3(i),err4(i))
end
diary off
```

The program uses the following program, named polyeval, to evaluate polynomials. The method used in the program is discussed in Section 1.3.

```
function value = polyeval(x,alpha,coeff,n);
%
% function value = polyeval(x,alpha,coeff,n)
%
% Evaluate a Taylor polynomial at the points given in x, with
% alpha the point of expansion of the Taylor polynomial, and
% with n the degree of the polynomial.  The coefficients are to
% be given in coeff; and it is assumed there are n+1 entries in
% coeff with coeff(1) the constant term in the polynomial

value = coeff(n+1)*ones(size(x));
z = x-alpha;
for i = n:-1:1
  value = coeff(i) + z.*value;
end
```

PROBLEMS

1. Using (1.9), compare $\log(x)$ and its Taylor polynomials of degrees 1, 2, and 3 in the manner of Table 1.1. Do this on the interval $\left[\frac{1}{2}, \frac{3}{2}\right]$.

2. Produce the linear and quadratic Taylor polynomials for the following cases. Graph the function and these Taylor polynomials.

   (a) $f(x) = \sqrt{x}$, $a = 1$ **(b)** $f(x) = \sin(x)$, $a = \pi/4$

   (c) $f(x) = e^{\cos(x)}$, $a = 0$ **(d)** $f(x) = \log(1 + e^x)$, $a = 0$

3. Produce a general formula for the degree n Taylor polynomials for the following functions, all using $a = 0$ as the point of approximation.

   (a) $1/(1 - x)$ **(b)** $\sin(x)$ **(c)** $\sqrt{1+x}$

   (d) $\cos(x)$ **(e)** $(1 + x)^{1/3}$

4. Does $f(x) = \sqrt[3]{x}$ have a Taylor polynomial approximation of degree 1 based on expanding about $x = 0$? $x = 1$? Explain and justify your answers.

5. Use the Taylor polynomials of degrees 1, 2, and 3 for the function $f(x) = \sqrt{1+x}$, obtained in Problem 3(c), to compute approxmimate values of $\sqrt{0.9}$, $\sqrt{1.1}$, $\sqrt{1.5}$, $\sqrt{2.0}$ and compare them with the exact values to 8 or more digits.

6. Repeat Problem 5, but with $f(x) = \sin x$ and $x = 0.01,\ 0.1,\ 0.5,\ 1.0$.

7. Compare $f(x) = \sin(x)$ with its Taylor polynomials of degrees 1, 3, and 5 on the interval $-\pi/2 \le x \le \pi/2$; $a = 0$. Produce a table in the manner of Table 1.1.

8. Let $f(x) = e^x$; recall the formulas (1.7) and (1.8) for $p_n(x; 0)$ and $p_n(x; 1)$, respectively. Compare $p_n(x; 0)$ and $p_n(x; 1)$ to e^x on the interval $[-1, 2]$ for $n = 1, 2, 3$. Discuss your results.

9. **(a)** Produce the Taylor polynomials of degrees 1, 2, 3, and 4 for $f(x) = e^{-x}$, with $a = 0$ the point of approximation.

(b) Using the Taylor polynomials for e^t, substitute $t = -x$ to obtain polynomial approximations for e^{-x}. Compare with the results in (a).

10. **(a)** Produce the Taylor polynomials of degrees 1, 2, 3, 4 for $f(x) = e^{x^2}$ with $a = 0$ the point of approximation.

(b) Using the Taylor polynomials for e^t, substitute $t = x^2$ to obtain polynomial approximations for e^{x^2}. Compare with the results in (a).

11. The quotient

$$g(x) = \frac{e^x - 1}{x}$$

is undefined for $x = 0$. Approximate e^x by using Taylor polynomials of degrees 1, 2, and 3, in turn, to determine a natural definition of $g(0)$.

12. The quotient

$$g(x) = \frac{\log(1 + x)}{x}$$

is undefined for $x = 0$. Approximate $\log(1 + x)$ using Taylor polynomials of degrees 1, 2, and 3, in turn, to determine a natural definition of $g(0)$.

13. **(a)** As an alternative to the linear Taylor polynomial, construct a linear polynomial $q(x)$, satisfying

$$q(a) = f(a), \qquad q(b) = f(b)$$

for given points a and b.

(b) Apply this to $f(x) = e^x$ with $a = 0$ and $b = 1$. For $0 \leq x \leq 1$, numerically compare $q(x)$ with the linear Taylor polynomial of this section.

14. For $f(x) = e^x$, construct a cubic polynomial $q(x)$ for which

$$\begin{aligned} q(0) &= f(0), & q(1) &= f(1) \\ q'(0) &= f'(0), & q'(1) &= f'(1) \end{aligned}$$

Numerically compare it to e^x and the Taylor polynomial $p_3(x)$ of (1.6) for $0 \leq x \leq 1$.

Hint: Write $q(x) = b_0 + b_1 x + b_2 x^2 + b_3 x^3$. Determine b_0 and b_1 from the conditions at $x = 0$. Then obtain a linear system of two equations for the remaining coefficients b_2 and b_3.

1.2. THE ERROR IN TAYLOR'S POLYNOMIAL

To make practical use of the Taylor polynomial approximation to $f(x)$, we need to know its accuracy. The following theorem gives the main tool for estimating this accuracy. We present it without proof, since it is given in most calculus texts.

Theorem 1.2.1 (Taylor's remainder) Assume that $f(x)$ has $n+1$ continuous derivatives on an interval $\alpha \le x \le \beta$, and let the point a belong to that interval. For the Taylor polynomial $p_n(x)$ of (1.6), let $R_n(x) \equiv f(x) - p_n(x)$ denote the remainder in approximating $f(x)$ by $p_n(x)$. Then

$$R_n(x) = \frac{(x-a)^{n+1}}{(n+1)!} f^{(n+1)}(c_x), \qquad \alpha \le x \le \beta \tag{1.11}$$

with c_x an unknown point between a and x.

Example 1.2.2 Let $f(x) = e^x$ and $a = 0$. The Taylor polynomial is given in (1.7). From the above theorem, the approximation error is given by

$$e^x - p_n(x) = \frac{x^{n+1}}{(n+1)!} e^c, \qquad n \ge 0 \tag{1.12}$$

with c between 0 and x. From this formula, we can prove that for each fixed x, the error tends to 0 as $n \to \infty$; this should be intuitively clear when $|x| \le 1$. Also from the formula, it appears that for each fixed value of n, the error becomes larger as x moves away from 0. To illustrate this graphically, it is good to graph the errors $e^x - p_n(x)$ rather than simply graphing the function e^x and the polynomials $p_n(x)$, in contrast to what was done in Section 1.1. This is illustrated in Figure 1.4 for degrees $n = 1, 2, 3, 4$ on the interval $[-1, 1]$. The graph illustrates the results stated above. ■

Example 1.2.3 As a special case of (1.12), let $x = 1$. Then from (1.7)

$$e \approx p_n(1) = 1 + 1 + \frac{1}{2!} + \frac{1}{3!} + \cdots + \frac{1}{n!}$$

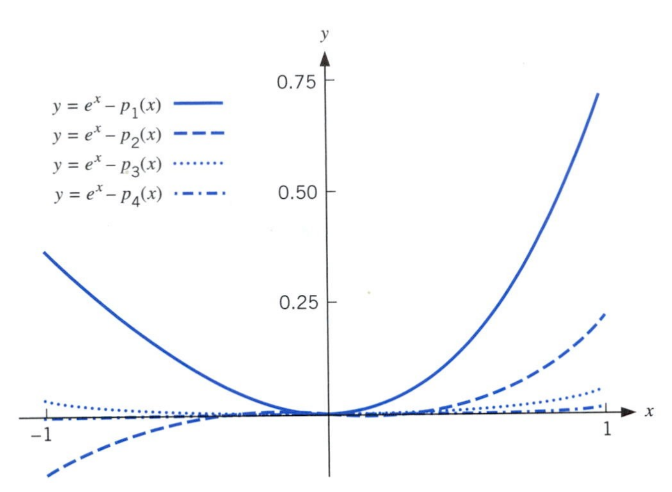

Figure 1.4. Errors in Taylor polynomial approximations to e^x

and from (1.12)

$$e - p_n(1) = R_n(1) = \frac{e^c}{(n+1)!}, \qquad 0 < c < 1$$

Since $e < 3$, we can bound $R_n(1)$ as follows:

$$\frac{1}{(n+1)!} \leq R_n(1) \leq \frac{e}{(n+1)!} < \frac{3}{(n+1)!}$$

This uses the inequality $e^0 \leq e^c \leq e^1$. As an actual numerical example, suppose we want to approximate e by $p_n(1)$ with

$$R_n(1) \leq 10^{-9}$$

Since we only know an upper bound for $R_n(1)$, we can obtain the desired error by making the upper bound satisfy

$$\frac{3}{(n+1)!} \leq 10^{-9}$$

This is true when $n \geq 12$; thus, $p_{12}(1)$ is a sufficiently accurate approximation to e. ■

The formulas (1.6) and (1.11) can be used to form approximations and remainder formulas for most of the standard functions encountered in undergraduate mathematics.

For later reference, we give some of the more important ones.

$$e^x = 1 + x + \frac{x^2}{2!} + \cdots + \frac{x^n}{n!} + \frac{x^{n+1}}{(n+1)!}e^c \tag{1.13}$$

$$\begin{aligned} \sin(x) = x - \frac{x^3}{3!} + \frac{x^5}{5!} - \cdots + (-1)^{n-1}\frac{x^{2n-1}}{(2n-1)!} \\ + (-1)^n \frac{x^{2n+1}}{(2n+1)!}\cos(c) \end{aligned} \tag{1.14}$$

$$\begin{aligned} \cos(x) = 1 - \frac{x^2}{2!} + \frac{x^4}{4!} - \cdots + (-1)^n \frac{x^{2n}}{(2n)!} \\ + (-1)^{n+1}\frac{x^{2n+2}}{(2n+2)!}\cos(c) \end{aligned} \tag{1.15}$$

$$\frac{1}{1-x} = 1 + x + x^2 + \cdots + x^n + \frac{x^{n+1}}{1-x}, \qquad x \neq 1 \tag{1.16}$$

$$\begin{aligned} (1+x)^\alpha = 1 + \binom{\alpha}{1}x + \binom{\alpha}{2}x^2 + \cdots + \binom{\alpha}{n}x^n \\ + \binom{\alpha}{n+1}x^{n+1}(1+c)^{\alpha-n-1} \end{aligned} \tag{1.17}$$

In this last formula, α is any real number. The coefficients $\binom{\alpha}{k}$ are called *binomial coefficients* and are defined by

$$\binom{\alpha}{k} = \frac{\alpha(\alpha-1)\cdots(\alpha-k+1)}{k!}, \qquad k = 1, 2, 3, \ldots, \qquad \binom{\alpha}{0} = 1$$

In all of the formulas, except (1.16), the point c is between 0 and x. The proof of (1.16) is taken up in Problem 10.

Example 1.2.4 Approximate $\cos(x)$ for $|x| \leq \pi/4$, with an error of no greater than 10^{-5}. Since the point c in the remainder of (1.15) is unknown, we consider the worst possible case and make it satisfy the desired error bound:

$$|R_{2n+1}(x)| \leq \frac{x^{2n+2}}{(2n+2)!} \leq 10^{-5}, \qquad \text{for } |x| \leq \pi/4$$

This uses $|\cos(c)| \leq 1$. For this inequality to be true, we must have

$$\frac{(\pi/4)^{2n+2}}{(2n+2)!} \leq 10^{-5}$$

which is satisfied when $n \geq 3$. The desired approximation is

$$\cos(x) \approx 1 - \frac{x^2}{2!} + \frac{x^4}{4!} - \frac{x^6}{6!} \quad \blacksquare$$

Many Taylor polynomials and remainder terms are not created directly from (1.6) and (1.11). Instead, the above standard Taylor formulas are manipulated to obtain Taylor approximations for other functions. As an example, consider constructing a Taylor polynomial approximation to $f(x) = e^{-x^2}$ about $x = 0$. Begin by replacing x by t in (1.13), as below:

$$e^t = 1 + t + \frac{t^2}{2!} + \cdots + \frac{t^n}{n!} + \frac{t^{n+1}}{(n+1)!}e^c$$

with c an unknown number between 0 and t. This is valid for all real numbers t. Now replace t by $-x^2$, yielding

$$e^{-x^2} = 1 - x^2 + \frac{x^4}{2!} - \frac{x^6}{3!} + \cdots + \frac{(-1)^n x^{2n}}{n!} + \frac{(-1)^{n+1} x^{2n+2}}{(n+1)!}e^c \tag{1.18}$$

with c an unknown number satisfying $-x^2 \leq c \leq 0$. This gives the Taylor polynomial approximation to e^{-x^2} of degree $2n$ (and also of degree $2n + 1$). If you attempt to construct this Taylor approximation directly, as in the manner of Section 1.1, then the derivatives of e^{-x^2} quickly become unmanageable.

As a somewhat more complicated example of an indirect construction of a Taylor polynomial approximation, we derive an approximation for $\log(1 - t)$. Begin by integrating (1.16) from 0 to t:

$$\int_0^t \frac{dx}{1-x} = \int_0^t (1 + x + x^2 + \cdots + x^n)\,dx + \int_0^t \frac{x^{n+1}}{1-x}\,dx$$

$$-\log(1-t) = t + \frac{1}{2}t^2 + \frac{1}{3}t^3 + \cdots + \frac{1}{n+1}t^{n+1} + \int_0^t \frac{x^{n+1}}{1-x}\,dx$$

$$\log(1-t) = -\left(t + \frac{1}{2}t^2 + \cdots + \frac{1}{n+1}t^{n+1}\right) - \int_0^t \frac{x^{n+1}}{1-x}\,dx \tag{1.19}$$

This is valid for $0 \leq t < 1$. The remainder term can be simplified by applying the integral mean value theorem (Theorem A.8 in Appendix A) to obtain

$$\int_0^t \frac{x^{n+1}}{1-x}\,dx = \frac{1}{1-c}\int_0^t x^{n+1}\,dx = \left(\frac{1}{1-c}\right)\frac{t^{n+2}}{n+2}$$

for some c between 0 and t. This entire argument is also valid for $-1 \leq t < 0$.

Summarizing this important case, we obtain

$$\log(1-t) = -\left(t + \frac{1}{2}t^2 + \cdots + \frac{1}{n+1}t^{n+1}\right) - \left(\frac{1}{1-c_t}\right)\frac{t^{n+2}}{n+2} \qquad (1.20)$$

with c_t an unknown number between 0 and t; and this is valid for $-1 \le t < 1$.

Notation 1.2.5 When we speak of bounding the error in some quantity, say, bounding the error $f(x) - p_n(x)$, we mean to find a number M_1 for which

$$|f(x) - p_n(x)| \le M_1$$

When we say to bound the error on an interval $\alpha \le x \le \beta$, we mean to find a number M_2 for which

$$\max_{\alpha \le x \le \beta} |f(x) - p_n(x)| \le M_2$$

Most examples are not as "nice" as Example 1.2.3, and we generally must use absolute values in dealing with error. ■

1.2.1 Infinite Series

By rearranging the terms in (1.16), we obtain the sum of a finite geometric series or progression.

$$1 + x + x^2 + \cdots + x^n = \frac{1 - x^{n+1}}{1 - x}, \qquad x \ne 1 \qquad (1.21)$$

Letting $n \to \infty$ in (1.16) when $|x| < 1$, we obtain the infinite *geometric series*

$$\frac{1}{1-x} = 1 + x + x^2 + x^3 + \cdots = \sum_{j=0}^{\infty} x^j, \qquad |x| < 1 \qquad (1.22)$$

In general, we say an infinite series

$$\sum_{j=0}^{\infty} c_j$$

is *convergent* if the partial sums

$$S_n = \sum_{j=0}^{n} c_j, \qquad n \ge 0$$

form a convergent sequence. This means that

$$S = \lim_{n \to \infty} S_n$$

exists, and we then write

$$S = \sum_{j=0}^{\infty} c_j \tag{1.23}$$

For the infinite series in (1.22) with $x \neq 1$, the partial sums are given by (1.21)

$$S_n = \frac{1 - x^{n+1}}{1 - x}$$

When $|x| < 1$, the sequence $\{S_n\}$ clearly has the limiting value $1/(1 - x)$, and therefore we say the infinite series is convergent to this value. When $|x| > 1$, the sequence $\{S_n\}$ clearly does not converge to any limiting value; we say the infinite geometric series diverges in this case. What happens with $|x| = 1$?

Assume $f(x)$ has derivatives of any order at $x = a$. The infinite series

$$\sum_{j=0}^{\infty} \frac{(x-a)^j}{j!} f^{(j)}(a)$$

is called the *Taylor series expansion* of the function $f(x)$ about the point $x = a$. The partial sum

$$\sum_{j=0}^{n} \frac{(x-a)^j}{j!} f^{(j)}(a)$$

is simply the Taylor polynomial $p_n(x)$. The sequence $\{p_n(x)\}$ has the limit $f(x)$ if the error tends to zero as $n \to \infty$

$$\lim_{n \to \infty} [f(x) - p_n(x)] = 0$$

In this case, we can write

$$f(x) = \sum_{j=0}^{\infty} \frac{(x-a)^j}{j!} f^{(j)}(a) \tag{1.24}$$

As examples, we can show that the error terms in (1.13)–(1.17) and (1.20) tend to zero as $n \to \infty$ for suitable values of x. This yields the following Taylor expansions:

$$e^x = \sum_{j=0}^{\infty} \frac{x^j}{j!}, \qquad -\infty < x < \infty \tag{1.25}$$

$$\sin x = \sum_{j=0}^{\infty} \frac{(-1)^j x^{2j+1}}{(2j+1)!}, \qquad -\infty < x < \infty \tag{1.26}$$

$$\cos x = \sum_{j=0}^{\infty} \frac{(-1)^j x^{2j}}{(2j)!}, \qquad -\infty < x < \infty \tag{1.27}$$

$$(1+x)^{\alpha} = \sum_{j=0}^{\infty} \binom{\alpha}{j} x^j, \qquad -1 < x < 1 \tag{1.28}$$

$$\log(1-t) = -\sum_{j=0}^{\infty} \frac{t^j}{j}, \qquad -1 \le x < 1 \tag{1.29}$$

In the case (1.28), the series may converge for other values of x, although that depends on α.

Infinite series of the form

$$\sum_{j=0}^{\infty} a_j (x-a)^j \tag{1.30}$$

are called *power series*. Taylor's formula is one way of obtaining such series, but they also arise in other ways. Their convergence can be examined directly without recourse to Taylor's error formula. We give two important theorems used in examining their convergence.

Theorem 1.2.6 Assume the series (1.30) converges for some value x_0. Then the series (1.30) converges also for all values of x satisfying $|x - a| < |x_0 - a|$.

We outline the basis of a proof of this result in Problem 24.

Theorem 1.2.7 For the series (1.30), assume that the limit

$$R = \lim_{n \to \infty} \left| \frac{a_{n+1}}{a_n} \right|$$

exists. Then for x satisfying $|x - a| < 1/R$, the series (1.30) converges to a limit $S(x)$. When $R = 0$, the series (1.30) converges for any real number x.

As an example, let us examine the convergence of the power series in formula (1.27). Letting $t = x^2$, we obtain the series

$$\sum_{j=0}^{\infty} \frac{(-1)^j t^j}{(2j)!} \tag{1.31}$$

Applying Theorem 1.2.7 with

$$a_j = \frac{(-1)^j}{(2j)!}$$

we find $R = 0$. So the series (1.31) converges for any value of t, and then the series in formula (1.27) converges for any value of x.

For more information on these two theorems and on the general area of infinite series, consult any introductory-level calculus textbook.

PROBLEMS

1. Bound the error in using $p_3(x)$ to approximate e^x on $[-1, 1]$, with $p_3(x)$ and its remainder given in (1.13). Compare it to the results given in Table 1.1.

2. Find the degree 2 Taylor polynomial for $f(x) = e^x \sin(x)$, about the point $a = 0$. Bound the error in this approximation when $-\pi/4 \le x \le \pi/4$.

3. Find linear and quadratic Taylor polynomial approximations to $f(x) = \sqrt[3]{x}$ about the point $a = 8$. Bound the error in each of your approximations on the interval $8 \le x \le 8 + \delta$ with $\delta > 0$. Obtain an actual numerical bound on the interval $[8, 8.1]$.

4. **(a)** Bound the error in the approximation

$$\sin(x) \approx x$$

for $-\pi/4 \le x \le \pi/4$.

(b) Since this is a good approximation for small values of x, also consider the "percentage error"

$$\frac{\sin(x) - x}{\sin(x)} \approx \frac{\sin(x) - x}{x}$$

Bound the absolute value of the latter quantity for $-\delta \le x \le \delta$. Pick δ to make the absolute value of the percentage error less than 1%.

5. How large should the degree $2n - 1$ be chosen in (1.14) to have

$$|\sin(x) - p_{2n-1}(x)| \le 0.001$$

for all $-\pi/2 \le x \le \pi/2$? Check your result by evaluating the resulting $p_{2n-1}(x)$ at $x = \pi/2$.

6. Let $p_n(x)$ be the Taylor polynomial of degree n of the function $f(x) = \log(1 - x)$ about $a = 0$. How large should n be chosen to have $|f(x) - p_n(x)| \le 10^{-4}$ for $-\frac{1}{2} \le x \le \frac{1}{2}$? For $-1 \le x \le \frac{1}{2}$?

7. Write out $p_4(x)$ for (1.17) with $\alpha = \frac{1}{2}$. Bound the error if $0 \le x \le \frac{1}{2}$.

8. How large should n be chosen in (1.13) to have

$$\left|e^x - p_n(x)\right| \le 10^{-5}, \qquad -1 \le x \le 1$$

9. Use Taylor polynomials with remainder term to evaluate the following limits:

(a) $\displaystyle\lim_{x \to 0} \frac{1 - \cos(x)}{x^2}$ **(b)** $\displaystyle\lim_{x \to 0} \frac{\log(1 + x^2)}{2x}$

(c) $\displaystyle\lim_{x \to 0} \frac{\log(1 - x) + xe^{x/2}}{x^3}$

Hint: Use Taylor polynomials for the standard functions [e.g., $\cos(t)$, $\log(1 + t)$, and e^t] to obtain polynomial approximations to the numerators of these fractions; and then simplify the results.

10. Verify (1.16).

Hint: Multiply both sides by $1 - x$ and simplify.

11. Show

$$(1 + t)^n = \sum_{j=0}^{n} \binom{n}{j} t^j$$

12. Rewrite $f(x) = 1 + x^6$ in the form

$$f(x) = \sum_{j=0}^{6} a_j (x - 1)^j$$

Hint: Expand $f(x)$ as a Taylor polynomial of degree 6 about $x = 1$. Using this, what are the coefficients $\{a_j\}$? What is the error $f(x) - p_6(x)$ in this case?

13. Evaluate $I = \displaystyle\int_0^1 \frac{e^x - 1}{x}\, dx$ within an accuracy of 10^{-6}.

Hint: Replace e^x by a general Taylor polynomial approximation plus its remainder.

14. **(a)** Obtain a Taylor polynomial with remainder for $f(t) = 1/(1 + t^2)$, about $a = 0$.

Hint: Substitute $x = -t^2$ into (1.16).

(b) Obtain a Taylor polynomial with remainder for $g(x) = \tan^{-1} x$. Do this by integrating the result in (a) and using

$$\tan^{-1}(x) = \int_0^x \frac{dt}{1 + t^2}$$

15. Define $f(x) = \int_0^x \frac{1 - e^{-t}}{t}\, dt$. Find a Taylor polynomial approximation of degree 2 for $f(x)$. Give a formula for the approximation error, bounding it on the interval $0 \le x \le \delta$ with $\delta > 0$. What is the error bound for $\delta = 0.1$?

16. Define $f(x) = \int_0^x \frac{\log(1+t)}{t}\, dt$.

(a) Give a Taylor polynomial approximation to $f(x)$ about $x = 0$.

(b) Bound the error in the degree n approximation for $|x| \le 1/2$.

(c) Find n so as to have a Taylor approximation with an error of at most 10^{-7} on $[-\frac{1}{2}, \frac{1}{2}]$.

17. Define $f(x) = \frac{\sqrt{\pi}}{2} + \int_0^x e^{-t^2}\, dt$.

(a) Using (1.13), give a Taylor polynomial approximation to $f(x)$ about $x = 0$.

(b) Bound the error in the degree n approximation for $|x| \le 1$.

(c) Find n so as to have a Taylor approximation with an error of at most 10^{-7} on $[-1, 1]$.

18. Define $f(x) = \frac{1}{x} \int_0^x \frac{dt}{1+t^3}$. Find a Taylor polynomial approximation to $f(x)$ with the degree of the approximation being degree 3 or larger. Give an error formula for your approximation. Estimate $f(0.1)$ and bound the error.

Hint: Begin with a Taylor series approximation for $1/(1-u)$.

19. Find the Taylor polynomial about $x = 0$ for

$$f(x) = \log\left(\frac{1+x}{1-x}\right), \qquad -1 < x < 1$$

Hint: Write

$$f(t) = \log(1+t) - \log(1-t)$$

$$f'(t) = \frac{1}{1+t} + \frac{1}{1-t} = \frac{2}{1-t^2}$$

Expand $1/(1-t^2)$ by applying (1.16) with $x = t^2$. Obtain

$$f'(t) = 2[1 + t^2 + t^4 + \cdots + t^{2n}] + \frac{2t^{2n+2}}{1-t^2}, \qquad -1 < t < 1.$$

Evaluate

$$f(x) = \int_0^x f'(t)\, dt$$

to obtain a Taylor polynomial approximation to $f(x)$, including an error term.

20. **(a)** Using Problem 19, give a Taylor polynomial approximation to $\log(2)$ and bound the error.

Hint: What is the needed value of x in order to use the result of Problem 19?

(b) Consider evaluating $\log(z)$ for $\frac{1}{2} \le z \le 1$. Use Problem 19 to give a way of calculating $\log(z)$. Bound the error.

21. **(a)** Consider evaluating π by using

$$\pi = 4 \tan^{-1}(1)$$

Using the results of Problem 14, how many terms would be needed in the Taylor approximation $\tan^{-1}(x) \approx p_{2n-1}(x)$ to calculate π with an accuracy of 10^{-10}? Is this a practical method of evaluating π?

(b) Suggest another computation of π by using the series for $\tan^{-1}(x)$, giving a more rapidly convergent method. (There is a large research literature on finding series that converge rapidly to π.)

22. Define $h(x) = f(x)g(x)$. Let the Taylor polynomials of degree n for $f(x)$ and $g(x)$ be given by

$$p_n(x) = \sum_{i=0}^{n} a_i x^i, \qquad q_n(x) = \sum_{j=0}^{n} b_j x^j$$

Let $r_n(x)$ be obtained by first multiplying $p_n(x)q_n(x)$ and then dropping all terms of degree greater than n.

(a) For $n = 2$, show that the Taylor polynomial of degree 2 for $h(x)$ equals $r_2(x)$.

(b) For general $n \ge 1$, show that the Taylor polynomial of degree n for $h(x)$ equals $r_n(x)$.

Hint: For repeated differentiation of the product $f(x)g(x)$, use the *Leibniz formula*:

$$\frac{d^k}{dx^k}[f(x)g(x)] = \sum_{j=0}^{k} \binom{k}{j} f^{(j)}(x) g^{(k-j)}(x)$$

23. Recall the definition of convergence for the infinite series

$$S = \sum_{j=0}^{\infty} c_j$$

of (1.23). Show that convergence implies that

$$\lim_{j \to \infty} c_j = 0$$

24. Consider the proof of Theorem 1.2.6. Since the series

$$\sum_{j=0}^{\infty} a_j (x_0 - a)^j$$

is assumed to converge for $x = x_0$, the result of Problem 23 implies that

$$\lim_{j\to\infty} a_j (x_0 - a)^j = 0$$

Consequently, there is a constant $c > 0$ for which

$$\left|a_j (x_0 - a)^j\right| \le c, \qquad j \ge 0$$

Introduce the partial sums

$$S_n = \sum_{j=0}^{n} a_j (x - a)^j, \qquad n \ge 0$$

and define

$$r = \left|\frac{x-a}{x_0 - a}\right|$$

and $r < 1$ is an assumption of the theorem.

(a) Show that the partial sums $\{S_n\}$ satisfy

$$|S_m - S_n| \le c\frac{r^{n+1}}{1-r}, \qquad \text{all } m > n \ge 0$$

This implies that

$$\lim_{n,m\to\infty} |S_m - S_n| = 0$$

and therefore the sequence $\{S_n\}$ forms what is called a *Cauchy sequence*. From a theorem of higher analysis, $\{S_n\}$ being a Cauchy sequence is sufficient to conclude that

$$S = \lim_{n\to\infty} S_n$$

exists, thus showing for the infinite series of (1.30) that

$$S = \sum_{j=0}^{\infty} a_j (x - a)^j, \qquad |x - a| < |x_0 - a|$$

is convergent.

(b) Show that the error in the partial sums satisfies

$$|S - S_n| \le \frac{cr^{n+1}}{1-r}$$

This is often used to obtain error bounds for the approximation

$$S \approx S_n$$

in which the infinite series S is approximated by its partial sum S_n.

1.3. POLYNOMIAL EVALUATION

The evaluation of a polynomial would appear to be a straightforward task. It is not, and to illustrate the possibilities, we consider the evaluation of

$$p(x) = 3 - 4x - 5x^2 - 6x^3 + 7x^4 - 8x^5$$

From a programmer's perspective, the simplest method of evaluation is to compute each term independently of the remaining terms. More precisely, the term cx^k is computed in a program by

$$c * x\hat{\ }k \quad \text{or} \quad c * x{*}{*}k$$

depending on which computer language is being used. This requires k multiplications with most compilers, although more "intelligent" compilers may produce a more efficient code. With this approach, there will be

$$1 + 2 + 3 + 4 + 5 = 15 \text{ multiplications}$$

in the evaluation of $p(x)$.

The second method of evaluation is more efficient. We compute each power of x by multiplying x with the preceding power of x, as

$$x^3 = x(x^2), \qquad x^4 = x(x^3), \qquad x^5 = x(x^4) \tag{1.32}$$

Thus, each term cx^k takes two multiplications for $k > 1$. The resulting evaluation of $p(x)$ uses

$$1 + 2 + 2 + 2 + 2 = 9 \text{ multiplications}$$

a considerable savings over the first method, especially with higher-degree polynomials.

The third method is called *nested multiplication*. With it, we write and evaluate $p(x)$ in the form

$$p(x) = 3 + x(-4 + x(-5 + x(-6 + x(7 - 8x))))$$

The number of multiplications is only 5, an additional saving over the second method. The nested multiplication method is the preferred evaluation procedure; and its advantage increases as the degree of the polynomial becomes larger.

Consider the general polynomial of degree n

$$p(x) = a_0 + a_1 x + \cdots + a_n x^n, \qquad a_n \neq 0 \tag{1.33}$$

If we use the second method, with the powers of x being computed as in (1.32), then the number of multiplications in the evaluation of $p(x)$ equals $2n - 1$. For the nested multiplication method, write and evaluate $p(x)$ in the form

$$p(x) = a_0 + x(a_1 + x(a_2 + \cdots + x(a_{n-1} + a_n x)\ldots)) \tag{1.34}$$

This uses only n multiplications, a savings of about 50% over the second method. All methods use n additions. The use of nested multiplication is implemented in the MATLAB program given at the end of Section 1.1.

Example 1.3.1 Evaluate the Taylor polynomial $p_5(x)$ for $\log(x)$ about $a = 1$. A general formula is given in (1.19), with t replaced by $-(x - 1)$. From it,

$$p_5(x) = (x-1) - \tfrac{1}{2}(x-1)^2 + \tfrac{1}{3}(x-1)^3 - \tfrac{1}{4}(x-1)^4 + \tfrac{1}{5}(x-1)^5$$

Let $w = x - 1$ and write

$$p_5(x) = w\left(1 + w\left(-\tfrac{1}{2} + w\left(\tfrac{1}{3} + w\left(-\tfrac{1}{4} + \tfrac{1}{5}w\right)\right)\right)\right)$$

In a computer program, you would store the coefficients in decimal form, to an accuracy consistent with the arithmetic of the machine. ■

We give a more formal algorithm for (1.34) because of its connection to some other topics. Suppose we want to evaluate $p(x)$ at some number z. Define a sequence of coefficients b_i as follows:

$$\begin{aligned} b_n &= a_n \\ b_{n-1} &= a_{n-1} + z b_n \\ b_{n-2} &= a_{n-2} + z b_{n-1} \\ &\ \vdots \\ b_0 &= a_0 + z b_1 \end{aligned} \tag{1.35}$$

Then

$$p(z) = b_0 \tag{1.36}$$

The coefficients b_j are the successive computations within a matching pair of parentheses in (1.34). b_{n-1} is the innermost computation, and b_0 is the final one. When looked at in this manner, the nested multiplication method is called *Horner's method*; it is closely related to *synthetic division*, which is discussed in many elementary algebra texts.

Using the coefficients of (1.35), define the polynomial

$$q(x) = b_1 + b_2 x + b_3 x^2 + \cdots + b_n x^{n-1} \tag{1.37}$$

It can be shown that

$$p(x) = b_0 + (x - z)q(x) \tag{1.38}$$

Thus, $q(x)$ is the quotient from dividing $p(x)$ by $x - z$, and b_0 is the remainder. The proof of (1.38) is left as Problem 7 for the reader. This result is used in connection with polynomial rootfinding methods to reduce the degree of a polynomial when a root z has been found, since then $b_0 = 0$ and $p(x) = (x - z)q(x)$.

1.3.1 An Example Program

We conclude this section by giving a MATLAB function for evaluating Taylor polynomial approximations for the function

$$\text{Sint}\, x = \frac{1}{x}\int_0^x \frac{\sin(t)}{t}\,dt, \qquad x \neq 0 \tag{1.39}$$

with $\text{Sint}(0) = 1$. This function is called a *sine integral*. A graph of $\text{Sint}\, x$ is given in Figure 1.5 for $-6 \le x \le 6$.

We begin by deriving a particular Taylor polynomial approximation on $[-1, 1]$ to $\text{Sint}\, x$, requiring that its maximum error on $[-1, 1]$ be bounded by 5×10^{-9}, a fairly arbitrary choice. Begin by using the standard series for $\sin(t)$, given by (1.14) with x replaced by t. First consider the case $x > 0$. Dividing by t and integrating over $[0, x]$ we get

$$\begin{aligned}
\text{Sint}\, x &= \frac{1}{x}\int_0^x \left[1 - \frac{t^2}{3!} + \frac{t^4}{5!} - \cdots + (-1)^{n-1}\frac{t^{2n-2}}{(2n-1)!}\right] dt + R_{2n-2}(x) \\
&= 1 - \frac{x^2}{3!3} + \frac{x^4}{5!5} - \cdots + (-1)^{n-1}\frac{x^{2n-2}}{(2n-1)!(2n-1)} + R_{2n-2}(x)
\end{aligned} \tag{1.40}$$

$$R_{2n-2}(x) = \frac{1}{x}\int_0^x (-1)^n \frac{t^{2n}}{(2n+1)!}\cos(c_t)\,dt$$

The point c_t is between 0 and t. Since $|\cos(c_t)| \le 1$,

$$|R_{2n-2}(x)| \le \frac{1}{x}\int_0^x \frac{t^{2n}}{(2n+1)!}\,dt = \frac{x^{2n}}{(2n+1)!(2n+1)} \tag{1.41}$$

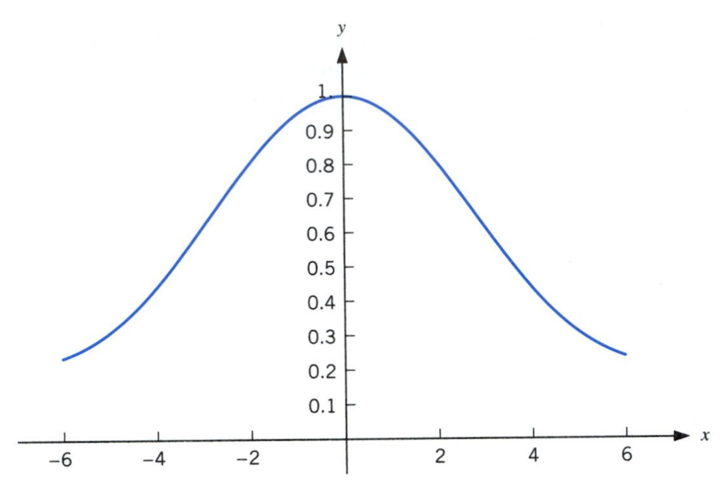

Figure 1.5. The sine integral Sint(x) of (1.39)

It is easy to see that this bound is also valid for $x < 0$. As required, choose the degree so that

$$|R_{2n-2}(x)| \le 5 \times 10^{-9} \tag{1.42}$$

From (1.41),

$$\max_{|x|\le 1} |R_{2n-2}(x)| \le \frac{1}{(2n+1)!(2n+1)}$$

We choose n so that this upper bound is itself bounded by 5×10^{-9}. This is true if $2n + 1 \ge 11$, i.e., $n \ge 5$, and then (1.42) is satisfied. The polynomial we use is

$$p(x) = 1 - \frac{x^2}{3!3} + \frac{x^4}{5!5} - \frac{x^6}{7!7} + \frac{x^8}{9!9}, \qquad -1 \le x \le 1 \tag{1.43}$$

and as an approximation to Sint x on $[-1, 1]$, its error satisfies (1.42).

The above polynomial $p(x)$ is an *even function*, meaning that $p(-x) = p(x)$ for all values of x; and even polynomials contain only even powers of x (cf. Problem 3). Then we can evaluate $p(x)$ more efficiently by evaluating the degree 4 polynomial $g(u)$ obtained by using $u = x^2$ in $p(x)$

$$g(u) = 1 - \frac{u}{18} + \frac{u^2}{600} - \frac{u^3}{35{,}280} + \frac{u^4}{3{,}265{,}920} \tag{1.44}$$

where the denominators of (1.43) have been calculated explicitly.

MATLAB PROGRAM: *Taylor polynomials for* Sint *x*. The Taylor polynomials $p_n(x)$ about 0 for Sint x are even polynomials. Proceeding in analogy with the reduction of (1.43) to (1.44), we can reduce by approximately one-half the number of multiplications needed to evaluate it. We give here a MATLAB program `plot_sint` to evaluate Taylor approximations of Sint x. The Taylor polynomials are of four different degrees n (given here as 2, 4, 6, and 8), and the evaluation is on a user-specified interval $[0, b]$.

```
% TITLE: Plot Taylor polynomials for the ''sine integral''
% about x = 0.
%
% This plots several Taylor polynomials and their errors
% for increasing degrees.  The particular function being
% approximated is Sint(x) on [0,b], with x = 0 the point of
% expansion for creating the Taylor polynomials.  We plot
% the Taylor polynomials for several degrees, which can
% be altered by the user.  Note that the function being
% plotted is symmetric about x=0.
%
% The Taylor polynomials in this case contain only terms
% of even degree.  Such polynomials are called '' even
% polynomials,'' and they are symmetric about x=0.
%
% TO THE STUDENT: run this program for various input values
% of '' b.''  Also, change the values given in the vector
% '' degree'' given below, to experiment with different
% degrees of Taylor polynomial approximations.

% Initialize
b = input('Give the number b defining the interval [0,b] ');
h = b/200;
x = 0:h:b;
max_degree = 20;

% Produce the Taylor coefficients for the ''sine integral''
% function ''Sint.''
c = sint_tay(max_degree);

% Specify the four values of degree to be considered.  They must
% all be even, and they must be less than or equal to max_degree.
degree = [2, 4, 6, 8];
if max(degree) > max_degree
  fprintf('Some value of degree is greater than max_degree =...
          %2.0f\n',max_degree)
  return
```

```
end

% Initialize the array to contain the polynomial values.  Row #i
% is to contain the values for the polynomial of degree=degree(i).
p = zeros(4,length(x));

% Calculate the Taylor polynomials
for i = 1:4
  p(i,:)  = poly_even(x,c,degree(i));
end

% Initialize for plotting the Taylor polynomials
hold off
clf
axis([0,b,0,1])
hold on

% Plot the Taylor polynomials
plot(x,p(1,:),x,p(2,:),':',x,p(3,:),'--',x,p(4,:),'-.')
plot([0,b],[0,0])
plot([0 0],[0 1])
title('Taylor approximations of Sint(x)')
text(1.025*b,0,'x')
text(0,1.03,'y')
legend(strcat('degree = ',int2str(degree(1))),...
       strcat('degree = ',int2str(degree(2))),...
       strcat('degree = ',int2str(degree(3))),...
       strcat('degree = ',int2str(degree(4))))
```

The program uses the following program, named `sint_tay`, to evaluate the Taylor coefficients of (1.40) for Sint x.

```
function coeff=sint_tay(n)
%
% function coeff = sint_tay(n)
%
% Evaluate the coefficients of the Taylor approximation
% of degree n which approximates the '' sine integral.''
% The input variable n must be an even integer.  The
% output vector coeff will have length m+1 where m=n/2.
% This is because only the even ordered coefficients are
% nonzero, and this must be considered in evaluating the
% associated Taylor polynomial.

m = double(int32(n/2));
```

```
if n ~= 2*m
  disp('Error in poly_even(x,coeff,n):')
  disp('The parameter n must be an even integer.')
end

coeff = ones(m+1,1);
sign = 1;
fact = 1;

for i = 2:m+1
  sign = -sign;
  d = 2*i-1;
  fact = fact*(d-1)*d;
  coeff(i) = sign/(fact*d);
end
```

The program `plot_sint` also uses the following program, named `poly_even`, to evaluate the even polynomials approximating Sint x. It is a variation on the program `polyeval` discussed earlier in this section, using the simplification illustrated in (1.44).

```
function value = poly_even(x,coeff,n);
%
% function value = poly_even(x,coeff,n)
%
% Evaluate an ''even'' Taylor polynomial at the points given
% in x, with n the degree of the polynomial.  The coefficients
% are to be given in coeff.  It is assumed that the numbers in
% coeff are the nonzero coefficients of the even-ordered
% terms in the polynomial.  The input parameter n must be an
% even integer.

m = double(int32(n/2));
if n ~= 2*m
  disp('Error in poly_even(x,coeff,n):')
  disp('The parameter n must be an even integer.')
end

xsq = x.*x;
value = coeff(m+1)*ones(size(x));

for i = m:-1:1
  value = coeff(i) + xsq.*value;
end
```

PROBLEMS

1. (a) Implement the function `plot_sint` on your computer, and use it to compare different degrees of approximation on the intervals [0, 1], [0, 2], [0, 3], and [0, 4]. Comment as best as you can on the accuracy of the approximations.

   (b) Modify the program by allowing for larger degrees of approximation, for example, in the program use

   ```
   degree = [4, 8, 12, 16]
   ```

   Repeat part (a) and also consider longer intervals $[0, b]$.

2. Repeat the process of creating a polynomial approximation to $\operatorname{Sint} x$ with an error tolerance of 10^{-9}, but now use the interval $-2 \le x \le 2$.

3. (a) Let the polynomial $p(x)$ be an *even function*, meaning that $p(-x) = p(x)$ for all x of interest. Show this implies that the coefficients are zero for all terms of odd degree.

   (b) Let the polynomial $p(x)$ be an odd function, meaning that $p(-x) = -p(x)$ for all x of interest. Show this implies that the coefficients are zero for all terms of even degree.

   (c) Let $p(x) = a_0 + a_1 x + a_2 x^2 + a_5 x^5$. Give conditions on the coefficients $\{a_0, a_1, a_2, a_5\}$ so that $p(x)$ is even. Repeat with $p(x)$ being odd.

4. In analogy with $\operatorname{Sint} x$, prepare a MATLAB program for evaluating

$$\operatorname{Cint} x = \frac{1}{x} \int_0^x \frac{1 - \cos(t)}{t^2}\, dt, \qquad -1 \le x \le 1$$

   Use the same error tolerance as in (1.42) for $\operatorname{Sint} x$. For comparison purposes, $\operatorname{Cint}(1) \doteq 0.486385376235$. Prepare a graph of $\operatorname{Cint} x$ on $[-1, 1]$.

5. The error function

$$\operatorname{erf} x = \frac{2}{\sqrt{\pi}} \int_0^x e^{-t^2}\, dt$$

   is useful in the theory of probability. Find its Taylor polynomial so that the error is bounded by 10^{-9} for $|x| \le b$ for a given $b > 0$. Show how to evaluate the Taylor polynomial efficiently. Draw a graph of the polynomial on $[-b, b]$. Use the values $b = 1, 3, 5$.

6. Suppose $p(x) = 4x^7 - 3x^6 - 2x^5 + x^4 + x^3 - 1$ is divided by $x - 1$. What is the remainder? What is the quotient?

7. Show (1.38). Compute the quantity $b_0 + (x - z)q(x)$ by substituting (1.37), and collect together common powers of x. Then simplify those coefficients by using (1.35). It may be easier to initially restrict the proof to a low degree, say, $n = 3$.

8. Show $p'(z) = q(z)$ in (1.38).

9. Evaluate

$$p(x) = 1 - \frac{x^3}{3!} + \frac{x^6}{6!} - \frac{x^9}{9!} + \frac{x^{12}}{12!} - \frac{x^{15}}{15!}$$

as efficiently as possible. How many multiplications are necessary? Assume all coefficients have been computed and stored for later use.

10. Show how to evaluate the function

$$f(x) = 2e^{4x} - e^{3x} + 5e^x + 1$$

efficiently.

Hint: Consider letting $z = e^x$.

11. For $f(x) = e^x$, find a Taylor approximation that is in error by at most 10^{-7} on $[-1, 1]$. Using this approximation, write a function program to evaluate e^x. Compare it to the standard value of e^x obtained from the MATLAB function `exp(x)`; calculate the difference between your approximation and `exp(x)` at 21 evenly spaced points in $[-1, 1]$.

CHAPTER 2

ERROR AND COMPUTER ARITHMETIC

Much of numerical analysis is concerned with how to solve a problem numerically, that is, how to develop a sequence of numerical calculations that will give a satisfactory answer. Part of this process is the consideration of the errors that arise in these calculations, whether from errors in arithmetic operations or from some other source. Throughout this book, as we look at the numerical solution of various problems, we will simultaneously consider the errors involved in whatever computational procedure is being used. In this chapter, we give a few general results on error.

We begin in Section 2.1 by considering the representation of numbers within present-day computers and some consequences of it. Section 2.2 contains basic definitions regarding the idea of error and some important examples of it, and Section 2.3 discusses the propagation of error in arithmetic calculations. Section 2.4 examines errors involved in some summation procedures and illustrates the importance of the definition of basic machine arithmetic operations in numerical calculations.

2.1. FLOATING-POINT NUMBERS

Numbers must be stored in computers and arithmetic operations must be performed on these numbers. Most computers have two ways of storing numbers, in integer format and in floating-point format. The integer format is relatively straightforward, and we will not consider it here. The floating-point format is a more general format allowing storage of numbers that are not integers, and in this section, we define floating-point format. Most computers use the binary number system, and an introductory discussion of it is given in Appendix E. We will discuss the most popular floating-point format being used currently for binary computers, but we begin with floating-point format for decimal numbers.

To simplify the explanation of the floating-point representation of a number, let us first consider a nonzero number x written in the decimal system. It can be written in a unique way as

$$x = \sigma \cdot \bar{x} \cdot 10^e \tag{2.1}$$

where $\sigma = +1$ or -1, e is an integer, and $1 \leq \bar{x} < 10$. These three quantities are called the *sign*, *exponent*, and *significand*, respectively, of the representation (2.1). As an example,

$$124.62 = (1.2462) \cdot 10^2$$

with the sign $\sigma = +1$, the exponent $e = 2$, and the significand $\bar{x} = 1.2462$. The format of (2.1) is often called *scientific notation* in high school texts on mathematics and science. Note that the *significand* is also called the *mantissa* in many texts.

The decimal floating-point representation of a number x is basically that given in (2.1), with limitations on the number of digits in $\bar{x}$ and on the size of e. For example, suppose we limit the number of digits in $\bar{x}$ to four and the size of e to between -99 and $+99$. We say that a computer with such a representation has a four-digit decimal floating-point arithmetic. As a corollary to the limitation on the length of $\bar{x}$, we cannot guarantee to store accurately more than the first four digits of a number, and even the fourth digit may need to be changed by rounding (which later we define more precisely). Some popular hand calculators use ten-digit decimal floating-point arithmetic. Because decimal arithmetic is more intuitive for most people, we occasionally illustrate ideas using decimal floating-point arithmetic rather than binary floating-point arithmetic.

Now consider a number x written in binary format. In analogy with (2.1), we can write

$$x = \sigma \cdot \bar{x} \cdot 2^e \tag{2.2}$$

where $\sigma = +1$ or -1, e is an integer, and $\bar{x}$ is a binary fraction satisfying

$$(1)_2 \leq \bar{x} < (10)_2 \tag{2.3}$$

Table 2.1. Storage of IEEE single precision floating-point format

$\underbrace{b_1}_{\sigma}$	$\underbrace{b_2 b_3 \ldots b_9}_{E}$	$\underbrace{b_{10} b_{11} \ldots b_{32}}_{\bar{x}}$

In decimal, $1 \leq \bar{x} < 2$. For example, if $x = (11011.0111)_2$, then $\sigma = +1$, $e = 4 = (100)_2$, and $\bar{x} = (1.10110111)_2$. Note that for $x \neq 0$, the digit to the left of the binary point in $\bar{x}$ is guaranteed to be 1.

The floating-point representation of a binary number x consists of (2.2) with a restriction on the number of binary digits in $\bar{x}$ and on the size of e. The allowable number of binary digits in $\bar{x}$ is called the *precision* of the binary floating-point representation of x. The IEEE floating-point arithmetic standard is the format for floating-point numbers used in almost all present-day computers. For example, Intel processors all use this standard. With this standard, the *IEEE single precision floating-point representation* of x has a precision of 24 binary digits and the exponent e is limited by $-126 \leq e \leq 127$.

$$x = \sigma \cdot (1.a_1 a_2 \ldots a_{23}) \cdot 2^e \tag{2.4}$$

In binary,

$$-(1111110)_2 \leq e \leq (1111111)_2$$

The *IEEE double precision floating-point representation* of x has a precision of 53 binary digits and the exponent e is limited by $-1022 \leq e \leq 1023$.

$$x = \sigma \cdot (1.a_1 a_2 \ldots a_{52}) \cdot 2^e \tag{2.5}$$

Single precision floating-point format uses four bytes (32 bits), and the storage scheme for it is sketched in Table 2.1. The sign σ is stored in bit b_1 ($b_1 = 0$ for $\sigma = +1$ and $b_1 = 1$ for $\sigma = -1$). Define $E = e + 127$. Rather than e, we store the positive binary integer E in bits b_2 through b_9. The bits $a_1 a_2 \ldots a_{23}$ of $\overline{x}$ are stored in bits b_{10} through b_{32}. The leading binary digit 1 in $\overline{x}$ is not stored in the floating-point representation when the number x is stored in memory; but the leading 1 is inserted into $\overline{x}$ when a floating-point number x is brought out of memory and into an arithmetic register for further use (or something equivalent is done). This leads to the need for a special representation of the number $x = 0$; it is stored as $E = 0$ with $\sigma = 0$ and $b_{10} b_{11} \ldots b_{32} = (00 \ldots 0)_2$. A list of all possible single precision floating-point numbers is given in Table 2.2. This and Table 2.4 are adapted from the excellent text of Overton [27, Tables 4.1, 4.2].

The double precision format uses eight bytes (64 bits), and how it is stored is sketched in Table 2.3. The sign σ is stored in bit b_1 ($b_1 = 0$ for $\sigma = +1$ and $b_1 = 1$ for $\sigma = -1$). Define $E = e + 1023$. Rather than e, we store the positive binary integer E in bits b_2 through b_{12}. The bits $a_1 a_2 \ldots a_{52}$ of $\overline{x}$ are stored in bits b_{13} through b_{64}. The leading binary digit 1 in $\overline{x}$ is not stored in the floating-point representation when the number x is stored in memory; but the leading 1 is inserted into $\overline{x}$ when a floating-point number x is brought out of memory and into an arithmetic register for further use.

Table 2.2. IEEE single precision format

$\pm$	$c_1 \ldots c_8$	$a_1 a_2 \ldots a_{23}$

$E = (c_1 \ldots c_8)_2$	x
$(00000000)_2 = (0)_{10}$	$\pm (0.a_1 a_2 \ldots a_{23})_2 \cdot 2^{-126}$
$(00000001)_2 = (1)_{10}$	$\pm (1.a_1 a_2 \ldots a_{23})_2 \cdot 2^{-126}$
$(00000010)_2 = (2)_{10}$	$\pm (1.a_1 a_2 \ldots a_{23})_2 \cdot 2^{-125}$
$\vdots$	$\vdots$
$(01111111)_2 = (127)_{10}$	$\pm (1.a_1 a_2 \ldots a_{23})_2 \cdot 2^{0}$
$(10000000)_2 = (128)_{10}$	$\pm (1.a_1 a_2 \ldots a_{23})_2 \cdot 2^{1}$
$\vdots$	$\vdots$
$(11111101)_2 = (253)_{10}$	$\pm (1.a_1 a_2 \ldots a_{23})_2 \cdot 2^{126}$
$(11111110)_2 = (254)_{10}$	$\pm (1.a_1 a_2 \ldots a_{23})_2 \cdot 2^{127}$
$(11111111)_2 = (255)_{10}$	$\pm\infty$ if $a_1 = \cdots = a_{23} = 0$; *NaN* otherwise

Table 2.3. Storage of IEEE double precision floating-point format

$\underbrace{b_1}_{\sigma}$	$\underbrace{b_2 b_3 \ldots b_{12}}_{E}$	$\underbrace{b_{13} b_{14} \ldots b_{64}}_{\bar{x}}$

This leads to the need for a special representation of the number $x = 0$; it is stored as $E = 0$ with $\sigma = 0$ and $b_{13} b_{14} \ldots b_{64} = (00 \ldots 0)_2$. A list of all possible double precision floating-point numbers is given in Table 2.4.

In addition, there are also representations for ∞ and $-\infty$, a computer version of $\pm$infinity. Thus, $\frac{1}{0} = \infty$ in this floating-point arithmetic. There is also a representation for *NaN*, meaning "not a number." Floating-point operations such as $\frac{0}{0}$ lead to an answer of *NaN*, rather than causing the program to stop abruptly with some possibly cryptic error message. The program can then test for *NaN* and respond to its presence as needed. There are additional nuances to the IEEE standard. When the leading 1 is missing in (2.4) and (2.5), corresponding to $E = 0$ in Tables 2.2 and 2.4, we refer to the floating-point format as *unnormalized*. It carries less precision in the significand than the *normalized format* of (2.4) and (2.5), and we ignore it for the most part in this text.

Example 2.1.1 MATLAB can be used to generate the entries in Table 2.4. Execute the command

```
format hex
```

This will cause all subsequent numerical output to the screen to be given in hexadecimal format (base 16). For example, listing the number 7 results in an output of

Table 2.4. IEEE double precision format

$\pm$	$c_1 \ldots c_{11}$	$a_1 a_2 \ldots a_{52}$

$E = (c_1 \ldots c_{11})_2$	x
$(00000000000)_2 = (0)_{10}$	$\pm (0.a_1 a_2 \ldots a_{52})_2 \cdot 2^{-1022}$
$(00000000001)_2 = (1)_{10}$	$\pm (1.a_1 a_2 \ldots a_{52})_2 \cdot 2^{-1022}$
$(00000000010)_2 = (2)_{10}$	$\pm (1.a_1 a_2 \ldots a_{52})_2 \cdot 2^{-1021}$
$\vdots$	$\vdots$
$(01111111111)_2 = (1023)_{10}$	$\pm (1.a_1 a_2 \ldots a_{52})_2 \cdot 2^{0}$
$(10000000000)_2 = (1024)_{10}$	$\pm (1.a_1 a_2 \ldots a_{52})_2 \cdot 2^{1}$
$\vdots$	$\vdots$
$(11111111101)_2 = (2045)_{10}$	$\pm (1.a_1 a_2 \ldots a_{52})_2 \cdot 2^{1022}$
$(11111111110)_2 = (2046)_{10}$	$\pm (1.a_1 a_2 \ldots a_{52})_2 \cdot 2^{1023}$
$(11111111111)_2 = (2047)_{10}$	$\pm\infty$ if $a_1 = \cdots = a_{52} = 0$; *NaN* otherwise

Table 2.5. Conversion of hexadecimal to binary

Hex digit	Binary equivalent
0	0000
1	0001
2	0010
$\vdots$	$\vdots$
9	1001
a	1010
$\vdots$	$\vdots$
f	1111

$$401c000000000000 \tag{2.6}$$

The 16 hexadecimal digits are $\{0, 1, 2, 3, 4, 5, 6, 7, 8, 9, a, b, c, d, e, f\}$. To obtain the binary representation, convert each hexadecimal digit to a four-digit binary number using Table 2.5. Replace each hexadecimal digit with its binary equivalent. For the above number, we obtain the binary expansion

$$0100\,0000\,0001\,1100\,0000 \ldots 0000 \tag{2.7}$$

for the number 7 in IEEE double precision floating-point format. ■

2.1.1 Accuracy of Floating-Point Representation

Consider how accurately a number can be stored in the floating-point representation. This is measured in various ways, with the *machine epsilon* being the most popular. The machine epsilon for any particular floating-point format is the difference between 1 and the next larger number that can be stored in that format. In single precision IEEE format, the next larger binary number is

$$1.00000000000000000000001 \tag{2.8}$$

with the final binary digit 1 in position 23 to the right of the binary point. Thus, the *machine epsilon in single precision IEEE format* is 2^{-23}. As an example, it follows that the number $1 + 2^{-24}$ cannot be stored exactly in IEEE single precision format. From

$$2^{-23} \doteq 1.19 \times 10^{-7} \tag{2.9}$$

we say that IEEE single precision format can be used to store approximately 7 decimal digits of a number x when it is written in decimal format. In a similar fashion, the *machine epsilon in double precision IEEE format* is $2^{-52} = 2.22 \times 10^{-16}$; IEEE double precision format can be used to store approximately 16 decimal digits of a number x. In MATLAB, the machine epsilon is available as the constant named `eps`.

As another way to measure the accuracy of a floating-point format, we look for the largest integer M having the property that any integer x satisfying $0 \le x \le M$ can be stored or represented exactly in floating-point form. Since x is an integer, we will have to convert it to a binary number $\bar{x}$ with a positive exponent e in (2.2). If n is the number of binary digits in the significand, then it is fairly easy to convince oneself that all integers less than or equal to

$$(1.11\ldots 1)_2 \cdot 2^{n-1}$$

can be stored exactly, where this significand contains n binary digits, all 1. This is the integer composed of n consecutive 1's; and using (1.21), it equals $2^n - 1$. In addition, $2^n = (1.0\ldots 0)_2 \cdot 2^n$ also stores exactly. But there are not enough digits in the significand to store $2^n + 1$, as this would require $n + 1$ binary digits in the significand. With an n digit binary floating-point representation,

$$M = 2^n \tag{2.10}$$

Any positive integer $\le M$ can be represented exactly in this floating-point representation.

In the IEEE single precision format,

$$M = 2^{24} = 16777216$$

and all 7-digit decimal integers will store exactly. In IEEE double precision format, the number

$$M = 2^{53} \doteq 9.0 \times 10^{15}$$

and thus all 15-digit decimal integers and most 16 digit ones will store exactly.

2.1.2 Rounding and Chopping

Let the significand in the floating-point representation contain n binary digits. If the number x in (2.2) has a significand $\bar{x}$ that requires more than n binary bits, then it must be shortened when x is stored in the computer. Currently, this is usually done in one of two ways, depending on the computer or the options selected within the IEEE standard for floating-point arithmetic. The simplest method is to simply truncate or *chop* $\bar{x}$ to n binary digits, ignoring the remaining digits. The second method is to *round* $\bar{x}$ to n digits, based on the size of the part of $\bar{x}$ following digit n. More precisely, if digit $n+1$ is zero, chop $\bar{x}$ to n digits; otherwise, chop $\bar{x}$ to n digits and add 1 to the last digit of the result. Regardless of whether chopping or rounding is being used, we will denote the *machine floating-point version* of a number x by $\text{fl}(x)$.

It can be shown that the number $\text{fl}(x)$ can be written in the form

$$\text{fl}(x) = x \cdot (1 + \epsilon) \tag{2.11}$$

with ϵ a small number, dependent on x. Since ϵ is small, this says that $\text{fl}(x)$ is a slight perturbation of x. If chopping is used,

$$-2^{-n+1} \leq \epsilon \leq 0 \tag{2.12}$$

and if rounding is used,

$$-2^{-n} \leq \epsilon \leq 2^{-n} \tag{2.13}$$

The most important characteristics of chopping are: (1) The worst possible error is twice as large as when rounding is used; and (2) the sign of the error $x - \text{fl}(x)$ is the same as the sign x. This last characteristic is the worst of the two. In many calculations it will lead to no possibility of cancellation of errors; examples are given later in Section 2.4. With rounding, the worst possible error is only one-half as large as for chopping. More important, the error $x - \text{fl}(x)$ is negative for half of the cases and positive for the other half. This leads to much better error propagation behavior in calculations involving many arithmetic operations.

For single precision IEEE floating-point arithmetic, several variants of rounding and chopping are available. The method of chopping described above is called "rounding towards zero," and with it we have

$$-2^{-23} \leq \epsilon \leq 0 \tag{2.14}$$

Most users of the IEEE standard use as a default the rounding described preceding (2.11). There are $n = 24$ digits in the significand; and with standard rounding,

$$-2^{-24} \leq \epsilon \leq 2^{-24} \tag{2.15}$$

The corresponding results for double precision IEEE floating-point arithmetic are

$$\begin{aligned} -2^{-52} &\leq \epsilon \leq 0 && \text{chopping} \\ -2^{-53} &\leq \epsilon \leq 2^{-53} && \text{rounding} \end{aligned} \tag{2.16}$$

2.1.3 Consequences for Programming of Floating-Point Arithmetic

Numbers that have finite decimal expansions may have infinite binary expansions. For example,

$$(0.1)_{10} = (0.000110011001100110011\ldots)_2$$

and therefore $(0.1)_{10}$ cannot be represented exactly in binary floating-point arithmetic. Consider the following MATLAB code on a binary machine, illustrating a possible problem:

```
x = 0.0
while x ~= 1.0
    x = x + 0.1
    disp([x,sqrt(x)])
end
```

This code forms an infinite loop. Clearly, the programmer intended it to stop when $x = 1.0$; but because the decimal number 0.1 is not represented exactly in the computer, the number 1.0 is never attained exactly. A different kind of looping test should be used, one not as sensitive to rounding or chopping errors.

In languages with both single precision and double precision, such as *Fortran* and *C*, it is important to specify double precision constants correctly. As an example in *Fortran*, consider the statement

```
PI = 3.14159265358979
```

and suppose that PI has been declared as a double precision variable. Even though the number on the right side is correct to double precision accuracy, it will not compile as such with many compilers. Rather, the number on the right side will be rounded as a single precision constant during the compilation phase; and then at run time, it will have zeros appended to extend it to double precision. Instead, one should write

```
PI = 3.14159265358979D0
```

which is a double precision constant. As a related example, consider

PI = 4.0*ATAN(1.0)

again with PI declared as a double precision variable. When executed, this statement will create an approximation to π with single precision accuracy, and then it will be extended to a double precision constant by appending zeros. To obtain a value for PI with double precision accuracy as in regard to the value of π, use

PI = 4.0D0*ATAN(1.0D0)

As another example, consider

H = 0.1

with H having been declared double precision. This will result in a value for H with only single precision accuracy to the decimal number 0.1. Instead, use

H = 0.1D0

In MATLAB, all the computations are done automatically in double precision. We do not need to pay attention to the format of specified double precision constants. In other words, with MATLAB we can focus more on the computational procedures.

PROBLEMS

1. Using MATLAB and proceeding as in the example of (2.6)–(2.7), find the binary double precision IEEE floating-point expressions for the following numbers:

   (a) 8 **(b)** 12 **(c)** 1.5
   (d) 0.5 **(e)** 1.25

2. Some microcomputers in the past used a binary floating-point format with 8 bits for the exponent e and 1 bit for the sign σ. The significand $\bar{x}$ contained 31 bits, $1 \leq \bar{x} < 2$, with no hiding of the leading bit 1. The arithmetic also used rounding. To determine the accuracy of the representation, find the machine epsilon [described preceding (2.8)] and the number M of (2.10). Also find bounds for the number ϵ of (2.11).

3. Write a program in the language of your choice to implement the following algorithm (given in MATLAB). Run it on one or more computers to which you have access. Comment on your results.

```
power = 1.0;
b = 1.0;
while b ~= 0.0
  power = power/2;
```

```
        a = 1.0 + power;
        b = a - 1.0;
        disp([power,a,b])
    end
    disp(['Unit round = ',num2str(2*power)])
```

4. Predict the output of the following section of code if it is run on a binary computer that uses chopping. More precisely, estimate the number of times the programmer probably intended for this loop to be executed; and state whether this intended behavior is what actually occurs. Would the outcome be any different if the statement "$X = X + H$" was replaced by "$X = I * H$"?

```
I = 0
X = 0.0
H = 0.1
WHILE X < 1.0
   I = I + 1
   X = X + H
   disp([I,X])
end
```

5. The following MATLAB program produced the given output. Explain the results.

```
x = 0.0
while x < 1.0
   x = x + 0.1;
   disp([x, sqrt(x)])
end
```

x	0.1	0.2	0.3	0.4	0.5	0.6
$\sqrt{x}$	0.3162	0.4472	0.5477	0.6325	0.7071	0.7746
x	0.7	0.8	0.9	1.0	1.1	
$\sqrt{x}$	0.8367	0.8944	0.9487	1.0000	1.0488	

6. Let $x > 0$ satisfy (2.2). Consider a computer using a positive binary floating-point representation with n bits of precision in the significand [e.g., $n = 24$ in (2.2)]. Assume that *chopping* (rounding toward zero) is used in going from a number x outside the computer to its floating-point approximation $\text{fl}(x)$ inside the computer.

(a) Show that

$$0 \le x - \text{fl}(x) \le 2^{e-n+1}$$

(b) Show that $x \ge 2^e$, and use this to show

$$\frac{x - \text{fl}(x)}{x} \le 2^{-n+1}$$

(c) Let

$$\frac{x - \text{fl}(x)}{x} = -\epsilon$$

and then solve for $\text{fl}(x)$. What are the bounds on ϵ? (This result extends to $x < 0$, with the assumption of $x > 0$ being used to simplify the algebra.)

7. Let $x > 0$ satisfy (2.2). Consider a computer using a positive binary floating-point representation with n bits of precision in the significand [e.g., $n = 24$ in (2.2)]. Assume that *rounding* is used in going from a number x outside the computer to its floating-point approximation $\text{fl}(x)$ inside the computer.

(a) Show that

$$-2^{e-n} \le x - \text{fl}(x) \le 2^{e-n}$$

(b) Show that $x \ge 2^e$, and use this to show

$$\frac{|x - \text{fl}(x)|}{x} \le 2^{-n}$$

(c) Let

$$\frac{x - \text{fl}(x)}{x} = -\epsilon$$

and then solve for $\text{fl}(x)$. What are the bounds on ϵ? (This result extends to $x < 0$, with the assumption of $x > 0$ being used to simplify the algebra.)

2.2. ERRORS: DEFINITIONS, SOURCES, AND EXAMPLES

The *error* in a computed quantity is defined as

$$\text{Error} = \text{true value} - \text{approximate value}$$

The quantity thus defined is also called the *absolute error*. The *relative error* is a measure of the error in relation to the size of the true value being sought:

$$\text{Relative error} = \frac{\text{error}}{\text{true value}}$$

This gives the size of the error in proportion to the true value being approximated. To simplify the notation when working with these quantities, we will usually denote the true and approximate values of a quantity x by x_T and x_A, respectively. Then we write

$$\text{Error}(x_A) = x_T - x_A$$

$$\text{Rel}(x_A) = \frac{x_T - x_A}{x_T}$$

As an illustration, consider the well-known approximation

$$\pi \doteq \frac{22}{7}$$

Here $x_T = \pi = 3.14159265\ldots$ and $x_A = 22/7 = 3.1428571\ldots$

$$\text{Error}\left(\frac{22}{7}\right) = \pi - \frac{22}{7} \doteq -0.00126$$

$$\text{Rel}\left(\frac{22}{7}\right) = \frac{\pi - (22/7)}{\pi} \doteq -0.000402$$

Another example of error measurement is given by the Taylor remainder (1.11).

The notion of the relative error is a more intrinsic error measure. For instance, suppose the exact distance between two cities is $d_T^{(1)} = 100$ km and the measured distance is $d_A^{(1)} = 99$ km. Then

$$\text{Error}\left(d_A^{(1)}\right) = d_T^{(1)} - d_A^{(1)} = 1\text{ km}$$

$$\text{Rel}\left(d_A^{(1)}\right) = \frac{\text{Error}\left(d_A^{(1)}\right)}{d_T^{(1)}} = 0.01 \equiv 1\%$$

Now suppose the exact distance between a certain two stores is $d_T^{(2)} = 2$ km and it is estimated to be $d_A^{(2)} = 1$ km. Then

$$\text{Error}\left(d_A^{(2)}\right) = d_T^{(2)} - d_A^{(2)} = 1\text{ km}$$

$$\text{Rel}\left(d_A^{(2)}\right) = \frac{\text{Error}\left(d_A^{(2)}\right)}{d_T^{(2)}} = 0.5 \equiv 50\%$$

In both cases the errors are equal. But obviously $d_A^{(1)}$ is a much more accurate estimate of $d_T^{(1)}$ than is $d_A^{(2)}$ as an estimate of $d_T^{(2)}$, and this is reflected in the sizes of the two relative errors.

An idea related to relative error is that of *significant digits*. For a number x_A, the number of its leading digits that are correct relative to the corresponding digits in the true value x_T is called the number of significant digits in x_A. For a more precise definition, and assuming the numbers are written in decimal form, calculate the magnitude of the error, $|x_T - x_A|$. If this error is less than or equal to five units in the $(m+1)^{\text{th}}$ digit of x_T, counting rightward from the first nonzero digit, then we say x_A has, at least, m significant digits of accuracy relative to x_T.

Example 2.2.1

(a) $x_A = 0.222$ has three digits of accuracy relative to $x_T = \frac{2}{9}$.

(b) $x_A = 23.496$ has four digits of accuracy relative to $x_T = 23.494$.

(c) $x_A = 0.02138$ has just two digits of accuracy relative to $x_T = 0.02144$.

(d) $x_A = \frac{22}{7}$ has three digits of accuracy relative to $x_T = \pi$. ■

It can be shown that if

$$\left|\frac{x_T - x_A}{x_T}\right| \le 5 \cdot 10^{-m-1} \tag{2.17}$$

then x_A has m significant digits with respect to x_T. Most people find it easier to measure relative error than significant digits; and in some textbooks, satisfaction of (2.17) is used as the definition of x_A having m significant digits of accuracy.

2.2.1 Sources of Error

Imagine solving a scientific-mathematical problem, and suppose this involves a computational procedure. Errors will usually be involved in this process, often of several different kinds. We sometimes think of errors as divided into "original errors" and "consequences of errors." We will give a rough classification of the kinds of original errors that might occur.

(E1) Modeling Errors Mathematical equations are used to represent physical reality, a process that is called *mathematical modeling*. This modeling introduces error into the description of the real-world problem that you are trying to solve.

For example, the simplest model for population growth is given by

$$N(t) = N_0 e^{kt} \tag{2.18}$$

where $N(t)$ equals the population at time t, and N_0 and k are positive constants. For some stages of growth of a population, when it has unlimited resources, this can be an accurate model. But more often, it will overestimate the actual population for large t. For example, it accurately models the growth of U.S. population over the period of $1790 \le t \le 1860$, with $k = 0.02975$ and $N_0 = 3{,}929{,}000 \times e^{-1790k}$; but it considerably overestimates the actual population in 1870.

Another example arises in studying the spread of rubella measles. We have the following model for the spread of the infection in a population, subject to certain assumptions:

$$\frac{ds(t)}{dt} = -a\,s(t)\,i(t)$$
$$\frac{di(t)}{dt} = a\,s(t)\,i(t) - b\,i(t)$$
$$\frac{dr(t)}{dt} = b\,i(t)$$

In this, $s(t)$, $i(t)$, and $r(t)$ refer, respectively, to the proportions of a total population at time t that are *susceptible*, *infectious*, and *removed* (from the susceptible and infectious pool of people). All variables are functions of time t. The constants can be taken as

$$a = \frac{6.8}{11}, \qquad b = \frac{1}{11}$$

The same model works for some other diseases (e.g., flu), with a suitable change of the constants a and b. Again, this is a useful approximation of reality.

The error in a mathematical model falls outside the scope of numerical analysis, but it is still an error with respect to the solution of the overall scientific problem of interest.

(E2) Blunders and Mistakes These errors are familiar to almost everyone. In the precomputer era, blunders generally consisted of isolated arithmetic errors, and elaborate check schemes were used to detect them. Today, the mistakes are more likely to be programming errors. To detect these errors, it is important to have some way of checking the accuracy of the program output. When first running the program, use cases for which you know the correct answer. With a complex program, break it into small subprograms, each of which can be tested separately. And when you believe the entire program is correct and are running it for cases of interest, maintain a watchful eye as to whether the output is reasonable.

(E3) Physical Measurement Errors Many problems involve physical data, and these data contain observational error. For example, the speed of light in a vacuum is

$$c = (2.997925 + \epsilon) \cdot 10^{10} \text{ cm/sec}, \qquad |\epsilon| \le 0.000003 \tag{2.19}$$

Because the physical data contain an error, calculations based on the data will contain the effect of this observational error. Numerical analysis cannot remove the error in the data, but it can look at its propagated effect in a calculation. Numerical analysis also can suggest the best form for a calculation that will minimize the propagated effect of the errors in the data. Propagation of errors in arithmetic calculations is considered in Section 2.3.

(E4) Machine Representation and Arithmetic Errors These occur when using computers or calculators, as for example with the rounding or chopping errors discussed in Section 2.1. These errors are inevitable when using floating-point arithmetic; and they form the main source of error with some problems, for example, the solution of systems of linear equations. In Section 2.4, we will analyze the effect of rounding errors for some summation procedures.

(E5) Mathematical Approximation Errors These are the major forms of error in which we are interested in the following chapters. To illustrate this type of error, consider evaluating the integral

$$I = \int_0^1 e^{-x^2}\, dx$$

There is no antiderivative for e^{-x^2} in terms of elementary functions and, thus, the integral cannot be evaluated explicitly. Instead, we approximate the integral with a quantity that can be computed. For example, use the Taylor approximation

$$e^{-x^2} \approx 1 - x^2 + \frac{x^4}{2!} - \frac{x^6}{3!} + \frac{x^8}{4!}$$

Then

$$I \approx \int_0^1 \left(1 - x^2 + \frac{x^4}{2!} - \frac{x^6}{3!} + \frac{x^8}{4!}\right) dx \tag{2.20}$$

which can be evaluated easily. The error in (2.20) is called a *mathematical approximation error*; some authors call it a *truncation error* or *discretization error*. Such errors arise when we have a problem that we cannot solve exactly and that we approximate with a new problem that we can solve. We try to estimate the size of these mathematical approximation errors and to make them sufficiently small. For the above approximate integral (2.20), apply the earlier example (1.18).

We will now describe three important errors that occur commonly in practice. They are sometimes considered as sources of error, but we think of them as deriving from the sources (E1) to (E5) described above. Also, the size of these errors can usually be minimized by using a properly chosen form for the computation being carried out.

2.2.2 Loss-of-Significance Errors

The idea of *loss of significant digits* is best understood by looking at some examples. Consider first the evaluation of

$$f(x) = x[\sqrt{x+1} - \sqrt{x}\,] \tag{2.21}$$

for an increasing sequence of values of x. The results of using a six-digit decimal calculator are shown in Table 2.6. As x increases, there are fewer digits of accuracy in the computed value of $f(x)$.

Table 2.6. Values of (2.21)

x	Computed $f(x)$	True $f(x)$
1	0.414210	0.414214
10	1.54340	1.54347
100	4.99000	4.98756
1000	15.8000	15.8074
10,000	50.0000	49.9988
100,000	100.000	158.113

To better understand what is happening, look at the individual steps of the calculation when $x = 100$. On the calculator,

$$\sqrt{100} = 10.0000, \quad \sqrt{101} = 10.0499$$

The first value is exact, and the second value is correctly rounded to six significant digits of accuracy. Next,

$$\sqrt{x+1} - \sqrt{x} = \sqrt{101} - \sqrt{100} = 0.0499000 \tag{2.22}$$

while the true value should be 0.0498756. The calculation (2.22) has a *loss-of-significance error*. Three digits of accuracy in $\sqrt{x+1} = \sqrt{101}$ were canceled by subtraction of the corresponding digits in $\sqrt{x} = \sqrt{100}$. The loss of accuracy was a by-product of the form of $f(x)$ and the finite precision six-digit decimal arithmetic being used.

For this particular $f(x)$, there is a simple way to reformulate it so as to avoid the loss-of-significance error. Consider (2.21) as a fraction with a denominator of 1, and multiply numerator and denominator by $\sqrt{x+1} + \sqrt{x}$, obtaining

$$f(x) = x\frac{\sqrt{x+1} - \sqrt{x}}{1} \cdot \frac{\sqrt{x+1} + \sqrt{x}}{\sqrt{x+1} + \sqrt{x}} = \frac{x}{\sqrt{x+1} + \sqrt{x}} \tag{2.23}$$

The latter expression will not have any loss-of-significance errors in its evaluation. On our six-digit decimal calculator, (2.23) gives

$$f(100) = 4.98756$$

the correct answer to six digits.

As a second example, consider evaluating

$$f(x) = \frac{1 - \cos(x)}{x^2} \tag{2.24}$$

for a sequence of values of x approaching 0. The results of using a popular 10-digit decimal hand calculator are shown in Table 2.7. To understand the loss of accuracy, look

Table 2.7. Values of (2.24) on a 10-digit calculator

x	Computed $f(x)$	True $f(x)$
0.1	0.4995834700	0.4995834722
0.01	0.4999960000	0.4999958333
0.001	0.5000000000	0.4999999583
0.0001	0.5000000000	0.4999999996
0.00001	0.0	0.5000000000

at the individual steps of the calculation when $x = 0.01$. First, on the calculator

$$\cos(0.01) = 0.9999500004$$

This has nine significant digits of accuracy, and it is off in the tenth digit by two units. Next, compute

$$1 - \cos(0.01) = 0.0000499996$$

This has only five significant digits, with four digits being lost in the subtraction. Division by $x^2 = 0.0001$ gives the entry in the table.

To avoid the loss of significant digits, we use another formulation for $f(x)$, avoiding the subtraction of nearly equal quantities. By using the Taylor approximation in (1.15), we get

$$\cos(x) = 1 - \frac{x^2}{2!} + \frac{x^4}{4!} - \frac{x^6}{6!} + R_6(x)$$

$$R_6(x) = \frac{x^8}{8!}\cos(\xi)$$

with ξ an unknown number between 0 and x. Therefore,

$$\begin{aligned} f(x) &= \frac{1}{x^2}\left\{1 - \left[1 - \frac{x^2}{2!} + \frac{x^4}{4!} - \frac{x^6}{6!} + R_6(x)\right]\right\} \\ &= \frac{1}{2!} - \frac{x^2}{4!} + \frac{x^4}{6!} - \frac{x^6}{8!}\cos(\xi) \end{aligned}$$

Thus, $f(0) = \frac{1}{2}$. And for $|x| \le 0.1$,

$$\left|\frac{x^6}{8!}\cos(\xi)\right| \le \frac{10^{-6}}{8!} \doteq 2.5 \cdot 10^{-11} \tag{2.25}$$

Hence,

$$f(x) \approx \frac{1}{2!} - \frac{x^2}{4!} + \frac{x^4}{6!}, \qquad |x| \le 0.1$$

with an accuracy given by (2.25). This gives a much better way of evaluating $f(x)$ for small values of x.

When two nearly equal quantities are subtracted, leading significant digits will be lost. Sometimes this is easily recognized, as in the above two examples (2.21) and (2.24), and then ways can usually be found to avoid the loss of significance. More often, the loss of significance will be subtle and difficult to detect. One common situation is in calculating sums containing a number of terms, as when using a Taylor polynomial to approximate a function $f(x)$. If the value of the sum is relatively small when compared with some of the terms being summed, then there are probably some significant digits of accuracy being lost in the summation process.

As an example of this last phenomenon, consider using the Taylor series approximation (1.13) for e^x to evaluate e^{-5}:

$$e^{-5} = 1 + \frac{(-5)}{1!} + \frac{(-5)^2}{2!} + \frac{(-5)^3}{3!} + \frac{(-5)^4}{4!} + \cdots \tag{2.26}$$

Imagine using a computer with four-digit decimal floating-point arithmetic, so that the terms of this series must all be rounded to four significant digits. In Table 2.8, we give these terms, along with the associated numerical sum of these terms through the given degree. The true value of e^{-5} is 0.006738, to four significant digits, and this is quite different from the final sum in the table. Also, if (2.26) is calculated to much higher precision for terms of degree ≤ 25, then the correct value of e^{-5} is obtained to four digits.

In this example, the terms become relatively large, but they are then added to form a much smaller number e^{-5}. This means there are loss-of-significance errors in the calculation of the sum. The rounding error in the term of degree 3 is of the same magnitude as the error in the final answer in the table. To avoid the loss of significance in this case is quite easy. Either use

$$e^{-5} = \frac{1}{e^5}$$

and form e^5 with a series not involving cancellation of positive and negative terms; or preferably, simply form $e^{-1} = 1/e$ and multiply it by itself four times to form e^{-5}. With most other series, there is usually not such a simple solution.

GENERAL OBSERVATION: Suppose a sequence of numbers is being summed in order to obtain an answer S. If S is much smaller in magnitude than some of the terms being summed, then S is likely to contain a loss of significance error.	(2.27)

Table 2.8. Calculation of (2.26) using four-digit decimal arithmetic

Degree	Term	Sum	Degree	Term	Sum
0	1.000	1.000	13	−0.1960	−0.04230
1	−5.000	−4.000	14	0.7001E − 1	0.02771
2	12.50	8.500	15	−0.2334E − 1	0.004370
3	−20.83	−12.33	16	0.7293E − 2	0.01166
4	26.04	13.71	17	−0.2145E − 2	0.009518
5	−26.04	−12.33	18	0.5958E − 3	0.01011
6	21.70	9.370	19	−0.1568E − 3	0.009957
7	−15.50	−6.130	20	0.3920E − 4	0.009996
8	9.688	3.558	21	−0.9333E − 5	0.009987
9	−5.382	−1.824	22	0.2121E − 5	0.009989
10	2.691	0.8670	23	−0.4611E − 6	0.009989
11	−1.223	−0.3560	24	0.9607E − 7	0.009989
12	0.5097	0.1537	25	−0.1921E − 7	0.009989

2.2.3 Noise in Function Evaluation

Consider evaluating a function $f(x)$ for all points x in some interval $a \leq x \leq b$. If the function is continuous, then the graph of this function is a continuous curve. Next, consider the evaluation of $f(x)$ on a computer using floating-point arithmetic with rounding or chopping. Arithmetic operations (e.g., additions and multiplications) cause errors in the evaluation of $f(x)$, generally quite small ones. If we look very carefully at the graph of the computed values of $f(x)$, it will no longer be a continuous curve, but instead a "fuzzy" curve reflecting the errors in the evaluation process.

Example 2.2.2 To illustrate these comments, we evaluate $f(x) = (x-1)^3$. We do so in the nested form

$$f(x) = -1 + x\,(3 + x\,(-3 + x)) \tag{2.28}$$

using MATLAB. The arithmetic in MATLAB uses IEEE double precision numbers and standard rounding. Figure 2.1 contains the graph of the computed values of $f(x)$ for $0 \leq x \leq 2$, and it appears to be a smooth continuous curve. Next we look at a small segment of this curve, for $0.99998 \leq x \leq 1.00002$. The plot of the computed values of $f(x)$ is given in Figure 2.2, for 81 evenly spaced values of x in [0.99998, 1.00002]. Note that the graph of $f(x)$ does not appear to be taken from a continuous curve, but rather, it is a narrow "fuzzy band" of seemingly random values. This is true of all parts of the computed curve of $f(x)$, but it becomes evident only when you look at the curve very closely. ■

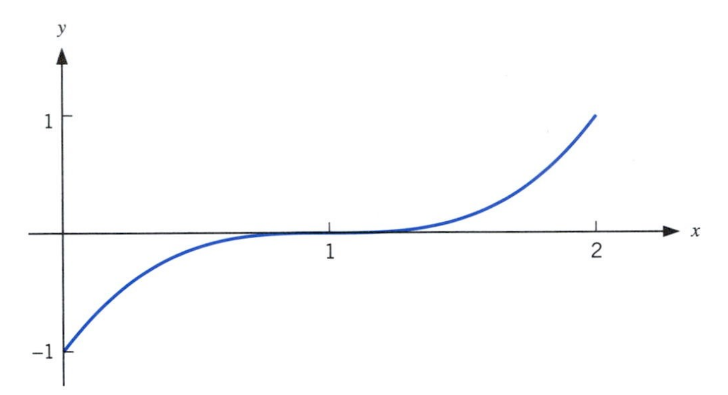

Figure 2.1. $f(x) = x^3 - 3x^2 + 3x - 1$

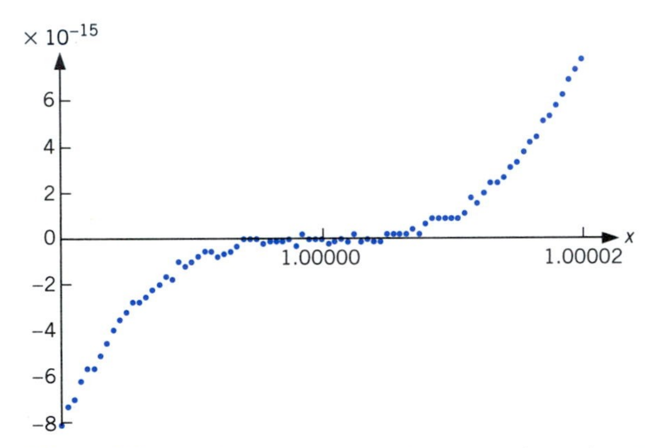

Figure 2.2. A detailed graph of $f(x) = x^3 - 3x^2 + 3x - 1$ near $x = 1$

Functions $f(x)$ on a computer should be thought of as a fuzzy band of random values, with the vertical thickness of the band quite small in most cases. The implications of this are minimal in most instances, but there are situations where it is important. For example, a rootfinding program might consider (2.28) to have a very large number of roots in [0.99998, 1.00002] on the basis of the many sign changes, as shown in Figure 2.2.

2.2.4 Underflow and Overflow Errors

From the definition of floating-point numbers given in Section 2.1, there are upper and lower limits for the magnitudes of the numbers that can be expressed in floating-point form. Attempts to create numbers that are too small lead to what are called *underflow* errors. The default option on most computers is to set the number to zero and to then proceed with the computation.

For example, consider evaluating

$$f(x) = x^{10} \tag{2.29}$$

for x near 0. When using IEEE single precision arithmetic, the smallest nonzero positive number expressible in *normalized* floating-point format is

$$m = 2^{-126} \doteq 1.18 \times 10^{-38} \tag{2.30}$$

See Table 2.2 with $E = 1$ and $(a_1 a_2 \ldots a_{23})_2 = (00\ldots0)_2$. Recall that *unnormalized* floating-point single precision format refers to (2.4) and Table 2.2 with $E = 0$. Thus, $f(x)$ will be set to zero if

$$\begin{aligned} x^{10} &< m \\ |x| &< \sqrt[10]{m} \doteq 1.61 \times 10^{-4} \\ -0.000161 < x &< 0.000161 \end{aligned}$$

Similar results will be valid for other floating-point formats, with the actual bound dependent on the exponent range of the floating-point representation. Some programs also allow the use of *unnormalized* floating-point numbers (e.g., MATLAB), and then the allowable size of a number is smaller, although with less precision in the significand.

Attempts to use numbers that are too large for the floating-point format will lead to *overflow* errors, and these are generally fatal errors on most computers. With the IEEE floating-point format, overflow errors can be carried along as having a value of $\pm\infty$ or *NaN*, depending on the context. Usually, an overflow error is an indication of a more significant problem or error in the program and the user needs to be aware of such errors.

In a few situations, it is possible to eliminate an overflow error by just reformulating the expression being evaluated. For example, consider evaluating

$$z = \sqrt{x^2 + y^2}$$

If x or y is very large, then $x^2 + y^2$ might create an overflow error, even though z might be within the floating-point range of the machine. To avoid this, let

$$z = \begin{cases} |x|\sqrt{1 + (y/x)^2}, & 0 \le |y| \le |x| \\ |y|\sqrt{1 + (x/y)^2}, & 0 \le |x| \le |y| \end{cases}$$

In both cases, the argument of the square root is of the form $1 + w^2$ with $|w| \le 1$. This calculation will not cause any overflow error, except when z is too large to be expressed in the floating-point format being used.

PROBLEMS

1. Calculate the error, relative error, and number of significant digits in the following approximations $x_A \approx x_T$:

(a) $x_T = 28.254$, $x_A = 28.271$ **(b)** $x_T = 0.028254$, $x_A = 0.028271$

(c) $x_T = e$, $x_A = 19/7$ **(d)** $x_T = \sqrt{2}$, $x_A = 1.414$

(e) $x_T = \log(2)$, $x_A = 0.7$

2. **(a)** In the population growth model of (2.18), $N(t) = N_0 e^{kt}$, give a physical meaning to the constant N_0.

(b) Show that $N(t)$ satisfies

$$\frac{N(t+1)}{N(t)} = \text{constant}$$

Thus, $N(t+1) = (\text{constant}) \cdot N(t)$, and the population increases by a constant ratio with every increase in t of 1 unit. Is this reasonable physically?

3. A slightly more sophisticated model for population growth is given by

$$N(t) = \frac{N_c}{1 + e^{-bt}}, \qquad t \geq 0$$

for some positive constants N_c and b. Identify a major difference between this model and the one in Problem 2. What is the physical meaning of the constant N_c?

Hint: Look at the behavior of each model as t becomes larger.

4. Bound the error in (2.20), using the remainder formula for the Taylor polynomial being used.

5. In some situations, loss-of-significance errors can be avoided by rearranging the function being evaluated, as was done with $f(x)$ in (2.23). Do something similar for the following cases, in some cases using trigonometric identities. In all but case (b), assume x is near 0.

(a) $\dfrac{1 - \cos(x)}{x^2}$ **(b)** $\log(x+1) - \log(x)$, x large

(c) $\sin(a + x) - \sin(a)$ **(d)** $\sqrt[3]{1+x} - 1$

(e) $\dfrac{\sqrt{4+x} - 2}{x}$

6. Use Taylor polynomial approximations to avoid the loss-of-significance errors in the following formulas when x is near 0:

(a) $\dfrac{e^x - 1}{x}$ **(b)** $\dfrac{1 - e^{-x}}{x}$ **(c)** $\dfrac{e^x - e^{-x}}{2x}$

(d) $\dfrac{\log(1-x) + xe^{x/2}}{x^3}$ **(e)** $\dfrac{1 - (1-x)^{\sqrt{2}-1}}{x}$

(f) $\dfrac{x - \sin(x)}{x^3}$ **(g)** $\dfrac{x - \sin(x)}{\tan(x)}$ **(h)** $\dfrac{x + \log(1-x)}{x^2}$

7. Consider the identity

$$\int_0^x \sin(xt)\, dt = \frac{1 - \cos(x^2)}{x}$$

Explain the difficulty in using the right-hand fraction to evaluate this expression when x is close to zero. Give a way to avoid this problem and be as precise as possible.

8. Repeat Problem 7 with the identity

$$f(x) = \frac{1}{x}\int_0^x e^{-xt} dt = \frac{1 - e^{-x^2}}{x^2}, \qquad x \neq 0$$

9. **(a)** Solve the equation $x^2 - 26x + 1 = 0$ using the quadratic formula. Use five-digit decimal arithmetic to find numerical values for the roots of the equation; for example, you will need $\sqrt{168} \doteq 12.961$. Identify any loss-of-significance error that you encounter.

(b) Find both roots accurately by using only five-digit decimal arithmetic.

Hint: Use

$$13 - \sqrt{168} = \frac{1}{13 + \sqrt{168}}$$

10. Repeat Problem 9 for the equation $x^2 - 40x + 1 = 0$, using $\sqrt{399} \doteq 19.975$.

11. Discuss the possible loss-of-significance error that may be encountered in solving the quadratic equation $ax^2 + bx + c = 0$. How might that loss-of-significance error be avoided?

12. Computationally, examine the accuracy of the identity

$$\sin(x) = \sqrt{1 - \cos^2(x)}, \qquad 0 \le x \le \frac{\pi}{2}$$

for values of x near to 0. Take values of x approaching 0 and examine the relative error in the computed values of $\sqrt{1 - \cos^2(x)}$, comparing them by using the values of $\sin(x)$ computed directly.

13. Find an accurate value of

$$f(x) = \sqrt{1 + \frac{1}{x}} - 1$$

for large values of x. Calculate

$$\lim_{x \to \infty} x f(x)$$

14. Consider evaluating $\cos(x)$ for large x by using the Taylor approximation

$$\cos(x) \approx 1 - \frac{x^2}{2!} + \cdots + (-1)^n \frac{x^{2n}}{(2n)!}$$

To see the difficulty involved in using this approximation, use it to evaluate $\cos(2\pi) = 1$:

$$\cos(2\pi) \approx 1 - \frac{(2\pi)^2}{2!} + \frac{(2\pi)^4}{4!} - \cdots + (-1)^n \frac{(2\pi)^{2n}}{(2n)!}$$

Assume that we are using decimal arithmetic, say, a four-digit decimal floating-point arithmetic with rounding. If $2n = 20$, then the error in this approximation is 0.00032 and, thus, the polynomial should be sufficiently accurate relative to the floating-point arithmetic being used. Now evaluate each term and round it to four significant digits, to find its exact floating-point representation. For example, the first three such terms are 1.000, -19.74, 64.94. Having found these 11 terms, add them up exactly. How closely do they approximate $\cos(2\pi) = 1.000$? Explain the source of the inaccuracy.

15. Repeat the example of Example 2.2.2 for the function $f(x) = (x - 1)^3$, but use

$$f(x) = x^3 - 3x^2 + 3x - 1$$

rather than the nested form of (2.28). Note any differences, if any.

16. Evaluate the following polynomials $p(x)$ on the given intervals. Evaluate $p(x)$ in the given form and in the nested form. Use a fairly large number of points x in the given interval. Both sets of function values will contain "noise," but generally they will be different because of the different formulations being used for $p(x)$. Subtract the nested values from the corresponding direct values, and print these numbers on the given intereval. This gives a composite of the noise from the two ways of evaluating $p(x)$, and it gives some idea of the size of the noise. Plot the values of x versus $p(x)$, and plot x versus the composite noise.

(a) $x^4 - 5.4x^3 + 10.56x^2 - 8.954x + 2.7951$, $0.9 \le x \le 1.3$

(b) $x^5 + 0.9x^4 - 1.62x^3 - 1.458x^2 + 0.6561x + 0.59049$, $-1.1 \le x \le -0.6$

17. Repeat Example 2.2.2 and Problem 15, using the polynomial

$$p(x) = x^4 - 5.4x^3 + 10.56x^2 - 8.954x + 2.7951$$

Look at values of $p(x)$ for x around 1.1.

18. To generate an overflow error on your computer, write a program to repeatedly square a number $x > 1$ and print the result. Eventually, you will exceed your machine's exponent limit for floating-point numbers.

19. For what values of x does x^{10} underflow using IEEE double precision *normalized* floating-point arithmetic. When does it overflow?

2.3. PROPAGATION OF ERROR

If a calculation is done with numbers that contain an error, then the resultant answer will be affected by these errors. We will look first at the effect of using such numbers with the ordinary arithmetic operations, and later we will consider the effect on more general function evaluations. Let x_A and y_A denote the numbers used in the calculation, and let x_T and y_T be the corresponding true values. We wish to bound

$$E = (x_T \omega y_T) - (x_A \omega y_A) \tag{2.31}$$

where ω denotes one of the operations "+," "−," "·," or "÷." The error E is called the *propagated error*.

The first technique used to bound E is known as *interval arithmetic*. Suppose we know bounds for $x_T - x_A$ and $y_T - y_A$. Then using these bounds and $x_A \omega y_A$, we look for an interval guaranteed to contain $x_T \omega y_T$.

Example 2.3.1 Let $x_A = 3.14$ and $y_A = 2.651$ be correctly rounded from x_T and y_T, to the number of digits shown. Then

$$|x_A - x_T| \le 0.005, \qquad |y_A - y_T| \le 0.0005$$

or, equivalently,

$$3.135 \le x_T \le 3.145, \qquad 2.6505 \le y_T \le 2.6515 \tag{2.32}$$

For the operation of addition

$$x_A + y_A = 5.791 \tag{2.33}$$

For the true value, use (2.32) to obtain the bounding interval

$$\begin{aligned} 3.135 + 2.6505 \le x_T + y_T \le 3.145 + 2.6515 \\ 5.7855 \le x_T + y_T \le 5.7965 \end{aligned} \tag{2.34}$$

To obtain a bound for the propagated error, subtract (2.33) from (2.34) to get

$$-0.0055 \le (x_T + y_T) - (x_A + y_A) \le 0.0055$$

With division,

$$\frac{x_A}{y_A} = \frac{3.14}{2.651} \doteq 1.184459 \tag{2.35}$$

Also,

$$\frac{3.135}{2.6515} \le \frac{x_T}{y_T} \le \frac{3.145}{2.6505}$$

Dividing the fractions and rounding to seven digits, we obtain

$$1.182350 \le \frac{x_T}{y_T} \le 1.186569 \tag{2.36}$$

For the error,

$$-0.002109 \le \frac{x_T}{y_T} - 1.184459 \le 0.002110 \quad \blacksquare$$

This technique of obtaining an interval that is guaranteed to contain the true answer is called *interval arithmetic*. It is a useful technique, and it has been implemented on computers, both using software and hardware. But for extended calculations, interval arithmetic must be implemented with a great deal of care or else it will lead to predicted error bounds that are far in excess of the true error. Much progress has been made in developing practical implementations of interval analysis, but it is not yet a widely used technique for the control or bounding of errors in practical computations. That seems likely to change in the future.

Propagated error in multiplication We compare $x_T y_T$ and $x_A y_A$. The relative error in $x_A y_A$ as compared to $x_T y_T$ is

$$\text{Rel}(x_A y_A) = \frac{x_T y_T - x_A y_A}{x_T y_T}$$

Let $x_T = x_A + \epsilon$, $y_T = y_A + \eta$. Then

$$\begin{aligned}
\text{Rel}(x_A y_A) &= \frac{x_T y_T - (x_T - \epsilon)(y_T - \eta)}{x_T y_T} \\
&= \frac{\eta x_T + \epsilon y_T - \epsilon \eta}{x_T y_T} \\
&= \frac{\epsilon}{x_T} + \frac{\eta}{y_T} - \left(\frac{\epsilon}{x_T}\right)\left(\frac{\eta}{y_T}\right) \\
&= \text{Rel}(x_A) + \text{Rel}(y_A) - \text{Rel}(x_A)\,\text{Rel}(y_A)
\end{aligned} \tag{2.37}$$

When both $\text{Rel}(x_A)$ and $\text{Rel}(y_A)$ are small in size compared with 1,

$$\text{Rel}(x_A y_A) \approx \text{Rel}(x_A) + \text{Rel}(y_A) \tag{2.38}$$

This shows that relative errors propagate slowly with multiplication, a very desirable property. The propagated error in using division is examined in Problem 9, and it has similarly nice properties.

Propagated error in addition and subtraction The operations of addition and subtraction need not be as well-behaved. We can have small values of $\text{Rel}(x_A)$ and $\text{Rel}(y_A)$, with $\text{Rel}(x_A \pm y_A)$ much larger; and Problems 9 and 10 of Section 2.2 can be used as illustrations of this. With reference to Problem 9, let

$$x_T = x_A = 13$$
$$y_T = \sqrt{168}, \qquad y_A = 12.961$$

For the relative errors,

$$\text{Rel}(x_A) = 0, \qquad \text{Rel}(y_A) \doteq 0.0000371$$
$$\text{Error}(x_A - y_A) \doteq -0.0004814$$
$$\text{Rel}(x_A - y_A) \doteq -0.0125$$

This source of error is connected closely to loss-of-significance errors.

Total calculational error When using floating-point arithmetic on a computer, the calculation of $x_A \omega y_A$ involves an additional rounding or chopping error, just as in the example (2.35). The computed value of $x_A \omega y_A$ will involve the propagated error plus a rounding or chopping error. To be more precise, let $\hat{\omega}$ denote the complete operation as carried out on the computer, including any rounding or chopping. Then the total error is given by

$$(x_T \omega y_T) - (x_A \hat{\omega} y_A) = [(x_T \omega y_T) - (x_A \omega y_A)] + [(x_A \omega y_A) - (x_A \hat{\omega} y_A)] \tag{2.39}$$

The first term on the right is the propagated error; the second term is the error in computing $x_A \omega y_A$.

When using IEEE arithmetic with the basic arithmetic operations, we have

$$x_A \hat{\omega} y_A = \text{fl}(x_A \omega y_A) \tag{2.40}$$

This says that the arithmetic calculation $x_A \omega y_A$ is to be carried out exactly and then rounded or chopped to standard floating-point length. For the error, use (2.11) to write

$$\text{fl}(x_A \omega y_A) = (1 + \epsilon)(x_A \omega y_A)$$

with ϵ bounded as in (2.12)–(2.16). Combining this with (2.40), we get

$$\begin{aligned}(x_A \omega y_A) - (x_A \hat{\omega} y_A) &= -\epsilon(x_A \omega y_A) \\ \frac{(x_A \omega y_A) - (x_A \hat{\omega} y_A)}{x_A \omega y_A} &= -\epsilon\end{aligned} \tag{2.41}$$

Thus, the process of rounding or chopping introduces a relatively small new error into $x_A \hat{\omega} y_A$ as compared with $x_A \omega y_A$.

2.3.1 Propagated Error in Function Evaluation

Consider evaluating $f(x)$ at the approximate value x_A rather than at x_T. Then consider how well does $f(x_A)$ approximate $f(x_T)$? There will also be an additional error introduced in the actual evaluation of $f(x_A)$, resulting in the noise phenomena described in Section 2.2. But we will be concerned here with just comparing the exact value of $f(x_A)$ with the exact value of $f(x_T)$.

Assume $f'(x)$ exists and is continuous for $a \le x \le b$. Using the mean value theorem (Theorem A.4, equation A.3) of Appendix A, we get

$$f(x_T) - f(x_A) = f'(c)(x_T - x_A) \tag{2.42}$$

with c an unknown point between x_A and x_T. Since x_A and x_T are generally very close together, we have

$$f(x_T) - f(x_A) \approx f'(x_T)(x_T - x_A) \approx f'(x_A)(x_T - x_A) \tag{2.43}$$

In most instances, it is better to consider the relative error

$$\text{Rel}(f(x_A)) \approx \frac{f'(x_T)}{f(x_T)}(x_T - x_A) = \frac{f'(x_T)}{f(x_T)} x_T \,\text{Rel}(x_A) \tag{2.44}$$

Example 2.3.2 In chemistry, one studies the ideal gas law

$$PV = nRT$$

in which R is a constant for all gases. In the MKS measurement system,

$$R = 8.3143 + \epsilon, \qquad |\epsilon| \le 0.0012$$

To see the effect of this uncertainty in R in a sample computation, consider evaluating T, assuming $P = V = n = 1$. Then

$$T = \frac{1}{R}$$

Define

$$f(x) = \frac{1}{x}$$

and evaluate it at $x = R$. Letting

$$x_T = R, \qquad x_A = 8.3143$$

we assume

$$|x_T - x_A| \le 0.0012$$

For the error

$$E = \frac{1}{R} - \frac{1}{8.3143} \tag{2.45}$$

we have from (2.43) that

$$|E| = |f(x_T) - f(x_A)| \approx \left|f'(x_A)\right| |x_T - x_A| \le \left(\frac{1}{x_A^2}\right)(0.0012) \doteq 1.74 \times 10^{-5}$$

For the relative error,

$$\left|\frac{E}{f(x_T)}\right| = \left|\frac{f(x_T) - f(x_A)}{f(x_T)}\right| \le \frac{1.74 \times 10^{-5}}{0.1202} \doteq 0.000144$$

Thus, the uncertainty in R has resulted in a relatively small error in the computed value

$$\frac{1}{R} \doteq \frac{1}{8.3143}$$ ■

Example 2.3.3 We apply (2.42) to the evaluation of

$$f(x) = b^x$$

where b is a positive constant. Using

$$f'(x) = (\log b)b^x$$

formulas (2.43) and (2.44) yield

$$\begin{aligned} b^{x_T} - b^{x_A} &\approx (\log b)b^{x_T}(x_T - x_A) \\ \text{Rel}(b^{x_A}) &\approx (\log b)x_T \,\text{Rel}(x_A) \end{aligned} \tag{2.46}$$

This example is also of interest for another reason. If the quantity

$$K = (\log b)x_T \tag{2.47}$$

is large in size, then the relative error in b^{x_A} will be much larger than the relative error in x_A. For example, if $\text{Rel}(x_A) = 10^{-7}$ and $K = 10^4$, then $\text{Rel}(b^{x_A}) \doteq 10^{-3}$, a significant decrease in the accuracy of b^{x_A} as compared with x_A, independent of how b^{x_A} is actually computed. The quantity in (2.47) is called a *condition number*. It relates the relative accuracy of the input to a problem (x_A in this case) to the relative accuracy of the output (namely, b^{x_A}). For a more formal definition, see Atkinson (1989, p. 86). We will return to the concept of condition number at other points in this text. ■

PROBLEMS

1. Let all of the numbers given below be correctly rounded to the number of digits shown. For each calculation, determine the smallest interval in which the result, using true instead of rounded values, must lie.

   (a) $1.1062 + 0.947$ **(b)** $23.46 - 12.753$

   (c) $(2.747)(6.83)$ **(d)** $8.473/0.064$

2. Use interval arithmetic to bound the error E of (2.45), just as was done following (2.34) and (2.36). Compare with the result given following (2.45).

3. Referring to Section 2.1, find the bounds for the ϵ in (2.41) for using the IEEE arithmetic standard.

4. Find bounds for the error and relative error in approximating $\sin(\sqrt{2})$ by $\sin(1.414)$.

5. In the following function evaluations $f(x_A)$, assume the numbers x_A are correctly rounded to the number of digits shown. Bound the error $f(x_T) - f(x_A)$ and the relative error in $f(x_A)$ using (2.43).

   (a) $\cos(1.473)$ **(b)** $\tan^{-1}(2.62)$ **(c)** $\log(1.4712)$

   (d) $e^{2.653}$ **(e)** $\sqrt{0.0425}$

6. For the function $f(x) = \sqrt{x}, x > 0$, estimate $f(x_T) - f(x_A)$ and $\text{Rel}(f(x_A))$. For what values of x_T, if any, is there a possible problem with loss of accuracy?

7. Bound

$$\int_0^{\pi} \frac{t^2}{1+t^4}\,dt - \int_0^{22/7} \frac{t^2}{1+t^4}\,dt$$

   Hint: Define

$$f(x) = \int_0^{x} \frac{t^2}{1+t^4}\,dt$$

   Apply (2.43) to $f(\pi) - f(22/7)$.

8. Let $f(x) = \tan x$. Let $x_T \approx x_A$, with $-\frac{1}{2}\pi < x_T, x_A < \frac{1}{2}\pi$ and $|x_T - x_A|$ quite small. Relate the relative error in x_A (in relation to x_T) to the relative error in $\tan x_A$ (in relation to $\tan x_T$). What happens in this relationship for the case that $x_T, x_A \approx \frac{1}{2}\pi$?

9. Following the method used in deriving (2.37)–(2.38), show that

$$\begin{aligned} \text{Rel}(x_A/y_A) &= \frac{\text{Rel}(x_A) - \text{Rel}(y_A)}{1 - \text{Rel}(y_A)} \\ &\approx \text{Rel}(x_A) - \text{Rel}(y_A) \end{aligned}$$

with the latter holding true for cases where $\text{Rel}(y_A)$ is small compared with 1.

10. To illustrate (2.46) and (2.47), compare π^{100} to $\pi^{100.1}$. Calculate these directly, as accurately as you can. (Most calculators carry enough decimal places to obtain sufficiently accurate answers for these exponentials.) Then calculate $\text{Rel}(\pi^{100.1})$ directly and using (2.46). Also give the condition number K of (2.47).

11. Let $f(x) = (x-1)(x-2)\ldots(x-n)$. Note that $f(1) = 0$. Estimate $f(1 + 10^{-4})$ by using (2.43) with $x_T = 1$, for $n = 2, 3, \ldots, 12$. Comment on the implications of this for finding the roots of $f(x)$, say, for the case $n = 8$.

Hint: Do not calculate $f'(x)$ by first multiplying out $f(x)$. Instead, use the product rule for derivatives to evaluate $f'(x)$, and then obtain $f'(1)$.

2.4. SUMMATION

There are many situations in which sums of a fairly large number of terms must be calculated. We will study the errors introduced when doing summation on a computer and look at ways to minimize the error in the final computed sum.

Let the sum be denoted by

$$S = a_1 + a_2 + \cdots + a_n = \sum_{j=1}^{n} a_j \tag{2.48}$$

where each a_j is a floating-point number. Adding these values in the machine amounts to calculating a sequence of $n - 1$ additions, each of which will probably involve a rounding or chopping error. More precisely, define

$$S_2 = \text{fl}(a_1 + a_2)$$

the floating-point version of $a_1 + a_2$ [recall the use of $\text{fl}(x)$, introduced in the paragraph preceding (2.11)]. Next, define

$$
\begin{aligned}
S_3 &= \mathrm{fl}(a_3 + S_2) \\
S_4 &= \mathrm{fl}(a_4 + S_3) \\
&\vdots \\
S_n &= \mathrm{fl}(a_n + S_{n-1})
\end{aligned}
$$

S_n is the computed version of S.

From (2.11),

$$
\begin{aligned}
S_2 &= (a_1 + a_2)(1 + \epsilon_2) \\
S_3 &= (a_3 + S_2)(1 + \epsilon_3) \\
&\vdots \\
S_n &= (a_n + S_{n-1})(1 + \epsilon_n)
\end{aligned}
\tag{2.49}
$$

If we assume IEEE arithmetic is used, each ϵ_j satisfies the bounds of (2.12)–(2.16). The choice depends on whether chopped or rounded arithmetic is used, and it also depends on whether single or double precision IEEE arithmetic is used.

The terms in (2.49) can be combined, manipulated, and estimated to give

$$
\begin{aligned}
S - S_n \approx -a_1(\epsilon_2 + \cdots + \epsilon_n) - a_2(\epsilon_2 + \cdots + \epsilon_n) - a_3(\epsilon_3 + \cdots + \epsilon_n) \\
-a_4(\epsilon_4 + \cdots + \epsilon_n) - \cdots - a_n\epsilon_n
\end{aligned}
\tag{2.50}
$$

If we look carefully at the formula and try to minimize the total error $S - S_n$, the following appears to be a reasonable strategy: Arrange the terms $a_1, a_2, \ldots, a_n$ before summing so that they are increasing in size

$$
|a_1| \le |a_2| \le |a_3| \le \cdots \le |a_n| \tag{2.51}
$$

Then the terms on the right side of (2.50) with the largest number of ϵ_j's are multiplied by the smaller values among the a_j's. This should make $S - S_n$ smaller, without much additional cost in most cases. This is a reasonable strategy, although there are better ones (which are also more complicated to implement).

Example 2.4.1 Define the terms a_j of the sum S as follows: Convert the fraction $1/j$ to a decimal fraction, round it to four significant digits, and let this be a_j. To make more clear the errors in the calculation of S, we use a decimal machine that has four digits in the significand of a floating-point number. Adding S from the largest term to the smallest term is denoted by "LS" in Tables 2.9 and 2.10; and adding from smallest to largest is denoted by "SL." The column "True" gives the true sum S, rounded to four digits. Table 2.9 gives the summation results for a decimal machine that uses chopping in all of its floating-point operations. Table 2.10 gives the analogous results for a machine using rounding.

In both tables, it is clear that the strategy of summing S from the smallest term to the largest is superior to the opposite procedure of summing from the largest term to the

Table 2.9. Calculating S on a machine using chopping

n	True	SL	Error	LS	Error
10	2.929	2.928	0.001	2.927	0.002
25	3.816	3.813	0.003	3.806	0.010
50	4.499	4.491	0.008	4.479	0.020
100	5.187	5.170	0.017	5.142	0.045
200	5.878	5.841	0.037	5.786	0.092
500	6.793	6.692	0.101	6.569	0.224
1000	7.486	7.284	0.202	7.069	0.417

Table 2.10. Calculating S on a machine using rounding

n	True	SL	Error	LS	Error
10	2.929	2.929	0	2.929	0
25	3.816	3.816	0	3.817	−0.001
50	4.499	4.500	−0.001	4.498	0.001
100	5.187	5.187	0	5.187	0
200	5.878	5.878	0	5.876	0.002
500	6.793	6.794	−0.001	6.783	0.010
1000	7.486	7.486	0	7.449	0.037

smallest. However, with the machine that rounds, it takes a fairly large number of terms before the order of summation makes any essential difference. ■

2.4.1 Rounding versus Chopping

A more important difference in the errors shown in the tables is that which exists between rounding and chopping. Rounding results in a far smaller error in the calculated sum than does chopping. To understand why this happens, return to the formula (2.50). As a typical case from it, consider the first term on the right side:

$$T \equiv -a_1(\epsilon_2 + \cdots + \epsilon_n) \tag{2.52}$$

Assume that we are using rounding with the four-digit decimal machine of the above example. In analogy with the derivation of (2.11)–(2.16) for our decimal machine, we know that all ϵ_j satisfy

Table 2.11. Calculation of (2.56): rounding versus chopping

n	True	Rounding Error	Chopping Error
10	2.92896825	$-1.76\text{E}-7$	$3.01\text{E}-7$
50	4.49920534	$7.00\text{E}-7$	$3.56\text{E}-6$
100	5.18737752	$-4.12\text{E}-7$	$6.26\text{E}-6$
500	6.79282343	$-1.32\text{E}-6$	$3.59\text{E}-5$
1000	7.48547086	$8.88\text{E}-8$	$7.35\text{E}-5$

$$-0.0005 \le \epsilon_j \le 0.0005 \tag{2.53}$$

Rounding errors can usually be treated as random in nature, subject to this bounding interval. Thus, the positive and negative values of the ϵ_j's in (2.52) will tend to cancel, and the sum T will be nearly zero. By using advanced methods from probability theory, it can be shown that (2.52) is very likely to satisfy

$$|T| \le (1.49)(0.0005) \cdot \sqrt{n}\,|a_1|$$

The value of T is proportional to $\sqrt{n}$. Thus, (2.52) tends to be small until n becomes quite large, and the same is true of the total error on the right of (2.50).

For our decimal machine with chopping, (2.53) is replaced by

$$-0.001 \le \epsilon_j \le 0 \tag{2.54}$$

and the errors are all of one sign. Again, the chopping errors will vary randomly in this interval. But now the average value of the ϵ_j's will be -0.0005, the middle of the interval; and the likely value of (2.52) will be

$$-a_1(n-1)(-0.0005) \tag{2.55}$$

Thus, T is proportional to n, whereas the corresponding result for the case of rounding was that T was proportional to $\sqrt{n}$; and n increases more rapidly than $\sqrt{n}$. Thus, the error (2.52) and (2.50) will grow much more rapidly when chopping is used rather than rounding.

Example 2.4.2 We use a binary computer for which both rounding and chopping are available, with a single precision accuracy of six to seven decimal digits. To illustrate the difference between rounding and chopping, consider evaluating

$$S = \sum_{j=1}^{n} \frac{1}{j} \tag{2.56}$$

in single precision arithmetic. For this calculation, errors occur in both the calculation of the floating-point form of $1/j$ and in the summation process. Table 2.11 contains the errors for the two modes of calculation. The value in column "True" was calculated by using double precision arithmetic to evaluate (2.56). Also, all sums were performed from the smallest term to the largest. ■

2.4.2 A Loop Error

An important example of accumulated errors is the computation of independent variables in a loop computation (e.g., *DO*-loops in *Fortran* and *for*-loops in MATLAB). Suppose we wish to calculate

$$x = a + jh \tag{2.57}$$

for $j = 0, 1, 2, \ldots, n$ for given $h > 0$. This is then used in a further computation, often to evaluate some function $f(x)$. The question we wish to consider here is whether x should be computed as in (2.57) or by using the statement

$$x = x + h \tag{2.58}$$

in the loop, having initially set $x = a$ before beginning the loop. These are mathematically equivalent ways to compute x, but they are usually not computationally equivalent.

The difficulty with computing x arises generally when h does not have a finite binary expansion that can be stored in the given floating-point significand, for example, $h = 0.1$. The computation (2.57) will involve two arithmetic operations and, thus, only two chopping or rounding errors, for each value of x. In contrast, the repeated use of (2.58) will involve a succession of additions, in fact, j of them for the x of (2.57). As x increases in size, the use of (2.58) involves a larger number of rounding or chopping errors, leading to a different quantity than in (2.57). Thus, (2.57) is usually the preferred way to evaluate x.

To provide a specific example, we give a program to compute the value of x in the two possible ways. We use $a = 0$ and $h = 0.1$. To check the accuracy, we also compute the true desired value of x by using a double precision computation. This program was run on a binary computer using IEEE arithmetic, and the results for selected $x_j = jh$ are shown in Table 2.12. There is a significant improvement with (2.57) over (2.58).

2.4.3 Calculation of Inner Products

A sum of the form

$$S = a_1 b_1 + a_2 b_2 + \cdots + a_n b_n = \sum_{j=1}^{n} a_j b_j \tag{2.59}$$

Table 2.12. Evaluation of $x = j \cdot h$, $h = 0.1$

j	x	Error Using (2.57)	Error Using (2.58)
10	1	1.49E − 8	−1.04E − 7
20	2	2.98E − 8	−2.09E − 7
30	3	4.47E − 8	7.60E − 7
40	4	5.96E − 8	1.73E − 6
50	5	7.45E − 8	2.46E − 6
60	6	8.94E − 8	3.43E − 6
70	7	1.04E − 7	4.40E − 6
80	8	1.19E − 7	5.36E − 6
90	9	1.34E − 7	2.04E − 6
100	10	1.49E − 7	−1.76E − 6

is called a *dot product* or *inner product*. Typically, the terms a_j and b_j are components of vectors A and B, and S is the inner product of A and B. Such sums occur quite often in solving certain kinds of problems, particularly those involving systems of simultaneous linear equations.

If we calculate S in single precision, then there will be a single precision rounding error for each multiplication and each addition. Thus, there will be $2n - 1$ single precision rounding errors involved in calculating S. The consequences of these errors can be analyzed in the manner of (2.50), and we could derive an optimal strategy for the calculation of (2.59). Instead, we will look at a simpler alternative, using the double precision arithmetic of the computer.

Convert each a_j and b_j to double precision by extending their significands with zeros. Multiply them in double precision and sum them in double precision; when done, round the answer to single precision to obtain the calculated value of S. For machines with IEEE arithmetic, this procedure is a simple and rapid way to obtain more accurate inner products in single precision computations, and there need be no increase in storage space for the arrays A and B. The accuracy is improved, since there will be only one single precision rounding error, regardless of the size of n. When the main calculations are already in double precision, some type of extended precision arithmetic is needed. The IEEE floating-point arithmetic standard contains such an extended precision arithmetic.

MATLAB uses only double precision IEEE arithmetic, and so it is not suitable to illustrate the ideas described here. Following is a function subprogram *SUMPRD*, written in *Fortran*. In it, the elements of A and B are converted to double precision by using the built-in function *DBLE*, which converts numbers to their equivalent forms in double precision by appending a suitable number of the digit 0.

```
      REAL FUNCTION SUMPRD(A,B,N)
C
C     THIS CALCULATES THE INNER PRODUCT
C
C                I=N
C     SUMPRD = SUM A(I)*B(I)
C                I=1
C
C     THE PRODUCTS AND SUMS ARE DONE IN DOUBLE
C     PRECISION, AND THE FINAL RESULT IS CONVERTED
C     BACK TO SINGLE PRECISION.
C
      REAL A(*), B(*)
      DOUBLE PRECISION DSUM
C
      DSUM = 0.0D0
      DO I=1,N
        DSUM = DSUM + DBLE(A(I))*DBLE(B(I))
      END DO
      SUMPRD = DSUM
      RETURN
      END
```

PROBLEMS

1. Write a computer program to evaluate

$$S = \sum_{j=1}^{n} a_j$$

for arbitrary n, and apply it to the series given below. Do the calculation by both the methods LS and SL. For LS, compute and sum the terms from the largest term first to the smallest term last. For SL, do the calculation in the reverse order. Calculate a true value for S using the given answer, and compare it with the values obtained by LS and SL.

(a) $$\sum_{j=1}^{n} \frac{1}{j(j+1)} = \frac{n}{n+1}$$

(b) $$\sum_{j=1}^{n} \frac{1}{j(j+2)} = \frac{3}{4} - \frac{2n+3}{2(n+1)(n+2)}$$

(c) $$\sum_{j=1}^{n} \frac{1}{\sqrt{j(j+1)}[\sqrt{j}+\sqrt{j+1}]} = \frac{n}{\sqrt{n+1}[\sqrt{n+1}+1]}$$

Use $n = 10, 50, 100, 500, 1000, 5000$.

2. Repeat Problem 1 for the sums of the geometric series

$$\sum_{j=0}^{n} x^j = \frac{1 - x^{n+1}}{1 - x}$$

with $x = 0.01, 0.1, 0.5, 0.9,$ and 0.99. Comment on the numerical results for different values of x.

3. Consider using the partial sum

$$S_n(x) = \sum_{j=0}^{n} (-1)^j \frac{x^{2j+1}}{(2j+1)!}$$

to approximate $\sin x$. For $x = 0.1$, 1, and 10, calculate $S_n(x)$ by both methods LS and SL, and compare the results against $\sin x$. Use $n = 10, 100, 1000$.

4. Derive the formula (2.50) for the cases $n = 2$, 3, and 4 from the formulas (2.49).

   Hint: If the ϵ_i's are small, as indicated in (2.12)–(2.16), then $\epsilon_i \epsilon_j$ is very small compared with ϵ_i and can therefore be neglected when added to ϵ_i.

5. Implement the *Fortran* subprogram *SUMPRD* given in this section. Apply it to the sums of Problem 1, regarding the sum as an inner product of two arrays. For (a), write

$$A = \left[1, \frac{1}{2}, \frac{1}{3}, \ldots, \frac{1}{n}\right]$$
$$B = \left[\frac{1}{2}, \frac{1}{3}, \ldots, \frac{1}{n+1}\right]$$

   Also, compute the same inner product by using only single precision arithmetic and compare it to your answer obtained by using *SUMPRD*.

6. Consider evaluating $p = a_1 a_2 \cdots a_n$ with $a_i = \mathrm{fl}(a_i), i = 1, 2, \ldots, n$. Define

$$p_2 = \mathrm{fl}(a_1 a_2), \qquad p_3 = \mathrm{fl}(p_2 a_3), \qquad \ldots, \qquad p_n = \mathrm{fl}(p_{n-1} a_n)$$

   Using the type of argument applied in deriving (2.50), derive an estimate for $p_n - p$ and $\mathrm{Rel}(p_n)$, showing the effect of the rounding or chopping errors that occur in forming $p_2, \ldots, p_n$.

CHAPTER 3

ROOTFINDING

Calculating the roots of an equation

$$f(x) = 0 \tag{3.1}$$

is a common problem in applied mathematics. We will explore some simple numerical methods for solving this equation and also will consider some possible difficulties. We begin by giving a simple example that arises in financial planning.

Example 3.0.1 When planning for retirement at some distant time, we invest money so as to have a satisfactory income during our retirement years. A simple version of this process leads to the idea of an *annuity*, and we briefly discuss it here.

Assume that an amount of P_{in} is deposited into a savings account at the beginning of each of N_{in} time periods. For example, if done monthly, then multiply the number of years until retirement by 12 to obtain N_{in}. Following the N_{in} time periods of saving contributions, assume an amount of P_{out} is withdrawn from the account for each of

N_{out} time periods. The first withdrawal occurs at the beginning of time period $N_{in} + 1$. During these $N_{in} + N_{out}$ time periods, the account is to earn interest at a compound rate of r per time period. For example, if the interest on the account is compounded monthly at an annual rate of 6%, then $r = 0.06/12 = 0.005$. After the final withdrawal, we assume the account has been drawn down to zero. What is the relationship between P_{in}, P_{out}, N_{in}, N_{out}, and r?

At the end of period N_{in}, the amount in the account will be

$$S_{in} \equiv P_{in}\{(1+r)^{N_{in}} + \cdots + (1+r)\} = P_{in}(1+r)\frac{(1+r)^{N_{in}} - 1}{r}$$

At the beginning of time period $N_{in} + 1$, we withdraw the first of N_{out} payments, each of amount P_{out}; and at the beginning of period $N_{in} + N_{out}$, the final payment is withdrawn and the account is assumed to be empty. The amount left after the first withdrawal is $S_{in} - P_{out}$; and after N_{out} withdrawals, with compounding of interest for amounts held in the account for intermediate periods, the account contains

$$(1+r)^{N_{out}-1} S_{in} - P_{out}\frac{(1+r)^{N_{out}} - 1}{r} = 0$$

By combining with the formula for S_{in} and simplifying, we obtain

$$f(r) \equiv P_{in}[(1+r)^{N_{in}} - 1] - P_{out}[1 - (1+r)^{-N_{out}}] = 0 \tag{3.2}$$

Given any four of the quantities P_{in}, P_{out}, N_{in}, N_{out}, and r, find the fifth one. This is straightforward in all instances but one: Finding r is a rootfinding problem for which we must use a numerical method. We will return later to this in the problems. ■

The function $f(x)$ of the equation (3.1) will usually have at least one continuous derivative, and often we will have some estimate of the root that is being sought. By using this information, most numerical methods for (3.1) compute a sequence of increasingly accurate estimates of the root. These methods are called *iteration methods*. We will study three different methods in the first three sections of this chapter; and in Section 3.4, we give a general theory for one-point iteration methods. Section 3.5 considers difficulties that occur in solving (3.1) for some special types of functions $f(x)$.

3.1. THE BISECTION METHOD

In this chapter, we assume that $f(x)$ is a function that is real-valued and that x is a real variable. Suppose that $f(x)$ is continuous on an interval $a \le x \le b$ and that

$$f(a)f(b) < 0 \tag{3.3}$$

Then $f(x)$ changes sign on $[a, b]$, and $f(x) = 0$ has at least one root on the interval. The simplest numerical procedure for finding a root is to repeatedly halve the interval $[a, b]$, keeping the half on which $f(x)$ changes sign. This procedure is called the *bisection method*. It is guaranteed to converge to a root, denoted here by α.

To be more precise in our definition, suppose that we are given an interval $[a, b]$ satisfying (3.3) and an error tolerance $\epsilon > 0$. Then the bisection method consists of the following steps:

B1. Define $c = (a + b)/2$.

B2. If $b - c \leq \epsilon$, then accept c as the root and stop.

B3. If $\text{sign}[f(b)] \cdot \text{sign}[f(c)] \leq 0$, then set $a = c$.
Otherwise, set $b = c$. Return to step B1.

The interval $[a, b]$ is halved with each loop through steps B1 to B3. The test B2 will be satisfied eventually, and with it the condition $|\alpha - c| \leq \epsilon$ will be satisfied. Justification is given in (3.7), which is presented in the discussion that follows. Notice that in step B3, we test the sign of $\text{sign}[f(b)] \cdot \text{sign}[f(c)]$ rather than that of $f(b)f(c)$ in order to avoid the possibility of underflow or overflow in the multiplication of $f(b)$ and $f(c)$.

Example 3.1.1 Find the largest root of

$$f(x) \equiv x^6 - x - 1 = 0 \tag{3.4}$$

accurate to within $\epsilon = 0.001$. With a graph, it is easy to check that $1 < \alpha < 2$. We choose $a = 1$, $b = 2$; then $f(a) = -1$, $f(b) = 61$, and (3.3) is satisfied. The results of the algorithm B1 to B3 are shown in Table 3.1. The entry n indicates that the associated row corresponds to iteration number n of steps B1 to B3. ■

Table 3.1. Bisection Method for (3.4)

n	a	b	c	$b - c$	$f(c)$
1	1.0000	2.0000	1.5000	0.5000	8.8906
2	1.0000	1.5000	1.2500	0.2500	1.5647
3	1.0000	1.2500	1.1250	0.1250	−0.0977
4	1.1250	1.2500	1.1875	0.0625	0.6167
5	1.1250	1.1875	1.1562	0.0312	0.2333
6	1.1250	1.1562	1.1406	0.0156	0.0616
7	1.1250	1.1406	1.1328	0.0078	−0.0196
8	1.1328	1.1406	1.1367	0.0039	0.0206
9	1.1328	1.1367	1.1348	0.0020	0.0004
10	1.1328	1.1348	1.1338	0.00098	−0.0096

3.1.1 Error Bounds

Let a_n, b_n, and c_n denote the nth computed values of a, b, and c, respectively. Then easily we get

$$b_{n+1} - a_{n+1} = \frac{1}{2}(b_n - a_n), \qquad n \geq 1 \tag{3.5}$$

and it is straightforward to deduce that

$$b_n - a_n = \frac{1}{2^{n-1}}(b - a), \qquad n \geq 1 \tag{3.6}$$

where $b - a$ denotes the length of the original interval with which we started. Since the root α is in either the interval $[a_n, c_n]$ or $[c_n, b_n]$, we know that

$$|\alpha - c_n| \leq c_n - a_n = b_n - c_n = \frac{1}{2}(b_n - a_n) \tag{3.7}$$

This is the error bound for c_n that is used in step B2 of the earlier algorithm. Combining it with (3.6), we obtain the further bound

$$|\alpha - c_n| \leq \frac{1}{2^n}(b - a) \tag{3.8}$$

This shows that the iterates c_n converge to α as $n \to \infty$.

To see how many iterations will be necessary, suppose we want to have

$$|\alpha - c_n| \leq \epsilon$$

This will be satisfied if

$$\frac{1}{2^n}(b - a) \leq \epsilon$$

Taking logarithms of both sides, we can solve this to give

$$n \geq \frac{\log\left(\dfrac{b-a}{\epsilon}\right)}{\log 2} \tag{3.9}$$

For our example (3.4), this results in

$$n \geq \frac{\log\left(\dfrac{1}{0.001}\right)}{\log 2} \doteq 9.97$$

Thus, we must have $n = 10$ iterates, exactly the number computed.

There are several advantages to the bisection method. The principal one is that the method is guaranteed to converge. In addition, the error bound, given in (3.7), is guaranteed to decrease by one-half with each iteration. Many other numerical methods have variable rates of decrease for the error, and these may be worse than the bisection method for some equations. The principal disadvantage of the bisection method is that it generally converges more slowly than most other methods. For functions $f(x)$ that have a continuous derivative, other methods are usually faster. These methods may not always converge; when they do converge, however, they are almost always much faster than the bisection method.

MATLAB PROGRAM: ***An implementation of the bisection method.*** A function `bisect` implementing the bisection method is given below. The comment statements at the beginning of the programs should be self-explanatory with regard to the purpose and use of the program. The form of the program is quite standard. The function f is given as an *internal function*, meaning that it is not recognizable to MATLAB programs stored in files separate from this one. The program involves an internal printing of the intermediate steps in the bisection method. This step can be omitted. At the conclusion of the program, we print the final value of the root and the error bound for it.

Lacking in the program is any attempt to check whether the given error tolerance `ep` is realistic for the length of the significand in the computer arithmetic being used. This was left out to keep the program simple, but a realistic rootfinding program in a computer library would need to include such a check. In MATLAB, such a test can be constructed using the given machine constant eps.

```
function [root,error_bound] = bisect(a0,b0,ep,max_iterate)
%
% function bisect(a0,b0,ep,max_iterate)
%
% This is the bisection method for solving an equation f(x)=0.
%
% The function f is defined below by the user.  The function f is
% to be continuous on the interval [a0,b0], and it is to be of
% opposite signs at a0 and b0.  The quantity ep is the error
% tolerance.  The parameter max_iterate is an upper limit on the
% number of iterates to be computed.
%
% This program guarantees ep as an error bound for the computed
% root provided:  (1) the restrictions on the given function f
% and the initial [a0,b0] are satisfied; (2) ep is not too small
% when the machine epsilon is taken into account; and (3) the
% number of iterates computed is at most max_iterate.  Only
% some of these conditions are checked in the program!
%
% For the given function f(x), an example of a calling sequence
```

```
% might be the following:
%       [root,error_bound] = bisect(1,1.5,1.0E-6,10)
%
% The following is printed for each iteration the values of
%       count, a, b, c, f(c), (b-a)/2
% with c the current iterate and (b-a)/2 the error bound for c.
% The variable count is the index of the current iterate.  Tap
% the carriage return to continue with the iteration.

if a0 >= b0
  disp('a0 < b0 is not true.  Stop!')
  return
end

format short e
a = a0; b = b0;
fa = f(a); fb = f(b);

if sign(fa)*sign(fb) > 0
  disp('f(a0) and f(b0) are of the same sign.  Stop!')
  return
end

c = (a+b)/2;
it_count = 0;
fprintf('\n  it_count  a  b  c  f(c)  b-c\n')
while b-c > ep & it_count < max_iterate
  it_count = it_count + 1;
  fc = f(c);
% Internal print of bisection method.  Tap the carriage
% return key to continue the computation.
  iteration = [it_count a b c fc b-c]
  if sign(fb)*sign(fc) <= 0
    a = c;
    fa = fc;
  else
    b = c;
    fb = fc;
  end
  c = (a+b)/2;
  pause
end

format long
root = c
```

```
format short e
error_bound = b-c

%%%%%%%%%%%%%%%%%%%%%%%%%%%%%%%
function value = f(x)
%
% function to define equation for rootfinding problem.

value = x.^6 - x - 1;
```

PROBLEMS

1. Use the bisection method with a hand calculator or computer to find the indicated roots of the following equations. Use an error tolerance of $\epsilon = 0.0001$.

(a) The real root of $x^3 - x^2 - x - 1 = 0$.

(b) The root of $x = 1 + 0.3\cos(x)$.

(c) The smallest positive root of $\cos(x) = \frac{1}{2} + \sin(x)$.

(d) The root of $x = e^{-x}$.

(e) The smallest positive root of $e^{-x} = \sin(x)$.

(f) The real root of $x^3 - 2x - 2 = 0$.

(g) All real roots of $x^4 - x - 1 = 0$.

2. To help determine the roots of $x = \tan(x)$, graph $y = x$ and $y = \tan(x)$, and look at the intersection points of the two curves.

(a) Find the smallest nonzero positive root of $x = \tan(x)$, with an accuracy of $\epsilon = 0.0001$.

Note: The desired root is greater than $\pi/2$.

(b) Solve $x = \tan(x)$ for the root that is closest to $x = 100$.

3. Consider equation (3.2) with $P_{\text{in}} = 1000$, $P_{\text{out}} = 20{,}000$, $N_{\text{in}} = 30$, and $N_{\text{out}} = 20$. Find r with an accuracy of $\epsilon = 0.0001$.

4. Show that for any real constants c and d, the equation $x = c + d\cos(x)$ has at least one root.

Hint: Find an interval $[a, b]$ on which $f(x) = x - c - d\cos(x)$ changes sign.

5. To use the bisection method, implement the routine `bisect` or write another program of your own design. Use it to solve the equations in Problems 1 and 2 with an accuracy of $\epsilon = 10^{-5}$.

6. Using the bisection method and a graph of $f(x)$, find all roots of

$$f(x) = 32x^6 - 48x^4 + 18x^2 - 1$$

The true roots are

$$\cos\left[(2j-1)\frac{\pi}{12}\right], \qquad j = 1, 2, \ldots, 6$$

7. Using the program of Problem 5, solve the equation

$$f(x) \equiv x^3 - 3x^2 + 3x - 1 = 0$$

with an accuracy of $\epsilon = 10^{-6}$. Experiment with different ways of evaluating $f(x)$; for example, use (i) the given form, (ii) reverse its order, and (iii) the nested form

$$f(x) = -1 + x\,(3 + x\,(-3 + x))$$

Try various initial intervals $[a, b]$, for example, [0, 1.5], [0.5, 2.0], and [0.5, 1.1]. Explain the results. Note that $\alpha = 1$ is the only root of $f(x) = 0$.

8. The polynomial

$$f(x) = x^4 - 5.4x^3 + 10.56x^2 - 8.954x + 2.7951$$

has a root α in [1, 1.2]. Repeat Problem 7, but vary the intervals $[a, b]$ to reflect the location of the root α.

9. Let the initial interval used in the bisection method have length $b - a = 3$. Find the number of midpoints c_n that must be calculated with the bisection method to obtain an approximate root within an error tolerance of 10^{-9}.

10. Consider the equation $e^{-x} = \sin x$. Find an interval $[a, b]$ that contains the smallest positive root. Estimate the number of midpoints c needed to obtain an approximate root that is accurate within an error tolerance of 10^{-10}.

11. Let α be the smallest positive root of

$$f(x) \equiv 1 - x + \sin x = 0$$

Find an interval $[a, b]$ containing α and for which the bisection method will converge to α. Then estimate the number of iterates needed to find α within an accuracy of 5×10^{-8}.

12. Let α be the unique root of

$$x = \frac{3}{1 + x^4}$$

Find an interval $[a, b]$ containing α and for which the bisection method will converge to α. Then estimate the number of iterates needed to find α within an accuracy of 5×10^{-8}.

13. Let α be the largest root of

$$f(x) \equiv e^x - x - 2 = 0$$

Find an interval $[a, b]$ containing α and for which the bisection method will converge to α. Then estimate the number of iterates needed to find α within an accuracy of 5×10^{-8}.

14. Imagine finding a root α satisfying $1 < \alpha < 2$. If you are using a binary computer with m binary digits in its significand, what is the smallest error tolerance that makes sense in finding an approximation to α? If the original interval $[a, b] = [1, 2]$, how many interval halvings are needed to find an approximation c_n to α with the maximum accuracy possible for this computer?

15. Let $f(x) = 1 - zx$ for some $z > 0$. Solving $f(x) = 0$ is equivalent to calculating $1/z$, thus doing a division.

(a) Give an interval $[a, b]$, or a way to calculate it, guaranteed to contain $1/z$. Do not use division in calculating or defining $[a, b]$.

(b) Assume $1 < z < 2$. By using some interval enclosing $1/z$, give the number of subdivisions n needed to obtain an estimate of $1/z$ within an accuracy of 2^{-25}.

(c) For a general $z > 0$, consider calculating $1/z$ in the single precision arithmetic of the IEEE standard floating-point arithmetic. Using the bisection method, give a way to calculate $1/z$ to the full accuracy of the arithmetic.

3.2. NEWTON'S METHOD

Consider the sample graph of $y = f(x)$ shown in Figure 3.1. The root α occurs where the graph crosses the x-axis. We will usually have an estimate of α, and it will be denoted here by x_0. To improve on this estimate, consider the straight line that is tangent to the graph at the point $(x_0, f(x_0))$. If x_0 is near α, this tangent line should be nearly coincident with the graph of $y = f(x)$ for points x about α. Then the root of the tangent line should nearly equal α. This root is denoted here by x_1.

To find a formula for x_1, consider the equation of the line tangent to the graph of $y = f(x)$ at $(x_0, f(x_0))$. It is simply the graph of $y = p_1(x)$ for the linear Taylor polynomial

$$p_1(x) = f(x_0) + f'(x_0)(x - x_0)$$

By definition, x_1 is the root of $p_1(x)$. Solving

$$f(x_0) + f'(x_0)(x_1 - x_0) = 0$$

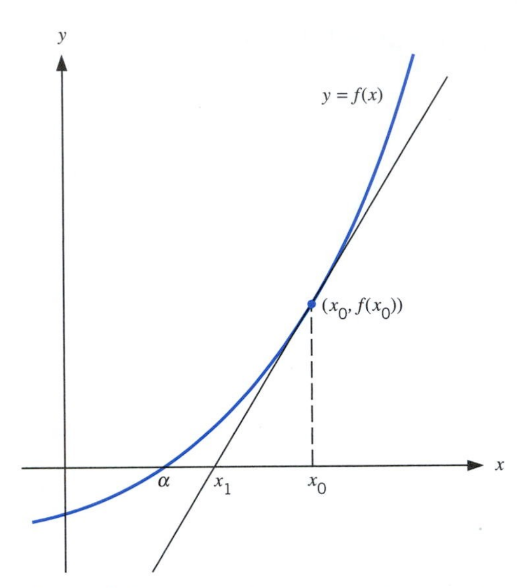

Figure 3.1. The schematic for Newton's method

leads to

$$x_1 = x_0 - \frac{f(x_0)}{f'(x_0)} \tag{3.10}$$

Since x_1 is expected to be an improvement over x_0 as an estimate of α, this entire procedure can be repeated with x_1 as the initial guess. This leads to the new estimate

$$x_2 = x_1 - \frac{f(x_1)}{f'(x_1)}$$

Repeating this process, we obtain a sequence of numbers $x_1, x_2, x_3, \ldots$ that we hope will approach the root α. These numbers are called iterates, and they are defined recursively by the following general *iteration formula*:

$$x_{n+1} = x_n - \frac{f(x_n)}{f'(x_n)}, \qquad n = 0, 1, 2, \ldots \tag{3.11}$$

This is *Newton's method* for solving $f(x) = 0$. It is also called the *Newton–Raphson method.*

Example 3.2.1 Using Newton's method, solve equation (3.4), which was used earlier in Example 3.1.1 to illustrate the bisection method. Here,

$$f(x) = x^6 - x - 1, \qquad f'(x) = 6x^5 - 1$$

Table 3.2. Newton's Method for $x^6 - x - 1 = 0$

n	x_n	$f(x_n)$	$x_n - x_{n-1}$	$\alpha - x_{n-1}$
0	1.5	8.89E + 1		
1	1.30049088	2.54E + 1	−2.00E − 1	−3.65E − 1
2	1.18148042	5.38E − 1	−1.19E − 1	−1.66E − 1
3	1.13945559	4.92E − 2	−4.20E − 2	−4.68E − 2
4	1.13477763	5.50E − 4	−4.68E − 3	−4.73E − 3
5	1.13472415	7.11E − 8	−5.35E − 5	−5.35E − 5
6	1.13472414	1.55E − 15	−6.91E − 9	−6.91E − 9

and the iteration is given by

$$x_{n+1} = x_n - \frac{x_n^6 - x_n - 1}{6x_n^5 - 1}, \qquad n \geq 0 \tag{3.12}$$

We use an initial guess of $x_0 = 1.5$. The results are shown in Table 3.2. The column "$x_n - x_{n-1}$" is an estimate of the error $\alpha - x_{n-1}$; justification for this is given later in the section.

The true root is $\alpha \doteq 1.134724138$, and x_6 equals α to nine significant digits. Compare this with the earlier results shown in Table 3.1 for the bisection method. Observe that Newton's method may converge slowly at first. As the iterates come closer to the root, however, the speed of convergence increases, as is shown in the table. ■

Example 3.2.2 We now consider a procedure that was used to carry out division on some early computers. These computers had hardware arithmetic for addition, subtraction, and multiplication, but division had to be implemented by software. The following iteration (3.14) is also used on some present-day supercomputers.

Suppose we want to form a/b. It can be done by multiplying a and $1/b$, with the latter produced approximately by using Newton's method. A necessary part of this is that we need to find $1/b$. To do this, we solve

$$f(x) \equiv b - \frac{1}{x} = 0 \tag{3.13}$$

where we assume $b > 0$. The root is $\alpha = 1/b$. The derivative is

$$f'(x) = \frac{1}{x^2}$$

and Newton's method is given by

$$x_{n+1} = x_n - \frac{b - \dfrac{1}{x_n}}{\dfrac{1}{x_n^2}}$$

Simplifying, we get

$$x_{n+1} = x_n(2 - bx_n), \qquad n \geq 0 \tag{3.14}$$

This involves only multiplication and subtraction and, thus, there is no difficulty in implementing it on the computers discussed previously. The initial guess, of course, should be chosen with $x_0 > 0$. For the error, it can be shown that

$$\text{Rel}(x_{n+1}) = [\text{Rel}(x_n)]^2, \qquad n \geq 0 \tag{3.15}$$

where

$$\text{Rel}(x_n) = \frac{\alpha - x_n}{\alpha}$$

the relative error when considering x_n as an approximation to $\alpha = 1/b$. From (3.15), we must have

$$|\text{Rel}(x_0)| < 1$$

Otherwise, the error in x_n will not decrease to zero as n increases. This condition means

$$-1 < \frac{\dfrac{1}{b} - x_0}{\dfrac{1}{b}} < 1$$

and this reduces to the equivalent condition

$$0 < x_0 < \frac{2}{b} \tag{3.16}$$

The iteration (3.14) converges to $\alpha = 1/b$ if and only if the initial guess x_0 satisfies (3.16). Figure 3.2 shows a sample case. From Figure 3.2, it is fairly easy to justify (3.16), since if it is violated, the calculated value of x_1 and all further iterates would be negative.

The result (3.15) shows that the convergence is very rapid, once we have a somewhat accurate initial guess. For example, suppose $|\text{Rel}(x_0)| = 0.1$, which corresponds to a

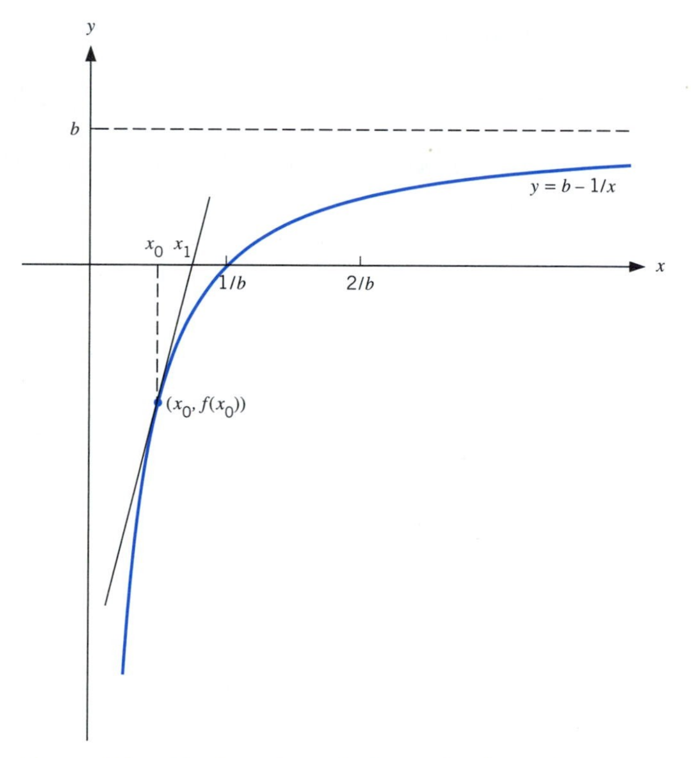

Figure 3.2. The iterative solution of $b - 1/x = 0$

10% error in x_0. Then from (3.15),

$$\begin{aligned} &\text{Rel}(x_1) = 10^{-2}, \qquad &\text{Rel}(x_2) = 10^{-4}, \\ &\text{Rel}(x_3) = 10^{-8}, \qquad &\text{Rel}(x_4) = 10^{-16} \end{aligned} \tag{3.17}$$

Thus, x_3 or x_4 should be sufficiently accurate for most purposes. ■

3.2.1 Error Analysis

Assume $f(x)$ has at least two continuous derivatives for all x in some interval about the root α. Further assume that

$$f'(\alpha) \neq 0 \tag{3.18}$$

This says that the graph of $y = f(x)$ is not tangent to the x-axis when the graph intersects it at $x = \alpha$. The case in which $f'(\alpha) = 0$ is treated in Section 3.5. Also, note that combining (3.18) with the continuity of $f'(x)$ implies that $f'(x) \neq 0$ for all x near α.

Use Taylor's theorem to write

$$f(\alpha) = f(x_n) + (\alpha - x_n)f'(x_n) + \tfrac{1}{2}(\alpha - x_n)^2 f''(c_n)$$

with c_n an unknown point between α and x_n. Note that $f(\alpha) = 0$ by assumption, and then divide $f'(x_n)$ to obtain

$$0 = \frac{f(x_n)}{f'(x_n)} + \alpha - x_n + (\alpha - x_n)^2 \frac{f''(c_n)}{2f'(x_n)}$$

From (3.11), the first term on the right side is $x_n - x_{n+1}$, and we have

$$0 = x_n - x_{n+1} + \alpha - x_n + (\alpha - x_n)^2 \frac{f''(c_n)}{2f'(x_n)}$$

Solving for $\alpha - x_{n+1}$, we have

$$\alpha - x_{n+1} = (\alpha - x_n)^2 \left[\frac{-f''(c_n)}{2f'(x_n)}\right] \tag{3.19}$$

This formula says that the error in x_{n+1} is nearly proportional to the square of the error in x_n. When the initial error is sufficiently small, this shows that the error in the succeeding iterates will decrease very rapidly, just as in (3.17). Formula (3.19) can also be used to give a formal mathematical proof of the convergence of Newton's method, but we omit it.

Example 3.2.3 For the earlier iteration (3.12), $f''(x) = 30x^4$. If we are near the root α, then

$$\frac{-f''(c_n)}{2f'(x_n)} \approx \frac{-f''(\alpha)}{2f'(\alpha)} = \frac{-30\alpha^4}{2(6\alpha^5 - 1)} \doteq -2.42$$

Thus for the error in (3.12),

$$\alpha - x_{n+1} \approx -2.42(\alpha - x_n)^2 \tag{3.20}$$

This explains the rapid convergence of the final iterates in Table 3.2. For example, consider the case of $n = 3$, with $\alpha - x_3 \doteq -4.73\text{E} - 3$. Then (3.20) predicts

$$\alpha - x_4 \doteq -2.42(4.73\text{E} - 3)^2 \doteq -5.42\text{E} - 5$$

which compares well to the actual error of $\alpha - x_4 \doteq -5.35\text{E} - 5$. ■

If we assume that the iterate x_n is near the root α, the multiplier on the right of (3.19) can be written as

$$\frac{-f''(c_n)}{2f'(x_n)} \approx \frac{-f''(\alpha)}{2f'(\alpha)} \equiv M \tag{3.21}$$

Thus,

$$\alpha - x_{n+1} \approx M(\alpha - x_n)^2, \qquad n \geq 0 \tag{3.22}$$

Multiply both sides by M to get

$$M(\alpha - x_{n+1}) \approx [M(\alpha - x_n)]^2 \tag{3.23}$$

Assuming that all of the iterates are near α, then inductively we can show that

$$M(\alpha - x_n) \approx [M(\alpha - x_0)]^{2^n}, \qquad n \geq 0$$

Since we want $\alpha - x_n$ to converge to zero, this says that we must have

$$|M(\alpha - x_0)| < 1$$

$$|\alpha - x_0| < \frac{1}{|M|} = \left|\frac{2f'(\alpha)}{f''(\alpha)}\right| \tag{3.24}$$

If the quantity $|M|$ is very large, then x_0 will have to be chosen very close to α to obtain convergence. In such situations, the bisection method is probably an easier method to use. An example of this situation is given in Problem 5.

The choice of x_0 can be very important in determining whether Newton's method will converge. Unfortunately, there is no single strategy that is always effective in choosing x_0. In most instances, a choice of x_0 arises from the physical situation that led to the rootfinding problem. In other instances, graphing $y = f(x)$ will probably be needed, possibly combined with the bisection method for a few iterates.

3.2.2 Error Estimation

We are computing a sequence of iterates x_n, and we would like to estimate their accuracy to know when to stop the iteration. To estimate $\alpha - x_n$, note that, since $f(\alpha) = 0$, we have

$$f(x_n) = f(x_n) - f(\alpha) = f'(\xi_n)(x_n - \alpha)$$

for some ξ_n between x_n and α, by the mean-value theorem. Solving for the error, we obtain

$$\alpha - x_n = \frac{-f(x_n)}{f'(\xi_n)} \approx \frac{-f(x_n)}{f'(x_n)}$$

provided that x_n is so close to α that $f'(x_n) \doteq f'(\xi_n)$. From (3.11), this becomes

$$\alpha - x_n \approx x_{n+1} - x_n \tag{3.25}$$

This is the standard error estimation formula for Newton's method, and it is usually fairly accurate. However, this formula is not valid if $f'(\alpha) = 0$, a case that is discussed in Section 3.5 of this chapter.

Example 3.2.4 Consider the error in the entry x_3 of Table 3.2.

$$\begin{aligned} \alpha - x_3 &\doteq -4.73\mathrm{E}-3 \\ x_4 - x_3 &\doteq -4.68\mathrm{E}-3 \end{aligned} \tag{3.26}$$

This illustrates the accuracy of (3.25) for that case. ■

MATLAB PROGRAM: *An implementation of Newton's method.* The function `newton` is an implementation of Newton's method, with error estimation and a safeguard against an infinite loop. The input parameter `max_iterate` is an upper limit on the number of iterates to be computed; this prevents the occurrence of an infinite loop.

```
function root = newton(x0,error_bd,max_iterate)
%
% function newton(x0,error_bd,max_iterate)
%
% This is Newton's method for solving an equation f(x) = 0.
%
% The functions f(x) and deriv_f(x) are given below.
% The parameter error_bd is used in the error test for the
% accuracy of each iterate.  The parameter max_iterate
% is an upper limit on the number of iterates to be
% computed.  An initial guess x0 must also be given.
%
% For the given function f(x), an example of a calling sequence
% might be the following:
%       root = newton(1,1.0E-12,10)
%
% The program prints the iteration values
%       iterate_number, x, f(x), deriv_f(x), error
% The value of x is the most current initial guess, called
% previous_iterate here, and it is updated with each iteration.
% The value of error is
%       error = newly_computed_iterate - previous_iterate
% and it is an estimated error for previous_iterate.
% Tap the carriage return to continue with the iteration.
```

```
format short e
error = 1;
it_count = 0;
fprintf('\n  it_count  x  f(x)  df(x)  error  \n')

while abs(error) > error_bd & it_count < max_iterate
  fx = f(x0);
  dfx = deriv_f(x0);
  if dfx == 0
    disp('The derivative is zero.  Stop')
    return
  end
  x1 = x0 - fx/dfx;
  error = x1 - x0;
% Internal print of the Newton method.  Tap the carriage
% return key to continue the computation.
  iteration = [it_count x0 fx dfx error]
  pause
  x0 = x1;
  it_count = it_count + 1;
end

if it_count >= max_iterate
  disp('The number of iterates calculated exceeded')
  disp('max_iterate.  An accurate root was not')
  disp('calculated.')
else
  format long
  root = x1
  format short
end

%%%%%%%%%%%%%%%%%%%%%%%%%%%%
function value = f(x)
%
% function to define equation for rootfinding problem.
%
value = x.^6 - x - 1;

%%%%%%%%%%%%%%%%%%%%%%%%%%%%
function value = deriv_f(x)
%
% Derivative of function defining equation for
% rootfinding problem.
%
value = 6*x.^5 - 1;
```

PROBLEMS

1. Carry out the Newton iteration (3.12) with the two initial guesses $x_0 = 1.0$ and $x_0 = 2.0$. Compare the results with Table 3.2.

2. Using Newton's method, find the roots of the equations in Problem 1, Section 3.1 (in this chapter). Use an error tolerance of $\epsilon = 10^{-6}$.

3. **(a)** On most computers, the computation of $\sqrt{a}$ is based on Newton's method. Set up the Newton iteration for solving $x^2 - a = 0$, and show that it can be written in the form

$$x_{n+1} = \frac{1}{2}\left(x_n + \frac{a}{x_n}\right), \qquad n \geq 0$$

   (b) Derive the error and relative error formulas

$$\sqrt{a} - x_{n+1} = -\frac{1}{2x_n}\left(\sqrt{a} - x_n\right)^2$$

$$\text{Rel}(x_{n+1}) = -\frac{\sqrt{a}}{2x_n}[\text{Rel}(x_n)]^2$$

   Hint: Apply (3.19).

   (c) For x_0 near $\sqrt{a}$, the last formula becomes

$$\text{Rel}(x_{n+1}) \approx -\frac{1}{2}[\text{Rel}(x_n)]^2, \qquad n \geq 0$$

   Assuming $\text{Rel}(x_0) = 0.1$, use this formula to estimate the relative error in x_1, x_2, x_3, and x_4.

4. Give Newton's method for finding $\sqrt[m]{a}$, with $a > 0$ and m a positive integer. Apply it to finding $\sqrt[m]{2}$ for $m = 3, 4, 5, 6, 7, 8$, say, to six significant digits.

   Hint: Solve $x^m - a = 0$.

5. **(a)** Repeat Problem 2 of Section 3.1, for finding roots of $x = \tan(x)$. Use an error tolerance of $\epsilon = 10^{-6}$.

   (b) The root near 100 will be difficult to find by using Newton's method. To explain this, compute the quantity M of (3.21), and use it in the condition (3.24) for x_0.

6. The equation

$$f(x) \equiv x + e^{-Bx^2}\cos(x) = 0, \qquad B > 0$$

   has a unique root, and it is in the interval $(-1, 0)$. Use Newton's method to find it as accurately as possible. Use values of $B = 1, 5, 10, 25, 50$. Among your choices of x_0, choose $x_0 = 0$, and explain the behavior observed in the iterates for the larger values of B.

   Hint: Draw a graph of $f(x)$ to better understand the behavior of the function.

7. Check the accuracy of the error approximation (3.25) for all entries in Table 3.2, as was done in (3.26).

8. Use the iteration (3.14) to compute $1/3$. Use $x_0 = 0.2$. Give the error in x_1, x_2, x_3, x_4.

9. Derive formula (3.15). Using it, show $\text{Rel}(x_n) = [\text{Rel}(x_0)]^{2^n}$.

Hint: Use

$$\text{Rel}(x_{n+1}) = \frac{\alpha - x_{n+1}}{\alpha} = 1 - bx_{n+1}$$

Replace x_{n+1} using (3.14), and then compare the result to $[\text{Rel}(x_n)]^2$.

10. Solve the equation

$$x^3 - 3x^2 + 3x - 1 = 0$$

on a computer and use Newton's method. Recalling Problem 7 of Section 3.1, experiment with the choice of initial guess x_0. Also experiment with different ways of evaluating $f(x)$ and $f'(x)$. Note any unusual behavior in the iteration.

11. Repeat Problem 10 with the equation

$$x^4 - 5.4x^3 + 10.56x^2 - 8.954x + 2.7951 = 0$$

Look for the root α located in $[1, 1.2]$.

12. Recall the material of Section 1.3 on the nested evaluation of polynomials. In particular, recall (1.35)–(1.38) and Problem 8 in that section. Using this, write a Newton program for finding the roots of polynomials $p(x)$, employing this earlier material to efficiently evaluate p and p'. Apply it to each of the following polynomial equations, finding its largest positive root:

(a) $x^3 - x^2 - x - 1 = 0$. **(b)** $32x^6 - 48x^4 + 18x^2 - 1 = 0$.

Note that the polynomial $q(x)$ referred to in (1.37)–(1.38) satisfies $p(x) = (x - \alpha)q(x)$ at the root α. Thus, $q(x)$ can be used to obtain the remaining roots of $p(x) = 0$. This is called *polynomial deflation*.

13. Consider applying Newton's method to find the root $\alpha = 0$ of $\sin(x) = 0$. Find an interval $[-r, r]$ for which the Newton iterates will converge to α, for any choice of x_0 in $[-r, r]$. Make r as large as possible.

Hint: Draw a graph of $y = \sin(x)$ and graphically interpret the placement of x_0 and x_1.

3.3. SECANT METHOD

The Newton method is based on approximating the graph of $y = f(x)$ with a tangent line and on then using the root of this straight line as an approximation to the root α of $f(x)$. From this perspective, other straight-line approximations to $y = f(x)$ would also lead to methods for approximating a root of $f(x)$. One such straight-line approximation leads to the *secant method.*

Assume that two initial guesses to α are known and denote them by x_0 and x_1. They may occur on opposite sides of α, as in Figure 3.3, or on the same side of α, as in Figure 3.4. The two points $(x_0, f(x_0))$ and $(x_1, f(x_1))$, on the graph of $y = f(x)$, determine a straight line, called a *secant line*. This line is an approximation to the graph of $y = f(x)$, and its root x_2 is an approximation of α.

To derive a formula for x_2, we proceed in a manner similar to that used to derive Newton's method: Find the equation of the line and then find its root x_2. The equation of the line is given by

$$y = p(x) \equiv f(x_1) + (x - x_1) \cdot \frac{f(x_1) - f(x_0)}{x_1 - x_0}$$

Solving $p(x_2) = 0$, we obtain

$$x_2 = x_1 - f(x_1) \cdot \frac{x_1 - x_0}{f(x_1) - f(x_0)}$$

Having found x_2, we can drop x_0 and use x_1, x_2 as a new set of approximate values for α. This leads to an improved value x_3; and this process can be continued indefinitely.

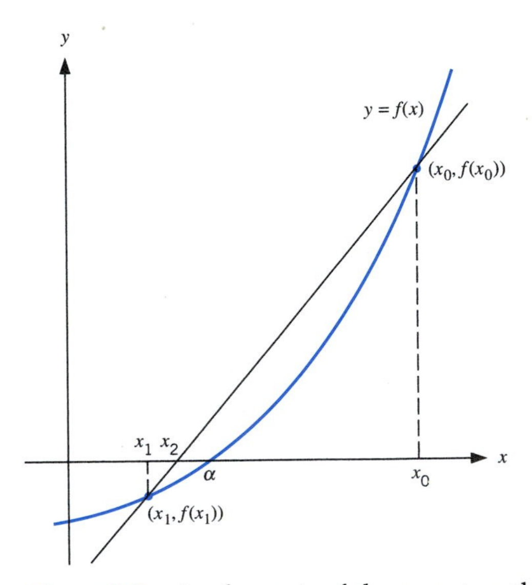

Figure 3.3. A schematic of the secant method: $x_1 < \alpha < x_0$

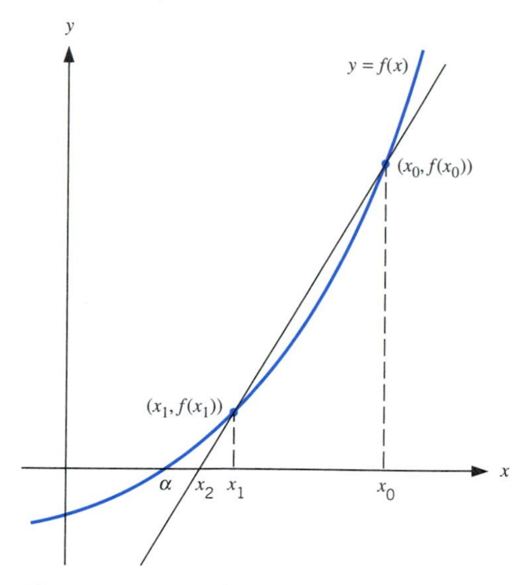

Figure 3.4. A schematic of the secant method: $\alpha < x_1 < x_0$

Doing so, we obtain the general iteration formula

$$x_{n+1} = x_n - f(x_n) \cdot \frac{x_n - x_{n-1}}{f(x_n) - f(x_{n-1})}, \qquad n \geq 1 \tag{3.27}$$

This is the *secant method*. It is called a two-point method, since two approximate values are needed to obtain an improved value. The bisection method is also a two-point method, but the secant method will almost always converge faster than bisection.

Example 3.3.1 We solve the equation

$$f(x) \equiv x^6 - x - 1 = 0$$

which was used previously as an example for both the bisection and Newton methods. The results are given in Table 3.3, including the quantity $x_n - x_{n-1}$ as an estimate of $\alpha - x_{n-1}$. The iterate x_8 equals α rounded to nine significant digits. As with the Newton method (3.12) for this equation, the initial iterates do not converge rapidly. But as the iterates become closer to α, the speed of convergence increases. ■

3.3.1 Error Analysis

By using techniques from calculus and some algebraic manipulation, it is possible to show that the iterates x_n of (3.27) satisfy

Table 3.3. Secant Method for $x^6 - x - 1 = 0$

n	x_n	$f(x_n)$	$x_n - x_{n-1}$	$\alpha - x_{n-1}$
0	2.0	61.0		
1	1.0	−1.0	−1.0	
2	1.01612903	−9.15E − 1	1.61E − 2	1.35E − 1
3	1.19057777	6.57E − 1	1.74E − 1	1.19E − 1
4	1.11765583	−1.68E − 1	−7.29E − 2	−5.59E − 2
5	1.13253155	−2.24E − 2	1.49E − 2	1.71E − 2
6	1.13481681	9.54E − 4	2.29E − 3	2.19E − 3
7	1.13472365	−5.07E − 6	−9.32E − 5	−9.27E − 5
8	1.13472414	−1.13E − 9	4.92E − 7	4.92E − 7

$$\alpha - x_{n+1} = (\alpha - x_n)(\alpha - x_{n-1})\left[\frac{-f''(\xi_n)}{2f'(\zeta_n)}\right] \tag{3.28}$$

The unknown number ζ_n is between x_n and x_{n-1}, and the unknown number ξ_n is between the largest and the smallest of the numbers α, x_n, and x_{n-1}. The error formula closely resembles the Newton error formula (3.19). This should be expected, since the secant method can be considered as an approximation of Newton's method, based on using

$$f'(x_n) \approx \frac{f(x_n) - f(x_{n-1})}{x_n - x_{n-1}} \tag{3.29}$$

Check that the use of this in the Newton formula (3.11) will yield (3.27).

The formula (3.28) can be used to obtain the further error result that if x_0 and x_1 are chosen sufficiently close to α, then we have convergence and

$$\lim_{n\to\infty} \frac{|\alpha - x_{n+1}|}{|\alpha - x_n|^r} = \left|\frac{f''(\alpha)}{2f'(\alpha)}\right|^{r-1} \equiv c \tag{3.30}$$

where $r = (\sqrt{5} + 1)/2 \doteq 1.62$. Thus,

$$|\alpha - x_{n+1}| \approx c\,|\alpha - x_n|^{1.62} \tag{3.31}$$

as x_n approaches α. Compare this with the Newton estimate (3.22), in which the exponent is 2 rather than 1.62. Thus, Newton's method converges more rapidly than the secant method. Also, the constant c in (3.31) plays the same role as M in (3.22), and they are related by

$$c = |M|^{r-1}$$

The restriction (3.24) on the initial guess for Newton's method can be replaced by a similar one for the secant iterates, but we omit it. Finally, the result (3.31) can be used to justify the error estimate

$$\alpha - x_{n-1} \approx x_n - x_{n-1} \tag{3.32}$$

for iterates x_n that are sufficiently close to the root.

Example 3.3.2 For the iterate x_5 in Table 3.3,

$$\begin{aligned} \alpha - x_5 &\doteq 2.19\text{E}-3 \\ x_6 - x_5 &\doteq 2.29\text{E}-3 \end{aligned} \tag{3.33}$$

■

MATLAB PROGRAM: ***An implementation of the secant method.*** The function `secant` is an implementation of the secant method, with error estimation and a safeguard against an infinite loop. The input parameter `max_iterate` is an upper limit on the number of iterates to be computed; this prevents the occurrence of an infinite loop.

```
function root = secant(x0,x1,error_bd,max_iterate)
%
% function secant(x0,x1,error_bd,max_iterate)
%
% This implements the secant method for solving an
% equation f(x) = 0.
%
% The parameter error_bd is used in the error test for the
% accuracy of each iterate.  The parameter max_iterate is
% an upper limit on the number of iterates to be computed.
% Two initial guesses, x0 and x1, must also be given.
%
% For the given function f(x), an example of a calling
% sequence might be the following:
%       root = secant(x0,x1,1.0E-12,10)
% The function f(x) is given below.
%
% The program prints the iteration values
%       iterate_number, x, f(x), error
% The value of x is the most current initial guess, called
% previous_iterate here, and it is updated with each
% iteration.  The value of error is
%       error = newly_computed_iterate - previous_iterate
% and it is an estimated error for previous_iterate.
% Tap the carriage return to continue with the iteration.

format short e
```

```
error = 1;
fx0 = f(x0);
it_count = 0;
iteration = [it_count x0 fx0]

while abs(error) > error_bd & it_count <= max_iterate
  it_count = it_count + 1;
  fx1 = f(x1);
  if fx1 - fx0 == 0
    disp('f(x1) = f(x0); Division by zero; Stop')
    return
  end
  x2 = x1 - fx1*(x1-x0)/(fx1-fx0);
  error = x2 - x1;
% Internal print of secant method.  Tap the carriage
% return key to continue the computation.
  iteration = [it_count x1 fx1 error]
  pause
  x0 = x1;
  x1 = x2;
  fx0 = fx1;
end

if it_count > max_iterate
  disp('The number of iterates calculated exceeded')
  disp('max_iterate.  An accurate root was not')
  disp('calculated.')
else
  format long
  root = x2
  format short
end

%%%%%%%%%%%%%%%%%%%%%%%%%%%%%
function value = f(x)
%
% function to define equation for rootfinding problems.
%
value = x.^6 - x - 1;
```

3.3.2 Comparison of Newton and Secant Methods

From the foregoing discussion, Newton's method converges more rapidly than the secant method. Thus, Newton's method should require fewer iterations to attain a given error tolerance. However, Newton's method requires two function evaluations per iteration, that of $f(x_n)$ and $f'(x_n)$. And the secant method requires only one evaluation, $f(x_n)$, if it is programed carefully to retain the value of $f(x_{n-1})$ from the preceding iteration. Thus, the secant method will require less time per iteration than the Newton method.

The decision as to which method should be used will depend on the factors just discussed, including the difficulty or expense of evaluating $f'(x_n)$; and it will depend on intangible human factors, such as convenience of use. Newton's method is very simple to program and to understand; but for many problems with a complicated $f'(x)$, the secant method will probably be faster in actual running time on a computer.

General Remarks The derivations of both the Newton and secant methods illustrate a general principle of numerical analysis. When trying to solve a problem for which there is no direct or simple method of solution, approximate it by another problem that you can solve more easily. In both cases, we have replaced the solution of $f(x) = 0$ with the solution of a much simpler rootfinding problem for a linear equation. This is another example of General Observation (1.10) given near the end of Section 1.1. The nature of the approximation being used also leads to the following observation:

GENERAL OBSERVATION: When dealing with problems involving differentiable functions $f(x)$, move to a nearby problem by approximating each such $f(x)$ with a linear function.	(3.34)

The linearization of mathematical problems is common throughout applied mathematics and numerical analysis.

3.3.3 The MATLAB Function fzero

MATLAB contains the rootfinding routine `fzero` that uses ideas involved in the bisection method and the secant method. As with many MATLAB programs, there are several possible calling sequences. The command

```
root=fzero(f_name,[a,b])
```

produces a root within $[a, b]$, where it is assumed that $f(a)f(b) \leq 0$. The command

```
root=fzero(f_name,x0)
```

tries to find a root of the function near $x0$. The default error tolerance is the maximum precision of the machine, although this can be changed by the user. This is an excellent rootfinding routine, combining guaranteed convergence with high efficiency.

PROBLEMS

1. Using the secant method, find the roots of the equations in Problem 1 of Section 3.1. Use an error tolerance of $\epsilon = 10^{-6}$.

2. Solve equation (3.2) with $P_{\text{in}} = 1000$, $P_{\text{out}} = 20{,}000$, $N_{\text{in}} = 30$, and $N_{\text{out}} = 20$. Find r with an accuracy of $\epsilon = 0.0001$.

3. Solve Problem 6 of Section 3.2, using the secant method. As one choice of initial guesses, use $x_0 = -1$, $x_1 = 0$.

4. Continuing Example 3.3.1, experimentally confirm the error estimate (3.31). For this purpose, first compute the constant c by the formula (3.30). Then for $n = 2, \dots, 7$, compute and compare both sides of the error estimate (3.31).

5. Using the secant method, repeat Problem 7 of Section 3.1 and Problem 10 of Section 3.2.

6. Using the secant method, repeat Problem 11 of Section 3.2.

7. As a partial step toward showing (3.28), use algebraic manipulation to show

$$\alpha - x_{n+1} = -(\alpha - x_n)(\alpha - x_{n-1})\frac{f[x_{n-1}, x_n, \alpha]}{f[x_{n-1}, x_n]}$$

where

$$f[a, b] = \frac{f(b) - f(a)}{b - a}$$

$$f[a, b, c] = \frac{f[b, c] - f[a, b]}{c - a}$$

These quantities are called *Newton divided differences*, and they are discussed in Section 4.1. The formula (4.24), applied to the above error formula, leads to (3.28).

8. Write formula (3.28) as

$$\alpha - x_{n+1} \approx M(\alpha - x_n)(\alpha - x_{n-1}), \qquad n \geq 0$$

with M defined in (3.21). Then multiply both sides by M, obtaining

$$|M(\alpha - x_{n+1})| \approx |M(\alpha - x_n)|\,|M(\alpha - x_{n-1})|$$

Let $B_n = |M(\alpha - x_n)|$, $n \geq 0$. To have x_n converge to α, we must have B_n converge to 0. The above formula yields

$$B_{n+1} \approx B_n B_{n-1}$$

For simplicity, assume $B_0 = B_1 = \delta$.

(a) Compute approximate values of $B_2, B_3, B_4, B_5, B_6, B_7$ in terms of δ.

(b) If we write $B_n = \delta^{q_n}$, $n \geq 0$, give a formula for q_{n+1} in terms of q_{n-1} and q_n. What are q_0 and q_1?

(c) Experimentally, confirm that

$$q_n \approx \frac{1}{\sqrt{5}} r^{n+1} \doteq \frac{1}{\sqrt{5}} (1.618)^{n+1}$$

for larger values of n, say, $n \geq 4$. The number r was used in (3.30). The numbers $\{q_n\}$ are called *Fibonacci numbers*. The number r is called the *golden mean* (some authors define the golden mean to be $r - 1$). For a more detailed derivation, see Atkinson (1989, pp. 68–69).

(d) Using (c), show that

$$\frac{B_{n+1}}{B_n^r}$$

is approximately constant. Find the constant. This result can be used to construct a proof of (3.30).

3.4. FIXED POINT ITERATION

The Newton method (3.11) and the secant method (3.27) are examples of one-point and two-point iteration methods, respectively. In this section, we give a more general introduction to iteration methods, presenting a general theory for one-point iteration formulas.

As a motivational example, consider solving the equation

$$x^2 - 5 = 0 \tag{3.35}$$

for the root $\alpha = \sqrt{5} \doteq 2.2361$. We give four iteration methods to solve this equation.

I1. $x_{n+1} = 5 + x_n - x_n^2$

I2. $x_{n+1} = \dfrac{5}{x_n}$

I3. $x_{n+1} = 1 + x_n - \dfrac{1}{5} x_n^2$

I4. $x_{n+1} = \dfrac{1}{2}\left(x_n + \dfrac{5}{x_n}\right)$

All four iterations have the property that if the sequence $\{x_n \mid n \geq 0\}$ has a limit α, then α is a root of (3.35). For each equation, check this as follows: Replace x_n and x_{n+1} by α, and then show that this implies $\alpha = \pm\sqrt{5}$. In Table 3.4, we give the iterates x_n for these four iteration methods. To explain these numerical results, we present a general theory for one-point iteration formulas.

Table 3.4. Iterations I1 to I4

n	x_n : I1	x_n : I2	x_n : I3	x_n : I4
0	2.5	2.5	2.5	2.5
1	1.25	2.0	2.25	2.25
2	4.6875	2.5	2.2375	2.2361
3	−12.2852	2.0	2.2362	2.2361

The iterations I1 to I4 all have the form

$$x_{n+1} = g(x_n) \tag{3.36}$$

for appropriate continuous functions $g(x)$. For example, with I1, $g(x) = 5 + x - x^2$. If the iterates x_n converge to a point α, then

$$\lim_{n\to\infty} x_{n+1} = \lim_{n\to\infty} g(x_n)$$
$$\alpha = g(\alpha)$$

Thus, α is a solution of the equation $x = g(x)$, and α is called a *fixed point* of the function g.

In this section, a general theory is given to explain when the iteration $x_{n+1} = g(x_n)$ will converge to a fixed point of g. We begin with a lemma on the existence of solutions of $x = g(x)$.

Lemma 3.4.1 Let $g(x)$ be a continuous function on an interval $[a, b]$, and suppose g satisfies the property

$$a \le x \le b \quad \Longrightarrow \quad a \le g(x) \le b \tag{3.37}$$

Then the equation $x = g(x)$ has at least one solution α in the interval $[a, b]$.

Proof. Define the function $f(x) = x - g(x)$. It is continuous for $a \le x \le b$. Moreover, $f(a) \le 0$ and $f(b) \ge 0$. By the intermediate value theorem (see Appendix A), there must be a point x in $[a, b]$ for which $f(x) = 0$. We usually denote this value of x by α. ■

See Figure 3.5 for a graphical interpretation of the solution of $x = g(x)$. The solutions α are the x-coordinates of the intersection points of the graphs of $y = x$ and $y = g(x)$.

Theorem 3.4.2 (Contraction mapping theorem) Assume $g(x)$ and $g'(x)$ are continuous for $a \le x \le b$, and assume g satisfies (3.37). Further assume that

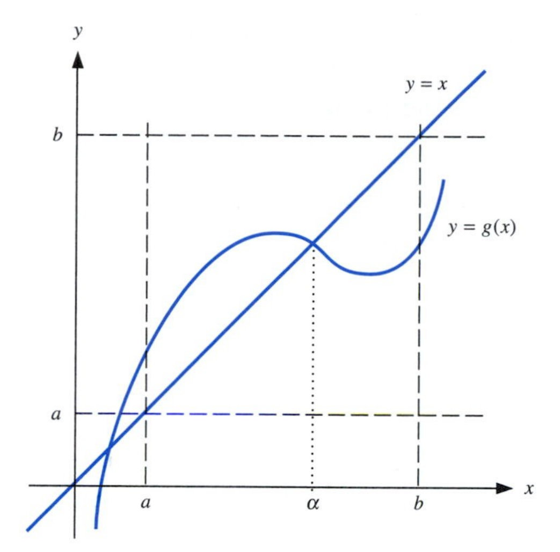

Figure 3.5. An example of Lemma 3.4.1

$$\lambda \equiv \max_{a \le x \le b} \left| g'(x) \right| < 1 \tag{3.38}$$

Then

S1. There is a unique solution α of $x = g(x)$ in the interval $[a, b]$.

S2. For any initial estimate x_0 in $[a, b]$, the iterates x_n will converge to α.

S3.

$$|\alpha - x_n| \le \frac{\lambda^n}{1 - \lambda} |x_0 - x_1|, \qquad n \ge 0 \tag{3.39}$$

S4.

$$\lim_{n \to \infty} \frac{\alpha - x_{n+1}}{\alpha - x_n} = g'(\alpha) \tag{3.40}$$

Thus for x_n close to α,

$$\alpha - x_{n+1} \approx g'(\alpha)(\alpha - x_n) \tag{3.41}$$

Proof. There is some useful information in the proof, so we go through most of the details of it. Note first that the hypotheses on g allow us to use Lemma 3.4.1 to assert the existence of at least one solution to $x = g(x)$. In addition, using the mean value theorem (see Appendix A), we have that for any two points w and z in $[a, b]$,

$$g(w) - g(z) = g'(c)(w - z)$$

for some c between w and z. By using (3.38), we obtain

$$\begin{aligned} |g(w) - g(z)| &= |g'(c)|\,|w - z| \\ &\le \lambda\,|w - z| \qquad a \le w, z \le b \end{aligned} \tag{3.42}$$

S1. Suppose there are two solutions, denoted by α and β. Then $\alpha = g(\alpha)$ and $\beta = g(\beta)$. By subtracting these, we find that

$$\alpha - \beta = g(\alpha) - g(\beta)$$

Take absolute values and use (3.42):

$$\begin{aligned} |\alpha - \beta| &\le \lambda\,|\alpha - \beta| \\ (1 - \lambda)\,|\alpha - \beta| &\le 0 \end{aligned}$$

Since $\lambda < 1$, we must have $\alpha = \beta$; and thus, the equation $x = g(x)$ has only one solution in the interval $[a, b]$.

S2. From the assumption (3.37), it can be shown that for any initial guess x_0 in $[a, b]$, the iterates x_n will all remain in $[a, b]$. For example, if $a \le x_0 \le b$, then (3.37) implies $a \le g(x_0) \le b$. Since $x_1 = g(x_0)$, this shows x_1 is in $[a, b]$. Repeat the argument to show that $x_2 = g(x_1)$ is in $[a, b]$, and continue the argument inductively.
To show that the iterates converge, subtract $x_{n+1} = g(x_n)$ from $\alpha = g(\alpha)$, obtaining

$$\begin{aligned} \alpha - x_{n+1} &= g(\alpha) - g(x_n) \\ &= g'(c_n)(\alpha - x_n) \end{aligned} \tag{3.43}$$

for some c_n between α and x_n. Using the assumption (3.38), we get

$$|\alpha - x_{n+1}| \le \lambda\,|\alpha - x_n|, \qquad n \ge 0 \tag{3.44}$$

Inductively, we can then show that

$$|\alpha - x_n| \le \lambda^n\,|\alpha - x_0|, \qquad n \ge 0 \tag{3.45}$$

Since $\lambda < 1$, the right side of (3.45) goes to zero as $n \to \infty$, and this then shows that $x_n \to \alpha$ as $n \to \infty$.

S3. Use (3.44) with $n = 1$ to obtain

$$\begin{aligned} |\alpha - x_0| &\le |\alpha - x_1| + |x_1 - x_0| \\ &\le \lambda\,|\alpha - x_0| + |x_1 - x_0| \\ (1 - \lambda)\,|\alpha - x_0| &\le |x_1 - x_0| \\ |\alpha - x_0| &\le \frac{1}{1 - \lambda}\,|x_1 - x_0| \end{aligned} \tag{3.46}$$

Combine this with (3.45) to conclude the derivation of (3.39).

S4. Use (3.43) to write

$$\lim_{n\to\infty} \frac{\alpha - x_{n+1}}{\alpha - x_n} = \lim_{n\to\infty} g'(c_n)$$

Each c_n is between α and x_n, and $x_n \to \alpha$, by S2. Thus, $c_n \to \alpha$. Combine this with the continuity of the function $g'(x)$ to obtain

$$\lim_{n\to\infty} g'(c_n) = g'(\alpha)$$

thus proving (3.40). ■

We need a more precise way to deal with the concept of the speed of convergence of an iteration method. We say that a sequence $\{x_n \mid n \geq 0\}$ converges to α with an *order of convergence* $p \geq 1$ if

$$|\alpha - x_{n+1}| \leq c\,|\alpha - x_n|^p, \qquad n \geq 0$$

for some constant $c \geq 0$. The cases $p = 1$, $p = 2$, and $p = 3$ are referred to as *linear convergence*, *quadratic convergence*, and *cubic convergence*, respectively. Newton's method usually converges quadratically; and the secant method has order of convergence $p = (1 + \sqrt{5})/2$. For linear convergence, we make the additional requirement that $c < 1$; as otherwise, the error $\alpha - x_n$ need not converge to zero.

If $|g'(\alpha)| < 1$ in the preceding theorem, then formula (3.44) shows that the iterates x_n are linearly convergent. If in addition, $g'(\alpha) \neq 0$, then formula (3.41) proves the convergence is exactly linear, with no higher order of convergence being possible. In this case, we call the value of $|g'(\alpha)|$ the linear rate of convergence.

In practice, Theorem 3.4.2 is seldom used directly. The main reason is that it is difficult to find an interval $[a, b]$ for which (3.37) is satisfied. Instead, we look for a way to use the theorem in a practical way. The key idea is the result (3.43), which shows how the iteration error behaves when the iterates x_n are near α.

Corollary 3.4.3 Assume that $g(x)$ and $g'(x)$ are continuous for some interval $c < x < d$, with the fixed point α contained in this interval. Moreover, assume that

$$|g'(\alpha)| < 1 \tag{3.47}$$

Then, there is an interval $[a, b]$ around α for which the hypotheses, and hence also the conclusions, of Theorem 3.4.2 are true. And if to the contrary, $|g'(\alpha)| > 1$, then the iteration method $x_{n+1} = g(x_n)$ will not converge to α. [When $|g'(\alpha)| = 1$, no conclusion can be drawn; and even if convergence were to occur, the method would be far too slow for the iteration method to be practical.]

The proof of this is taken up in Problem 10. Using this result, we can examine the iteration methods I1 to I4. Recall $\alpha = \sqrt{5}$.

I1. $g(x) = 5 + x - x^2$, $g'(x) = 1 - 2x$, $g'(\alpha) = 1 - 2\sqrt{5} < -1$. Thus, the iteration I1 will not converge to $\sqrt{5}$.

I2. $g(x) = 5/x$, $g'(x) = -5/x^2$, $g'(\alpha) = -1$. We cannot conclude that the iteration converges or diverges. But from Table 3.4, it is clear that the iterates will not converge to α.

I3. $g(x) = 1 + x - \frac{1}{5}x^2$, $g'(x) = 1 - \frac{2}{5}x$, $g'(\alpha) = 1 - \frac{2}{5}\sqrt{5} \doteq 0.106$. From the corollary, the iteration will converge. And from (3.40),

$$|\alpha - x_{n+1}| \approx 0.106\,|\alpha - x_n|$$

when x_n is close to α. The errors decrease by approximately a factor of 0.1 with each iteration.

I4. $g(x) = \frac{1}{2}(x + 5/x)$, $g'(x) = \frac{1}{2}(1 - 5/x^2)$, $g'(\alpha) = 0$. Thus, the condition for convergence is easily satisfied. Note that this is Newton's method for computing $\sqrt{5}$.

It is often difficult to know how to convert a rootfinding problem $f(x) = 0$ into a fixed point problem $x = g(x)$ that leads to a convergent method. One such process is given in Problem 9, and it makes essential use of Corollary 3.4.3.

The possible behavior of the fixed point iterates x_n is shown graphically in Figure 3.6, for various sizes of $g'(\alpha)$. To see the convergence, consider the case of $x_1 = g(x_0)$, the height of the graph of $y = g(x)$ at x_0. We bring the number x_1 back to the x-axis by using the line $y = x$ and the height $y = x_1$. We continue this with each iterate, obtaining a stairstep behavior when $g'(\alpha) > 0$. When $g'(\alpha) < 0$, the iterates oscillate around the fixed point α, as can be seen in Figure 3.6.

Example 3.4.4 In Table 3.5, we give results from the iteration I3, along with more information on the convergence of the iterates. The errors are given, along with the ratios

$$r_n = \frac{\alpha - x_n}{\alpha - x_{n-1}} \tag{3.48}$$

Empirically, the values of r_n converge to $g'(\alpha) \doteq 0.105573$, which agrees with (3.40). ■

3.4.1 Aitken Error Estimation and Extrapolation

With the formula (3.41), it is possible to estimate the error in the iterates x_n and to accelerate their convergence. Let $g'(\alpha)$ be denoted by λ. Then we assume that (3.41) holds true for all n of interest, and we write it with n replaced by $n - 1$:

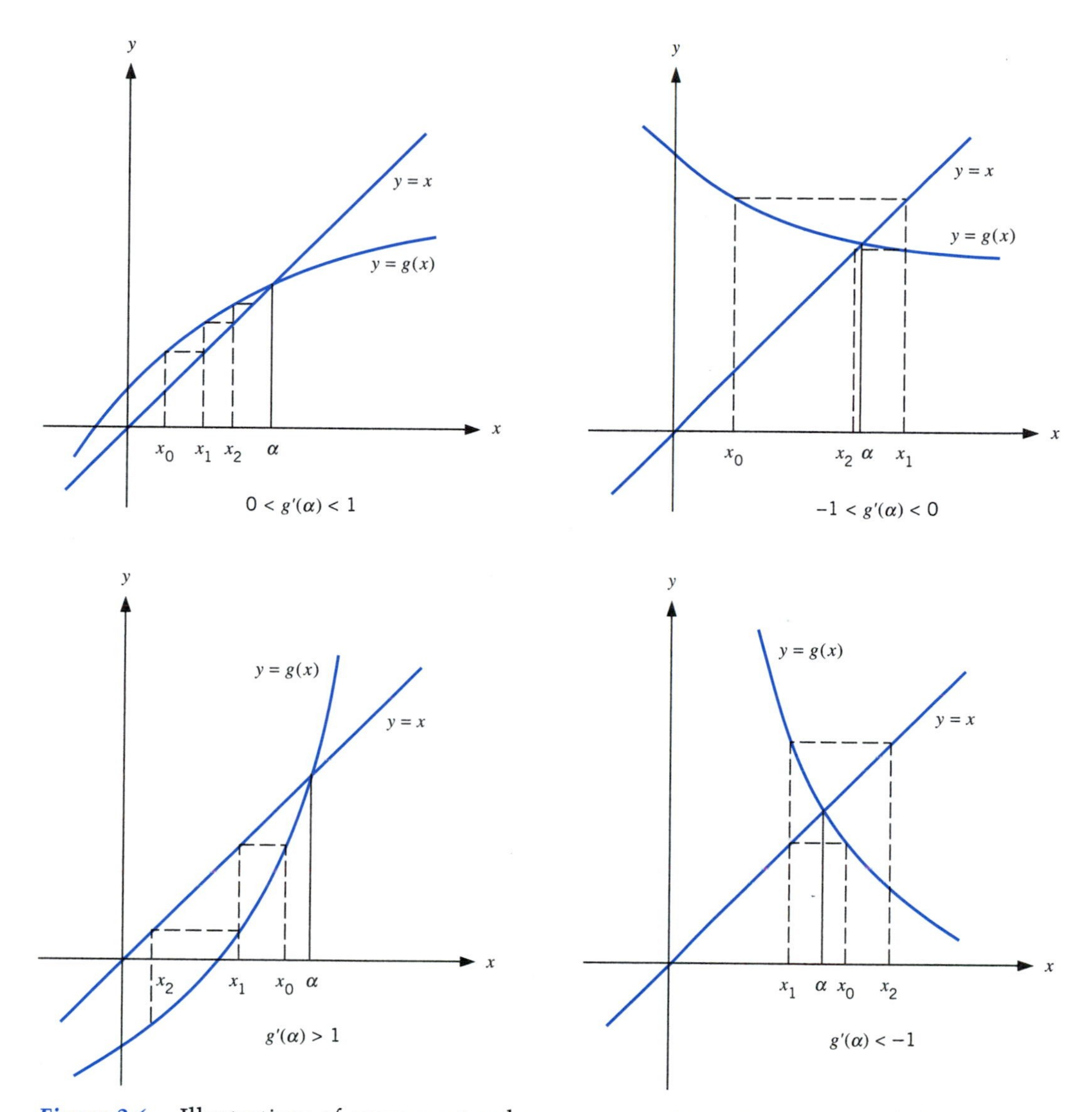

Figure 3.6. Illustrations of convergent and nonconvergent sequences $x_{n+1} = g(x_n)$

$$\alpha - x_n \approx \lambda(\alpha - x_{n-1}) \tag{3.49}$$

Solving for α and putting it in a computationally convenient form, we have

$$\alpha \approx x_n + \frac{\lambda}{1-\lambda}(x_n - x_{n-1}) \tag{3.50}$$

We need an estimate of λ. It cannot be calculated from its definition, since that requires knowing the solution α. The same is true of the ratios in (3.48). To estimate λ,

Table 3.5. The Iteration $x_{n+1} = 1 + x_n - \frac{1}{5}x_n^2$

n	x_n	$\alpha - x_n$	r_n
0	2.5	−2.64E − 1	
1	2.25	−1.39E − 2	0.0528
2	2.2375	−1.43E − 3	0.1028
3	2.23621875	−1.51E − 4	0.1053
4	2.23608389	−1.59E − 5	0.1055
5	2.23606966	−1.68E − 6	0.1056
6	2.23606815	−1.77E − 7	0.1056
7	2.23606800	−1.87E − 8	0.1056

we use the ratios

$$\lambda_n = \frac{x_n - x_{n-1}}{x_{n-1} - x_{n-2}}, \qquad n \geq 2 \tag{3.51}$$

To see that this should be an increasingly good estimate of λ as $n \to \infty$, write it as

$$\lambda_n = \frac{g(x_{n-1}) - g(x_{n-2})}{x_{n-1} - x_{n-2}} = g'(c_n)$$

with some c_n satisfying $\min\{x_{n-1}, x_{n-2}\} \leq c_n \leq \max\{x_{n-1}, x_{n-2}\}$. The last equality follows from the mean value theorem. Since $x_n \to \alpha$ as $n \to \infty$, we also have that $c_n \to \alpha$; and by the continuity of $g'(x)$, $g'(c_n) \to g'(\alpha) = \lambda$. Thus, $\lambda_n \to \lambda$ as $n \to \infty$. Combining (3.50) and (3.51), we obtain

$$\alpha \approx x_n + \frac{\lambda_n}{1 - \lambda_n}(x_n - x_{n-1}) \tag{3.52}$$

This is called *Aitken's extrapolation formula*. Writing it in the equivalent form

$$\alpha - x_n \approx \frac{\lambda_n}{1 - \lambda_n}(x_n - x_{n-1}) \tag{3.53}$$

gives *Aitken's error estimate*. This formula can be used to estimate the error in the original iterates $\{x_n\}$, or formula (3.52) can be used to create a more rapidly convergent sequence.

Example 3.4.5 We repeat the example on the iteration I3. Table 3.6 contains the differences $x_n - x_{n-1}$, the ratios λ_n, and the estimated error from (3.53), given in the column labeled Estimate. Compare the column Estimate with the error column in Table 3.5. ■

General Remarks There are a number of reasons to perform theoretical error analyses of a numerical method. We want to better understand the method, when it will

Table 3.6. The Iteration $x_{n+1} = 1 + x_n - \frac{1}{5}x_n^2$ and Aitken Error Estimation

n	x_n	$x_n - x_{n-1}$	λ_n	Estimate
0	2.5			
1	2.25	−2.50E − 1		
2	2.2375	−1.25E − 2	0.0500	−6.58E − 4
3	2.23621875	−1.28E − 3	0.1025	−1.46E − 4
4	2.23608389	−1.35E − 4	0.1053	−1.59E − 5
5	2.23606966	−1.42E − 5	0.1055	−1.68E − 6
6	2.23606815	−1.50E − 6	0.1056	−1.77E − 7
7	2.23606800	−1.59E − 7	0.1056	−1.87E − 8

perform well, when it will perform poorly, and perhaps, when it may not work at all. With a mathematical proof, we convince ourselves of the correctness of a numerical method under precisely stated hypotheses on the problem being solved. Finally, we often can improve on the performance of a numerical method. The use of Theorem 3.4.2 to obtain the Aitken extrapolation formula of (3.52) is an illustration of the following:

> GENERAL OBSERVATION:
> By understanding the behavior of the error in a numerical method, it is often possible to improve on that method and to obtain another more rapidly convergent method. (3.54)

We will illustrate this at other points in the text.

3.4.2 Higher-Order Iteration Formulas

The convergence formula (3.41) gives less information in the case $g'(\alpha) = 0$, although the convergence is clearly quite good. To improve on the results in Theorem 3.4.2, consider the Taylor expansion of $g(x_n)$ about α, assuming that $g(x)$ is twice continuously differentiable:

$$g(x_n) = g(\alpha) + (x_n - \alpha)g'(\alpha) + \frac{1}{2}(x_n - \alpha)^2 g''(c_n) \tag{3.55}$$

with c_n between x_n and α. Using $x_{n+1} = g(x_n)$, $\alpha = g(\alpha)$, and $g'(\alpha) = 0$, we have

$$x_{n+1} = \alpha + \frac{1}{2}(x_n - \alpha)^2 g''(c_n)$$

$$\alpha - x_{n+1} = -\frac{1}{2}(\alpha - x_n)^2 g''(c_n) \tag{3.56}$$

$$\lim_{n\to\infty} \frac{\alpha - x_{n+1}}{(\alpha - x_n)^2} = -\frac{1}{2}g''(\alpha) \tag{3.57}$$

If $g''(\alpha) \neq 0$, then this formula shows that the iteration $x_{n+1} = g(x_n)$ is of *order* 2 or is *quadratically convergent*.

If also $g''(\alpha) = 0$, and perhaps also some higher-order derivatives are zero at α, then expand the Taylor series through higher-order terms in (3.55), until the final error term contains a derivative of g that is nonzero at α. This leads to methods with an order of convergence greater than 2.

As an example, consider Newton's method as a fixed-point iteration:

$$x_{n+1} = g(x_n), \qquad g(x) = x - \frac{f(x)}{f'(x)} \tag{3.58}$$

Then,

$$g'(x) = \frac{f(x)f''(x)}{[f'(x)]^2}$$

and if $f'(\alpha) \neq 0$, then

$$g'(\alpha) = 0$$

Similarly, it can be shown that $g''(\alpha) \neq 0$ if moreover, $f''(\alpha) \neq 0$. If we use (3.57), these results show that Newton's method is of order 2, provided that $f'(\alpha) \neq 0$ and $f''(\alpha) \neq 0$.

PROBLEMS

1. **(a)** Calculate the first six iterates in the iteration

$$x_{n+1} = 1 + 0.3\sin(x_n)$$

with $x_0 = 1$. Choose other initial guesses x_0 and repeat this calculation.

(b) Find an interval $[a, b]$ satisfying the hypotheses of Theorem 3.4.2.

Hint: For $g(x) = 1 + 0.3\sin(x)$, let

$$a = \min_{-\infty<x<\infty} g(x), \qquad b = \max_{-\infty<x<\infty} g(x)$$

(c) Prepare a table in the same manner as Table 3.5 in the preceding discussion. The true solution is $\alpha \doteq 1.28809131321184$.

2. Repeat Problem 1 for the iteration $x_{n+1} = 0.5/(1 + x_n^2)$. The true solution is $\alpha \doteq 0.423853799069783$.

3. How many solutions are there to the equation $x = e^{-x}$? Will the iteration $x_{n+1} = e^{-x_n}$ converge for suitable choices of x_0? Calculate the first six iterates when $x_0 = 0$.

4. Repeat Problem 3 with $x_{n+1} = 1 + \tan^{-1}(x_n)$.

5. Show that for any constants c and d, $|d| < 1$, the equation $x = c + d\cos(x) \equiv g(x)$ has a unique solution α. In addition, show that the iteration $x_{n+1} = c + d\cos(x_n)$ will converge to α. Bound the rate of convergence.

6. Convert the equation $x^2 - 5 = 0$ to the fixed-point problem

$$x = x + c(x^2 - 5) \equiv g(x)$$

with c a nonzero constant. Determine the possible values of c to ensure convergence of

$$x_{n+1} = x_n + c(x_n^2 - 5)$$

to $\alpha = \sqrt{5}$.

7. What are the solutions α, if any, of the equation $x = \sqrt{1 + x}$? Does the iteration $x_{n+1} = \sqrt{1 + x_n}$ converge to any of these solutions (assuming x_0 is chosen sufficiently close to α)?

8. Which of the following iterations will converge to the indicated α, provided x_0 is chosen sufficiently close to α? If it does converge, determine the convergence order.

(a) $x_{n+1} = \dfrac{15x_n^2 - 24x_n + 13}{4x_n}, \quad \alpha = 1.$

(b) $x_{n+1} = \frac{3}{4}x_n + 1/x_n^3, \quad \alpha = \sqrt{2}.$

9. Consider the rootfinding problem $f(x) = 0$ with root α, with $f'(x) \neq 0$. Convert it to the fixed-point problem

$$x = x + cf(x) \equiv g(x)$$

with c a nonzero constant. How should c be chosen to ensure rapid convergence of

$$x_{n+1} = x_n + cf(x_n)$$

to α (provided that x_0 is chosen sufficiently close to α)? Apply your way of choosing c to the rootfinding problem $x^3 - 5 = 0$.

10. Prove Corollary 3.4.3. To do this, note first that combining $|g'(\alpha)| < 1$ and the continuity of $g'(x)$ shows that for some $r > 0$,

$$|\alpha - x| \leq r \text{ implies } |g'(\alpha) - g'(x)| \leq \frac{1 - |g'(\alpha)|}{2}$$

Examine the interval $[a, b] = [\alpha - r, \alpha + r]$ and the size of $g'(x)$ on this interval.

11. The iteration $x_{n+1} = 2 - (1+c)x_n + cx_n^3$ will converge to $\alpha = 1$ for some values of c (provided that x_0 is chosen sufficiently close to α). Find the values of c for which convergence occurs. For what values of c, if any, will the convergence be quadratic?

12. Consider the equation $x = g_c(x) \equiv cx(1-x)$, with c a nonzero constant. This equation has two solutions, and we let α_c denote the nonzero solution. What is α_c? For what values of c will the iteration $x_{n+1} = g_c(x_n)$ converge to α_c (provided that x_0 is chosen sufficiently close to α)?

Note: This equation $x = g_c(x)$ is called the *logistic equation*, and it and the associated iteration $x_{n+1} = g_c(x_n)$ have recently been of great interest as an example in the *mathematical theory of chaos*. To observe some of the behavior that has been of interest, slowly increase the value of c past the interval found earlier in the problem, keeping $c < 4$. Observe the behavior of the iterates over large number values of n. For a more extensive discussion, see Ian Stewart, *Does God Play Dice?*, Blackwell Ltd. Publishers, Oxford, 1989, p. 155.

13. Use Aitken's error estimation formula (3.53) to estimate the error $\alpha - x_2$ in the following iterations:

(a) $x_{n+1} = e^{-x_n}, \qquad x_0 = 0.57$

(b) $x_{n+1} = \dfrac{0.5}{1 + x_n^4}, \qquad x_0 = 0.48$

(c) $x_{n+1} = 1 + 0.5\sin(x_n), \qquad x_0 = 1.5$

14. For slowly convergent sequences, the Aitken extrapolation formula (3.52) can greatly accelerate the convergence. Use the following algorithm:

$$\begin{aligned}
x_1 &= g(x_0) \\
x_2 &= g(x_1) \\
x_3 &= \text{Aitken extrapolate of } x_0, x_1, \text{ and } x_2 \\
x_4 &= g(x_3) \\
x_5 &= g(x_4) \\
x_6 &= \text{Aitken extrapolate of } x_3, x_4, \text{ and } x_5.
\end{aligned}$$

Continue this process in the same manner. Apply it to the following iterations:

(a) $x_{n+1} = 2e^{-x_n}, \qquad x_0 = 0.8$

(b) $x_{n+1} = \dfrac{0.9}{1 + x_n^4}, \qquad x_0 = 0.75$

(c) $x_{n+1} = 6.28 + \sin(x_n), \qquad x_0 = 6$

15. Show that (3.52) can be rewritten as

$$\alpha \approx x_n - \frac{(x_n - x_{n-1})^2}{(x_n - x_{n-1}) - (x_{n-1} - x_{n-2})}$$

16. Compute (3.57) for Newton's method (3.58). Compare your result with (3.19) in Section 3.2.

17. Derive the generalization of (3.57) when $\alpha = g(\alpha)$, $g'(\alpha) = g''(\alpha) = 0$, and $g^{(3)}(\alpha) \neq 0$.

18. What is the order of convergence of the iteration

$$x_{n+1} = \frac{x_n(x_n^2 + 3a)}{3x_n^2 + a}$$

as it converges to the fixed point $\alpha = \sqrt{a}$?

3.5. ILL-BEHAVING ROOTFINDING PROBLEMS

We will examine two classes of problems for which the methods of Sections 3.1 to 3.4 do not perform well. Often there is little that a numerical analyst can do to improve these problems, but one should be aware of their existence and of the reason for their ill-behavior.

We begin with functions that have a *multiple root*. The root α of $f(x)$ is said to be of multiplicity m if

$$f(x) = (x - \alpha)^m h(x) \tag{3.59}$$

for some continuous function $h(x)$ with $h(\alpha) \neq 0$, m a positive integer. If we assume that $f(x)$ is sufficiently differentiable, an equivalent definition is that

$$f(\alpha) = f'(\alpha) = \cdots = f^{(m-1)}(\alpha) = 0, \qquad f^{(m)}(\alpha) \neq 0 \tag{3.60}$$

A root of multiplicity $m = 1$ is called a *simple root* [recall (3.18) in Section 3.2].

Example 3.5.1 (a) $f(x) = (x-1)^2(x+2)$ has two roots. The root $\alpha = 1$ has multiplicity 2, and $\alpha = -2$ is a simple root.

(b) $f(x) = x^3 - 3x^2 + 3x - 1$ has $\alpha = 1$ as a root of multiplicity 3. To see this, note that

$$f(1) = f'(1) = f''(1) = 0, \qquad f'''(1) = 6$$

The result follows from (3.60).

(c) $f(x) = 1 - \cos(x)$ has $\alpha = 0$ as a root of multiplicity $m = 2$. To see this, write

$$f(x) = x^2 \left[\frac{2\sin^2(x/2)}{x^2}\right] \equiv x^2 h(x)$$

with $h(0) = \frac{1}{2}$. The function $h(x)$ is continuous for all x. ■

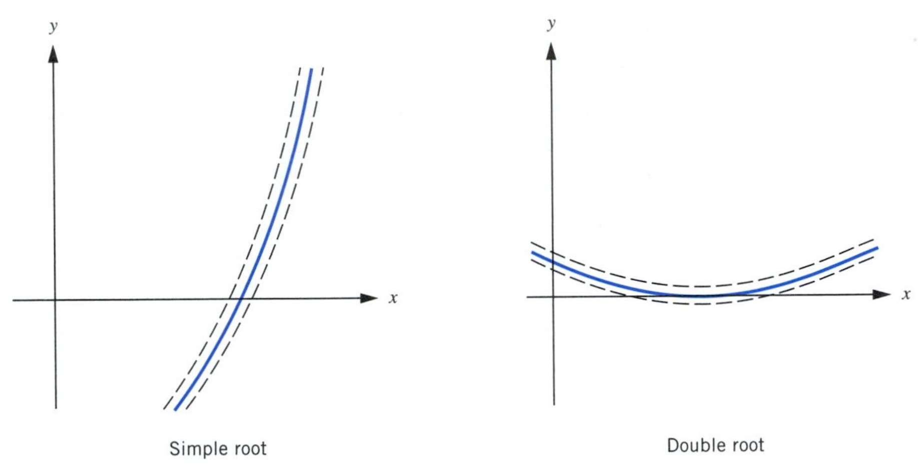

Figure 3.7. The interval of uncertainty in evaluation of a root

When the Newton and secant methods are applied to the calculation of a multiple root α, the convergence of $\alpha - x_n$ to zero is much slower than it would be for a simple root. In addition, there is a large *interval of uncertainty* as to where the root actually lies, because of the noise in evaluating $f(x)$.

The large interval of uncertainty for a multiple root is the most serious problem associated with numerically finding such a root. In Figure 2.2 of Chapter 2, we illustrate the noise in evaluating $f(x) = (x-1)^3$, which has $\alpha = 1$ as a root of multiplicity 3. That graph also illustrates the large interval of uncertainty in finding α. To further illustrate the difference in the intervals of uncertainty between simple roots and multiple roots, see Figure 3.7. The dashed lines drawn about each side of the graph of $y = f(x)$ are meant to give the outer limits on the noise in evaluating $f(x)$, and in both graphs we use the same outer limits on the noise (this is the vertical distance for any x, and they are the same). The intersection of the band of noise with the x-axis shows the interval in which the root α may be located, and it is much larger with the multiple root.

Example 3.5.2 To illustrate the effect of a multiple root on a rootfinding method, we use Newton's method to calculate the root $\alpha = 1.1$ of

$$\begin{aligned} f(x) &= (x-1.1)^3(x-2.1) \\ &= 2.7951 + x(-8.954 + x(10.56 + x(-5.4 + x))) \end{aligned} \tag{3.61}$$

The computer used is decimal with six digits in the significand, and it uses rounding. The function $f(x)$ is evaluated in the nested form of (3.61), and $f'(x)$ is evaluated similarly. The results are given in Table 3.7. The column "Ratio" gives the values of

$$\frac{\alpha - x_n}{\alpha - x_{n-1}} \tag{3.62}$$

Table 3.7. Newton's Method for (3.61)

n	x_n	$f(x_n)$	$\alpha - x_n$	Ratio
0	0.800000	0.03510	0.300000	
1	0.892857	0.01073	0.207143	0.690
2	0.958176	0.00325	0.141824	0.685
3	1.00344	0.00099	0.09656	0.681
4	1.03486	0.00029	0.06514	0.675
5	1.05581	0.00009	0.04419	0.678
6	1.07028	0.00003	0.02972	0.673
7	1.08092	0.0	0.01908	0.642

and we can see that these values equal about $\frac{2}{3}$. The iteration is linearly convergent with a rate of $\frac{2}{3}$. ■

It is possible to show that when we use Newton's method to calculate a root of multiplicity m, the ratios (3.62) will approach

$$\lambda = \frac{m-1}{m}, \qquad m \geq 1 \tag{3.63}$$

(This is left to Problem 3.) Thus as x_n approaches α,

$$\alpha - x_n \approx \lambda(\alpha - x_{n-1}) \tag{3.64}$$

and the error decreases at about the constant rate. In our example, $\lambda = \frac{2}{3}$, since the root has multiplicity $m = 3$, which corresponds to the values in the last column of the table. The error formula (3.64) implies a much slower rate of convergence than is usual for Newton's method. With any root of multiplicity $m \geq 2$, the number $\lambda \geq \frac{1}{2}$; thus, the bisection method is always at least as fast as Newton's method for multiple roots. Of course, m must be an odd integer to have $f(x)$ change sign at $x = \alpha$, thus permitting the bisection method to be applied.

A further observation from Table 3.7 is that the iterate x_7 is an exact root of $f(x)$ in the computer, even though it is very far from $\alpha = 1.1$. This is explained by the rounding errors that occur in the evaluation of $f(x)$. The resulting noise in evaluating $f(x)$ leads to a fairly large interval in which the root α might lie, just as illustrated earlier in Figure 3.7.

The only way to obtain accurate values for multiple roots is to analytically remove the multiplicity, obtaining a new function for which α is a simple root. Otherwise, there will be a large interval of uncertainty for the location of the root. To remove the multiplicity, first use Newton's method to determine the multiplicity m of α, using the results (3.63) and (3.64), together with the approximation

$$\lambda \approx \lambda_n \equiv \frac{x_n - x_{n-1}}{x_{n-1} - x_{n-2}} \tag{3.65}$$

(See Problem 3.) Once λ is found, we can determine m from (3.63). Then analytically we can calculate

$$F(x) = f^{(m-1)}(x)$$

From (3.59), it can be shown that $F(x)$ will have α as a simple root. Solve $F(x) = 0$ to find α accurately.

Example 3.5.3 Differentiate (3.61) twice to obtain the new rootfinding problem

$$f''(x) \equiv 21.12 - 32.4x + 12x^2 = 0$$

This equation has $\alpha = 1.1$ as a simple root, and Newton's method will converge rapidly to a very accurate value. Use the final computed value x_7 from Table 3.7 as an initial guess for the desired root α of $f''(x) = 0$. ■

3.5.1 Stability of Roots

With most functions $f(x)$, if a small error is made in calculating the function, then the root will change by a correspondingly small amount. However, there are a number of functions for which this is not true. With those functions, very small errors in evaluating $f(x)$ will lead to very large changes in the roots of the function. Finding the root of such a function is called an *ill-conditioned* or *unstable problem.* We give a well-known example of such a function, and then we return to an analysis of unstable rootfinding problems.

Example 3.5.4 Define

$$\begin{aligned} f(x) &= (x-1)(x-2)(x-3)(x-4)(x-5)(x-6)(x-7) \\ &= x^7 - 28x^6 + 322x^5 - 1960x^4 + 6769x^3 - 13{,}132x^2 + 13{,}068x - 5040 \end{aligned} \tag{3.66}$$

Change the coefficient of x^6 from -28 to -28.002, and call the new function $F(x)$. The change in the coefficient is relatively small

$$|\text{Rel}(28.002)| = \frac{0.002}{28} \doteq 7.14 \times 10^{-5} \tag{3.67}$$

The roots of $f(x)$ are clearly $\{1, 2, \ldots, 7\}$. The roots of $F(x)$ are given in Table 3.8, correctly rounded to eight significant digits. Some of these roots are far from the corresponding roots of $f(x)$, even though the change from $f(x)$ to $F(x)$ is relatively small. A similar change in some of the other coefficients of $f(x)$ will lead to the same kind of behavior in the roots. ■

Table 3.8. Roots of $f(x)$ and $F(x)$

Root of $f(x)$	Root of $F(x)$	Error
1	1.0000028	$-2.8E-6$
2	1.9989382	$1.1E-3$
3	3.0331253	$-3.3E-2$
4	3.8195692	0.18
5	$5.4586758 + 0.54012578i$	$-0.46 - 0.54i$
6	$5.4586758 - 0.54012578i$	$0.54 + 0.54i$
7	7.2330128	-0.23

To talk about the approximate evaluation of a function $f(x)$, introduce the perturbed function

$$F_\epsilon(x) = f(x) + \epsilon g(x) \tag{3.68}$$

The function $g(x)$ is assumed to be continuously differentiable, and ϵ is to be a reasonably small number. For small values of ϵ, the functions $F_\epsilon(x)$ and $f(x)$ will be nearly the same.

Example 3.5.5 For the preceding example (3.66), we would have

$$F_\epsilon(x) = f(x) + \epsilon g(x), \qquad g(x) = x^6, \qquad \epsilon = -0.002 \quad \blacksquare \tag{3.69}$$

The roots of $F_\epsilon(x)$ will depend on ϵ, and we denote such a root by $\alpha(\epsilon)$. The original root α of $f(x)$ is just $\alpha(0)$. To simplify our discussion, we will assume that $\alpha(0)$ is a simple root of $f(x)$ and, thus, $f'(\alpha(0)) \neq 0$. This discussion will be adequate for understanding the example (3.66). If we use these assumptions, it can be shown by more advanced mathematical results that a Taylor polynomial approximation can be used to estimate $\alpha(\epsilon)$ if ϵ is sufficiently small. We will use

$$\alpha(\epsilon) \approx \alpha(0) + \epsilon\alpha'(0) \tag{3.70}$$

Thus, we need to compute $\alpha'(0)$.

Since $\alpha(\epsilon)$ is a root of $F_\epsilon(x)$, we have

$$f(\alpha(\epsilon)) + \epsilon g(\alpha(\epsilon)) = 0 \tag{3.71}$$

for all small values of ϵ. Take the derivative of both sides of this equation, using ϵ as the variable for the differentiation. This yields

$$f'(\alpha(\epsilon))\alpha'(\epsilon) + g(\alpha(\epsilon)) + \epsilon g'(\alpha(\epsilon))\alpha'(\epsilon) = 0 \tag{3.72}$$

Substitute $\epsilon = 0$ to get

$$f'(\alpha(0))\alpha'(0) + g(\alpha(0)) = 0$$

and solve for $\alpha'(0)$

$$\alpha'(0) = -\frac{g(\alpha(0))}{f'(\alpha(0))} \tag{3.73}$$

Using this in (3.70), we get

$$\alpha(\epsilon) \approx \alpha(0) - \epsilon \frac{g(\alpha(0))}{f'(\alpha(0))} \tag{3.74}$$

for all sufficiently small values of ϵ. If the derivative $\alpha'(0)$ is very large in size, then the small change ϵ will be magnified greatly in its effect on the root.

Example 3.5.6 Consider the root $\alpha(0) = 4$ for the polynomial $f(x)$ of (3.66). Use the definition of $F_\epsilon(x)$ given in (3.69). Then

$$\begin{aligned} f'(4) &= (4-1)(4-2)(4-3)(4-5)(4-6)(4-7) = -36 \\ g(4) &= 4^6 = 4096 \\ \alpha'(0) &= -\frac{4096}{-36} \doteq 114 \\ \alpha(\epsilon) &\approx 4 + 114\epsilon \end{aligned} \tag{3.75}$$

This shows that the small change ϵ will be magnified greatly in its effect on the root $\alpha(0) = 4$ for $f(x)$. For the particular choice of (3.69),

$$\alpha(\epsilon) \doteq 4 + 114(-0.002) = 3.772$$

which is approximately the actual root 3.820 given in Table 3.8. For the results of Table 3.8 with $\alpha(0) = 5$ or 6, the estimation formula (3.74) cannot be valid. Since (3.74) predicts a real number as the perturbation, the complex perturbations in Table 3.8 could not have been predicted from it. The value of ϵ would have needed to have been smaller in order for (3.74) to be valid. ■

Finding the roots of a polynomial such as (3.66) is an unstable problem. There is not much that can be done with such a problem except to go to higher precision arithmetic. The main difficulty lies in the original formulation of the mathematical equation to be solved, and often there is another way to approach the problem that will prevent the unstable behavior.

PROBLEMS

1. Use Newton's method to calculate the roots of

$$f(x) = x^5 + 0.9x^4 - 1.62x^3 - 1.458x^2 + 0.6561x + 0.59049$$

Print out the iterates and the function values. Produce the ratios of (3.62) by using the approximation (3.65)

$$\frac{\alpha - x_n}{\alpha - x_{n-1}} \approx \frac{x_{n+1} - x_n}{x_n - x_{n-1}}$$

Repeat the problem for several choices of x_0. Make observations that seem important relative to the rootfinding problem.

Note: The above will first approach $\lambda = (m-1)/m$, as in (3.63), but they will then depart from it because of noise in the evaluation of $f(x)$ as x_n approaches the root.

2. Repeat Problem 1 for

$$f(x) = x^4 - 3.2x^3 + 0.96x^2 + 4.608x - 3.456$$

3. Use the fixed point iteration theory of Section 3.4 to derive the results (3.63) to (3.65). To aid with this, first write

$$x_{n+1} = g(x_n) \equiv x_n - \frac{f(x_n)}{f'(x_n)}$$

and use (3.59) to write

$$g(x) = x - \frac{(x-\alpha)h(x)}{h(x) + (x-\alpha)h'(x)}$$

Apply Corollary 3.4.3.

4. Do the calculation of $\alpha(\epsilon)$ for the roots $\alpha(0) = 3$ and 7, continuing Example 3.5.6.

5. Do the perturbation calculation for another change in (3.66). Change the coefficient of x^4 from -1960 to -1960.14. What is the relative perturbation in the coefficient? Calculate $\alpha(\epsilon)$ for $\alpha(0) = 3$ and $\alpha(0) = 5$.

6. Consider the problem of solving $x/(1+x) - 0.99 = 0$, calling its root α. Then let $\alpha(\epsilon)$ be the solution of $x/(1+x) - 0.99 + \epsilon = 0$.

(a) Using (3.74), estimate $\alpha(\epsilon) - \alpha$.

(b) Calculate $\alpha(\epsilon)$ directly, compute $\alpha(\epsilon) - \alpha$, and compare with (a). Comment on your results.

7. Consider the polynomial $f(x) = x^5 - 300x^2 - 126x + 5005$, which has a root $\alpha = 5$. Also consider the perturbed function

$$F_\epsilon(x) = f(x) + \epsilon x^5 = (1 + \epsilon)x^5 - 300x^2 - 126x + 5005$$

with ϵ being a small number. Letting $\alpha(\epsilon)$ denote the perturbed root of $F_\epsilon(x) = 0$ corresponding to $\alpha(0) = 5$, estimate $\alpha(\epsilon) - 5$. Is finding $\alpha = 5$ for $f(x) = 0$ an unstable rootfinding problem?

8. Newton's method is used to find a root of $f(x) = 0$. The first few iterates are shown in the following table, giving a very slow speed of convergence. What can be said about the root α to explain this convergence? Knowing $f(x)$, how would you find an accurate value for α?

n	x_n	$x_n - x_{n-1}$
0	0.75	
1	0.752710	0.00271
2	0.754795	0.00208
3	0.756368	0.00157
4	0.757552	0.00118
5	0.758441	0.000889

CHAPTER 4

INTERPOLATION AND APPROXIMATION

Most functions encountered in mathematics courses cannot be evaluated exactly, even though we usually handle them as if they were completely known quantities. The simplest and most important of these are $\sqrt{x}$, e^x, $\log(x)$, and the trigonometric functions; and there are many other functions that occur commonly in physics, engineering, and other disciplines.

In evaluating functions, by hand or using a computer, we are essentially limited to the elementary arithmetic operations $+$, $-$, $\times$, and $\div$. Combining these operations means that we can evaluate polynomials and *rational functions*, which are polynomials divided by polynomials. All other functions must be evaluated by using approximations based on polynomials or rational functions, including piecewise variants of them (e.g., spline functions). In this chapter we discuss polynomial approximations of functions. Rational functions generally give slightly more efficient approximations; but polynomials are adequate for most problems, their theory is much easier to work with, and therefore we limit our discussion to polynomials.

Interpolation is the process of finding and evaluating a function whose graph goes through a set of given points. The points may arise as measurements in a physical problem, or they may be obtained from a known function. The interpolating function is usually chosen from a restricted class of functions, and *polynomials* are the most commonly used class. In Section 4.1 we define the polynomial interpolation problem and give two formulas for constructing an interpolating polynomial. In Section 4.2 we analyze the error involved in polynomial interpolation. In the past few decades, much more use has been made of piecewise polynomial functions, and chief among these is the class of functions called *spline functions*. In Section 4.3 we introduce spline functions in the context of interpolation.

In Chapter 1, we studied the Taylor polynomial as a means to evaluate a given function $f(x)$. This is a relatively easy way to approximate most commonly used functions to any desired level of accuracy; and often the Taylor polynomial is the only direct method of approximating such functions. Nonetheless, a Taylor polynomial for $f(x)$ is usually a very inefficient approximation; if the approximation is to be used many times, then it should be replaced by a formula requiring less evaluation time. If a Taylor polynomial of some degree is being used, then there is usually another polynomial of much lower degree that will be of equal accuracy; and its lower degree will decrease both the evaluation time and the number of rounding errors in the evaluation process.

In the last four sections of the chapter, we consider the general problem of approximating a function $f(x)$ by a polynomial. This extends the ideas of Chapter 1 where Taylor polynomial approximations were discussed. In Section 4.4 we discuss the general problem of approximating a function using polynomials, and in Section 4.6 we use interpolation to construct an improvement on Taylor approximations. Section 4.5 contains an introduction to an important class of polynomials called the *Chebyshev polynomials*, and these are needed in the construction given in Section 4.6. The chapter concludes with Section 4.7, in which the idea of approximation in the sense of *least squares* is introduced.

4.1. POLYNOMIAL INTERPOLATION

Interpolation is used in a wide variety of ways. Originally, it was used widely to do interpolation in tables defining common mathematical functions; but that is a far less important use in the present day, due to the availability of computers and calculators. Interpolation is still used in the related problem of extending functions that are known only at a discrete set of points, and such problems occur frequently when numerically solving differential and integral equations. Next, interpolation is used to solve problems from the more general area of approximation theory. Interpolation is an important tool in producing computable approximations to commonly used functions. Moreover, to numerically integrate or differentiate a function, we often replace the function with a simpler approximating expression, and it is then integrated or differentiated. These simpler expressions are almost always obtained by interpolation. Also, some of the most widely used numerical methods for solving differential equations are obtained from interpolating approximations. Finally, interpolation is widely used in computer graphics, to produce smooth curves and surfaces when the geometric object of interest is given at only a discrete set of data points.

4.1.1 Linear Interpolation

We begin by considering the construction of a polynomial whose graph will pass through two given data points. This form of interpolation is called *linear interpolation*, and it is used here as an introduction to more general polynomial interpolation.

Given two points (x_0, y_0) and (x_1, y_1) with $x_0 \neq x_1$, draw a straight line through them, as in Figure 4.1. The straight line is the graph of the linear polynomial

$$P_1(x) = \frac{(x_1 - x)y_0 + (x - x_0)y_1}{x_1 - x_0} \tag{4.1}$$

The reader should check that the graph of this function is the straight line determined by (x_0, y_0) and (x_1, y_1). We say that this function interpolates the value y_i at the point x_i, $i = 0, 1$; or

$$P_1(x_i) = y_i, \qquad i = 0, 1$$

Example 4.1.1 Let the data points be $(1, 1)$ and $(4, 2)$. The polynomial $P_1(x)$ is given by

$$P_1(x) = \frac{(4 - x)(1) + (x - 1)(2)}{3} \tag{4.2}$$

The graph of $y = P_1(x)$ is shown in Figure 4.2, along with that of $y = \sqrt{x}$ from which the data points were taken. ■

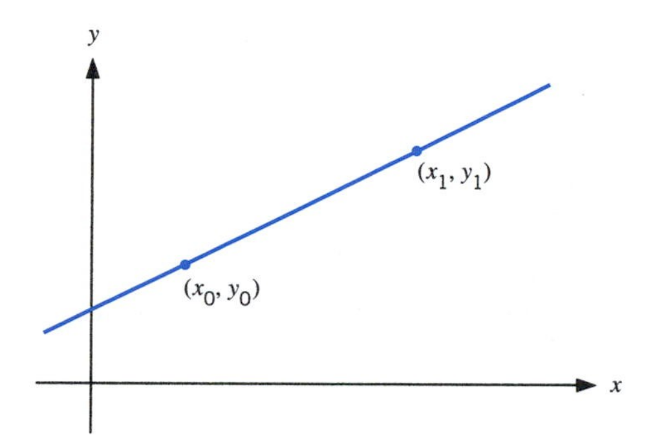

Figure 4.1. Linear interpolation

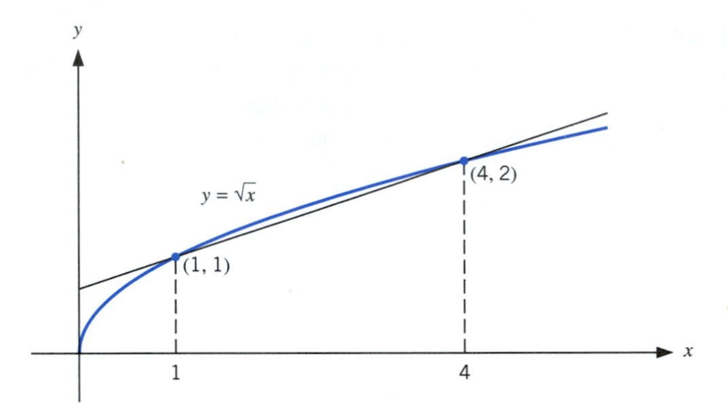

Figure 4.2. $y = \sqrt{x}$ and its linear interpolating polynomial (4.2)

Example 4.1.2 Obtain an estimate of $e^{0.826}$ using the function values

$$e^{0.82} \doteq 2.270500, \qquad e^{0.83} \doteq 2.293319$$

Denote $x_0 = 0.82$ and $x_1 = 0.83$. The polynomial $P_1(x)$ interpolating e^x at x_0 and x_1 is given by

$$P_1(x) = \frac{(0.83 - x)(2.270500) + (x - 0.82)(2.293319)}{0.01} \tag{4.3}$$

In particular,

$$P_1(0.826) = 2.2841914 \tag{4.4}$$

For comparison, the true value is

$$e^{0.826} \doteq 2.2841638 \tag{4.5}$$

to eight significant digits. ■

4.1.2 Quadratic Interpolation

Most data arise from graphs that are curved rather than straight. To better approximate such behavior, we look at polynomials of a degree greater than 1. Assume that three data points (x_0, y_0), (x_1, y_1), and (x_2, y_2) are given, with x_0, x_1, x_2 being distinct numbers. We construct the quadratic polynomial that passes through these points as follows:

$$P_2(x) = y_0 L_0(x) + y_1 L_1(x) + y_2 L_2(x) \tag{4.6}$$

with

$$
\begin{aligned}
L_0(x) &= \frac{(x - x_1)(x - x_2)}{(x_0 - x_1)(x_0 - x_2)} \\
L_1(x) &= \frac{(x - x_0)(x - x_2)}{(x_1 - x_0)(x_1 - x_2)} \\
L_2(x) &= \frac{(x - x_0)(x - x_1)}{(x_2 - x_0)(x_2 - x_1)}
\end{aligned}
\tag{4.7}
$$

Formula (4.6) is called *Lagrange's formula* for the quadratic interpolating polynomial; and the polynomials L_0, L_1, and L_2 are called the *Lagrange interpolation basis functions*.

Each polynomial $L_i(x)$ has degree 2 and, thus, $P_2(x)$ has degree ≤ 2. In addition,

$$
\begin{aligned}
L_i(x_j) &= 0, \qquad j \neq i \\
L_i(x_i) &= 1
\end{aligned}
$$

for $0 \leq i, j \leq 2$. These two statements are combined into the statement

$$L_i(x_j) = \delta_{ij}, \qquad 0 \leq i, j \leq 2 \tag{4.8}$$

where δ_{ij} is called the *Kronecker delta function*

$$\delta_{ij} = \begin{cases} 1, & i = j \\ 0, & i \neq j \end{cases}$$

With the use of these properties, we can easily show that $P_2(x)$ interpolates the data

$$P_2(x_i) = y_i, \qquad i = 0, 1, 2$$

Example 4.1.3 Construct $P_2(x)$ for the data points $(0, -1)$, $(1, -1)$, and $(2, 7)$. Then

$$P_2(x) = \frac{(x - 1)(x - 2)}{2}(-1) + \frac{x(x - 2)}{-1}(-1) + \frac{x(x - 1)}{2}(7) \tag{4.9}$$

Its graph is shown in Figure 4.3. ■

With linear interpolation, it was obvious that there was only one straight line passing through two given data points. But with three data points, it is less obvious that there is only one quadratic interpolating polynomial whose graph passes through the points. To see that there is only one such polynomial, we assume that there is a second one and then show that it must equal $P_2(x)$. Let $Q_2(x)$ denote another polynomial of degree ≤ 2 whose graph passes through (x_i, y_i), $i = 0, 1, 2$. Define

$$R(x) = P_2(x) - Q_2(x) \tag{4.10}$$

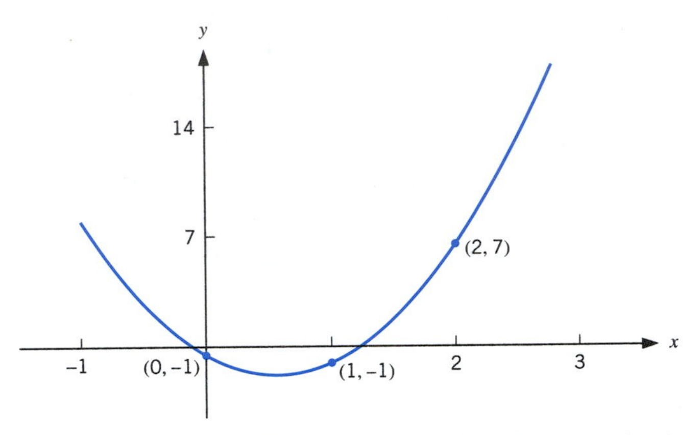

Figure 4.3. The quadratic interpolating polynomial (4.9)

Since P_2 and Q_2 both have degree ≤ 2, we also have $\deg(R) \leq 2$. In addition, using the interpolating property of both P_2 and Q_2, we get

$$\begin{aligned} R(x_i) &= P_2(x_i) - Q_2(x_i) \\ &= y_i - y_i = 0 \end{aligned}$$

for $i = 0, 1, 2$. Thus, $R(x)$ has three distinct roots x_0, x_1, and x_2, but its degree is ≤ 2. By the fundamental theorem of algebra [see B.1(e) of Appendix B], this is not possible unless $R(x) \equiv 0$, the zero polynomial. But from (4.10), that says $P_2(x) \equiv Q_2(x)$; and thus there is only one polynomial of degree ≤ 2 that satisfies

$$P_2(x_i) = y_i, \qquad i = 0, 1, 2 \tag{4.11}$$

Example 4.1.4 Calculate a quadratic interpolate to $e^{0.826}$ from the function values

$$e^{0.82} \doteq 2.270500, \qquad e^{0.83} \doteq 2.293319, \qquad e^{0.84} \doteq 2.316367$$

We choose $x_0 = 0.82$, $x_1 = 0.83$, $x_2 = 0.84$. Then it is straightforward (although tedious) to calculate

$$P_2(0.826) \doteq 2.2841639 \tag{4.12}$$

to eight digits. Comparison to the true answer $e^{0.826} \doteq 2.2841638$ shows that $P_2(0.826)$ is a significant improvement over $P_1(0.826) \doteq 2.2841914$. ■

4.1.3 Higher-Degree Interpolation

We now consider the general case. Assume that we are given $n+1$ data points $(x_0, y_0), \ldots, (x_n, y_n)$, with all of the x_i's distinct. The interpolating polynomial of degree $\leq n$ is given by

$$P_n(x) = y_0 L_0(x) + y_1 L_1(x) + \cdots + y_n L_n(x) \tag{4.13}$$

with each $L_i(x)$ a polynomial of degree n given by

$$L_i(x) = \frac{(x - x_0)\cdots(x - x_{i-1})(x - x_{i+1})\cdots(x - x_n)}{(x_i - x_0)\cdots(x_i - x_{i-1})(x_i - x_{i+1})\cdots(x_i - x_n)} \tag{4.14}$$

for $i = 0, 1, \ldots, n$. The denominator equals the value of the numerator at $x = x_i$. For the case $i = 0$, this formula becomes

$$L_0(x) = \frac{(x - x_1)\cdots(x - x_n)}{(x_0 - x_1)\cdots(x_0 - x_n)}$$

and $L_n(x)$ is defined analogously. Note that the function $L_i(x)$ defined in (4.14) depends on n. Since the value of n is usually clear from the context, we do not explicitly express the dependence of $L_i(x)$ on n.

From (4.14), it is easy to show that $L_i(x)$ satisfies

$$L_i(x_j) = \delta_{ij}, \qquad 0 \leq j \leq n \tag{4.15}$$

for $i = 0, 1, \ldots, n$. If we use (4.15), it follows by direct substitution of $x = x_j$ into (4.13) that

$$P_n(x_j) = y_j, \qquad j = 0, 1, \ldots, n \tag{4.16}$$

Formula (4.13) is called *Lagrange's formula* for the degree n interpolating polynomial.

The fact that there is only one polynomial satisfying (4.16) among all polynomials of degree $\leq n$ is proved in a manner analogous to that used with quadratic interpolation, following (4.10). This is left as Problem 11. Because of its importance, we state the results of this subsection as a formal mathematical theorem.

Theorem 4.1.5 Let $n \geq 0$, assume $x_0, x_1, \ldots, x_n$ are $n+1$ distinct numbers, and let $y_0, \ldots, y_n$ be $n+1$ given numbers, not necessarily distinct. Then, among all polynomials of degree $\leq n$, there is exactly one polynomial $P_n(x)$ that satisfies

$$P_n(x_i) = y_i, \qquad i = 0, 1, \ldots, n$$

Our primary use in this text for polynomial interpolation is to produce approximations to a given function $f(x)$. Polynomials can be easily evaluated, and it is also straightforward to integrate or differentiate them. Interpolation polynomials will be used in Section 4.6 and in Chapter 5 to produce approximate methods for evaluating, integrating, and differentiating arbitrary functions $f(x)$.

Another important use for interpolation is to find a *smooth curve* passing through a set of data points $\{(x_i, y_i) \mid i = 1, \ldots, N\}$; and generally, we want a curve that does not have any oscillations or ripples in it. The solution to this problem leads us to *spline functions*, taken up in Section 4.3; but the material on polynomial interpolation will still be useful in developing this new type of interpolation function.

We caution the reader that the degree of the interpolation polynomial $P_n(x)$ may be less than n. For purposes of illustration, suppose the three data points (x_i, y_i), $i = 0, 1, 2$, lie on a straight line. Then $P_2(x) \equiv P_1(x)$, where $P_1(x)$ is the linear interpolating polynomial determined by the points (x_0, y_0) and (x_1, y_1). The reason for this lies in the uniqueness result proven following (4.10). The polynomial $P_1(x)$ interpolates to all three data points, by our assumption on them and, thus, it must equal $P_2(x)$, by the uniqueness of $P_2(x)$. A similar argument holds true for higher-order polynomial interpolation.

4.1.4 Divided Differences

The Lagrange formula (4.13) is well-suited for many theoretical uses of interpolation, but it is less desirable when actually computing the value of an interpolating polynomial. As an example, knowing $P_2(x)$ does not lead to a less expensive way to evaluate $P_3(x)$, at least not in a simple manner. For this reason, we introduce an alternative and more easily calculable formulation for the interpolation polynomials $P_1(x), P_2(x), \ldots, P_n(x)$.

As a needed preliminary to this new formula for interpolation, we introduce a discrete version of the derivative of a function $f(x)$. Let x_0 and x_1 be distinct numbers, and define

$$f[x_0, x_1] = \frac{f(x_1) - f(x_0)}{x_1 - x_0} \tag{4.17}$$

This is called the *first-order divided difference* of $f(x)$. If $f(x)$ is differentiable on an interval containing x_0 and x_1, then the mean value theorem implies

$$f[x_0, x_1] = f'(c) \tag{4.18}$$

for some c between x_0 and x_1. This is one justification for thinking of $f[x_0, x_1]$ as being an analog of the derivative of $f(x)$. Also, if x_0 and x_1 are close together, then

$$f[x_0, x_1] \approx f'\left(\frac{x_0 + x_1}{2}\right) \tag{4.19}$$

which is usually a very accurate approximation. The analysis of this approximation is taken up in Problem 20.

Example 4.1.6 Let $f(x) = \cos(x)$, $x_0 = 0.2$, $x_1 = 0.3$. Then

$$f[x_0, x_1] = \frac{\cos(0.3) - \cos(0.2)}{0.3 - 0.2} \doteq -0.2473009 \tag{4.20}$$

For (4.19)

$$f'\left(\frac{x_0 + x_1}{2}\right) = -\sin(0.25) \doteq -0.2474040$$

and $f[x_0, x_1]$ is a very good approximation of this derivative. ■

We define higher-order divided differences recursively using lower-order ones. Let x_0, x_1, and x_2 be distinct real numbers, and define

$$f[x_0, x_1, x_2] = \frac{f[x_1, x_2] - f[x_0, x_1]}{x_2 - x_0} \tag{4.21}$$

This is called the *second-order divided difference.* For x_0, x_1, x_2, and x_3 distinct, define

$$f[x_0, x_1, x_2, x_3] = \frac{f[x_1, x_2, x_3] - f[x_0, x_1, x_2]}{x_3 - x_0} \tag{4.22}$$

the *third-order divided difference.* In general, let $x_0, x_1, \ldots, x_n$ be $n + 1$ distinct numbers, and define

$$f[x_0, \ldots, x_n] = \frac{f[x_1, \ldots, x_n] - f[x_0, \ldots, x_{n-1}]}{x_n - x_0} \tag{4.23}$$

This is the divided difference of order n, sometimes also called the *Newton divided difference.*

The relationship of the higher-order divided differences to the derivatives of corresponding order is given by the following theorem. Its proof is given in the next section.

Theorem 4.1.7 Let $n \geq 1$, and assume $f(x)$ is n times continuously differentiable on some interval $\alpha \leq x \leq \beta$. Let $x_0, x_1, \ldots, x_n$ be $n + 1$ distinct numbers in $[\alpha, \beta]$. Then

$$f[x_0, x_1, \ldots, x_n] = \frac{1}{n!} f^{(n)}(c) \tag{4.24}$$

for some unknown point c lying between the minimum and maximum of the numbers $x_0, \ldots, x_n$.

Example 4.1.8 Let $f(x) = \cos(x)$, $x_0 = 0.2$, $x_1 = 0.3$, $x_2 = 0.4$. Then $f[x_0, x_1]$ is given in (4.20), and

$$f[x_1, x_2] = \frac{\cos(0.4) - \cos(0.3)}{0.4 - 0.3} \doteq -0.3427550$$

From (4.21),

$$f[x_0, x_1, x_2] \doteq \frac{-0.3427550 - (-0.2473009)}{0.4 - 0.2} = -0.4772705 \tag{4.25}$$

For the case $n = 2$, (4.24) becomes

$$f[x_0, x_1, x_2] = \tfrac{1}{2} f''(c) \tag{4.26}$$

for some c between the minimum and maximum of x_0, x_1, and x_2. Taking $f''(x) = -\cos(x)$ with $x = x_1$, we have

$$\tfrac{1}{2} f''(0.3) = -\tfrac{1}{2}\cos(0.3) \doteq -0.4776682$$

which is nearly equal to the result in (4.25). ■

4.1.5 Properties of Divided Differences

The divided differences (4.23) have a number of special properties that can simplify work with them. First, let $(i_0, i_1, \ldots, i_n)$ denote a permutation (or rearrangement) of the integers $(0, 1, \ldots, n)$. Then it can be shown that

$$f[x_{i_0}, x_{i_1}, \ldots, x_{i_n}] = f[x_0, x_1, \ldots, x_n] \tag{4.27}$$

The original definition (4.23) seems to imply that the order of $x_0, x_1, \ldots, x_n$ will make a difference in the calculation of $f[x_0, \ldots, x_n]$; but (4.27) asserts that this is not true. The proof of (4.27) is nontrivial, and we will consider only the cases $n = 1$ and $n = 2$. For $n = 1$,

$$f[x_1, x_0] = \frac{f(x_0) - f(x_1)}{x_0 - x_1} = \frac{f(x_1) - f(x_0)}{x_1 - x_0} = f[x_0, x_1]$$

For $n = 2$, we can expand (4.21) to obtain

$$\begin{aligned} f[x_0, x_1, x_2] &= \frac{f(x_0)}{(x_0 - x_1)(x_0 - x_2)} \\ &\quad + \frac{f(x_1)}{(x_1 - x_0)(x_1 - x_2)} + \frac{f(x_2)}{(x_2 - x_0)(x_2 - x_1)} \end{aligned} \tag{4.28}$$

If we interchange values of x_0, x_1, and x_2, then the fractions on the right side will interchange their order, but the sum will remain the same. Try this with an interchange of x_0 and x_1 on the right-hand side of (4.27). We leave the derivation of (4.28) as Problem 22.

A second useful property is that the definitions (4.17), (4.21) to (4.23) can be extended to the case where some or all of the node points x_i are coincident, provided that $f(x)$ is sufficiently differentiable. For example, define

$$f[x_0, x_0] = \lim_{x_1 \to x_0} f[x_0, x_1] = \lim_{x_1 \to x_0} \frac{f(x_1) - f(x_0)}{x_1 - x_0}$$
$$f[x_0, x_0] = f'(x_0)$$

For an arbitrary $n \geq 1$, let all of the nodes in (4.24) approach x_0. This leads to the definition

$$f[x_0, x_0, \ldots, x_0] = \frac{1}{n!} f^{(n)}(x_0) \tag{4.29}$$

where the left-hand side denotes an order n divided difference, all of whose nodes are x_0.

For cases where only some of the nodes are coincident, we can use (4.27), (4.29), and (4.23) to extend the definition of divided difference. For example,

$$\begin{aligned} f[x_0, x_1, x_0] &= f[x_0, x_0, x_1] \\ &= \frac{f[x_0, x_1] - f[x_0, x_0]}{x_1 - x_0} = \frac{f[x_0, x_1] - f'(x_0)}{x_1 - x_0} \end{aligned} \tag{4.30}$$

Further properties of divided differences are explored in the problems at the end of the section.

MATLAB PROGRAM: *Evaluating divided differences.* Given a set of values $f(x_0), \ldots, f(x_n)$, we will often need to calculate the set of divided differences

$$f[x_0, x_1], f[x_0, x_1, x_2], \ldots, f[x_0, x_1, \ldots, x_n]$$

The MATLAB function given below can be used to this end, using the function call

```
divdif_y=divdif(x_nodes,y_values)
```

Note however that MATLAB does not allow zero subscripts, and therefore the vector input variables *x_nodes* and *y_values* will need to be defined as vectors containing $n + 1$ components. More precisely, when given the nodes $\{x_0, x_1, \ldots, x_n\}$ and the associated function values, we define the input vectors *x_nodes* and *y_values* as follows:

$$\begin{aligned} x\_nodes &= [x_0, x_1, \ldots, x_n] \\ x\_nodes(i) &= x_{i-1}, \quad i = 1, \ldots, n+1 \\ y\_values &= [f(x_0), f(x_1), \ldots, f(x_n)] \\ y\_values(i) &= f(x_{i-1}), \quad i = 1, \ldots, n+1 \end{aligned} \tag{4.31}$$

```
function divdif_f = divdif(t_nodes,f_values)
%
% This is a function,
```

```
%        divdif_f = divdif(t_nodes,f_values)
% It calculates the divided differences of the function
% values given in the vector f_values, which are the values of
% some function f(t) at the nodes given in t_nodes.  On exit,
%        divdif_f(i) = f[t_1,...,t_i], i=1,...,m
% with m the length of t_nodes.  The input values t_nodes and
% f_values are not changed by this program.

divdif_f = f_values;
m = length(t_nodes);
for i=2:m
  for j=m:-1:i
    divdif_f(j) = (divdif_f(j)-divdif_f(j-1)) ...
                        /(t_nodes(j)-t_nodes(j-i+1));
  end
end
```

This program is illustrated in Table 4.1.

Table 4.1. Values and Divided Differences of $\cos(x)$

i	x_i	$\cos(x_i)$	D_i
0	0.0	1.000000	$0.1000000\text{E}+1$
1	0.2	0.980067	$-0.9966711\text{E}-1$
2	0.4	0.921061	$-0.4884020\text{E}+0$
3	0.6	0.825336	$0.4900763\text{E}-1$
4	0.8	0.696707	$0.3812246\text{E}-1$
5	1.0	0.540302	$-0.3962047\text{E}-2$
6	1.2	0.362358	$-0.1134890\text{E}-2$

4.1.6 Newton's Divided Difference Interpolation Formula

From the discussion at the beginning of Subsection 4.1.4, it is clear that the Lagrange formula (4.13) is very inconvenient for actual calculations for a sequence of interpolation polynomials of increasing degree. Moreover, when computing with polynomials $P_n(x)$ of varying n, the calculation of $P_n(x)$ for a particular n is of little use in calculating an interpolation polynomial of higher degree. These problems are avoided by using another formula for $P_n(x)$, one using the divided differences of the data being interpolated.

Let $P_n(x)$ denote the polynomial interpolating $f(x_i)$ at x_i, for $i = 0, 1, \ldots, n$. Thus, $\deg(P_n) \le n$ and

$$P_n(x_i) = f(x_i), \qquad i = 0, 1, \ldots, n \tag{4.32}$$

Then the interpolation polynomials can be written as follows:

$$P_1(x) = f(x_0) + (x - x_0) f[x_0, x_1] \tag{4.33}$$

$$\begin{aligned} P_2(x) = f(x_0) &+ (x - x_0) f[x_0, x_1] \\ &+ (x - x_0)(x - x_1) f[x_0, x_1, x_2] \end{aligned} \tag{4.34}$$

$$\vdots$$

$$\begin{aligned} P_n(x) = f(x_0) &+ (x - x_0) f[x_0, x_1] + \cdots \\ &+ (x - x_0)(x - x_1) \cdots (x - x_{n-1}) f[x_0, x_1, \ldots, x_n] \end{aligned} \tag{4.35}$$

This is called *Newton's divided difference formula* for the interpolating polynomial. Note that for $k \geq 0$,

$$P_{k+1}(x) = P_k(x) + (x - x_0) \cdots (x - x_k) f[x_0, x_1, \ldots, x_k, x_{k+1}] \tag{4.36}$$

Thus, we can go from degree k to degree $k + 1$ with a minimum of calculation, once the divided difference coefficients have been computed.

We will consider only the proof of (4.33) and (4.34). For the first case, consider $P_1(x_0)$ and $P_1(x_1)$. Easily, $P_1(x_0) = f(x_0)$; and

$$\begin{aligned} P_1(x_1) &= f(x_0) + (x_1 - x_0) \left[\frac{f(x_1) - f(x_0)}{x_1 - x_0} \right] \\ &= f(x_0) + [f(x_1) - f(x_0)] = f(x_1) \end{aligned}$$

Thus, $\deg(P_1) \leq 1$, and it satisfies the interpolation conditions. By the uniqueness of polynomial interpolation (see Theorem 4.1.5), formula (4.33) is the linear interpolation polynomial to $f(x)$ at x_0, x_1.

For (4.34), note that

$$P_2(x) = P_1(x) + (x - x_0)(x - x_1) f[x_0, x_1, x_2] \tag{4.37}$$

It satisfies $\deg(P_2) \leq 2$; and for x_0, x_1,

$$P_2(x_i) = P_1(x_i) + 0 = f(x_i), \qquad i = 0, 1$$

Also,

$$\begin{aligned} P_2(x_2) &= f(x_0) + (x_2 - x_0) f[x_0, x_1] + (x_2 - x_0)(x_2 - x_1) f[x_0, x_1, x_2] \\ &= f(x_0) + (x_2 - x_0) f[x_0, x_1] + (x_2 - x_1)\{f[x_1, x_2] - f[x_0, x_1]\} \\ &= f(x_0) + (x_1 - x_0) f[x_0, x_1] + (x_2 - x_1) f[x_1, x_2] \\ &= f(x_0) + \{f(x_1) - f(x_0)\} + \{f(x_2) - f(x_1)\} \\ &= f(x_2) \end{aligned}$$

By the uniqueness of polynomial interpolation, this is the quadratic interpolating polynomial to $f(x)$ at $\{x_0, x_1, x_2\}$.

A general proof of the formula (4.35) is more complicated than we wish to consider here. See Atkinson (1989, Chapter 3) for a detailed derivation of (4.35), along with further properties of divided differences.

Example 4.1.9 Let $f(x) = \cos(x)$. Table 4.1 contains a set of nodes x_i, the function values $f(x_i)$, and the divided differences

$$D_i = f[x_0, \ldots, x_i], \qquad i \geq 0$$

These were computed using the MATLAB program `divdif` given above. Table 4.2 contains the values of $P_n(x)$ for $x = 0.1, 0.3, 0.5$ for various values of n. The true values $f(x)$ are given in the last row of the table. The calculations were done by using (4.35) as implemented in the MATLAB program `interp` given below. ■

The preceding example used evenly spaced points x_i. But, in general, the interpolation node points need not be evenly spaced to use the divided difference interpolation formula (4.35); nor need they be arranged in any particular order. In Section 4.6, we will use (4.35) with node points that are not evenly spaced.

To evaluate (4.35) in an efficient manner, we can use a variation on the nested multiplication algorithm of Section 1.3 in Chapter 1. First, write (4.35) as

$$\begin{aligned} P_n(x) = D_0 &+ (x - x_0)D_1 + (x - x_0)(x - x_1)D_2 \\ &+ \cdots + (x - x_0)\cdots(x - x_{n-1})D_n \end{aligned} \tag{4.38}$$

with

$$D_0 = f(x_0), \qquad D_i = f[x_0, \ldots, x_i] \qquad \text{for } i \geq 1$$

Table 4.2. Interpolation to $\cos(x)$

n	$P_n(0.1)$	$P_n(0.3)$	$P_n(0.5)$
1	0.9900333	0.9700999	0.9501664
2	0.9949173	0.9554478	0.8769061
3	0.9950643	0.9553008	0.8776413
4	0.9950071	0.9553351	0.8775841
5	0.9950030	0.9553369	0.8775823
6	0.9950041	0.9553365	0.8775825
True	0.9950042	0.9553365	0.8775826

Next, rewrite (4.38) in the nested form

$$P_n(x) = D_0 + (x - x_0)[D_1 + (x - x_1)[D_2 + \cdots + (x - x_{n-2})[D_{n-1} + (x - x_{n-1})D_n] \cdots] \tag{4.39}$$

For example,

$$P_3(x) = D_0 + (x - x_0)[D_1 + (x - x_1)[D_2 + (x - x_2)D_3]]$$

Using (4.39), we need only n multiplications to evaluate $P_n(x)$. The nested form (4.39) is more convenient with a fixed degree n. To compute a sequence of interpolation polynomials of increasing degree, it is more efficient to use the original form (4.35).

MATLAB PROGRAM: ***Evaluating the Newton divided difference form of the interpolation polynomial.*** To aid in constructing examples like those in Tables 4.1 and 4.2, we include the following interpolation program. Also, the program uses the Newton divided difference form of the interpolating polynomial, with the nested form given above. The program should be preceded by calling the function `divdif` that was given earlier in this section. As earlier with `divdif`, note that MATLAB does not allow a zero subscript; see (4.31).

```
function p_eval = interp(x_nodes,divdif_y,x_eval)
%
% This is a function
%       p_eval = interp(x_nodes,divdif_y,x_eval)
% It calculates the Newton divided difference form of
% the interpolation polynomial of degree m-1, where the
% nodes are given in x_nodes, m is the length of x_nodes,
% and the divided differences are given in divdif_y.  The
% points at which the interpolation is to be carried out
% are given in x_eval; and on exit, p_eval contains the
% corresponding values of the interpolation polynomial.

m = length(x_nodes);
p_eval = divdif_y(m)*ones(size(x_eval));
for i=m-1:-1:1
  p_eval = divdif_y(i) + (x_eval - x_nodes(i)).*p_eval;
end
```

PROBLEMS

1. Given the data points (0, 2), (1, 1), find the following:
   (a) The straight line interpolating this data.
   (b) The function $f(x) = a + be^x$ interpolating this data.
   Hint: Find a and b so that $f(0) = 2$, $f(1) = 1$.

(c) The function $f(x) = a/(b + x)$ interpolating this data.

In each case, graph the interpolating function.

2. (a) Find the function $P(x) = a + b\cos(\pi x) + c\sin(\pi x)$, which interpolates the data

x	0	0.5	1
y	2	5	4

(b) Find the quadratic polynomial interpolating this data.

In each instance, graph the interpolating function.

3. The following table was obtained in solving a differential equation. Using linear interpolation between adjacent nodes x_i, produce a continuous graph of this data on the interval $0 \le x \le 6$.

x_i	0.0	1.0	2.0	3.0	4.0	5.0	6.0
y_i	2.0000	2.1592	3.1697	5.4332	9.1411	14.406	21.303

4. Given $x_0 < x_1$ and x, define

$$\mu = \frac{x - x_0}{x_1 - x_0}$$

and let $P_1(x)$ interpolate the function $f(x)$ at the points x_0 and x_1. Show

$$P_1(x) = f(x_0) + \mu\,[f(x_1) - f(x_0)]$$

In the past, this formula was the most commonly used way of doing linear interpolation in tables of mathematical functions.

5. Write a computer program to do linear interpolation and to check its accuracy. Input x_0 and x_1 and then generate the data values using $y = e^x$ (or some other commonly available function). For a variety of values of x, both inside and outside $[x_0, x_1]$, compute $P_1(x)$, e^x, and their difference $E(x) = e^x - P_1(x)$. Plot the values of $E(x)$, to see how the error varies with x.

6. Show that the formula (4.9) simplifies to

$$P_2(x) = 4x^2 - 4x - 1$$

7. Produce the quadratic polynomial interpolating the data $\{(-2, -15), (-1, -8), (0, -3)\}$. Find the first zero of $P_2(x)$ to the right of $x = 0$. Does $P_2(x)$ have a maximum?

8. Using (4.6), find the polynomial $P_2(x)$ that interpolates the following data. In each case, simplify (4.6) as much as possible.

(a) $\{(0, 1), (1, 2), (2, 3)\}$

(b) $\{(0, 1), (1, 1), (2, 1)\}$

Comment on your results.

9. Let the nodes $\{x_0, x_1, x_2\}$ be evenly spaced with $x_1 - x_0 = h = x_2 - x_1$. Recalling the notation of Problem 4, define $\mu = (x - x_0)/h$. Let $P_2(x)$ be the polynomial of degree ≤ 2 that interpolates $f(x)$ at the nodes x_0, x_1, x_2. Show that

$$P_2(x) = f(x_0) + \mu\,[f(x_1) - f(x_0)] + \frac{\mu\,(\mu - 1)}{2}\{[f(x_2) - f(x_1)] - [f(x_1) - f(x_0)]\}$$

In the past, this formula was the most commonly used way of doing quadratic interpolation in tables of mathematical functions. The quantities $f(x_{i+1}) - f(x_i)$ are called forward differences of $f(x)$, and their use is explored further in Problem 33.

10. Write out the complete formula (4.13) for $P_3(x)$, including all four of the polynomials $L_0(x), L_1(x), L_2(x), L_3(x)$ for this case.

11. **(a)** Prove that there is only one polynomial $P_3(x)$ among all polynomials of degree ≤ 3 that satisfy the interpolating conditions

$$P_3(x_i) = y_i, \qquad i = 0, 1, 2, 3$$

where the x_i's are distinct.

Hint: Generalize the proof given in and following (4.10) for the uniqueness of $P_2(x)$.

(b) Give a proof of uniqueness for $P_n(x)$ in Theorem 4.1.5.

12. **(a)** For $n = 3$, explain why

$$L_0(x) + L_1(x) + L_2(x) + L_3(x) = 1$$

for all x.

Hint: It is unnecessary to actually multiply out and combine the functions $L_i(x)$ of (4.14). Use (4.13) with a suitable choice of $\{y_0, y_1, y_2, y_3\}$.

(b) Generalize part (a) to an arbitrary degree $n > 0$.

13. Continuing Problem 12 with degree of interpolation $n = 3$, explain why

$$x_0^j L_0(x) + x_1^j L_1(x) + x_2^j L_2(x) + x_3^j L_3(x) = x^j, \qquad j \leq 3$$

and

$$x_0^j L_0(x) + x_1^j L_1(x) + x_2^j L_2(x) + x_3^j L_3(x) \neq x^j, \qquad j > 3$$

14. **(a)** Let $P_2^{(0,2)}(x)$ denote the quadratic polynomial that interpolates the data $\{(x_0, y_0), (x_1, y_1), (x_2, y_2)\}$; let $P_2^{(1,3)}(x)$ denote the quadratic polynomial that interpolates the data $\{(x_1, y_1), (x_2, y_2), (x_3, y_3)\}$. Finally, let $P_3(x)$ denote the cubic polynomial interpolating the data $\{(x_0, y_0), (x_1, y_1), (x_2, y_2), (x_3, y_3)\}$. Show that

$$P_3(x) = \frac{(x_3 - x)P_2^{(0,2)}(x) + (x - x_0)P_2^{(1,3)}(x)}{x_3 - x_0}$$

(b) How might this be generalized to constructing $P_n(x)$, interpolating $\{(x_0, y_0), \ldots, (x_n, y_n)\}$, from interpolation polynomials of degree $n - 1$?

15. Find a polynomial $P(x)$ of degree ≤ 3 for which

$$P(0) = y_1, \qquad P(1) = y_2.$$
$$P'(0) = y_1', \qquad P'(1) = y_2'$$

with y_1, y_2, y_1', y_2' given constants. The resulting polynomial is called the *cubic Hermite interpolating polynomial.*

Hint: Write

$$P(x) = y_1 H_1(x) + y_1' H_2(x) + y_2 H_3(x) + y_2' H_4(x)$$

with H_1, H_2, H_3, H_4 cubic polynomials satisfying appropriate properties, in analogy with (4.15). For example, choose $H_1(x)$ to be a cubic polynomial that satisfies

$$H_1(0) = 1, \qquad H_1(1) = 0$$
$$H_1'(0) = 0, \qquad H_1'(1) = 0$$

16. As a generalized interpolation problem, find the quadratic polynomial $q(x)$ for which

$$q(0) = -1, \qquad q(1) = -1, \qquad q'(1) = 4$$

17. Find the solution to the interpolation problem of finding a polynomial $q(x)$ with $\deg(q) \leq 2$ and such that

$$q(x_0) = y_0, \qquad q(x_1) = y_1, \qquad q'(x_1) = y_1'$$

with $x_0 \neq x_1$.

Hint: Write $q(x) = y_0 M_0(x) + y_1 M_1(x) + y_1' M_2(x)$ where $\deg(M_i) \le 2$, $i = 0, 1, 2$ and each $M_i(x)$ satisfies suitable interpolating conditions at the points x_0 and x_1. For example, $M_0(x)$ should satisfy

$$M_0(x_0) = 1, \qquad M_0(x_1) = M_0'(x_1) = 0$$

18. Let $x_0 < x_1 < x_2$, let $\{y_0, y_1', y_2\}$ be given. Find the polynomial $q(x)$ with $\deg(q) \le 2$ and such that

$$q(x_0) = y_0, \qquad q'(x_1) = y_1', \qquad q(x_2) = y_2$$

What condition, if any, is required in order to have $q(x)$ be uniquely determined from the given interpolation conditions?

19. Let $x_0 = 0.85$, $x_1 = 0.87$, $x_2 = 0.89$. Using the values of e^{x_0}, e^{x_1}, e^{x_2} and $f(x) = e^x$, calculate $f[x_0, x_1]$, $f[x_1, x_2]$, and $f[x_0, x_1, x_2]$. Check the accuracy of the approximation (4.19).

20. Let $z = (x_0 + x_1)/2$, $h = (x_1 - x_0)/2$. To analyze the approximation (4.19), consider the error

$$E = f[x_0, x_1] - f'\left(\frac{x_0 + x_1}{2}\right) = \frac{f(z+h) - f(z-h)}{2h} - f'(z)$$

Expand $f(z+h)$ and $f(z-h)$ about z by using Taylor's theorem from Section 1.2, and include terms of degree ≤ 3 in the variable h. Use these expansions to show that

$$E \approx \frac{h^2}{6} f'''(z)$$

for small values of h. Thus, (4.19) is a good approximation when $x_1 - x_0$ is relatively small.

21. **(a)** Let $f(x)$ be a polynomial of degree m. For $x \ne x_0$, define

$$g_1(x) = f[x_0, x] = \frac{f(x) - f(x_0)}{x - x_0}$$

Show $g(x)$ is a polynomial of degree $m - 1$. This is another justification for regarding the divided difference as an analog of the derivative.

Hint: The numerator $f(x) - f(x_0)$ is a polynomial of degree m with a zero at $x = x_0$. Use the fundamental theorem of algebra, noting that x_0 is a root of $f(x) - f(x_0)$.

(b) For x_0, x_1, x distinct and for $f(x)$ a polynomial of degree m, define

$$g_2(x) = f[x_0, x_1, x]$$

Show $g_2(x)$ is a polynomial of degree $m - 2$.

22. Verify formula (4.28), thus showing the symmetry of $f[x_0, x_1, x_2]$ under permutations of the nodes.

23. Using (4.29) and the ideas used in obtaining (4.30), obtain a formula for calculating

(a) $f[x_0, x_0, x_1, x_1]$ **(b)** $f[x_0, x_0, x_0, x_1]$

24. Given the data below, find $f[x_0, x_1]$ and $f[x_0, x_1, x_2]$. Then calculate $P_1(0.15)$ and $P_2(0.15)$, the linear and quadratic interpolates evaluated at $x = 0.15$.

n	x_n	$f(x_n)$
0	0.1	0.2
1	0.2	0.24
2	0.3	0.30

25. Let $f(x) = 1/(1 + x)$ and let $x_0 = 0$, $x_1 = 1$, $x_2 = 2$. Calculate the divided differences $f[x_0, x_1]$ and $f[x_0, x_1, x_2]$. Using these divided differences, give the quadratic polynomial $P_2(x)$ that interpolates $f(x)$ at the given node points $\{x_0, x_1, x_2\}$. Graph the error $f(x) - P_2(x)$ on the interval $[0, 2]$.

26. **(a)** By using function program `divdif`, calculate the divided differences $D_0 = f(x_0)$, $D_1 = f[x_0, x_1], \ldots, D_5 = f[x_0, x_1, x_2, x_3, x_4, x_5]$, for $f(x) = e^x$. Use $x_0 = 0$, $x_1 = 0.2$, $x_2 = 0.4, \ldots, x_5 = 1.0$.

(b) Using the results of (a), calculate $P_j(x)$ for $x = 0.1, 0.3, 0.5$ and $j = 1, \ldots, 5$. Compare these results to the true values of e^x.

27. Repeat Problem 26 with $f(x) = \tan^{-1}(x)$.

28. The following data are taken from a polynomial $p(x)$ of degree ≤ 5. What is the polynomial and what is its degree?

x	-2	-1	0	1	2	3
$p(x)$	-5	1	1	1	7	25

29. Produce a program to check the accuracy of higher-order interpolation by using Newton's formula (4.35). For some $f(x)$ and some node points $x_0, x_1, \ldots, x_6$, use `divdif` to produce the divided differences

$$D_i = f[x_0, \ldots, x_i], \qquad i = 0, 1, \ldots, 6$$

Then evaluate $P_6(x)$ for a variety of values of x, and compare them to the true values of $f(x)$. Check the program by reproducing some of the results in Tables

4.1 and 4.2 for $f(x) = \cos(x)$. Then repeat the process with a nonpolynomial function $f(x)$ of your choice.

30. **(a)** In the linear formula (4.33), let $x_1 - x_0 = h$ and $\mu = (x - x_0)/(x_1 - x_0) = (x - x_0)/h$. Show that (4.33) reduces to the earlier formula of Problem 4.

(b) In the quadratic formula (4.34), let $x_1 - x_0 = x_2 - x_1 = h$ and $\mu = (x - x_0)/(x_1 - x_0)$. Show that (4.34) reduces to the earlier formula of Problem 9.

31. Let $f(x) = x^n$ for some integer $n \geq 0$. Let $x_0, x_1, \ldots, x_m$ be $m + 1$ distinct numbers. What is $f[x_0, x_1, \ldots, x_m]$ for $m = n$? For $m > n$?

32. Let $f(x) = e^x$. Show that $f[x_0, x_1, \ldots, x_m] > 0$ for all values of m and all distinct nodes $\{x_0, x_1, \ldots, x_m\}$.

33. When the nodes $\{x_j\}$ are evenly spaced, the formulas for the divided difference and the interpolation polynomial $P_n(x)$ take simpler forms. Let $x_j = x_0 + jh$ for $j = 0, \pm 1, \pm 2, \ldots$, and let $f_j \equiv f(x_j)$. Define the first-order *forward difference* of $f(x)$ by

$$\Delta f(x_j) \equiv \Delta f_j = f_{j+1} - f_j$$

Define the second-order forward difference by

$$\begin{aligned}\Delta^2 f(x_j) &= \Delta f_{j+1} - \Delta f_j = [f_{j+2} - f_{j+1}] - [f_{j+1} - f_j] \\ &= f_{j+2} - 2f_{j+1} + f_j\end{aligned}$$

Define higher-order forward differences by

$$\Delta^k f_j = \Delta^{k-1} f_{j+1} - \Delta^{k-1} f_j, \qquad k \geq 2$$

Show that

(a) $f[x_j, x_{j+1}] = \dfrac{1}{h} \Delta f_j$

(b) $f[x_j, x_{j+1}, x_{j+2}] = \dfrac{1}{2! h^2} \Delta^2 f_j$

(c) $f[x_j, x_{j+1}, \ldots, x_{j+k}] = \dfrac{1}{k! h^k} \Delta^k f_j, \qquad k \geq 1$

34. **(a)** Using (4.33) and Problem 33(a), derive the interpolation formula of Problem 4.

(b) Using (4.34) and Problem 33(b), derive the interpolation formula of Problem 9.

(c) Generalize (a) and (b) by applying Problem 33(c) to (4.35). Show that

$$P_n(x) = \sum_{k=0}^{n} \binom{\mu}{k} \Delta^k f_0, \qquad \mu = \frac{x - x_0}{h}$$

using the *binomial coefficient*

$$\binom{\mu}{k} = \frac{\mu(\mu-1)\cdots(\mu-k+1)}{k!}, \qquad \binom{\mu}{0} = 1$$

This is called *Newton's forward difference form of the interpolation polynomial.*

4.2. ERROR IN POLYNOMIAL INTERPOLATION

For a given function $f(x)$ defined on an interval $[a, b]$, let $P_n(x)$ denote the polynomial of degree $\leq n$ interpolating $f(x)$ at $n+1$ points $x_0, x_1, \ldots, x_n$ in $[a, b]$

$$P_n(x) = \sum_{j=0}^{n} f(x_j)L_j(x) \tag{4.40}$$

In this section, we consider carefully the error in polynomial interpolation, giving more precise information on its behavior as x and n vary. We begin with a formula for the error.

Theorem 4.2.1 Let $n \geq 0$, let $f(x)$ have $n+1$ continuous derivatives on $[a, b]$, and let $x_0, x_1, \ldots, x_n$ be distinct node points in $[a, b]$. Then

$$f(x) - P_n(x) = \frac{(x-x_0)(x-x_1)\cdots(x-x_n)}{(n+1)!} f^{(n+1)}(c_x) \tag{4.41}$$

for $a \leq x \leq b$, where c_x is an unknown point between the minimum and maximum of $x_0, x_1, \ldots, x_n$, and x.

We omit the proof, since it does not contain any useful additional information regarding interpolation. A sketch of the case $n = 1$, linear interpolation, is given in Problem 18.

Example 4.2.2 Take $f(x) = e^x$ on $[0, 1]$, and consider the error in linear interpolation to $f(x)$ using nodes x_0 and x_1 satisfying $0 \leq x_0 \leq x_1 \leq 1$. From (4.41),

$$e^x - P_1(x) = \frac{(x-x_0)(x-x_1)}{2} e^{c_x} \tag{4.42}$$

for some c_x between the minimum and maximum of x_0, x_1, and x. For this example, assume $x_0 < x < x_1$. Then we note the interpolation error is negative, and we write

$$e^x - P_1(x) = -\frac{(x_1-x)(x-x_0)}{2} e^{c_x}$$

This shows that the error is approximately a quadratic polynomial with roots at x_0 and x_1, provided that e^{c_x} is approximately constant for $x_0 < x < x_1$ (which is approximately true if $[x_0, x_1]$ is a short interval). Since $x_0 \le c_x \le x_1$, we have the upper and lower bounds

$$\frac{(x_1 - x)(x - x_0)}{2} e^{x_0} \le \left| e^x - P_1(x) \right| \le \frac{(x_1 - x)(x - x_0)}{2} e^{x_1}$$

To obtain a bound independent of x, use

$$\max_{x_0 \le x \le x_1} \frac{(x_1 - x)(x - x_0)}{2} = \frac{h^2}{8}, \qquad h = x_1 - x_0 \tag{4.43}$$

This follows easily by noting that $(x_1 - x)(x - x_0)$ is a quadratic with roots at x_0 and x_1 and thus its maximum value occurs midway between the roots. Substituting $x = (x_0 + x_1)/2$ yields the value $h^2/8$.

Noting that $e^{x_1} \le e$ on [0, 1], we have the bound

$$\left| e^x - P_1(x) \right| \le \frac{h^2 e}{8}, \qquad 0 \le x_0 \le x \le x_1 \le 1 \tag{4.44}$$

independent of x, x_0, and x_1. Recall Example 4.1.2 of Section 4.1. With $x = 0.826$ and $h = 0.01$, we have

$$\left| e^x - P_1(x) \right| \le \frac{(0.01)^2(2.72)}{8} = 0.0000340 \tag{4.45}$$

The actual error is -0.0000276, which satisfies this bound. ■

Example 4.2.3 Again let $f(x) = e^x$ on [0, 1], but consider the error in quadratic interpolation. Then

$$e^x - P_2(x) = \frac{(x - x_0)(x - x_1)(x - x_2)}{6} e^{c_x} \tag{4.46}$$

for some c_x between the minimum and maximum of x_0, x_1, x_2, and x. Assume x_0, x_1, and x_2 are evenly spaced, and let $h = x_1 - x_0 = x_2 - x_1$; assume that $0 \le x_0 < x < x_2 \le 1$. As before, we have

$$\left| e^x - P_2(x) \right| \le \left| \frac{(x - x_0)(x - x_1)(x - x_2)}{6} \right| e^1 \tag{4.47}$$

To proceed further, we use

$$\max_{x_0 \le x \le x_2} \left| \frac{(x - x_0)(x - x_1)(x - x_2)}{6} \right| = \frac{h^3}{9\sqrt{3}} \tag{4.48}$$

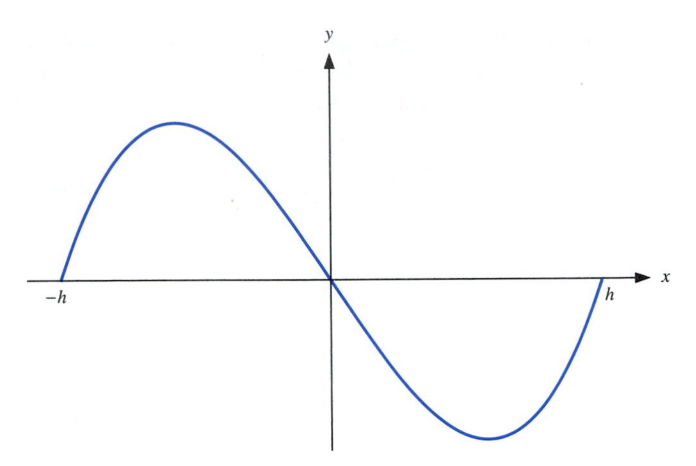

Figure 4.4. $y = w_2(x)$

This result is obtained by using elementary calculus in the following way. To simplify the calculations, we consider the special case

$$w_2(x) = \frac{(x+h)x(x-h)}{6} = \frac{x^3 - xh^2}{6} \tag{4.49}$$

Its graph is shown in Figure 4.4. The cubic polynomial $w_2(x)$ is a translation along the x-axis of the polynomial in (4.48), and it has the same shape and size. With (4.49), we look for the points x at which

$$0 = w_2'(x) = \frac{3x^2 - h^2}{6}$$

Then $x = \pm h/\sqrt{3}$, and

$$\left| w_2\left(\pm\frac{h}{\sqrt{3}}\right) \right| = \frac{h^3}{9\sqrt{3}}$$

proving (4.48). Using this in (4.47), we get

$$\left| e^x - P_2(x) \right| \le \frac{h^3 e}{9\sqrt{3}} \approx 0.174h^3 \tag{4.50}$$

For $h = 0.01$, $0 \le x \le 1$,

$$\left| e^x - P_2(x) \right| \le 1.74 \times 10^{-7} \tag{4.51}$$

Compare this with the earlier result (4.45) for linear interpolation. ■

4.2.1 Another Error Formula

Construct the interpolation formula $P_{n+1}(x)$ that interpolates to $f(x)$ at the $n + 2$ nodes $x_0, x_1, \ldots, x_n, x_{n+1}$. Using the Newton divided difference form gives us

$$P_{n+1}(x) = P_n(x) + (x - x_0)(x - x_1) \cdots (x - x_n) f[x_0, x_1, \ldots, x_n, x_{n+1}] \qquad (4.52)$$

with $P_n(x)$ interpolating $f(x)$ at $x_0, x_1, \ldots, x_n$. Using the interpolation property and letting $x = x_{n+1}$, we have

$$f(x_{n+1}) = P_n(x_{n+1}) + (x_{n+1} - x_0) \cdots (x_{n+1} - x_n) f[x_0, x_1, \ldots, x_n, x_{n+1}]$$

Regard x_{n+1} as a variable point, distinct from $x_0, \ldots, x_n$, and rename it t. Moving the P_n term to the left side of the equation, we have

$$f(t) - P_n(t) = (t - x_0)(t - x_1) \cdots (t - x_n) f[x_0, x_1, \ldots, x_n, t] \qquad (4.53)$$

This is the interpolation error in $P_n(t)$. From what was stated in the preceding section, if $f(x)$ is a sufficiently differentiable function, then the divided difference $f[x_0, x_1, \ldots, x_n, t]$ can be extended to t a node point. Then the right side of (4.53) will be zero when $t = x_0, x_1, \ldots, x_n$.

Comparing (4.53) with (4.41), with $x = t$, we have two formulas for the interpolation error:

$$\frac{\Psi_n(t)}{(n+1)!} f^{(n+1)}(c_t) = \Psi_n(t) f[x_0, x_1, \ldots, x_n, t]$$

with

$$\Psi_n(t) = (t - x_0)(t - x_1) \cdots (t - x_n)$$

Canceling $\Psi_n(t)$ from both sides, we have

$$\frac{1}{(n+1)!} f^{(n+1)}(c_t) = f[x_0, x_1, \ldots, x_n, t] \qquad (4.54)$$

with c_t between the minimum and maximum of $x_0, x_1, \ldots, x_n$, and t. Let $m = n + 1$ and $t = x_m$. Then (4.54) becomes

$$\frac{f^{(m)}(c)}{m!} = f[x_0, x_1, \ldots, x_{m-1}, x_m]$$

where

$$\min\{x_0, \ldots, x_m\} \le c \le \max\{x_0, \ldots, x_m\}$$

This proves Theorem 4.1.7 of Section 4.1.

Usually we use formula (4.41) to bound the interpolation error, but occasionally (4.53) is preferable. The derivation of error formulas for numerical integration methods often uses (4.53).

4.2.2 Behavior of the Error

When we consider the error formula (4.41) or (4.53), the polynomial

$$\Psi_n(x) = (x - x_0) \cdots (x - x_n) \tag{4.55}$$

is the most important quantity in determining the behavior of the error. We will examine its behavior for $x_0 \le x \le x_n$ when the node points $x_0, \ldots, x_n$ are evenly spaced.

For larger values of n, say, $n \ge 5$, the values of $\Psi_n(x)$ change greatly through the interval $x_0 \le x \le x_n$. The values in $[x_0, x_1]$ and $[x_{n-1}, x_n]$ become much larger than the values in the middle of $[x_0, x_n]$. This can be proved theoretically, but we only suggest the result by looking at the graph of $\Psi_n(x)$ when $n = 6$. This is given in Figure 4.5; note the relatively larger values in $[x_0, x_1]$ and $[x_5, x_6]$ as compared with the values in $[x_2, x_4]$. As n increases, this disparity also increases.

When considering $\Psi_n(x)$ as a part of the error formula (4.41) or (4.53) for $f(x) - P_n(x)$, these remarks imply that the interpolation error at x is likely to be smaller when it is near the middle of the node points. In practical interpolation problems, high-degree polynomial interpolation with evenly spaced nodes is seldom used because of these difficulties. However, we will learn in Section 4.6 that high-degree polynomial interpolation with a suitably chosen set of node points can be very useful in obtaining polynomial approximations to functions.

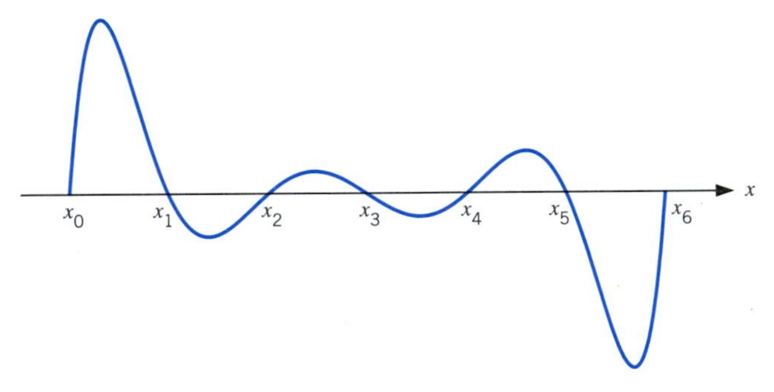

Figure 4.5. $y = \Psi_6(x)$

Example 4.2.4 Let $f(x) = \cos(x)$, $h = 0.2$, $n = 8$, and then interpolate at $x = 0.9$.

Case (i) $x_0 = 0.8$, $x_8 = 2.4$. Thus, $x = 0.9$ is in the first subinterval $[x_0, x_1]$. By direct calculation of $P_8(0.9)$,

$$\cos(0.9) - P_8(0.9) \doteq -5.51 \times 10^{-9}$$

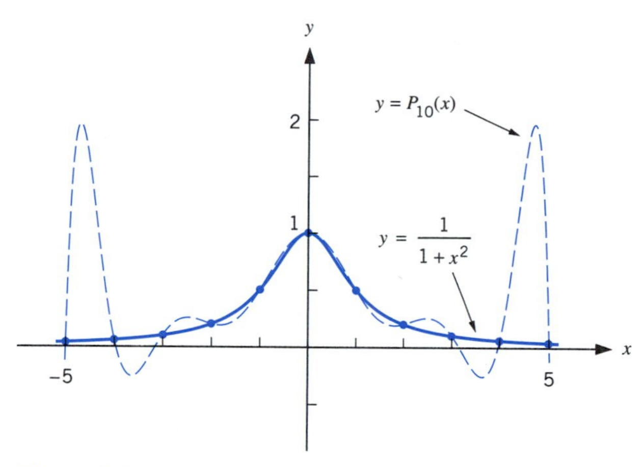

Figure 4.6. The interpolation to $1/(1+x^2)$

Case (ii) $x_0 = 0.2$, $x_8 = 1.8$. Thus, $x = 0.9$ is in the subinterval $[x_3, x_4]$, where x_4 is the midpoint for the interpolation interval $[x_0, x_8]$. By direct calculation

$$\cos(0.9) - P_8(0.9) \doteq 2.26 \times 10^{-10}$$

a factor of 24 smaller than in case (i). ■

Example 4.2.5 Let $f(x) = 1/(1+x^2)$ for $-5 \le x \le 5$. Let $n > 0$ be an even integer, and define $h = 10/n$

$$x_j = -5 + jh, \qquad j = 0, 1, 2, \ldots, n$$

Then it can be shown that for many points x in the interval $[-5, 5]$, the sequence of interpolating polynomials $\{P_n(x)\}$ does not converge to $f(x)$ as $n \to \infty$. To illustrate the poor approximation of $P_n(x)$ to $f(x)$ in this case, we give the graphs of $f(x)$ and $P_{10}(x)$ in Figure 4.6. The divergence of $P_n(x)$ from $f(x)$ for $|x| \ge 4$ becomes worse as n increases. ■

PROBLEMS

1. Consider interpolating $f(x) = \sin(x)$ from a table of values of the function f given at equally spaced values of x for $0 \le x \le 1.58$; the x entries are given in steps of $h = 0.01$.

   (a) Bound the error $f(x) - P_1(x)$ of linear interpolation in this table. The value of x is to satisfy $x_0 < x < x_1$, with x_0 and x_1 adjacent x entries in the table.

   (b) Bound the error $f(x) - P_2(x)$ of quadratic interpolation. The value of x is to satisfy $x_0 < x < x_2$, with x_0, x_1, and x_2 adjacent x entries in the table.

2. Repeat Problem 1 with $f(x) = \tan^{-1}(x)$ on $[0, 0.8]$.

3. Repeat Problem 1 with

$$f(x) = \int_0^x e^{-t^2}\,dt$$

for $0 \le x \le 1$.

4. Suppose a table of values of $f(x) = \sin(x)$, $0 \le x \le 1.58$, is to be constructed, with the values of $\sin(x)$ given with a spacing of h.

(a) If linear interpolation is used in this table, how small should h be in order for the interpolation error to be less than 10^{-6}? For notational assumptions, see Problem 1(a).

(b) If quadratic interpolation is used in this table, how small should h be in order for the interpolation error to be less than 10^{-6}? For notational assumptions, see Problem 1(b).

5. Repeat Problem 4, doing so for $f(x) = \log(x)$, $1 \le x \le 7$, with an interpolation error tolerance of 5×10^{-6}.

6. Repeat Problem 4, doing so for $f(x) = \tan^{-1}(x)$, $0 \le x \le 1$, with an interpolation error tolerance of 5×10^{-6}. Note that this allows a straightforward calculation of $\tan^{-1}(x)$ for any x since

$$\tan^{-1}(-x) = -\tan^{-1}(x)$$

$$\tan^{-1}\left(\frac{1}{x}\right) = \frac{\pi}{2} - \tan^{-1}(x), \qquad x > 0$$

7. Consider constructing a table of values of $f(x) = \sqrt{x}$ for $1 \le x \le 100$, with values of $f(x)$ given for $x = 0, h, 2h, \ldots$. Choose h so that when linear interpolation is used in this table, the error is bounded by 5×10^{-6}. Discuss and compare the linear interpolation error near $x = 1$ and $x = 100$.

8. Let $f(x) = x^4 + \sqrt{2}x^3 + \pi x$. Verify whether $f[0, 1, 2, 3, 4] = f[0, 1, \pi, e, -1]$.

9. Let $f(x) = a_0 + a_1 x + \cdots + a_n x^n$ be a polynomial of degree less than or equal to n, and let $\{x_0, x_1, \ldots, x_n\}$ be distinct points. What is the value of $f[x_0, x_1, \ldots, x_n]$?

10. Bound the error of cubic interpolation to $f(x) = e^x$ on $[0, 1]$ with evenly spaced node points.

Hint: Replace the bounding of

$$\Psi_3(x) = (x - x_0)(x - x_1)(x - x_2)(x - x_3), \qquad x_0 \le x \le x_3$$

with the bounding of its translate

$$\Psi_3(x) = \left(x + \tfrac{3}{2}h\right)\left(x + \tfrac{1}{2}h\right)\left(x - \tfrac{1}{2}h\right)\left(x - \tfrac{3}{2}h\right)$$

on the interval $-\frac{3}{2}h \le x \le \frac{3}{2}h$. This generalizes the method used in obtaining (4.48).

11. Consider using nodes

$$x_0 = a - \frac{h}{\sqrt{3}}, \qquad x_1 = a + \frac{h}{\sqrt{3}}$$

to linearly interpolate a function $f(x)$ on the interval $[a-h, a+h]$ for some real numbers a and h, $h > 0$. Calculate a bound for

$$\max_{a-h \le x \le a+h} |f(x) - P_1(x)|$$

Using the illustrative function $f(x) = e^x$, choose values for h and a and then draw a graph to illustrate what you are calculating. This problem has applications in solving differential and integral equations.

12. To visualize the change in the values of

$$\Psi_n(x) = (x - x_0)(x - x_1) \cdots (x - x_n)$$

as n increases and as x varies over $[x_0, x_n]$, graph the special case

$$\Psi_n(x) = x(x-1) \cdots (x-n)$$

for $0 \le x \le n$. Do this for $n = 3, 4, \ldots, 8$.

13. For an interval $[a, b]$, define $h = (b-a)/n$ for an integer $n > 0$. Define evenly spaced node points by

$$x_j = a + jh, \qquad j = 0, 1, \ldots, n$$

Thus, $x_0 = a$, $x_1 = a + h, \ldots, x_n = a + nh = b$. Consider the polynomial

$$\Psi_n(x) = (x - x_0)(x - x_1) \cdots (x - x_n)$$

and show

$$|\Psi_n(x)| \le n! h^{n+1}, \qquad a \le x \le b \tag{*}$$

Hint: Consider first a lower-order case such as $n = 2$ or 3. Also, consider separately the cases $x_0 < x < x_1$, $x_1 < x < x_2$, $\ldots$, $x_{n-1} < x < x_n$. With each case, bound the various factors $x - x_j$ by multiples of h.

14. Let $P_n(x)$ be the degree n polynomial interpolating to $f(x) = e^x$ on $[a, b] = [0, 1]$ using evenly spaced node points (as defined in Problem 13). Using the main result (*) shown in Problem 13, show that

$$\max_{0 \le x \le 1} \left| e^x - P_n(x) \right| \longrightarrow 0 \qquad \text{as } n \longrightarrow \infty$$

15. Repeat Problem 14 with $f(x) = \sin x$ on $[0, \pi]$.

16. Consider computing the interpolating polynomial $P_n(x)$ for the function

$$f(x) = \frac{1}{2 + \cos x}$$

with a uniformly spaced subdivision of the given interval $[a, b]$. Study the interpolating polynomial and its error for the intervals $[0, 2\pi]$ and $[-\pi, \pi]$, for varying n. Do so for $n = 10, 20, 30, 40$. Make observations on the error.

Hint: Graphs are useful, especially graphs of the interpolation error. Also, double precision arithmetic is insufficient for $n > 40$.

17. Study the error in using evenly spaced node points (as defined in Problem 13) to interpolate $f(x) = \tan^{-1} x$. Do so on the interval $[-a, a]$ with $a = 1, 2, 3, 4$. Construct $P_n(x)$ with $n = 3, 7, 11, 15, 19$. Estimate the maximum of $|f(x) - P_n(x)|$ on $[-a, a]$ for each value of a and n, and display these in a table. Also print out graphs of the error for as many cases as possible. Comment on your results.

18. Consider the proof of the error formula for linear interpolation

$$f(x) - P_1(x) = \frac{(x - x_0)(x - x_1)}{2} f''(c)$$

with $\min\{x_0, x_1, x\} \le c \le \max\{x_0, x_1, x\}$. From the construction of $P_1(x)$, the error formula is clearly true if $x = x_0$ or $x = x_1$. Thus, we consider only the case $x \ne x_0, x_1$. Introduce

$$E(t) = f(t) - P_1(t)$$

and

$$G(t) = E(t) - \frac{(t - x_0)(t - x_1)}{(x - x_0)(x - x_1)} E(x)$$

with t varying and x fixed.

(a) Show $G(x_0) = G(x_1) = G(x) = 0$.

(b) Using Rolle's theorem or the mean value theorem, show that $G'(t)$ has at least two zeros; and then show that $G''(t)$ has at least one root, calling it c.

(c) Calculate $E''(t)$ and $G''(t)$. Then evaluate $G''(c)$ and solve for $E(x)$ to conclude the derivation of the error formula. This derivation generalizes to a proof of the general result in Theorem 4.2.1 for any $n \geq 1$.

4.3. INTERPOLATION USING SPLINE FUNCTIONS

To motivate the definition and use of spline functions, we begin with the problem of interpolating the data shown in Table 4.3. The simplest method of interpolation is to connect the node points by straight-line segments; the resulting graph is shown in Figure 4.7. This is called *piecewise linear interpolation*, and the associated interpolating function is denoted by $l(x)$. It agrees with the data, but it has the disadvantage of not having a smooth graph. Most data will represent a smooth curved graph, one without the corners of $y = l(x)$. Consequently, we usually want to construct a smooth curve that interpolates the given data points, but one that follows the shape of $y = l(x)$.

The next choice of interpolation is to use polynomial interpolation. There are seven data points, and thus we consider the interpolating polynomial $P_6(x)$ of degree 6. Its graph is shown in Figure 4.8 (note the change in vertical scale), and it differs markedly from that of $y = l(x)$. Although it is a smooth graph, it is quite different from that of $y = l(x)$ between some of the interpolation node points, for example, on $0 \leq x \leq 1$.

Table 4.3. Interpolation Data Points

x	0	1	2	2.5	3	3.5	4
y	2.5	0.5	0.5	1.5	1.5	1.125	0

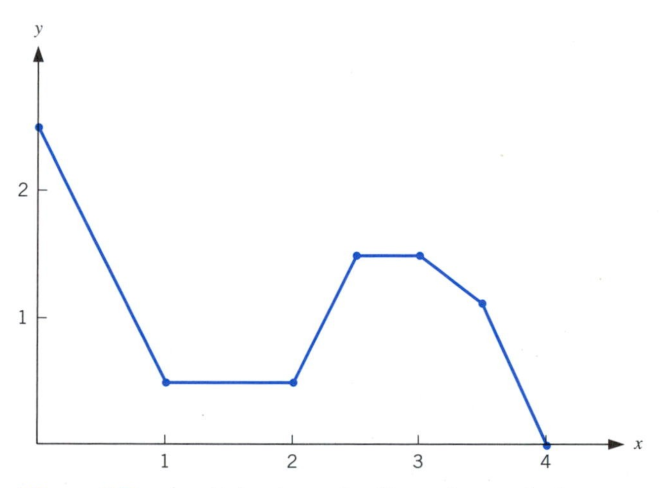

Figure 4.7. $y = l(x)$: piecewise linear interpolation

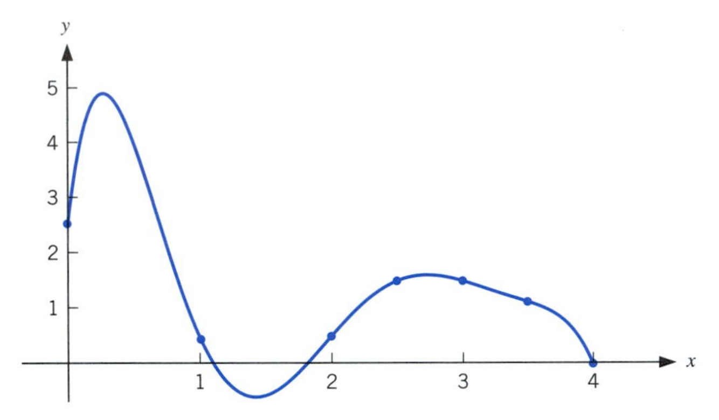

Figure 4.8. $y = P_6(x)$: polynomial interpolation

A third choice is to connect the data points of Table 4.3 by using a succession of quadratic interpolating polynomials. We denote this function on $0 \le x \le 4$ by $q(x)$, and we give its graph in Figure 4.9. On each of the subintervals $[0, 2]$, $[2, 3]$, and $[3, 4]$, $q(x)$ is the quadratic polynomial interpolating the data on that subinterval. The graph of $q(x)$ is somewhat smoother than $y = l(x)$, and follows it more closely than does $y = P_6(x)$. Nonetheless, there is still a problem at the points $x = 2$ and $x = 3$ where the graph has "corners"; the derivative $q'(x)$ is discontinuous at such points. We wish to find an interpolating function that is smooth and does not change too much between the node points, when it is compared to the graph of $y = l(x)$. For many applications, however, the interpolating function $q(x)$ may be completely adequate as, for example, with the numerical integration of the data.

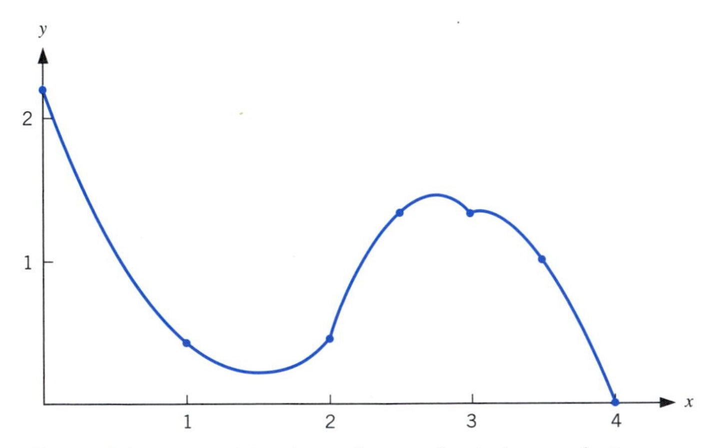

Figure 4.9. $y = q(x)$: piecewise quadratic interpolation

4.3.1 Spline Interpolation

To pose the problem more generally, suppose n data points (x_i, y_i), $i = 1, \ldots, n$, are given. For simplicity, assume that

$$x_1 < x_2 < \cdots < x_n \tag{4.56}$$

and let $a = x_1$, $b = x_n$. We seek a function $s(x)$ defined on $[a, b]$ that interpolates the data:

$$s(x_i) = y_i, \qquad i = 1, \ldots, n \tag{4.57}$$

For smoothness of $s(x)$, we require that $s'(x)$ and $s''(x)$ be continuous. In addition, we want the curve to follow the general shape given by the piecewise linear function connecting the data points (x_i, y_i), as illustrated in Figure 4.7. The standard way in which this has been done has been to ask that the derivative $s'(x)$ not change too rapidly between node points. This has been carried out by requiring the second derivative $s''(x)$ to be as small as possible and, more precisely, by requiring that

$$\int_a^b [s''(x)]^2 \, dx \tag{4.58}$$

be made as small as possible. This may not be a perfect mathematical realization of the idea of a smooth shape-preserving interpolation function for the data $\{(x_i, y_i)\}_{i=1}^n$, but it usually gives a very good interpolating function from a visual perspective.

There is a unique solution $s(x)$ to this problem, and it satisfies the following:

S1. $s(x)$ is a polynomial of degree ≤ 3 on each subinterval $[x_{j-1}, x_j]$, $j = 2, 3, \ldots, n$;

S2. $s(x)$, $s'(x)$, and $s''(x)$ are continuous for $a \leq x \leq b$;

S3. $s''(x_1) = s''(x_n) = 0$.

The function $s(x)$ is called the *natural cubic spline function* that interpolates the data $\{(x_i, y_i)\}$. We will give a method for constructing $s(x)$, and then apply it to the data in Table 4.3. Then, we will return to a more general discussion of cubic spline functions and interpolation with them.

4.3.2 Construction of the Interpolating Natural Cubic Spline

Introduce the variables $M_1, \ldots, M_n$ with

$$M_i \equiv s''(x_i), \qquad i = 1, 2, \ldots, n \tag{4.59}$$

We will express $s(x)$ in terms of the (unknown) values M_i; then, we will produce a system of linear equations from which the values M_i can be calculated.

Since $s(x)$ is cubic on each interval $[x_{j-1}, x_j]$, the function $s''(x)$ is linear on the interval. A linear function is determined by its values at two points, and we use

$$s''(x_{j-1}) = M_{j-1}, \qquad s''(x_j) = M_j \tag{4.60}$$

Then

$$s''(x) = \frac{(x_j - x)M_{j-1} + (x - x_{j-1})M_j}{x_j - x_{j-1}}, \qquad x_{j-1} \le x \le x_j \tag{4.61}$$

We will now form the second antiderivative of $s''(x)$ on $[x_{j-1}, x_j]$ and apply the interpolating conditions

$$s(x_{j-1}) = y_{j-1}, \qquad s(x_j) = y_j \tag{4.62}$$

After quite a bit of manipulation, this results in the cubic polynomial

$$\begin{aligned} s(x) = & \frac{(x_j - x)^3 M_{j-1} + (x - x_{j-1})^3 M_j}{6(x_j - x_{j-1})} + \frac{(x_j - x)y_{j-1} + (x - x_{j-1})y_j}{x_j - x_{j-1}} \\ & - \frac{1}{6}(x_j - x_{j-1})[(x_j - x)M_{j-1} + (x - x_{j-1})M_j] \end{aligned} \tag{4.63}$$

for $x_{j-1} \le x \le x_j$. It can be checked directly that the second derivative of this formula yields (4.61); and by direct substitution, (4.63) satisfies the interpolating conditions (4.62).

Formula (4.63) applies to each of the intervals $[x_1, x_2], \ldots, [x_{n-1}, x_n]$. The formulas for adjacent intervals $[x_{j-1}, x_j]$ and $[x_j, x_{j+1}]$ will agree at their common point $x = x_j$ because of the interpolating condition $s(x_j) = y_j$, which is common to the definitions. This implies that $s(x)$ will be continuous over the entire interval $[a, b]$. Similarly, formula (4.61) for $s''(x)$ implies that it is continuous on $[a, b]$.

To ensure the continuity of $s'(x)$ over $[a, b]$, the formulas for $s'(x)$ on $[x_{j-1}, x_j]$ and $[x_j, x_{j+1}]$ are required to give the same value at their common point $x = x_j$, for $j = 2, 3, \ldots, n-1$. After a great deal of simplification, this leads to the following system of linear equations:

$$\begin{aligned} & \frac{x_j - x_{j-1}}{6} M_{j-1} + \frac{x_{j+1} - x_{j-1}}{3} M_j + \frac{x_{j+1} - x_j}{6} M_{j+1} \\ & = \frac{y_{j+1} - y_j}{x_{j+1} - x_j} - \frac{y_j - y_{j-1}}{x_j - x_{j-1}}, \qquad j = 2, 3, \ldots, n-1 \end{aligned} \tag{4.64}$$

These $n - 2$ equations together with the earlier assumption (S3)

$$M_1 = M_n = 0 \tag{4.65}$$

leads to the values $M_1, \ldots, M_n$ and then to the interpolating function $s(x)$. The linear system (4.64) is called a *tridiagonal system*, and there are special methods for its solution.

One of these is given in Chapter 6, Section 6.4; and most computer centers will have special routines for tridiagonal systems of equations.

Example 4.3.1 Calculate the natural cubic spline interpolating the data

$$\left\{(1,1), \left(2, \tfrac{1}{2}\right), \left(3, \tfrac{1}{3}\right), \left(4, \tfrac{1}{4}\right)\right\} \tag{4.66}$$

The number of points is $n = 4$; and all $x_j - x_{j-1} = 1$. The system (4.64) becomes

$$\begin{aligned} \tfrac{1}{6}M_1 + \tfrac{2}{3}M_2 + \tfrac{1}{6}M_3 &= \tfrac{1}{3} \\ \tfrac{1}{6}M_2 + \tfrac{2}{3}M_3 + \tfrac{1}{6}M_4 &= \tfrac{1}{12} \end{aligned} \tag{4.67}$$

Together with (4.65), this yields

$$M_2 = \tfrac{1}{2}, \qquad M_3 = 0$$

Substituting into (4.63), we obtain

$$s(x) = \begin{cases} \tfrac{1}{12}x^3 - \tfrac{1}{4}x^2 - \tfrac{1}{3}x + \tfrac{3}{2}, & 1 \le x \le 2, \\ -\tfrac{1}{12}x^3 + \tfrac{3}{4}x^2 - \tfrac{7}{3}x + \tfrac{17}{6}, & 2 \le x \le 3, \\ -\tfrac{1}{12}x + \tfrac{7}{12}, & 3 \le x \le 4 \end{cases} \tag{4.68}$$

It is left as an exercise to compute $s'(x)$ and $s''(x)$ and to check that they are continuous. ■

Example 4.3.2 Calculate the natural cubic spline interpolating the data in Table 4.3. Since there are $n = 7$ points, the system (4.64) will contain five equations. The graph of the resulting function $s(x)$ is given in Figure 4.10; and for comparison we also include the graph of $y = l(x)$, the piecewise linear interpolant of Figure 4.7. The graph of $y = s(x)$ is generally quite similar to that of $q(x)$ in Figure 4.9. But with $s(x)$, the graph no longer contains the corners or discontinuous changes in slope that are present in the graph of $y = q(x)$ at $x = 2$ and 3. To the eye, the graph of $s(x)$ would seem a better interpolating function than $q(x)$ for the data of Table 4.3. ■

4.3.3 Other Interpolating Spline Functions

Up until this point, we have not considered the accuracy of the interpolating spline $s(x)$. This is satisfactory where only the data points are known, and we only want a smooth curve that looks correct to the eye. But often, we want the spline to interpolate a known function, and then we are also interested in accuracy.

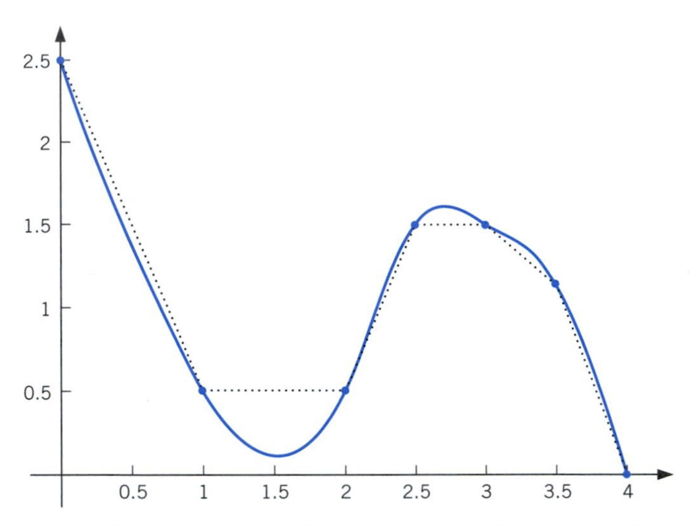

Figure 4.10. Natural cubic spline interpolation (solid line) and piecewise linear interpolation (dotted line)

Let $f(x)$ be given on $[a, b]$. We will consider the case where the interpolation of $f(x)$ is performed at evenly spaced values of x. Let $n > 1$,

$$h = \frac{b-a}{n-1}, \qquad x_j = a + (j-1)h, \qquad j = 1, 2, \ldots, n \tag{4.69}$$

Let $s_n(x)$ be the natural cubic spline interpolating $f(x)$ at $x_1, \ldots, x_n$. Then it can be shown that

$$\max_{a \le x \le b} |f(x) - s_n(x)| \le ch^2 \tag{4.70}$$

where c depends on $f''(a)$, $f''(b)$, and $\max_{a \le x \le b} \left| f^{(4)}(x) \right|$. The primary reason that the approximation $s_n(x)$ does not converge more rapidly (i.e., have an error bound with a higher power of h) is that $f''(x)$ is generally nonzero at $x = a$ and b, whereas $s_n''(a) = s_n''(b) = 0$ by definition. For functions $f(x)$ with $f''(a) = f''(b) = 0$, the right-hand side of (4.70) can be replaced by ch^4.

To improve on $s_n(x)$, we can look at other cubic spline functions $s(x)$ that interpolate $f(x)$. Referring back to the general node-point definition (4.56), we say that $s(x)$ is a *cubic spline* on $[a, b]$ if:

1. $s(x)$ is cubic on each subinterval $[x_{j-1}, x_j]$,
2. $s(x)$, $s'(x)$, and $s''(x)$ are all continuous on $[a, b]$.

In general, cubic splines are fairly smooth functions that are convenient to work with, and they have come to be widely used in the past two decades in computer graphics and in many areas of applied mathematics and statistics.

If $s(x)$ is again chosen to satisfy the earlier interpolating conditions of (4.57),

$$s(x_i) = y_i, \qquad i = 1, \ldots, n$$

then the representation formula (4.63) and the tridiagonal system (4.64) are still valid. This system contains $n - 2$ equations and the n unknowns $M_1, \ldots, M_n$. By replacing the endpoint conditions (4.65) with other conditions, we can obtain other interpolating cubic splines.

If the data are obtained by evaluating a function $f(x)$

$$y_i = f(x_i), \qquad i = 1, \ldots, n$$

then we choose endpoint conditions (or boundary conditions) for $s(x)$ that will result in a better approximation to $f(x)$. If possible, require

$$s'(x_1) = f'(x_1), \qquad s'(x_n) = f'(x_n) \tag{4.71}$$

or

$$s''(x_1) = f''(x_1), \qquad s''(x_n) = f''(x_n) \tag{4.72}$$

When combined with (4.63) and (4.64), either of these conditions leads to a unique interpolating spline $s(x)$, dependent on which of the conditions is chosen. In both instances, the right-hand side of (4.70) can be replaced by ch^4, where c depends on $\max_{a \le x \le b} \left| f^{(4)}(x) \right|$.

If the derivatives of $f(x)$ are not known, then extra interpolating end conditions can also be used to ensure that the error bound of (4.70) is proportional to h^4. In particular, suppose that

$$x_1 < z_1 < x_2, \qquad x_{n-1} < z_2 < x_n$$

and suppose that $f(z_1)$, $f(z_2)$ are known. Then use the formula for $s(x)$ in (4.63) and

$$s(z_1) = f(z_1), \qquad s(z_2) = f(z_2) \tag{4.73}$$

This will add two new equations to the system (4.64), one equation for M_1 and M_2, and a second equation for M_{n-1} and M_n. This form of spline is generally preferable to the interpolating natural cubic spline, and it is almost equally easy to produce. This is the default form of spline interpolation that is implemented in MATLAB, and we discuss it further below. The form of spline formed in this way is said to satisfy the *not-a-knot interpolation boundary conditions*. Figure 4.11 contains the graph of $l(x)$ and the "not-a-knot" interpolating cubic spline for the data given in Table 4.3 at the beginning of this section ($z_1 = 1$, $z_2 = 3.5$). Compare this graph and Figure 4.10.

Interpolating cubic spline functions have become a popular way to represent data analytically because they are relatively smooth (two continuous derivatives), they do

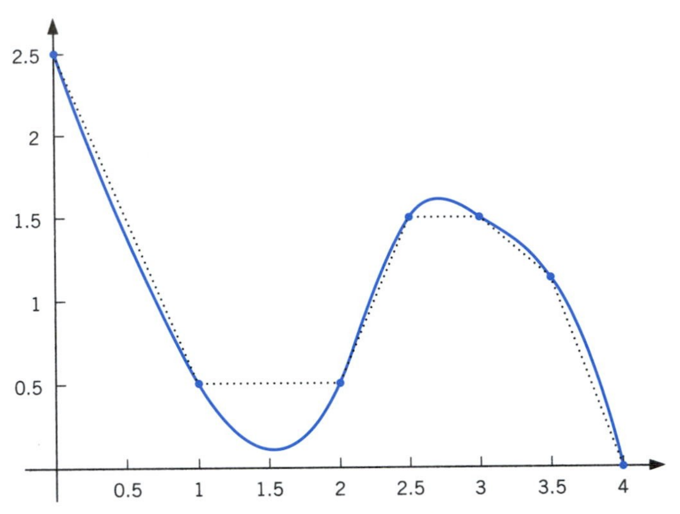

Figure 4.11. The not-a-knot interpolating spline (solid line) and piecewise linear interpolation (dotted line)

not have the rapid oscillation that sometimes occurs with higher-degree polynomial interpolation, and they are reasonably easy to work with on a computer. They do not replace polynomials, but they are a very useful extension of them.

Most academic computer centers will have several programs involving cubic splines, for use in numerical integration, numerical differentiation, interpolation, and the curve-fitting of data. The interpolation programs will generally allow a variety of endpoint conditions, such as those given in (4.65) and (4.71) to (4.73). The student is well advised to use these packages rather than attempting to write his or her own programs.

4.3.4 The MATLAB Program `spline`

The standard MATLAB package contains the function `spline`. It has several possible calling sequences; the standard calling sequence is

```
y=spline(x_nodes,y_nodes,x)
```

This produces the cubic spline function $s(x)$ whose graph passes through the points $\{(\xi_i, \eta_i) \mid i = 1, \ldots, n\}$ with

$$(\xi_i, \eta_i) = (\texttt{x\_nodes(i)}, \texttt{y\_nodes(i)})$$

and n the length of `x_nodes` (and `y_nodes`). The *not-a-knot interpolation conditions* of (4.73) are used. The point (ξ_2, η_2) is the point $(z_1, f(z_1))$ of (4.73), and (ξ_{n-1}, η_{n-1}) is the point $(z_2, f(z_2))$. Following construction of the cubic spline $s(x)$, it is evaluated at the abscissae given in x, and these values are output from `spline` to be stored in y.

If one uses the statements

```
pp=spline(x_nodes,y_nodes)
[breaks,coefs,l,k,d]=unmkpp(pp)
```

then one can obtain the coefficients of a cubic polynomial representation of $s(x)$ on each subinterval determined by adjacent abscissae in `x_nodes`. More detailed information on these commands can be obtained from the MATLAB *help* command.

Example 4.3.3 Approximate the function $f(x) = e^x$ on the interval $[a, b] = [0, 1]$. For $n > 0$, define $h = 1/n$. We use two different choices for the interpolation nodes. First we use the choice

$$x_1 = 0, \ x_2 = h, \ x_3 = 2h, \ \ldots, x_{n+1} = nh = 1$$

and we call this "Choice #1." Using `spline`, we produce the cubic spline interpolant $s_{n,1}(x)$ to $f(x)$. With the not-a-knot interpolation conditions, the nodes x_2 and x_n are the points z_1 and z_2 of (4.73). For a general smooth function $f(x)$, it usually turns out that the magnitude of the error $f(x) - s_{n,1}(x)$ is largest around the endpoints of the interval of approximation. For that reason, we also try "Choice #2" in which two new nodes are inserted, the midpoints of the boundary subintervals $[0, h]$ and $[1 - h, 1]$. The nodes are now

$$x_1 = 0, \quad x_2 = \tfrac{1}{2}h, \quad x_3 = h, \quad x_4 = 2h, \ \ldots, x_{n+1} = (n-1)h,$$
$$x_{n+2} = 1 - \tfrac{1}{2}h, \ x_{n+3} = 1$$

Using `spline` results in a cubic spline function that we denote by $s_{n,2}(x)$. With the not-a-knot interpolation conditions, the nodes x_2 and x_{n+2} are the points z_1 and z_2 of (4.73). Generally, $s_{n,2}(x)$ is a more accurate approximation than is $s_{n,1}(x)$.

The cubic polynomials produced for $s_{n,2}(x)$ by `spline` for the intervals $[x_1, x_2]$ and $[x_2, x_3]$ are the same, and thus we can use the polynomial for $[0, \frac{1}{2}h]$ for the entire interval $[0, h]$; and an analogous situation is true on $[1 - h, h]$. In Table 4.4 we give the maximum errors

$$E_n^{(k)} = \max_{0 \le x \le 1} \left| f(x) - s_{n,k}(x) \right|$$

for both choices of interpolation node points. ■

Table 4.4. Cubic Spline Approximation to $f(x) = e^x$

n	$E_n^{(1)}$	Ratio	$E_n^{(2)}$	Ratio
5	1.01E − 4		1.11E − 5	
10	6.92E − 6	14.6	7.88E − 7	14.1
20	4.56E − 7	15.2	5.26E − 8	15.0
40	2.92E − 8	15.6	3.39E − 9	15.5

PROBLEMS

1. Consider the data points $\{(0, 1), (1, 1), (2, 5)\}$.
   (a) Find the piecewise linear interpolating function for the data.
   (b) Find the quadratic interpolating polynomial.
   (c) Find the natural cubic spline that interpolates the data.

   In all three cases, graph the interpolating functions for $0 \le x \le 2$.

2. Consider the data

x	1	2	3	4	5
y	3	1	2	3	2

   (a) Find the piecewise linear interpolating functions $l(x)$.
   (b) Find the cubic spline function $s(x)$ that interpolates the data and satisfies the not-a-knot boundary conditions (4.73). Note in this case that $n = 3$, $x_1 = 1$, $x_2 = 3$, $x_3 = 5$, $z_1 = 2$, $z_2 = 4$.

   Graph both $s(x)$ and $l(x)$ for $1 \le x \le 5$.

3. Consider the data

x	0	1/2	1	2	3
y	0	1/4	1	−1	−1

   (a) Find the piecewise linear interpolating function for the data.
   (b) Find the piecewise quadratic interpolating function.
   (c) Find the natural cubic spline that interpolates the data.
   (d) Find the not-a-knot interpolating cubic spline. When using (4.73), let $x_1 = 0$, $x_2 = 1$, $x_3 = 3$, and $z_1 = \frac{1}{2}$, $z_2 = 2$.

   Graph all four cases for $0 \le x \le 3$.

4. Consider these data:

x	0	1	2	2.5	3	4
y	1.4	0.6	1.0	0.65	0.6	1.0

   Repeat Problem 3. But note that the piecewise quadratic interpolation will require modification. Use quadratic interpolation on [0, 2] and [2, 3]. For $q(x)$ on [3, 4], construct the quadratic interpolating polynomial on [2.5, 4] and then use it on only [3, 4]. The linear system for the cubic spline function is best solved with a linear system solver like those given in Chapter 6, Section 6.4.

5. Use the MATLAB built-in function `spline` to interpolate the function $f(x) = 1/\left(1 + x^2\right)$, $-5 \le x \le 5$, from Example 4.2.5, with the following sets of x nodes:

(i) $\{-5, \ -2.5, \ 0, \ 2.5, \ 5\}$,

(ii) $\{-5, \ -3.5, \ -2, \ 0, \ 2, \ 3.5, \ 5\}$,

(iii) $\{-5, \ -4.5, \ -4, \ -3, \ -2, \ -1, \ 0, \ 1, \ 2, \ 3, \ 4, \ 4.5, \ 5\}$.

In each case, graph the spline function and the function $f(x)$. Compare to Figure 4.6.

6. Show that the boundary conditions (4.71) lead to the respective equations

$$\begin{aligned} \frac{x_2 - x_1}{3} M_1 + \frac{x_2 - x_1}{6} M_2 &= \frac{y_2 - y_1}{x_2 - x_1} - f'(x_1) \\ \frac{x_n - x_{n-1}}{6} M_{n-1} + \frac{x_n - x_{n-1}}{3} M_n &= f'(x_n) - \frac{y_n - y_{n-1}}{x_n - x_{n-1}} \end{aligned} \tag{4.74}$$

Hint: Differentiate (4.63) on $[x_1, x_2]$ and $[x_{n-1}, x_n]$, and then use (4.71).

7. **(a)** Solve for the cubic spline that interpolates the data in (4.66), with the addition of the boundary conditions (4.71),

$$s'(1) = -1, \qquad s'(4) = -\tfrac{1}{16}$$

Hint: Combine (4.74) in Problem 6 with (4.67), and use a linear system solver to solve your linear system of four equations.

(b) Compare both the natural spline (4.68) and the present spline to the function $f(x) = 1/x$ from which the data were generated. Calculate $f(x) - s(x)$ for a sampling of points in $1 \le x \le 4$.

8. **(a)** Solve for the cubic spline that interpolates the data in (4.66), with the addition of the boundary conditions (4.72),

$$s''(1) = 2, \qquad s''(4) = \tfrac{1}{32}$$

Hint: Combine $M_1 = 2$, $M_4 = \frac{1}{32}$ with the equations (4.67).

(b) Compare this spline and the natural cubic spline (4.68) to the function $f(x) = 1/x$ from which the data were generated.

9. Find the cubic spline that satisfies the conditions

$$s(0) = 0, \quad s(1) = 1, \quad s(2) = 2, \quad s'(0) = 0, \quad s''(2) = 2$$

Also graph it.

10. Is the following function a cubic spline on the interval $0 \le x \le 2$?

$$s(x) = \begin{cases} (x-1)^3, & 0 \le x \le 1 \\ 2(x-1)^3, & 1 \le x \le 2 \end{cases}$$

11. Define

$$s(x) = \begin{cases} -5 + 8x - 6x^2 + 2x^3, & 1 \le x \le 2 \\ 27 - 40x + 18x^2 - 2x^3, & 2 \le x \le 3 \end{cases}$$

Verify that $s(x)$ is a cubic spline function on $[1, 3]$. Is it a natural cubic spline function on this interval?

12. Define

$$s(x) = \begin{cases} 2x^3, & 0 \le x \le 1 \\ x^3 + 3x^2 - 3x + 1, & 1 \le x \le 2 \\ 9x^2 - 15x + 9, & 2 \le x \le 3 \end{cases}$$

Verify that $s(x)$ is a cubic spline function on $[0, 3]$. Is it a natural cubic spline function on this interval?

13. Define

$$s(x) = \begin{cases} x^3 - 3x^2 + 2x + 1, & 1 \le x \le 2, \\ -x^3 + 9x^2 - 22x + 17, & 2 \le x \le 3 \end{cases}$$

Is $s(x)$ a cubic spline function on $[1, 3]$? Is it a natural cubic spline function?

14. Is the following function a cubic spline on $[0, 3]$?

$$s(x) = \begin{cases} x^3, & 0 \le x \le 1 \\ 2x - 1, & 1 < x < 2 \\ 3x^2 - 9, & 2 \le x \le 3 \end{cases}$$

15. Define

$$s(x) = \begin{cases} x^3 + 2x^2 + 1, & 1 \le x \le 2 \\ -2x^3 + \beta x^2 - 36x + 25, & 2 \le x \le 3 \end{cases}$$

For a special value of β, $s(x)$ is a cubic spline function on $[1, 3]$. Find that value of β and then verify that $s(x)$ is a cubic spline function on $[1, 3]$. Is it a natural cubic spline function on this interval?

16. Is there a choice of coefficients $\{a, b, c, d\}$ for which the following function is a cubic spline?

$$s(x) = \begin{cases} (x+1)^3, & -2 \le x \le -1 \\ ax^3 + bx^2 + cx + d, & -1 < x < 1 \\ (x-1)^2, & 1 \le x \le 2 \end{cases}$$

17. **(a)** Formula (4.63) gives $s(x)$ on $x_{j-1} \le x \le x_j$. Write the corresponding formula when $x_j \le x \le x_{j+1}$.

(b) Form $s'(x)$ for $x_{j-1} \le x \le x_j$ and $x_j \le x \le x_{j+1}$.

(c) Set $x = x_j$ in both formulas for $s'(x)$, and require them to be equal. Derive (4.64).

18. Let $s(x)$ be a cubic spline with a single node c satisfying $a < c < b$. Suppose $s(x) = 0$ for $a \le x \le c$. Show that for some constant d,

$$s(x) = d\,(x - c)^3\,, \qquad c \le x \le b$$

19. **(a)** Let $[a, b]$ be a given interval and let $a < c < b$. Define

$$\sigma_c(x) = \begin{cases} 0, & a \le x \le c \\ (x - c)^3, & c \le x \le b \end{cases}$$

Show $\sigma_c(x)$ is a cubic spline function on $[a, b]$.

(b) Let $x_1 < x_2 < \cdots < x_n$, let $p(x)$ be an arbitrary polynomial of degree ≤ 3, and define

$$s(x) = \sum_{j=2}^{n-1} b_j \sigma_{x_j}(x) + p(x), \qquad x_1 \le x \le x_n$$

with $b_2, \ldots, b_{n-1}$ arbitrary constants. The definition uses the definition of $\sigma_c(x)$ in part (a). Show $s(x)$ is a cubic spline on $[a, b] \equiv [x_1, x_n]$. What are the points at which $s'''(x)$ is not continuous? This formula can be shown to represent all cubic splines on $[x_1, x_n]$; but it is not recommended for practical calculations because of ill-conditioning when solving for $p(x)$ and the coefficients $\{b_j\}_{j=2}^{n-1}$.

20. To study the accuracy of cubic spline interpolation, use a package program to construct interpolating splines to $y = e^x$ on $0 \le x \le 1$. Use evenly spaced nodes on [0, 1] to generate the data, say, with 10, 20, and 40 subdivisions. Then check the accuracy of the spline function at four times that many points. Do this for (a) the cubic natural interpolating spline function, and (b) the spline satisfying (4.71) or (4.72). Discuss your results. How does the error behave as the number of subdivisions is doubled?

4.4. THE BEST APPROXIMATION PROBLEM

In this section we look at the concept of best possible approximation. This is illustrated with improvements to the Taylor polynomials for $f(x) = e^x$.

To understand how much it is possible to improve on the Taylor series approximation, we look at the concept of best possible approximation. Let $f(x)$ be a given function that is continuous on some interval $a \le x \le b$. If $p(x)$ is a polynomial, then we are interested in measuring

$$E(p) = \max_{a \le x \le b} |f(x) - p(x)| \tag{4.75}$$

the maximum possible error in the approximation of $f(x)$ by $p(x)$ on the interval $[a, b]$. For each degree $n > 0$, define

$$\begin{aligned} \rho_n(f) &= \min_{\deg(p) \le n} E(p) \\ &= \min_{\deg(p) \le n} \left[\max_{a \le x \le b} |f(x) - p(x)| \right] \end{aligned} \tag{4.76}$$

This is the smallest possible value for $E(p)$ that can be attained with a polynomial of degree $\le n$. It is called the *minimax error*. It can be shown that there is a unique polynomial of degree $\le n$ for which the maximum error on $[a, b]$ is $\rho_n(f)$. This polynomial is called the *minimax polynomial approximation* of order n, and we denote it here by $m_n(x)$.

Example 4.4.1 Let $f(x) = e^x$ on $-1 \le x \le 1$, and consider linear polynomial approximations to $f(x)$. From Section 1.2, the linear Taylor polynomial is

$$t_1(x) = 1 + x \tag{4.77}$$

and by direct computation,

$$\max_{-1 \le x \le 1} \left| e^x - t_1(x) \right| \doteq 0.718 \tag{4.78}$$

By using methods not discussed here, the linear minimax polynomial to e^x on $[-1, 1]$ is

$$m_1(x) = 1.2643 + 1.1752x \tag{4.79}$$

and

$$\max_{-1 \le x \le 1} \left| e^x - m_1(x) \right| \doteq 0.279 \tag{4.80}$$

The graphs of these two linear polynomials are given in Figure 4.12, along with that of e^x. ■

Example 4.4.2 Again let $f(x) = e^x$ for $-1 \le x \le 1$. We compare the maximum errors (4.75) for both the degree n Taylor polynomial $t_n(x)$ and the degree n minimax polynomial approximation $m_n(x)$. These results are shown in Table 4.5 for several values of n. Note that the accuracy of $m_n(x)$ relative to that of $t_n(x)$ becomes greater as n increases. This is a

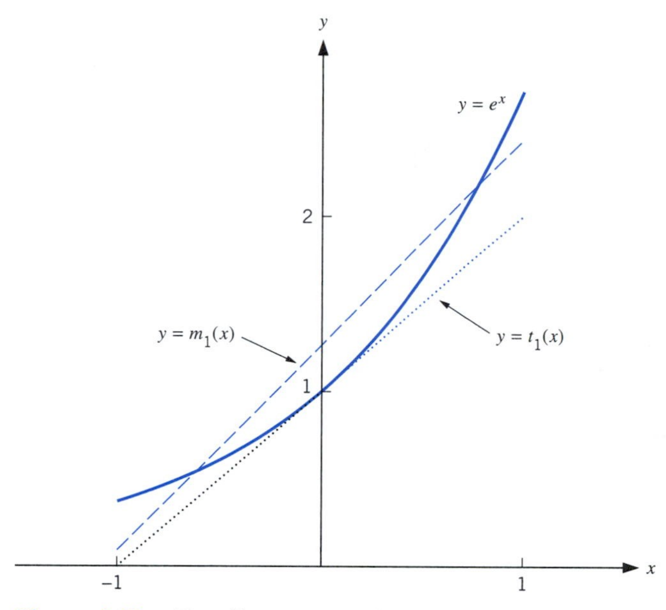

Figure 4.12. Two linear approximations to e^x

Table 4.5. Taylor and Minimax Errors for e^x on $[-1, 1]$

	Maximum Error in:	
n	$t_n(x)$	$m_n(x)$
1	7.18E − 1	2.79E − 1
2	2.18E − 1	4.50E − 2
3	5.16E − 2	5.53E − 3
4	9.95E − 3	5.47E − 4
5	1.62E − 3	4.52E − 5
6	2.26E − 4	3.21E − 6
7	2.79E − 5	2.00E − 7
8	3.06E − 6	1.11E − 8
9	3.01E − 7	5.52E − 10

general characteristic of minimax approximations. This example is not as dramatic as for many functions $f(x)$ because the Taylor series for e^x converges very rapidly.

To further examine the differences in the Taylor and minimax approximations, we present graphs of the errors when $n = 3$. We have

$$\begin{aligned} t_3(x) &= 1 + x + \tfrac{1}{2}x^2 + \tfrac{1}{6}x^3 \\ m_3(x) &= 0.994579 + 0.995668x + 0.542973x^2 + 0.179533x^3 \end{aligned} \tag{4.81}$$

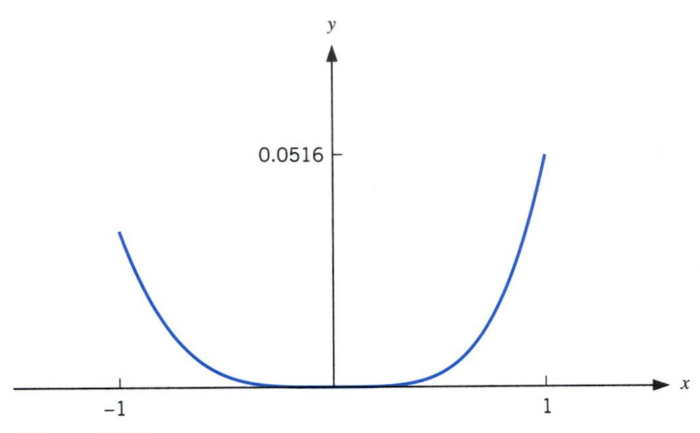

Figure 4.13. $e^x - t_3(x)$

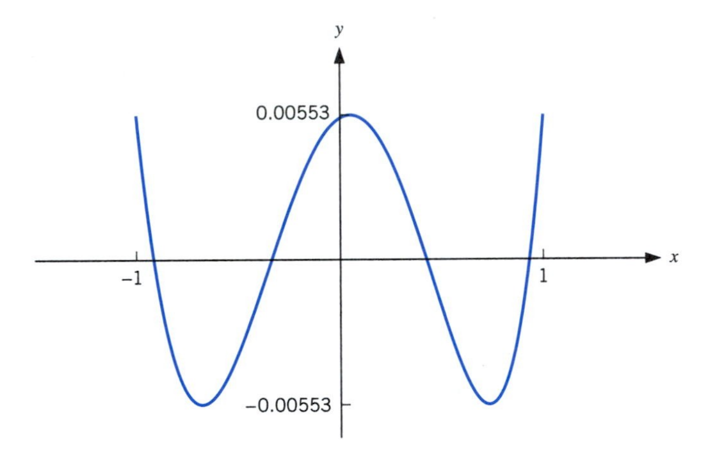

Figure 4.14. $e^x - m_3(x)$

The graphs are shown in Figures 4.13 and 4.14. Note that the vertical scales are quite different. ■

These examples illustrate several general properties of the minimax approximation $m_n(x)$ for approximating a function $f(x)$ on an interval $[a, b]$. First, $m_n(x)$ is usually a significant improvement on the Taylor polynomial $t_n(x)$. This means that to have an approximation with a given degree of accuracy, the minimax approximation will be of lower degree than the Taylor polynomial of equivalent accuracy, often much lower. We note that the actual degree of $m_n(x)$ may be less than n, just as is true with the Taylor polynomial $t_n(x)$, and we still refer to $m_n(x)$ as "the minimax approximation of degree n."

A second characteristic of $m_n(x)$ is that the larger values of the error $f(x) - m_n(x)$ are dispersed over the entire interval $[a, b]$. In comparison, the Taylor error $f(x) - t_n(x)$

is much smaller around the point of expansion than at other points of the interval $[a, b]$. In some sense, the smaller sizes for $f(x) - m_n(x)$ are obtained by distributing the error values more uniformly throughout the interval. An associated third characteristic is that the error $f(x) - m_n(x)$ is oscillatory on $[a, b]$, as illustrated in Figure 4.14. It can be shown that this error will change sign at least $n + 1$ times inside the interval $[a, b]$, and the sizes of the oscillations will be equal.

These properties can be used to construct $m_n(x)$, but it is still a process that is best left to an expert. Programs to do this are available in most computer center libraries. Our approach will be to use these properties to motivate the method given in Section 4.6, which gives a polynomial close to the minimax approximation, and is much easier to construct.

4.4.1 Accuracy of the Minimax Approximation

From the above examples for $f(x) = e^x$, it appears that $m_n(x)$ is very accurate for relatively small values of n. This can be made more precise for some commonly occurring functions such as e^x, $\cos(x)$, and others. Assume $f(x)$ has an infinite number of continuous derivatives on an interval $[a, b]$, and let $m_n(x)$ be the minimax approximation of degree n for $f(x)$ on $[a, b]$. Then the minimax error satisfies

$$\rho_n(f) \le \frac{[(b-a)/2]^{n+1}}{(n+1)!2^n} \max_{a \le x \le b} \left| f^{(n+1)}(x) \right| \tag{4.82}$$

A derivation of this bound follows from the results of Section 4.6. This error bound will not always become smaller with increasing n, but it will give a fairly accurate bound for many common functions $f(x)$.

Table 4.6. Bound on $\rho_n(e^x)$

n	Bound (4.82)	$\rho_n(f)$
1	6.80E − 1	2.79E − 1
2	1.13E − 1	4.50E − 2
3	1.42E − 2	5.53E − 3
4	1.42E − 3	5.47E − 4
5	1.18E − 4	4.52E − 5
6	8.43E − 6	3.21E − 6
7	5.27E − 7	2.00E − 7

Example 4.4.3 Let $f(x) = e^x$ for $-1 \le x \le 1$. Then (4.82) becomes

$$\rho_n(e^x) \le \frac{e}{(n+1)!2^n} \tag{4.83}$$

Table 4.6 gives these values for various n, along with the corresponding exact value for

$\rho_n(f)$ from Table 4.5. The computed value overestimates the true value by a factor of about 2.5, but this is still considered a fairly accurate estimator of $\rho_n(f)$. ■

PROBLEMS

1. Verify (4.78) and (4.80).

2. **(a)** For $f(x) = \tan^{-1}(x)$, calculate the Taylor approximations $t_1(x)$ and $t_3(x)$. Also find their maximum errors relative to $\tan^{-1}(x)$ on $[-1, 1]$.

   (b) The linear and cubic minimax polynomials for $f(x) = \tan^{-1}(x)$ on $[-1, 1]$ are, respectively,

$$m_1(x) = 0.833278x$$
$$m_3(x) = 0.97238588x - 0.19193797x^3$$

   Find their maximum errors on $[-1, 1]$.

   (c) Graph $f(x) - t_3(x)$ and $f(x) - m_3(x)$ on $[-1, 1]$.

3. With many functions $f(x)$, various identities can be used to reduce the interval of approximation from one of very large or infinite extent to a relatively small finite interval. For example, with $f(x) = e^x$, use

$$e^x = \frac{1}{e^{-x}} \qquad \text{if } x < 0$$

   Also, if $m \leq x < m + 1$ for some $m \geq 1$, then

$$e^x = e^m e^y, \qquad y = x - m, \qquad 0 \leq y < 1$$

   These two identities reduce the evaluation of e^x to that of ordinary multiplication, $e^m = e \cdot e \cdot \cdots \cdot e$, multiplying m terms together; and the evaluation of e^y is needed with y between 0 and 1.

   (a) With this as motivation, compute the bound (4.82) with $f(x) = e^x$ on $[0, 1]$ and let $n = 1, 2, \ldots, 7$. Compare these results to those given in Table 4.6.

   (b) To check the accuracy of (a), compute the exact error in $m_3(x)$ for the interval $[0, 1]$, with

$$m_3(x) = 0.9994552 + 1.0166023x + 0.4217030x^2 + 0.2799765x^3$$

   Plot the graph of $e^x - m_3(x)$ on $[0, 1]$.

4. In analogy with the interval reduction discussion in Problem 3, write a section of MATLAB code to reduce the evaluation of $\cos(x)$ to that of $\cos(y)$, where $0 \leq y \leq \pi/2$ for some y related to x.

5. **(a)** Compute the bound (4.82) for $f(x) = \cos(x)$, $0 \leq x \leq \pi/2$, for $n = 1, 2, \dots, 7$.

(b) To check the accuracy of (a), compute the exact error in $m_3(x) \approx \cos(x)$ on $[0, \pi/2]$, where

$$m_3(x) = 0.9986329 + 0.0296140x - 0.6008616x^2 + 0.1125060x^3$$

(c) Graph $\cos(x) - m_3(x)$ on $[0, \pi/2]$.

6. Explain why a natural interval for approximating $f(x) = \log x$ on a binary computer is $[a, b] = [1, 2]$. Give an algorithm for reducing the evaluation of $\log x$ for general $x > 0$ to that of evaluating $\log y$ for a suitable y in $[1, 2]$.

7. **(a)** For $f(x) = \log x$, bound the minimax approximation error $\rho_n(f)$ on $1 \leq x \leq 2$. Find a bound for each case $n \geq 1$, and have the bound converge to zero as $n \to \infty$.

(b) Find the Taylor polynomial $t_3(x)$ of degree 3 for $f(x) = \log x$ about $x = \frac{3}{2}$. What is its maximum error on $1 \leq x \leq 2$?

(c) For $f(x) = \log(x)$ and the interval $[1, 2]$,

$$m_3(x) = -1.492776 + 2.112632x - 0.729104x^2 + 0.109690x^3$$

Find $\rho_3(f)$ and graph $f(x) - m_3(x)$ on $[1, 2]$. Compare $t_3(x)$ and $m_3(x)$.

8. **(a)** Verify the identities

$$\tan^{-1}(-x) = -\tan^{-1}(x)$$

$$\tan^{-1}(x) = \frac{\pi}{2} - \tan^{-1}\left(\frac{1}{x}\right), \qquad x > 0$$

(b) With these, what smaller interval can be used to approximate the following?

$$f(x) = \tan^{-1}(x), \qquad -\infty < x < \infty$$

4.5. CHEBYSHEV POLYNOMIALS

We introduce a family of polynomials, the *Chebyshev polynomials*, that are used in many parts of numerical analysis and, more generally, in mathematics and physics. A few of their properties will be given in this section, and then they will be used in Section 4.6 to produce a polynomial approximation close to the minimax approximation.

For an integer $n \geq 0$, define the function

$$T_n(x) = \cos\left(n \cos^{-1} x\right), \qquad -1 \leq x \leq 1 \tag{4.84}$$

This may not appear to be a polynomial, but we will show it is a polynomial of degree n. To simplify the manipulation of (4.84), we introduce

$$\theta = \cos^{-1}(x) \quad \text{or} \quad x = \cos(\theta), \qquad 0 \leq \theta \leq \pi \tag{4.85}$$

Then

$$T_n(x) = \cos(n\theta) \tag{4.86}$$

Graphs of the Chebyshev polynomials $T_0(x), T_1(x), \ldots, T_4(x)$ are given in Figures 4.15 and 4.16.

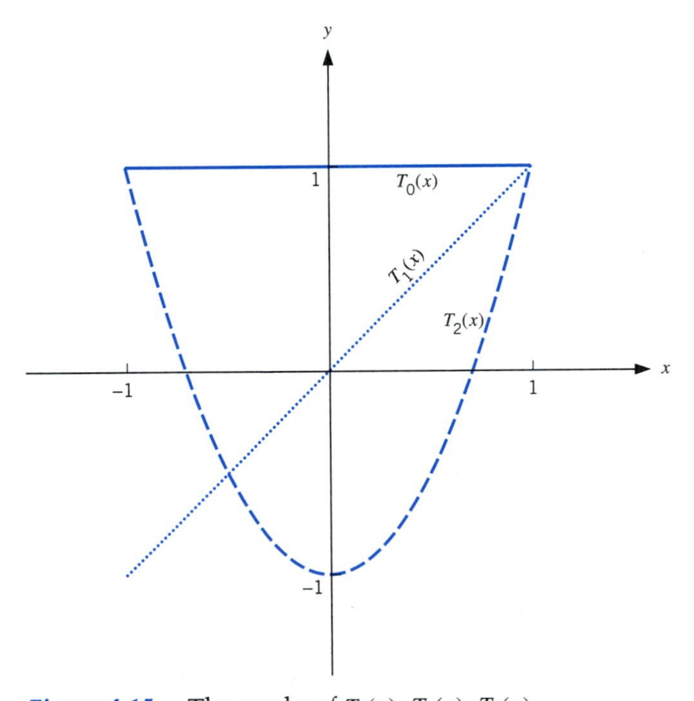

Figure 4.15. The graphs of $T_0(x)$, $T_1(x)$, $T_2(x)$

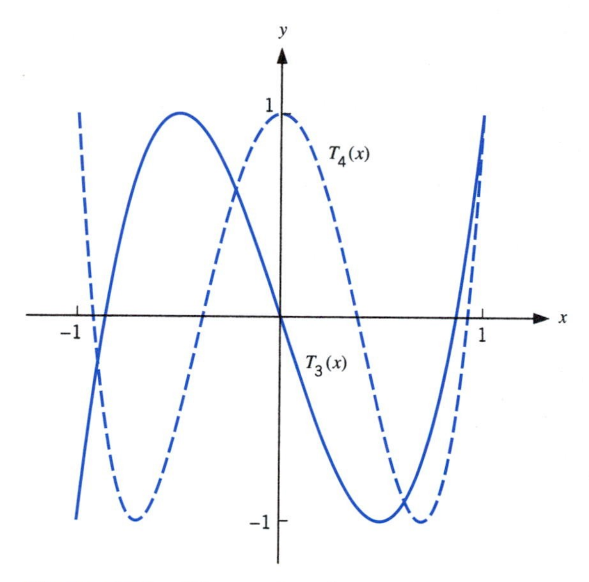

Figure 4.16. The graphs of $T_3(x)$, $T_4(x)$

Example 4.5.1 (a) Let $n = 0$. Then

$$T_0(x) = \cos(0 \cdot \theta) = 1$$

(b) Let $n = 1$. Then

$$T_1(x) = \cos(\theta) = x$$

(c) Let $n = 2$. Then using a common trigonometric identity, we obtain

$$T_2(x) = \cos(2\theta) = 2\cos^2(\theta) - 1 = 2x^2 - 1 \quad \blacksquare$$

4.5.1 The Triple Recursion Relation

Recall the trigonometric addition formulas

$$\cos(\alpha \pm \beta) = \cos(\alpha)\cos(\beta) \mp \sin(\alpha)\sin(\beta)$$

For any $n \geq 1$, apply these identities to get

$$\begin{aligned} T_{n+1}(x) &= \cos[(n+1)\theta] = \cos(n\theta + \theta) \\ &= \cos(n\theta)\cos(\theta) - \sin(n\theta)\sin(\theta) \\ T_{n-1}(x) &= \cos[(n-1)\theta] = \cos(n\theta - \theta) \\ &= \cos(n\theta)\cos(\theta) + \sin(n\theta)\sin(\theta) \end{aligned}$$

Add these two equations, and then use (4.85) and (4.86) to obtain

$$T_{n+1}(x) + T_{n-1}(x) = 2\cos(n\theta)\cos(\theta) = 2xT_n(x)$$
$$T_{n+1}(x) = 2xT_n(x) - T_{n-1}(x), \qquad n \geq 1 \tag{4.87}$$

This is called the *triple recursion relation* for the Chebyshev polynomials. It is often used in evaluating them, rather than using the explicit formula (4.84).

Example 4.5.2 (a) Let $n = 2$ in (4.87). Using the previously computed values of $T_1(x)$ and $T_2(x)$, we obtain

$$T_3(x) = 2xT_2(x) - T_1(x) = 2x(2x^2 - 1) - x$$
$$= 4x^3 - 3x$$

(b) Let $n = 3$. Then

$$T_4(x) = 2xT_3(x) - T_2(x) = 2x(4x^3 - 3x) - (2x^2 - 1)$$
$$= 8x^4 - 8x^2 + 1 \quad \blacksquare$$

4.5.2 The Minimum Size Property

Before stating the main result, we note that

$$|T_n(x)| \leq 1, \qquad -1 \leq x \leq 1 \tag{4.88}$$

for all $n \geq 0$. Also, note that

$$T_n(x) = 2^{n-1}x^n + \text{lower-degree terms}, \qquad n \geq 1 \tag{4.89}$$

The first result follows directly from the definition (4.84) using the bound $|\cos(n\theta)| \leq 1$ [cf. (4.86)]. The second can be proved by using mathematical induction and the recursion relation (4.87). Note that the earlier computations of T_1, T_2, T_3, T_4 are also examples of (4.89).

Introduce a modified version of $T_n(x)$

$$\widetilde{T}_n(x) = \frac{1}{2^{n-1}}T_n(x) = x^n + \text{lower-degree terms}, \qquad n \geq 1 \tag{4.90}$$

From (4.88) and (4.89),

$$\left|\widetilde{T}_n(x)\right| \leq \frac{1}{2^{n-1}}, \qquad -1 \leq x \leq 1, \qquad n \geq 1 \tag{4.91}$$

A polynomial whose highest-degree term has a coefficient of 1 is called a *monic polynomial.* Formula (4.91) states that the monic polynomial $\widetilde{T}_n(x)$ has size $1/2^{n-1}$ on $-1 \le x \le 1$, and this becomes smaller as the degree n increases. In comparison,

$$\max_{-1 \le x \le 1} |x^n| = 1$$

Thus, x^n is a monic polynomial whose size does not change with increasing n.

Theorem 4.5.3 Let $n \ge 1$ be an integer, and consider all possible monic polynomials of degree n. Then the degree n monic polynomial with the smallest maximum absolute value on $[-1, 1]$ is the modified Chebyshev polynomial $\widetilde{T}_n(x)$, and its maximum value on $[-1, 1]$ is $1/2^{n-1}$.

This theorem leads to a number of useful applications for Chebyshev polynomials, one of which we consider in the next section. A proof of the theorem is suggested in Problem 10.

PROBLEMS

1. Find $T_5(x)$ explicitly in polynomial form, and then graph it on $-1 \le x \le 1$.

2. Demonstrate as best you can why (4.89) is true.

3. From (4.84), we know that $|T_n(x)| \le 1$ on $[-1, 1]$ and $T_n(1) = 1$. Find a general formula for the points x at which $T_n(x) = \pm 1$. How many such points are there on $[-1, 1]$?

   Hint: Begin with a special case such as $n = 3$.

4. Evaluate $T_n(0.5)$ for $2 \le n \le 10$, without using the definitions (4.84) to (4.86). Use $T_0(0.5) = 1$, $T_1(0.5) = 0.5$.

5. Suppose that we require $T_0(x), T_1(x), \ldots, T_n(x)$ for a particular value of x. Do a count of the number of multiplications that are needed to produce these values if we use the triple recursion formula (4.87).

6. Let $q(x)$ be a polynomial of degree $\le n - 1$, and consider

   $$\max_{-1 \le x \le 1} |x^n - q(x)|$$

   What is the smallest possible value for this quantity? Solve for the $q(x)$ for which the smallest value is attained.

7. For $n, m \ge 0$ and $n \ne m$, show

   $$\int_{-1}^{1} \frac{T_n(x) T_m(x)}{\sqrt{1 - x^2}}\, dx = 0$$

   This is called the *orthogonality relation* for Chebyshev polynomials.

   Hint: Use (4.84) and the change of variable $x = \cos\theta$.

8. The functions

$$S_n(x) = \frac{1}{n+1} T'_{n+1}(x), \qquad n \ge 0$$

are called *Chebyshev polynomials of the second kind.*

(a) Calculate $S_0(x)$, $S_1(x)$, $S_2(x)$, $S_3(x)$.

(b) Show

$$S_n(x) = \frac{\sin(n+1)\theta}{\sin\theta} \qquad \text{with } x = \cos\theta$$

for $0 \le \theta \le \pi$.

(c) Use addition formulas for $\sin(\alpha \pm \beta)$ to produce a triple recursion formula for $S_{n+1}(x)$.

9. The monomials $\{x^j \mid j \ge 0\}$ can be expressed as combinations of the Chebyshev polynomials. Easily, $1 = T_0(x)$ and $x = T_1(x)$. Next,

$$x^2 = \tfrac{1}{2}[T_2(x) + 1] = \tfrac{1}{2}T_2(x) + \tfrac{1}{2}T_0(x)$$

(a) Proceeding similarly, express x^3 and x^4 as combinations of Chebyshev polynomials.

(b) Show that for each $n \ge 0$, it is possible to write x^n in the form

$$x^n = a_{n,n}T_n(x) + a_{n,n-1}T_{n-1}(x) + \cdots + a_{n,0}T_0(x) \tag{4.92}$$

for a suitable choice of coefficients $\{a_{n,j}\}_{j=0}^{n}$.

(c) Show

$$a_{n,0} = \frac{1}{\pi}\int_{-1}^{1} \frac{x^n}{\sqrt{1-x^2}}\,dx, \qquad a_{n,j} = \frac{2}{\pi}\int_{-1}^{1} \frac{x^n T_j(x)}{\sqrt{1-x^2}}\,dx$$

for $j > 0$.

Hint: Use Problem 7. Multiply both sides of (4.92) by $T_j(x)/\sqrt{1-x^2}$ and integrate over $[-1, 1]$.

10. To indicate the proof of Theorem 4.5.3, consider the case of degree $n = 3$.

(a) Using Problem 3, find the values of x for which

$$\widetilde{T}_3(x) = \pm\tfrac{1}{4}$$

Call these values $x_3 < x_2 < x_1 < x_0$.

(b) Assume there is another monic polynomial $q(x)$ for which

(i) $\max_{-1 \le x \le 1} |q(x)| < \dfrac{1}{2^{n-1}} = \dfrac{1}{4}$

(ii) $\deg(q) \le 3$.

(c) Evaluate the polynomial

$$R(x) = \widetilde{T}_3(x) - q(x)$$

at x_0, x_1, x_2, x_3. Show that

$$R(x_0) > 0, \qquad R(x_1) < 0, \qquad R(x_2) > 0, \qquad R(x_3) < 0$$

(d) Show $R(x)$ has degree ≤ 2 and that it must have at least three roots.

(e) Show $\widetilde{T}_3(x) \equiv q(x)$, contrary to the assumption (i) in (b). This proves that there is no smaller maximum on $[-1, 1]$ than $\frac{1}{4}$ for a monic polynomial of degree 3.

4.6. A NEAR-MINIMAX APPROXIMATION METHOD

Since we are looking for polynomial approximations to a given function $f(x)$, it would seem reasonable to consider using an interpolating polynomial. The most obvious choice is to choose an evenly spaced set of interpolation node points on the interval $a \le x \le b$ of interest. Unfortunately, this often gives an interpolating polynomial that is a very poor approximation to $f(x)$, for reasons we will not go into here. This was illustrated in Figure 4.6, Example 4.2.4 of Section 4.2. To consider interpolation in a more methodical way, we will examine it by means of the error formula (4.41), also from Section 4.2.

To simplify the presentation, we choose the special interval $-1 \le x \le 1$ as the approximation interval for $f(x)$, and we initially limit the degree of the approximating polynomial to $n = 3$. Let x_0, x_1, x_2, x_3 be the interpolation node points in $[-1, 1]$, and let $c_3(x)$ denote the polynomial of degree ≤ 3 that interpolates $f(x)$ at x_0, x_1, x_2, and x_3. Then from (4.41), the interpolation error is given by

$$f(x) - c_3(x) = \frac{(x - x_0)(x - x_1)(x - x_2)(x - x_3)}{4!} f^{(4)}(c_x) \tag{4.93}$$

for $-1 \le x \le 1$ and for some c_x in $[-1, 1]$. The nodes x_0, x_1, x_2, x_3 are to be chosen so that the maximum value of $|f(x) - c_3(x)|$ on $[-1, 1]$ is made as small as possible.

Looking at the right side of (4.93), we see that the only quantity we can use to influence the size of the error is the degree 4 polynomial

$$\omega(x) = (x - x_0)(x - x_1)(x - x_2)(x - x_3) \tag{4.94}$$

We want to choose the interpolation points x_0, x_1, x_2, x_3 so that

$$\max_{-1 \le x \le 1} |\omega(x)| \tag{4.95}$$

is made as small as possible.

If $\omega(x)$ is multiplied out, it is fairly easy to see that

$$\omega(x) = x^4 + \text{lower-degree terms}$$

This is a monic polynomial of degree 4. From Theorem 4.5.3 in the preceding section, the smallest possible value for (4.95) is obtained with

$$\omega(x) = \frac{T_4(x)}{2^3} = \frac{1}{8}(8x^4 - 8x^2 + 1) \tag{4.96}$$

and the smallest value of (4.95) is $1/2^3$ in this case.

The choice (4.96) defines implicitly the interpolation node points. From (4.94), the node points are the zeros of $\omega(x)$, and from (4.96), they must therefore be the zeros of $T_4(x)$. The node points could be calculated by numerically finding the roots of the right side of (4.96); but there is a simpler procedure, which is also much better to use when the polynomial degree becomes larger.

Look at the definition (4.84), (4.85). In our case,

$$T_4(x) = \cos(4\theta), \qquad x = \cos(\theta)$$

This is zero when

$$\begin{aligned} 4\theta &= \pm\frac{\pi}{2}, \pm\frac{3\pi}{2}, \pm\frac{5\pi}{2}, \pm\frac{7\pi}{2}, \ldots \\ \theta &= \pm\frac{\pi}{8}, \pm\frac{3\pi}{8}, \pm\frac{5\pi}{8}, \pm\frac{7\pi}{8}, \ldots \\ x &= \cos\left(\frac{\pi}{8}\right), \cos\left(\frac{3\pi}{8}\right), \cos\left(\frac{5\pi}{8}\right), \cos\left(\frac{7\pi}{8}\right), \cos\left(\frac{9\pi}{8}\right), \ldots \end{aligned} \tag{4.97}$$

using $\cos(-\theta) = \cos(\theta)$. The first four values of x are distinct, but the successive values repeat the first four values. Thus, when we evaluate (4.97), the nodes are approximately

$$\pm 0.382683, \qquad \pm 0.923880 \tag{4.98}$$

Example 4.6.1 Let $f(x) = e^x$ on $[-1, 1]$, and use the nodes (4.98) to produce the interpolating polynomials $c_3(x)$ of degree 3. Table 4.7 lists the nodes, function values, and divided differences needed for Newton's divided difference formula for the interpolating polynomial (4.35). ■

Table 4.7. Interpolation Data for $c_3(x)$

i	x_i	$f(x_i)$	$f[x_0, \ldots, x_i]$
0	0.923880	2.5190442	2.5190442
1	0.382683	1.4662138	1.9453769
2	−0.382683	0.6820288	0.7047420
3	−0.923880	0.3969760	0.1751757

Example 4.6.2 By evaluating $c_3(x)$ at a large number of points, we find that

$$\max_{-1\le x\le 1} \left|e^x - c_3(x)\right| \doteq 0.00666 \tag{4.99}$$

The graph of $e^x - c_3(x)$ is given in Figure 4.17. Compare this error to the corresponding value from Table 4.5

$$\rho_3(e^x) \doteq 0.00553$$

and compare the graph to that given in Figure 4.14 for $m_3(x) \approx e^x$. ■

The above construction of $c_3(x)$ generalizes to finding a degree n near-minimax approximation to $f(x)$ on $[-1, 1]$. The interpolation error is given by

$$f(x) - c_n(x) = \frac{(x - x_0) \cdots (x - x_n)}{(n + 1)!} f^{(n+1)}(c_x), \qquad -1 \le x \le 1 \tag{4.100}$$

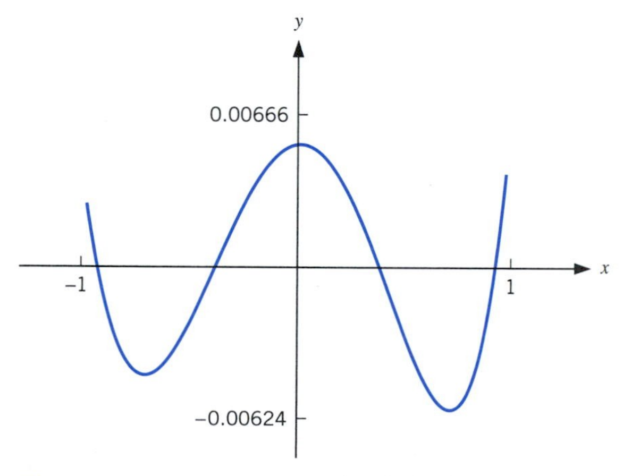

Figure 4.17. $e^x - c_3(x)$ with Chebyshev nodes

and we seek to minimize

$$\max_{-1 \le x \le 1} |(x - x_0) \cdots (x - x_n)| \tag{4.101}$$

The polynomial being minimized is monic of degree $n + 1$. And from Theorem 4.5.3, this minimum is attained by the monic polynomial

$$\frac{1}{2^n} T_{n+1}(x)$$

Thus, the interpolation nodes are the zeros of $T_{n+1}(x)$; and by the procedure leading to (4.97), they are given by

$$x_j = \cos\left(\frac{2j+1}{2n+2}\pi\right), \qquad j = 0, 1, \ldots, n \tag{4.102}$$

The near-minimax approximation $c_n(x)$ of degree n is obtained by interpolating to $f(x)$ at these $n + 1$ nodes on $[-1, 1]$.

Example 4.6.3 Let $f(x) = e^x$. Then the maximum error in $c_n(x)$ on $[-1, 1]$ is given in Table 4.8. For comparison, we also include the corresponding minimax errors $\rho_n(x)$. These figures show that for practical purposes, $c_n(x)$ is a satisfactory replacement for the minimax approximation $m_n(x)$. ■

MATLAB PROGRAM. We give a program for constructing and evaluating the near-minimax approximation $c_n(x)$ for a given function $f(x)$ on $[-1, 1]$. The polynomial $c_n(x)$ is written in the Newton divided difference form of the interpolating polynomial; and it is evaluated using the nested multiplication method described in (4.39) of Section 4.1. Note that the programs `divdif` and `interp` from Section 4.1 are used by the program.

Table 4.8. Near-Minimax Errors for e^x on $[-1, 1]$

n	$\max \lvert e^x - c_n(x)\rvert$	$\rho_n(e^x)$
1	3.72E − 1	2.79E − 1
2	5.65E − 2	4.50E − 2
3	6.66E − 3	5.53E − 3
4	6.40E − 4	5.47E − 4
5	5.18E − 5	4.52E − 5
6	3.62E − 6	3.21E − 6

The program uses $f(x) = e^x$, but this is easily changed to some other function by changing the function subprogram fcn. The maximum error

$$\max_{-1 \leq x \leq 1} \left| e^x - c_n(x) \right|$$

is estimated by evaluating the error at 501 evenly spaced points x in $[-1, 1]$. The program outputs the interpolation nodes, corresponding function values, and divided differences, so that $c_n(x)$ can be evaluated in a separately constructed program.

```
function [nodes, fcn_values, div_diff_fcn] = chebyshev_interp(n)
%
% This creates an interpolant of degree n to the function
% fcn(x) on [-1,1], which is given below by a function
% subprogram.  The nodes are the Chebyshev zeroes of the
% degree n+1 Chebyshev polynomial on [-1,1].  The program
% gives two plots:  first the true function and its
% interpolant, and second, the error in the interpolation.

% Create the nodes and associated divided differences.
h = pi/(2*(n+1));
nodes = cos(h*[1:2:2*n+1]);
fcn_values = fcn(nodes);
div_diff_fcn = divdif(nodes,fcn_values);

% Create the points at which the functions are to be
% graphed.
x_eval = -1:.002:1;
true_fcn = fcn(x_eval);
y_eval = interp(nodes,div_diff_fcn,x_eval);

% Create the window for the graph of the function
% and its interpolant.
m = min([min(true_fcn),min(y_eval)]);
M = max([max(true_fcn),max(y_eval)]);
axis([-1.1,1.1,m,M])
hold on

% Create the graph of the function and its interpolant.
plot(x_eval,true_fcn,'r')
plot(x_eval,y_eval,':')
legend('True function','Interpolant',0)
plot(nodes,fcn_values,'.','MarkerSize',6)
hold off
```

```
pause
clf

% Create the window for the graph of the error.
error = true_fcn - y_eval;
M = max(error);
m = min(error);
axis([-1.1,1.1,m,M])
hold on

% Create the graph of the error in the interpolant.
plot(x_eval,error,'r')
hold off

% Print the maximum error.
disp(['maximum error = ',num2str(max(abs(error)))])

function fval = fcn(x)
fval = exp(x);
```

For a practical problem, we would have $f(x)$ evaluated by some accurate, but inefficient method, often a Taylor polynomial. This approximation would be given in the function subprogram `fcn(x)`, to be used by the main program to produce a more efficient approximation. One other difficulty is that most functions do not have $[-1, 1]$ as the interval on which we wish to approximate them. This limitation is removed in Problem 4.

4.6.1 Odd and Even Functions

A word of warning needs to be given regarding the use of this algorithm. If $f(x)$ is what is called an even or odd function on $[-1, 1]$, then n should be chosen in a more restricted way. We say $f(x)$ is *even* if

$$f(-x) = f(x), \qquad \text{all } x \tag{4.103}$$

An example is $f(x) = \cos(x)$. Such functions have graphs that are symmetric about the y-axis. We say $f(x)$ is *odd* if

$$f(-x) = -f(x), \qquad \text{all } x \tag{4.104}$$

An example is $f(x) = \sin(x)$. Such functions are said to be symmetric about the origin.

For these two cases, choose n in the above algorithm as follows:

$$\begin{array}{l}\text{If } f(x) \text{ is odd, then choose } n \text{ even.}\\ \text{If } f(x) \text{ is even, then choose } n \text{ odd.}\end{array} \tag{4.105}$$

This will result in $c_n(x)$ having degree only $n-1$, but it will give an appropriate formula. The difficulty arises from the following. If a polynomial is odd as a function, as in (4.104), then $c_n(x)$ will have only odd degree terms; and if $c_n(x)$ is even, then it will have only even degree terms. The restriction (4.105) will take this difficulty into account in using the program while also giving good accuracy. See Atkinson (1989, Section 4.7) for a further discussion.

PROBLEMS

1. **(a)** Demonstrate that there are only four distinct values in (4.97).

   (b) Show that the zeros of $T_{n+1}(x)$ are given by (4.102).

2. Give the interpolation nodes for the linear near-minimax approximation of this section, for the interval $[-1, 1]$. Give the linear near-minimax approximation for $f(x) = e^x$ on $[-1, 1]$.

3. **(a)** Implement the program given in this section. Include a printout of either the divided differences and nodes or the errors; these can be used to generate a graph of the error function.

   (b) Calculate the cubic near-minimax approximation for $f(x) = \tan^{-1}(x)$ on $[-1, 1]$. Compare it to the cubic minimax approximation given in Problem 2(b) of Section 4.4.

   Note: Use $n = 4$ in the program, based on (4.105), because $\tan^{-1}(x)$ is an odd function on $[-1, 1]$.

4. Most functions do not have $[-1, 1]$ as the interval on which they are to be approximated. Suppose $g(t)$ is to be evaluated for $a \le t \le b$. Then define a new function $f(x)$ on $[-1, 1]$ by

   $$f(x) = g\left[\frac{(b+a)+x(b-a)}{2}\right], \qquad -1 \le x \le 1$$

   Here

   $$t = \tfrac{1}{2}[(b+a)+x(b-a)]$$

   represents a linear change of variable. We now approximate $f(x)$ on $[-1, 1]$.

   As a specific example, produce the cubic near-minimax approximation for $g(t) = e^t$, $0 \le t \le 1$. Compare this to the minimax approximation given in Problem 3(b) of Section 4.4.

5. Find the cubic near-minimax approximation to $g(t) = \cos(t)$, $0 \le t \le \pi/2$. Compare it to the minimax approximation given in Problem 5(b) of Section 4.4.

6. Find the maximum error in the degree n near-minimax approximation to $g(t) = \tan^{-1}(t), 0 \le t \le 1$. Do this for $n = 1, 2, \ldots, 6$.

7. Repeat Problem 6 with $g(t) = \log(t)$, $1 \le t \le 2$.

8. Repeat Problem 6 for

$$g(t) = \frac{1}{t} \int_0^t \frac{\sin(x)}{x}\, dx, \qquad -1 \le t \le 1$$

To initially evaluate $g(t)$, use a Taylor polynomial approximation, say, with an accuracy of 5×10^{-15} on the interval $-1 \le t \le 1$. Write a MATLAB function to evaluate this polynomial, following the example of Sint x in Section 1.3 of Chapter 1.

9. Repeat Problem 8 for the function

$$g(t) = \frac{1}{t} \int_0^t \frac{e^u - 1}{u}\, du, \qquad -1 \le t \le 1$$

10. Show that if a polynomial $p(x)$ is an even function, satisfying (4.103), then all odd degree terms in $p(x)$ will be zero. Extend this result to polynomials that are odd functions.

11. By using (4.100) and (4.101), along with Theorem 4.5.3, derive the bound (4.82) of Section 4.4 in the case $[a, b] = [-1, 1]$.

4.7. LEAST SQUARES APPROXIMATION

In the preceding section we gave a near-minimax polynomial approximation based on using polynomial interpolation at suitably chosen node points. Another approach is to seek an approximation with a small "average error" over the interval of approximation. If a function $f(x)$ is being approximated by a polynomial $p(x)$ over an interval $a \le x \le b$, then a convenient definition of the average error of the approximation is given by

$$E(p; f) \equiv \sqrt{\frac{1}{b-a} \int_a^b [f(x) - p(x)]^2\, dx} \tag{4.106}$$

This is also called the *root-mean-square-error* in the approximation of $f(x)$ by $p(x)$. We begin by illustrating its use in a relatively straightforward case. Note first that minimizing $E(p; f)$ for different choices of $p(x)$ is equivalent to minimizing

$$\int_a^b [f(x) - p(x)]^2\, dx \tag{4.107}$$

thus dispensing with the square root and multiplying fraction (although the minimums are generally different).

Example 4.7.1 Let $f(x) = e^x$, and let $p(x) = \alpha_0 + \alpha_1 x$, with α_0, α_1 arbitrary. We want to approximate $f(x)$ over the interval $[-1, 1]$. Thus, we want to choose α_0, α_1 so as to minimize the integral

$$g(\alpha_0, \alpha_1) \equiv \int_{-1}^{1} [e^x - \alpha_0 - \alpha_1 x]^2 \, dx \tag{4.108}$$

Expanding this, we obtain

$$g(\alpha_0, \alpha_1) = \int_{-1}^{1} \left\{ e^{2x} + \alpha_0^2 + \alpha_1^2 x^2 - 2\alpha_0 e^x - 2\alpha_1 x e^x + 2\alpha_0 \alpha_1 x \right\} dx \tag{4.109}$$

This can be integrated to give an expression of the form

$$g(\alpha_0, \alpha_1) = c_1 \alpha_0^2 + c_2 \alpha_1^2 + c_3 \alpha_0 \alpha_1 + c_4 \alpha_0 + c_5 \alpha_1 + c_6$$

with suitable constants $\{c_1, \ldots, c_6\}$ calculated using integration. This is a quadratic polynomial in the two variables α_0, α_1. Its minimum can be found by solving the simultaneous equations

$$\frac{\partial g}{\partial \alpha_0} = 0, \qquad \frac{\partial g}{\partial \alpha_1} = 0 \tag{4.110}$$

in which we are using the derivatives of g with respect to α_0 and α_1. In our case, it is simpler to return to (4.108) to differentiate, obtaining

$$2 \int_{-1}^{1} [e^x - \alpha_0 - \alpha_1 x]\,(-1)\, dx = 0$$

$$2 \int_{-1}^{1} [e^x - \alpha_0 - \alpha_1 x]\,(-x)\, dx = 0$$

This simplifies to

$$2\alpha_0 = \int_{-1}^{1} e^x dx = e - e^{-1}$$

$$\tfrac{2}{3}\alpha_1 = \int_{-1}^{1} x e^x dx = 2e^{-1}$$

$$\alpha_0 = \frac{e - e^{-1}}{2} \doteq 1.1752$$

$$\alpha_1 = 3e^{-1} \doteq 1.1036$$

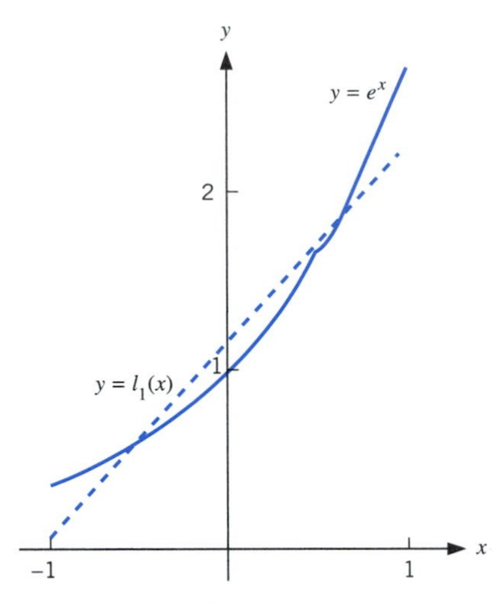

Figure 4.18. The linear least squares approximation to e^x

Table 4.9. Errors in Linear Approximations of e^x

Approximation	Maximum Error	Root-Mean-Square-Error
Taylor $t_1(x)$	0.718	0.246
Least squares $\ell_1(x)$	0.439	0.162
Chebyshev $c_1(x)$	0.372	0.184
Minimax $m_1(x)$	0.279	0.190

Using these values for α_0 and α_1, we denote the resulting linear approximation by

$$\ell_1(x) = \alpha_0 + \alpha_1 x$$

It is called the best linear approximation to e^x in the sense of least squares. For the error,

$$\max_{-1 \le x \le 1} \left| e^x - \ell_1(x) \right| \doteq 0.439$$

A graph of it is given in Figure 4.18. We compare four forms of linear approximation to e^x which have been studied in this chapter, giving the maximum errors and root-mean-square-errors for each in Table 4.9. ■

Return to the minimization of $E(p; f)$ in (4.106) for a general function $f(x)$ on a general interval $[a, b]$, and let $n \ge 0$ be a given integer. We seek a polynomial $p(x)$ of a degree less than or equal to n, that minimizes (4.107). Imitating the preceding example, we can write

$$p(x) = \alpha_0 + \alpha_1 x + \cdots + \alpha_n x^n \tag{4.111}$$

We define

$$g(\alpha_0, \alpha_1, \ldots, \alpha_n) \equiv \int_{-1}^{1} [f(x) - \alpha_0 - \alpha_1 x - \cdots - \alpha_n x^n]^2 \, dx \tag{4.112}$$

and we seek coefficients $\alpha_0, \alpha_1, \ldots, \alpha_n$ that minimize this integral. If the integral is expanded, it can be shown that $g(\alpha_0, \alpha_1, \ldots, \alpha_n)$ is a quadratic polynomial in the $n + 1$ variables $\alpha_0, \alpha_1, \ldots, \alpha_n$.

A minimizer for $g(\alpha_0, \alpha_1, \ldots, \alpha_n)$ can be found by invoking the conditions

$$\frac{\partial g}{\partial \alpha_i} = 0, \qquad i = 0, 1, \ldots, n$$

leading to a set of $n + 1$ equations that must be satisfied by a minimizing set $\alpha_0, \alpha_1, \ldots, \alpha_n$ for g. Manipulating this set of conditions leads to a simultaneous linear system. To better understand the form of the linear system, consider the special case of $[a, b] = [0, 1]$. Then the linear system is

$$\sum_{j=0}^{n} \frac{\alpha_j}{i + j + 1} = \int_0^1 x^i f(x) \, dx, \qquad i = 0, 1, \ldots, n \tag{4.113}$$

We will study the solution of simultaneous linear systems in Chapter 6. In Section 6.5 we will see that this linear system is "ill-conditioned" and difficult to solve accurately, even for moderately sized values of n such as $n = 5$. As a consequence, this is not a good approach to solving for a minimizer of $E(p; f)$ in (4.107).

4.7.1 Legendre Polynomials

A better approach to minimizing $E(p; f)$ requires the introduction of a special set of polynomials, the *Legendre polynomials*. They are defined as follows:

$$\begin{aligned} P_0(x) &= 1 \\ P_n(x) &= \frac{1}{n!2^n} \cdot \frac{d^n}{dx^n} \left[\left(x^2 - 1\right)^n \right], \qquad n = 1, 2, \ldots \end{aligned} \tag{4.114}$$

For example,

$$\begin{aligned} P_1(x) &= x \\ P_2(x) &= \tfrac{1}{2}\left(3x^2 - 1\right) \\ P_3(x) &= \tfrac{1}{2}\left(5x^3 - 3x\right) \\ P_4(x) &= \tfrac{1}{8}\left(35x^4 - 30x^2 + 3\right) \end{aligned} \tag{4.115}$$

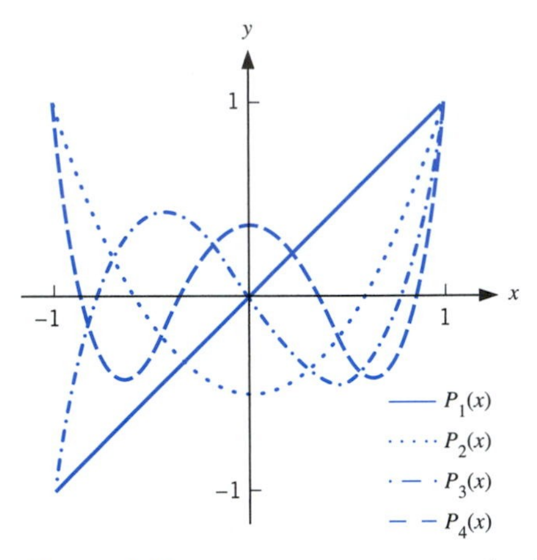

Figure 4.19. Legendre polynomials of degrees $1, 2, 3, 4$

Graphs of these are given in Figure 4.19. As we will see later, when a polynomial is expressed in terms of the Legendre polynomials, in contrast to the formula (4.111), the coefficients of the expression for the least squares approximation are directly determined.

The Legendre polynomials have many special properties, and they are widely used in numerical analysis and applied mathematics. We give some properties, but refer to Atkinson (1989, Chap. 4) for proofs. We first introduce the special notation

$$(f, g) = \int_a^b f(x)g(x)\,dx \tag{4.116}$$

for general functions $f(x)$ and $g(x)$.

Properties:

- Degree and normalization:

$$\deg P_n = n, \qquad P_n(1) = 1, \qquad n \geq 0$$

- Triple recursion relation:

$$P_{n+1}(x) = \frac{2n+1}{n+1} x P_n(x) - \frac{n}{n+1} P_{n-1}(x), \qquad n \geq 1 \tag{4.117}$$

- Orthogonality and size:

$$(P_i, P_j) = \begin{cases} 0, & i \neq j \\ \dfrac{2}{2j+1}, & i = j \end{cases} \tag{4.118}$$

- Zeros:

$$\text{All zeros of } P_n(x) \text{ are located in the interval } -1 < x < 1, \text{ and moreover, all zeros are simple roots of } P_n(x). \tag{4.119}$$

- Basis: Every polynomial $p(x)$ of degree $\leq n$ can be written in the form

$$p(x) = \sum_{j=0}^{n} \beta_j P_j(x) \tag{4.120}$$

with the choice of $\beta_0, \beta_1, \ldots, \beta_n$ uniquely determined from $p(x)$.

4.7.2 Solving for the Least Squares Approximation

We will solve the least squares approximation problem on only the interval $[-1, 1]$. Approximation problems on other intervals $[a, b]$ can be accomplished using a linear change of variable, as is discussed in Problem 4 of Section 4.6. Note that by using the definition in (4.116), the quantity in (4.107) that we are seeking to minimize can be written as

$$(f - p, f - p) \tag{4.121}$$

We begin by writing $p(x)$ in the form of (4.120). We substitute this into (4.121), obtaining

$$\widetilde{g}(\beta_0, \beta_1, \ldots, \beta_n) \equiv (f - p, f - p) = \left(f - \sum_{j=0}^{n} \beta_j P_j, f - \sum_{i=0}^{n} \beta_i P_i \right)$$

Using the property (4.118), we can expand this into the following:

$$\widetilde{g} = (f, f) - \sum_{j=0}^{n} \frac{(f, P_j)^2}{(P_j, P_j)} + \sum_{j=0}^{n} (P_j, P_j) \left[\beta_j - \frac{(f, P_j)}{(P_j, P_j)} \right]^2$$

Looking at this carefully, we see that it is smallest when

$$\beta_j = \frac{(f, P_j)}{(P_j, P_j)}, \qquad j = 0, 1, \ldots, n \tag{4.122}$$

and then the minimum for this choice of coefficients is

$$\widetilde{g} = (f, f) - \sum_{j=0}^{n} \frac{(f, P_j)^2}{(P_j, P_j)}$$

We call

$$\ell_n(x) = \sum_{j=0}^{n} \frac{(f, P_j)}{(P_j, P_j)} P_j(x) \tag{4.123}$$

the *least squares approximation of degree n* to $f(x)$ on $[-1, 1]$.

Example 4.7.2 We continue the preceding Example 4.7.1 in which $f(x) = e^x$ on $[-1, 1]$. We use (4.123) with $n = 3$. The coefficients $\{\beta_0, \beta_1, \beta_2, \beta_3\}$ are given in Table 4.10.

When combined with the formulas given in (4.115), we obtain

$$\ell_3(x) = 0.996294 + 0.997955x + 0.536722x^2 + 0.176139x^3$$

The graph of $e^x - \ell_3(x)$ is shown in Figure 4.20, and one can see that

$$\max_{-1 \le x \le 1} \left| e^x - \ell_3(x) \right| \doteq 0.0112$$

Table 4.10. Coefficients β_j for Cubic Least Squares Approximation to e^x

j	0	1	2	3
β_j	1.17520	1.10364	0.35781	0.07046

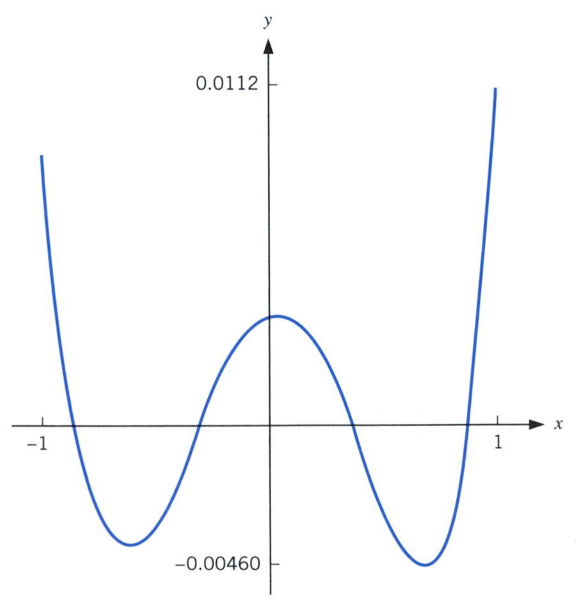

Figure 4.20. The error in the cubic least squares approximation to e^x

Table 4.11. Errors in Cubic Approximations of e^x

Approximation	Maximum Error	Root-Mean-Square-Error
Taylor $t_3(x)$	0.0516	0.0145
Least squares $\ell_3(x)$	0.0112	0.00334
Chebyshev $c_3(x)$	0.00666	0.00384
Minimax $m_3(x)$	0.00553	0.00388

We compare four forms of cubic approximation to e^x that have been studied in this chapter, giving the maximum errors and root-mean-square-errors for each in Table 4.11. ■

4.7.3 Generalizations of Least Squares Approximation

For various reasons, we generalize the concept of "average error" by considering

$$E(p; f) \equiv \sqrt{\frac{1}{c}\int_a^b w(x)\,[f(x) - p(x)]^2\,dx}, \qquad c = \int_a^b w(x)\,dx \tag{4.124}$$

and we seek to minimize this quantity. This is called *weighted least squares approximation*, with $w(x)$ called the *weight function*. The function $w(x)$ is assumed to satisfy the following assumptions:

A1. $w(x) > 0$ for $a < x < b$;

A2. For all integers $n \geq 0$,

$$\int_a^b w(x)\,|x|^n\,dx < \infty$$

With these assumptions, we can quickly generalize the development given above that used Legendre polynomials, provided we generalize the notation (4.116) to

$$(f, g)_w \equiv \int_a^b w(x) f(x) g(x)\,dx \tag{4.125}$$

There are generalizations of the Legendre polynomials that satisfy the crucial orthogonality property (4.118), and we give some examples in Problems 8 and 9. There is a general theory for such orthogonal polynomials; for an introduction, see Atkinson (1989, Chapter 4).

PROBLEMS

1. Find the linear least squares approximation to $f(x) = e^x$ on the interval [0, 1]. Use the direct method of Example 4.7.1, but on the interval [0, 1].

2. Find the linear least squares approximation to $f(x) = \sin x$ on the interval $[0, \frac{1}{2}\pi]$. Use the direct method of Example 4.7.1, but on the interval $[0, \frac{1}{2}\pi]$.

3. For the following functions on $[-1, 1]$, find their least squares approximations of degree $n \leq 4$, written in the form (4.123). The coefficients (f, P_j) can be found by either numerical integration (see Chapter 5) or by using a symbolic integrator such as *MAPLE* or *MATHEMATICA*. Also graph the errors $f(x) - \ell_n(x)$ on $[-1, 1]$.

   (a) $f(x) = \sin(\pi x)$;

   (b) $f(x) = \log\left(1 + x^2\right)$;

   (c) $f(x) = \tan^{-1} x$;

   (d) $f(x) = e^t,\ t = \frac{1}{2}(x + 1)$;

   (e) $f(x) = \cos\left[\frac{1}{4}\pi\,(x + 1)\right]$.

4. Show (4.118) for $0 \leq i, j \leq 2$.

5. Show (4.119) for $n = 1, 2, 3$.

6. Show that for each $n \geq 0$,

$$x^n = \sum_{j=0}^{n} c_{j,n} P_j(x) \tag{4.126}$$

   for appropriate choices of coefficients $\{c_{j,n}\}$. Begin by showing the cases of $n = 0$ and $n = 1$, obtaining the needed coefficients $\{c_{j,n}\}$. For general $n \geq 2$, note that you can solve for x^n in the formula (4.114), obtaining a formula involving $P_n(x)$ and $\{1, x, x^2, \ldots, x^{n-1}\}$. Then show that a recursive algorithm can be devised to obtain (4.126). For example, with $n = 2$, solve for x^2 in terms of $P_2(x)$, 1, and x. Use the formula (4.126) for the cases of $n = 0, 1$ to show (4.126) for $n = 2$. Repeat the process for $n = 3$, solving first for x^3 in terms of $P_3(x)$, 1, x, and x^2. Then use the formula (4.126) for the cases of $n = 0, 1, 2$ to show (4.126) for $n = 3$. Also, see the closely related Problem 9 of Section 4.5.

7. Using the results of Problem 6, show the basis property (4.120) for $n = 2$. Find the coefficients $\{\beta_j\}$ in (4.120) from the coefficients $\{a_j\}$ of $p(x)$ written in the standard form

$$p(x) = a_0 + a_1 x + a_2 x^2$$

   Also do the same procedure to obtain (4.120) for $n = 3$.

8. Let $w(x) = 1/\sqrt{1 - x^2}$ for $-1 < x < 1$. Show that the Chebyshev polynomials $\{T_k(x) : k \geq 0\}$ satisfy

$$\left(T_i, T_j\right)_w = 0 \quad \text{for} \quad i \neq j$$

   What is $\left(T_j, T_j\right)_w$ in this case?

9. Recall the definition of Chebyshev polynomials of the second kind, $S_n(x)$, from Problem 8, Section 4.5. Let $w(x) = \sqrt{1 - x^2}$ for $-1 \leq x \leq 1$. Show the orthogonality property

$$\left(S_i, S_j\right)_w = 0 \quad \text{for} \quad i \neq j$$

10. Repeat Problem 6, but do so with the Chebyshev polynomials of the second kind $\{S_k(x)\}_{k \geq 0}$ replacing the Legendre polynomials $\{P_k(x)\}_{k \geq 0}$.

CHAPTER 5

NUMERICAL INTEGRATION AND DIFFERENTIATION

The definite integral

$$I(f) = \int_a^b f(x)\,dx \tag{5.1}$$

is defined in calculus as a limit of what are called *Riemann sums*. It is then proved that

$$I(f) = F(b) - F(a) \tag{5.2}$$

where $F(x)$ is any antiderivative of $f(x)$; this is the *fundamental theorem of calculus*. Many integrals can be evaluated by using this formula, and a significant portion of most calculus textbooks is devoted to this approach. Nonetheless, most integrals cannot be evaluated by using (5.2) because most integrands $f(x)$ do not have antiderivatives expressible in terms of elementary functions. Examples of such integrals are

$$\int_0^1 e^{-x^2}\,dx, \qquad \int_0^{\pi} x^{\pi} \sin\left(\sqrt{x}\right) dx \tag{5.3}$$

Other methods are needed for evaluating such integrals.

In the first section of this chapter, we define two of the oldest and most popular numerical methods for approximating (5.1): the trapezoidal rule and Simpson's rule. Section 5.2 analyzes the error in using these methods and then obtains improvements on them. Section 5.3 gives another approach to numerical integration: Gaussian quadrature. It is more complicated in its origins than the Simpson and trapezoidal rules, but it is almost always much superior in accuracy for similar amounts of computation. The last section discusses numerical differentiation. Some simple methods are derived, and their sensitivity to rounding errors is analyzed.

5.1. THE TRAPEZOIDAL AND SIMPSON RULES

The central idea behind most formulas for approximating

$$I(f) = \int_a^b f(x)\,dx$$

is to replace $f(x)$ by an approximating function whose integral can be evaluated. In this section, we look at methods based on using both linear and quadratic interpolation.

Approximate $f(x)$ by the linear polynomial

$$P_1(x) = \frac{(b-x)f(a) + (x-a)f(b)}{b-a}$$

which interpolates $f(x)$ at a and b (see Figure 5.1). The integral of $P_1(x)$ over $[a, b]$ is the area of the shaded trapezoid shown in Figure 5.1; it is given by

$$T_1(f) = (b-a)\left[\frac{f(a)+f(b)}{2}\right] \tag{5.4}$$

This approximates the integral $I(f)$ if $f(x)$ is almost linear on $[a, b]$.

Example 5.1.1 Approximate the integral

$$I = \int_0^1 \frac{dx}{1+x} \tag{5.5}$$

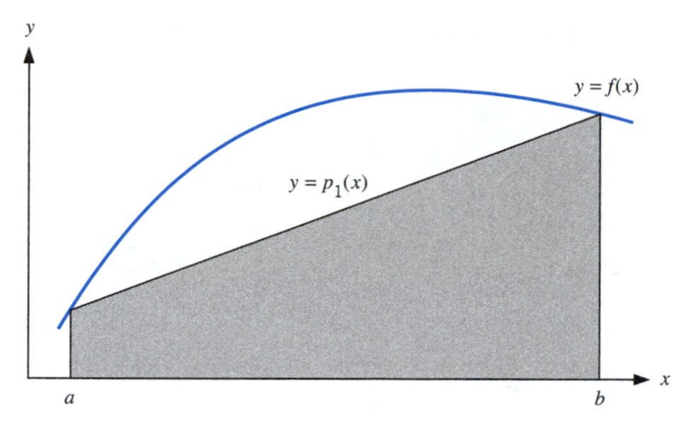

Figure 5.1. An illustration of the trapezoidal rule (5.4)

The true value is $I = \log(2) \doteq 0.693147$. Using (5.4), we obtain

$$T_1 = \frac{1}{2}\left[1 + \frac{1}{2}\right] = \frac{3}{4} = 0.75 \tag{5.6}$$

This is in error by

$$I - T_1 \doteq -0.0569 \quad \blacksquare \tag{5.7}$$

To improve on the approximation $T_1(f)$ in (5.4) when $f(x)$ is not a nearly linear function on $[a, b]$, break the interval $[a, b]$ into smaller subintervals and apply (5.4) on each subinterval. If the subintervals are small enough, then $f(x)$ will be nearly linear on each one. This idea is illustrated in Figure 5.2.

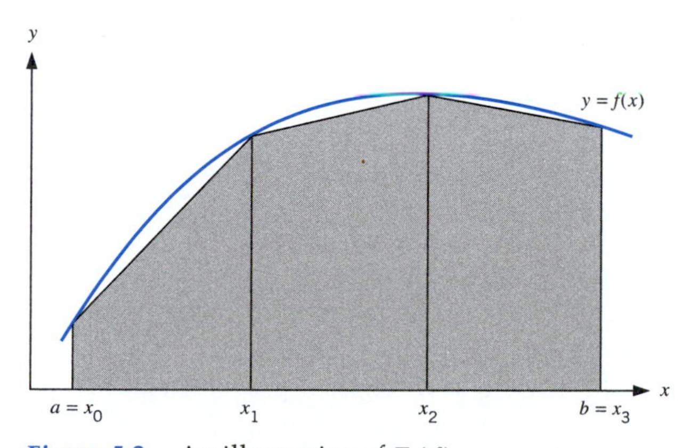

Figure 5.2. An illustration of $T_3(f)$

Example 5.1.2 Evaluate the preceding example by using $T_1(f)$ on two subintervals of equal length. For two subintervals,

$$I = \int_0^{1/2} \frac{dx}{1+x} + \int_{1/2}^{1} \frac{dx}{1+x}$$

$$I \doteq \frac{1}{2}\left[\frac{1+\frac{2}{3}}{2}\right] + \frac{1}{2}\left[\frac{\frac{2}{3}+\frac{1}{2}}{2}\right]$$

$$T_2 = \frac{17}{24} \doteq 0.70833 \tag{5.8}$$

$$I - T_2 \doteq -0.0152 \tag{5.9}$$

The error in T_2 is about $\frac{1}{4}$ of that given for T_1 in (5.7). ■

We will derive a general formula to simplify the calculations when using several subintervals of equal length. Let the number of subintervals be denoted by n, and let

$$h = \frac{b-a}{n}$$

be the length of each subinterval. The endpoints of the subintervals are given by

$$x_j = a + jh, \qquad j = 0, 1, \ldots, n$$

Then break the integral into n subintegrals

$$\begin{aligned} I(f) &= \int_a^b f(x)\,dx = \int_{x_0}^{x_n} f(x)\,dx \\ &= \int_{x_0}^{x_1} f(x)\,dx + \int_{x_1}^{x_2} f(x)\,dx + \cdots + \int_{x_{n-1}}^{x_n} f(x)\,dx \end{aligned} \tag{5.10}$$

Approximate each subintegral by using (5.4), noting that each subinterval $[x_{j-1}, x_j]$ has length h. Then

$$\begin{aligned} I(f) \approx h\left[\frac{f(x_0)+f(x_1)}{2}\right] + h\left[\frac{f(x_1)+f(x_2)}{2}\right] \\ + h\left[\frac{f(x_2)+f(x_3)}{2}\right] + \cdots + h\left[\frac{f(x_{n-1})+f(x_n)}{2}\right] \end{aligned}$$

The terms on the right can be combined to give the simpler formula

$$T_n(f) = h\left[\tfrac{1}{2}f(x_0) + f(x_1) + f(x_2) + \cdots + f(x_{n-1}) + \tfrac{1}{2}f(x_n)\right] \tag{5.11}$$

This is called the *trapezoidal numerical integration rule*. The subscript n gives the number of subintervals being used; and the points $x_0, x_1, \ldots, x_n$ are called the numerical integration *node points*.

Before giving some numerical examples of $T_n(f)$, we would like to discuss the choice of n. With a sequence of increasing values of n, $T_n(f)$ will usually be an increasingly accurate approximation of $I(f)$. But which sequence of values of n should be used? If n is doubled repeatedly, then the function values used in each $T_{2n}(f)$ will include all of the earlier function values used in the preceding $T_n(f)$. Thus, the doubling of n will ensure that all previously computed information is used in the new calculation, making the trapezoidal rule less expensive than it would be otherwise. To illustrate how function values are reused when n is doubled, consider $T_2(f)$ and $T_4(f)$.

$$T_2(f) = h\left[\frac{f(x_0)}{2} + f(x_1) + \frac{f(x_2)}{2}\right] \tag{5.12}$$

with

$$h = \frac{b-a}{2}, \qquad x_0 = a, \qquad x_1 = \frac{a+b}{2}, \qquad x_2 = b$$

Also,

$$T_4(f) = h\left[\frac{f(x_0)}{2} + f(x_1) + f(x_2) + f(x_3) + \frac{f(x_4)}{2}\right] \tag{5.13}$$

with

$$h = \frac{b-a}{4}, \qquad x_0 = a, \qquad x_1 = \frac{3a+b}{4}, \qquad x_2 = \frac{a+b}{2},$$

$$x_3 = \frac{a+3b}{4}, \qquad x_4 = b$$

In (5.13), only $f(x_1)$ and $f(x_3)$ need to be evaluated, as the other function values are known from (5.12). For this and other reasons, all of our examples of $T_n(f)$ are based on doubling n.

Example 5.1.3 We give calculations of $T_n(f)$ for three integrals:

$$I^{(1)} = \int_0^1 e^{-x^2}\,dx \doteq 0.746824132812427 \tag{5.14}$$

$$I^{(2)} = \int_0^4 \frac{dx}{1+x^2} = \tan^{-1}(4) \doteq 1.32581766366803 \tag{5.15}$$

$$I^{(3)} = \int_0^{2\pi} \frac{dx}{2+\cos(x)} = \frac{2\pi}{\sqrt{3}} \doteq 3.62759872846844 \tag{5.16}$$

Table 5.1. Examples of the Trapezoidal Rule

n	$I^{(1)}$		$I^{(2)}$		$I^{(3)}$	
	Error	Ratio	Error	Ratio	Error	Ratio
2	1.55E − 2		−1.33E − 1		−5.61E − 1	
4	3.84E − 3	4.02	−3.59E − 3	37.0	−3.76E − 2	14.9
8	9.59E − 4	4.01	5.64E − 4	−6.37	−1.93E − 4	195.0
16	2.40E − 4	4.00	1.44E − 4	3.92	−5.19E − 9	37,600.0
32	5.99E − 5	4.00	3.60E − 5	4.00	*	
64	1.50E − 5	4.00	9.01E − 6	4.00	*	
128	3.74E − 6	4.00	2.25E − 6	4.00	*	

The results are shown in Table 5.1. Only the errors $I(f) - T_n(f)$ are given, since this is the main quantity of interest in considering the speed with which $T_n(f)$ approaches $I(f)$. The column labeled "Ratio" gives the ratio of successive errors, the factor by which the error decreases when n is doubled.

From the table, the error in calculating $I^{(1)}$ and $I^{(2)}$ decreases by a factor of about 4 when n is doubled. The third example, of $I^{(3)}$, converges much more rapidly. The answers for $n = 32, 64$, and 128 were correct up to the limits due to rounding error on the computer (about 16 decimal digits), and this is denoted by $*$ in the table. An explanation of all of these results will be given in the next section. ■

MATLAB PROGRAM. We give a program for the trapezoidal rule. We calculate $T_n(f)$ for $n = n_0, 2n_0, 4n_0, \ldots, 256n_0$, with n_0 supplied by the user. When n is doubled to $2n$, all of the function values occurring in $T_n(f)$ are also used in computing $T_{2n}(f)$. We also allow for a variety of integrands in the function $f(x)$, with the user specifying the integrand through the subprogram f. The program comments explain the organization of the program.

```
function [integral,difference,ratio]=trapezoidal(a,b,n0,index_f)
%
% This uses the trapezoidal rule with n subdivisions to
% integrate the function f over the interval [a,b].  The
% values of n used are
%   n = n0,2*n0,4*n0,...,256*n0
% The value of n0 must be a positive integer.
% The corresponding numerical integrals are returned in the
% vector integral.  The differences of successive numerical
% integrals are returned in the vector difference:
%   difference(i) = integral(i)-integral(i-1), i=2,...,9
% The entries in ratio give the rate of decrease in these
% differences.
```

```
%
% In using this program, define the integrand using the
% function given below.  The parameter index_f allows the
% user to do calculations with multiple integrands.

% Initialize output vectors.
integral = zeros(9,1);
difference = zeros(9,1);
ratio = zeros(9,1);

% Initialize for trapezoidal rule.
sumend = (f(a,index_f) +f(b,index_f))/2;
sum = 0;

% Initialize for case of n0 > 2.
if(n0 > 2)
  h = (b-a)/n0;
  for i=2:2:n0-2
    sum = sum + f(a+i*h,index_f);
  end
end

% Calculate the numerical integrals, doing each
% by appropriately modifying the preceding case.
for i=1:9
  n = n0*2^(i-1);
  h = (b-a)/n;
  for k=1:2:n-1
    sum = sum + f(a+k*h,index_f);
  end
  integral(i) = h*(sumend + sum);
end

% Calculate the differences of the successive
% trapezoidal rule integrals and the ratio
% of decrease in these differences.
difference(2:9) = integral(2:9)-integral(1:8);
ratio(3:9) = difference(2:8)./difference(3:9);

function f_value = f(x,index)
%
% This defines the integrand.

switch index
case 1
```

```
 f_value = exp(-x.^2);
case 2
 f_value = 1 ./(1+x.^2);
case 3
 f_value = 1 ./(2+cos(x));
end
```

5.1.1 Simpson's Rule

To improve on $T_1(f)$ in (5.4), use quadratic interpolation to approximate $f(x)$ on $[a, b]$. Let $P_2(x)$ be the quadratic polynomial that interpolates $f(x)$ at a, $c = (a + b)/2$ and b. Using this to approximate $I(f)$, we get

$$I(f) \approx \int_a^b P_2(x)\,dx = \int_a^b \left[\frac{(x-c)(x-b)}{(a-c)(a-b)} f(a) + \frac{(x-a)(x-b)}{(c-a)(c-b)} f(c) + \frac{(x-a)(x-c)}{(b-a)(b-c)} f(b) \right] dx \tag{5.17}$$

This integral can be evaluated directly, but it is easier to first introduce $h = (b - a)/2$ and then to change the variable of integration. We will evaluate the first term to illustrate the general procedure. Let $u = x - a$. Then

$$\begin{aligned}
\int_a^b \frac{(x-c)(x-b)}{(a-c)(a-b)}\,dx &= \frac{1}{2h^2} \int_a^{a+2h} (x-c)(x-b)\,dx \\
&= \frac{1}{2h^2} \int_0^{2h} (u-h)(u-2h)\,du \\
&= \frac{1}{2h^2} \left[\frac{u^3}{3} - \frac{3}{2}u^2 h + 2h^2 u \right]_0^{2h} = \frac{h}{3}
\end{aligned}$$

The complete evaluation of (5.17) yields

$$S_2(f) = \frac{h}{3} \left[f(a) + 4f\left(\frac{a+b}{2}\right) + f(b) \right] \tag{5.18}$$

The method is illustrated in Figure 5.3.

Example 5.1.4 Use the earlier integral from (5.5)

$$I = \int_0^1 \frac{dx}{1+x}$$

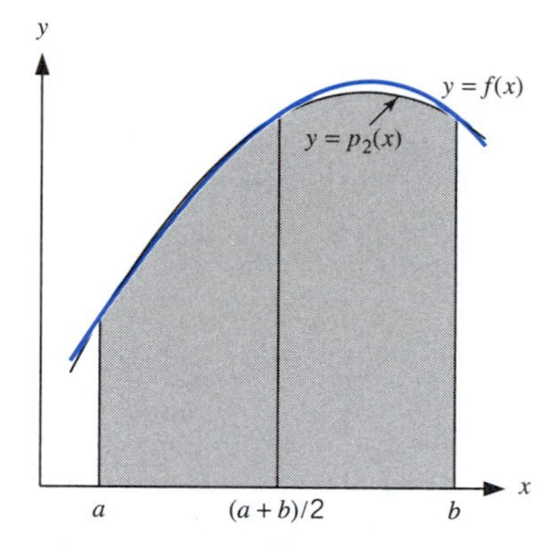

Figure 5.3. An illustration of Simpson's rule (5.18)

Then $h = (b - a)/2 = \frac{1}{2}$, and

$$S_2 = \frac{1/2}{3}\left[1 + 4\left(\frac{2}{3}\right) + \frac{1}{2}\right] = \frac{25}{36} \doteq 0.69444 \tag{5.19}$$

The error is

$$I - S_2 = \log(2) - S_2 \doteq -0.00130 \tag{5.20}$$

To compare this with the trapezoidal rule, use T_2 from (5.8), since the number of function evaluations is the same for both S_2 and T_2. The error in S_2 is smaller than that in (5.9) for T_2 by a factor of about 12, a significant increase in accuracy. ■

The rule $S_2(f)$ will be an accurate approximation to $I(f)$ if $f(x)$ is nearly quadratic on $[a, b]$. For other cases, proceed in the same manner as for the trapezoidal rule. Let n be an even integer, $h = (b - a)/n$, and define the evaluation points for $f(x)$ by

$$x_j = a + jh, \qquad j = 0, 1, \ldots, n \tag{5.21}$$

Follow the idea of (5.10), but break $[a, b] = [x_0, x_n]$ into larger subintervals, each containing three interpolation node points. Thus,

$$\begin{aligned} I(f) &= \int_a^b f(x)\,dx = \int_{x_0}^{x_n} f(x)\,dx \\ &= \int_{x_0}^{x_2} f(x)\,dx + \int_{x_2}^{x_4} f(x)\,dx + \cdots + \int_{x_{n-2}}^{x_n} f(x)\,dx \end{aligned}$$

Table 5.2. Examples of the Simpson Rule

n	$I^{(1)}$		$I^{(2)}$		$I^{(3)}$	
	Error	Ratio	Error	Ratio	Error	Ratio
2	−3.56E − 4		8.66E − 2		−1.26	
4	−3.12E − 5	11.4	3.95E − 2	2.2	1.37E − 1	−9.2
8	−1.99E − 6	15.7	1.95E − 3	20.3	1.23E − 2	11.2
16	−1.25E − 7	15.9	4.02E − 6	485.0	6.43E − 5	191.0
32	−7.79E − 9	16.0	2.33E − 8	172.0	1.71E − 9	37,600.0
64	−4.87E − 10	16.0	1.46E − 9	16.0	*	
128	−3.04E − 11	16.0	9.15E − 11	16.0	*	

Approximate each subintegral by (5.18). This yields

$$I(f) \approx \frac{h}{3}[f(x_0)+4f(x_1)+f(x_2)] + \frac{h}{3}[f(x_2)+4f(x_3)+f(x_4)] + \cdots + \frac{h}{3}[f(x_{n-2})+4f(x_{n-1})+f(x_n)]$$

If these terms are combined and simplified, we obtain the formula

$$S_n(f) = \frac{h}{3}[f(x_0)+4f(x_1)+2f(x_2)+4f(x_3)+2f(x_4)+\cdots + 2f(x_{n-2})+4f(x_{n-1})+f(x_n)] \tag{5.22}$$

This is called *Simpson's rule*, and it has been among the most popular numerical integration methods for more than two centuries. The index n gives the number of subdivisions used in defining the integration node points $x_0, \ldots, x_n$.

Example 5.1.5 We evaluate the integrals (5.14) to (5.16), which were used previously to illustrate the trapezoidal rule. The results of using Simpson's rule are given in Table 5.2. For integrals $I^{(1)}$ and $I^{(2)}$, the ratio by which the error decreases approaches 16. For integral $I^{(3)}$, the errors converge to zero much more rapidly. An explanation of these results is given in the next section. ■

MATLAB PROGRAM. We give a program for Simpson's rule. We calculate $S_n(f)$ for $n = n_0, 2n_0, 4n_0, \ldots, 256n_0$, with n_0 supplied by the user. When n is doubled to $2n$, all of the function values occurring in $S_n(f)$ are also used in computing $S_{2n}(f)$. We also allow for a variety of integrands in the function $f(x)$, with the user specifying the integrand through the subprogram f. The program comments explain the organization of the program.

```
function [integral,difference,ratio]=simpson(a,b,n0,index_f)
%
% This uses Simpson's rule with n subdivisions to integrate the
% function f over the interval [a,b].  The values of n used are
%   n = n0,2*n0,4*n0,...,256*n0
% The value of n0 MUST be a positive even integer.
% The corresponding numerical integrals are returned in the
% vector integral.  The differences of successive numerical
% integrals are returned in the vector difference:
%   difference(i) = integral(i)-integral(i-1), i=2,...,9
% The entries in ratio give the rate of decrease in these
% differences.
%
% In using this program, define the integrand using the
% function given below.  The parameter index_f allows the
% user to do calculations with multiple integrands.

% Initialize output vectors.
integral = zeros(9,1);
difference = zeros(9,1);
ratio = zeros(9,1);

% Initialize for Simpson integration.
sumend = f(a,index_f) + f(b,index_f);
sumodd = 0;
sumeven = 0;

% Initialize for case of n0 > 2.
if(n0 > 2)
  h = (b-a)/n0;
  for i=2:2:n0-2
    sumeven = sumeven + f(a+i*h,index_f);
  end
end

% Calculate the numerical integrals, doing each
% by appropriately modifying the preceding case.
for i=1:9
  n = (n0)*(2^(i-1));
  h = (b-a)/n;
  sumeven = sumeven + sumodd;
  sumodd = 0;
  for k=1:2:n-1
    sumodd = sumodd + f(a+k*h,index_f);
  end
```

```
  integral(i) = h*(sumend + 4*sumodd + 2*sumeven)/3;
end

% Calculate the differences of the successive
% Simpson rule integrals and the ratio
% of decrease in these differences.
difference(2:9) = integral(2:9)-integral(1:8);
ratio(3:9) = difference(2:8)./difference(3:9);

function f_value = f(x,index)
%
% This defines the integrand.

switch index
case 1
 f_value = exp(-x.^2);
case 2
 f_value = 1 ./(1+x.^2);
case 3
 f_value = 1 ./(2+cos(x));
end
```

PROBLEMS

1. Compute $T_4(f)$ and $S_4(f)$ for the integral I in (5.5). Compute the errors $I - T_4$ and $I - S_4$; compare them to the errors $I - T_2$ and $I - S_2$, respectively.

2. Using the program for the trapezoidal rule given in the text, prepare a table of values of $T_n(f)$ for $n = 2, 4, 8, \ldots, 512$ for the following integrals. Also find the errors and the ratios by which the errors decrease.

   (a) $\displaystyle\int_0^{\pi} e^x \cos(4x)\,dx = \frac{e^{\pi} - 1}{17}$

   (b) $\displaystyle\int_0^{1} x^{5/2}\,dx = \tfrac{2}{7}$

   (c) $\displaystyle\int_0^{5} \frac{dx}{1 + (x - \pi)^2} = \tan^{-1}(5 - \pi) + \tan^{-1}(\pi)$

   (d) $\displaystyle\int_{-\pi}^{\pi} e^{\cos(x)}\,dx \doteq 7.95492652101284$

   (e) $\displaystyle\int_0^{\pi/4} e^{\cos(x)}\,dx \doteq 1.93973485062365$

   (f) $\displaystyle\int_0^{1} \sqrt{x}\,dx = \tfrac{2}{3}$

3. Repeat Problem 2 using Simpson's rule.

4. Use the trapezoidal rule and Simpson's rule with $n = 4, 8, \ldots, 512$ to find approximate values of the area under the curve of $y = f(x)$ for the following functions $f(x)$ on the given intervals:

(a) $f(x) = e^{-x^2}, \quad 0 \le x \le 10$

(b) $f(x) = \tan^{-1}\left(1 + x^2\right), \quad 0 \le x \le 2$

(c) $f(x) = \sqrt{x}e^x, \quad 0 \le x \le 1$

5. Recall that the length of the curve represented by a function $y = f(x)$ on an interval $[a, b]$ is given by the integral

$$I(f) = \int_a^b \sqrt{1 + [f'(x)]^2}\, dx$$

Use the trapezoidal rule and Simpson's rule with $n = 4, 8, \ldots, 512$ to compute the lengths of the following curves:

(a) $f(x) = \sin(\pi x), \quad 0 \le x \le 1$

(b) $f(x) = e^x, \quad 0 \le x \le 1$

(c) $f(x) = e^{x^2}, \quad 0 \le x \le 1$

6. Experiment with computing numerically the integral

$$I = \int_0^b \frac{dx}{2 + \sin x} = \frac{2\sqrt{3}}{3}\left(\tan^{-1}\left(\frac{\sqrt{3}}{3}\left(2\tan\frac{b}{2} + 1\right)\right)\right) - \frac{\pi\sqrt{3}}{9}$$

for various intervals $[0, b]$. In particular, use $b = \frac{1}{2}\pi, \pi, 2\pi, 10$. Do so with both the trapezoidal rule T_n and Simpson's rule S_n, and use a number of values of n. This integral can be evaluated analytically, although some caution is needed in using the most common formulas; the above formula is accurate for only $-\frac{1}{2}\pi < b < \frac{1}{2}\pi$. Comment on your numerical results, including the rate of convergence of the numerical methods being used.

7. Calculate the last two integrals in (5.17), verifying (5.18).

8. **(a)** As another approximation to $I(f) = \int_a^b f(x)\, dx$, replace $f(x)$ by the constant $f[(a + b)/2]$ on the entire interval $a \le x \le b$. Show that this leads to the numerical integration formula

$$M_1(f) = (b - a) f\left(\frac{a + b}{2}\right)$$

Graphically illustrate this approximation.

(b) In analogy with the derivation of the trapezoidal rule (5.11) and Simpson's rule (5.22), generalize (a) to the numerical integration formula

$$M_n(f) = h[f(x_1) + f(x_2) + \cdots + f(x_n)]$$

where $h = (b - a)/n$ and

$$x_j = a + \left(j - \tfrac{1}{2}\right) h, \qquad j = 1, \ldots, n$$

This is called the *midpoint rule;* it is a popular alternative to the trapezoidal rule.

(c) For the integral, $I = \int_0^1 dx/(1 + x)$ of (5.5), calculate $M_1(f)$ and $M_2(f)$. Compare the errors to those for T_1 and T_2, given in (5.7) and (5.9).

9. Repeat Problem 2 using the midpoint rule.

10. **(a)** As another approximation to $I(f) = \int_a^b f(x)\, dx$, replace $f(x)$ by the degree 4 polynomial $P_4(x)$ interpolating $f(x)$ at the five points

$$x_j = a + jh, \qquad j = 0, 1, 2, 3, 4, \qquad h = \frac{b - a}{4}$$

Show that this leads to the approximating formula

$$B_4(f) = \frac{2h}{45}[7f(x_0) + 32f(x_1) + 12f(x_2) + 32f(x_3) + 7f(x_4)]$$

This is called *Boole's rule*, and it can be generalized to a larger number of subintervals in the same manner as was done for the Simpson and trapezoidal rules.

(b) Compute $B_4(f)$ for $I = \int_0^1 dx/(1 + x)$, and compare the results to those obtained in Problem 1.

11. The *degree of precision* of a numerical integration formula is defined as follows: If the formula has zero error when integrating any polynomial of degree $\leq r$, and if the error is nonzero for some polynomial of degree $r + 1$, then we say the formula has degree of precision equal to r. Show that the following rules have the indicated degree of precision r:

(a) $M_1(f) : r = 1$ (see Problem 8 for definition of $M_1(f)$)

(b) $T_1(f) : r = 1$

(c) $S_2(f) : r = 3$

(d) $B_4(f) : r = 5$ (see Problem 10 for definition of $B_4(f)$)

Are these results still valid when the various subscripts are replaced by n?

Hint: In your work, consider only $I = \int_0^b f(x)\,dx$, for suitable b, and $f(x) = 1, x, x^2$, etc.

12. Determine the degree of precision of the approximation

$$\int_0^1 f(x)\,dx \approx \frac{1}{4}f(0) + \frac{3}{4}f\left(\frac{2}{3}\right)$$

13. Let

$$I_h(f) = \frac{3h}{4}[f(0) + 3f(2h)]$$

What is the degree of precision of the approximation $I_h(f) \approx \int_0^{3h} f(x)\,dx$?
Hint: Consider $f(x) = 1, x, x^2, x^3$, etc.

14. Approximate $I(f) = \int_{-1}^{1} f(x)\,dx$ by replacing $f(x)$ with $P_1(x)$, the linear interpolant to $f(x)$ at $x = -\frac{1}{3}$ and $x = \frac{1}{3}$. Give the resulting numerical integration formula. What is its degree of precision?

15. Consider the approximation

$$I(f) = \int_{-1}^{1} f(x)\,dx \approx f(-\beta) + f(\beta)$$

for some β satisfying $0 < \beta \leq 1$. Show it has degree of precision greater than or equal to 1 for any such choice of β. Choose β to obtain a formula with degree of precision greater than 1. What is the degree of precision of this formula?

16. Approximate $I(f) = \int_0^{2h} f(x)\,dx$ by replacing $f(x)$ with $P_1(x)$, the linear interpolant to $f(x)$ at $x = 0$ and $x = h$. Give the resulting numerical integration formula. What is its degree of precision?

5.2. ERROR FORMULAS

In the preceding section, numerical results for all integrands but one showed a regular behavior in the error for both the trapezoidal and Simpson rules. To explain this regular behavior, we consider error formulas for these integration methods. These formulas will lead to a better understanding of the methods, showing both their weaknesses and strengths, and they will allow improvements of the methods. We begin by examining the error for the trapezoidal rule.

Theorem 5.2.1 Let $f(x)$ have two continuous derivatives on $[a, b]$, and let n be a positive integer. Then for the error in integrating

$$I(f) = \int_a^b f(x)\,dx$$

using the trapezoidal rule $T_n(f)$ of (5.11), we have

$$E_n^T(f) \equiv I(f) - T_n(f) = \frac{-h^2(b-a)}{12} f''(c_n) \tag{5.23}$$

The number c_n is some unknown point in $[a, b]$, and $h = (b-a)/n$.

The proof is omitted here, although part of it is given later. The formula (5.23) can be used to bound the error in $T_n(f)$, generally by bounding the term $|f''(c_n)|$ by its largest possible value on the interval $[a, b]$. This will be illustrated in the following example. Also note that the formula for $E_n^T(f)$ is consistent with the behavior of the errors observed in Table 5.1 for the integrals $I^{(1)}$ and $I^{(2)}$. When n is doubled, h is halved, and the term h^2 decreases by a factor of 4. This is exactly the factor observed in Table 5.1 for the decrease in the trapezoidal error.

Example 5.2.2 Recall the examples involving (5.5) in the preceding section, with

$$I = \int_0^1 \frac{dx}{1+x} = \log(2)$$

Here $f(x) = 1/(1+x)$, $[a, b] = [0, 1]$, and $f''(x) = 2/(1+x)^3$. Substituting into (5.23), we obtain

$$E_n^T(f) = -\frac{h^2}{12} f''(c_n), \qquad 0 \le c_n \le 1, \qquad h = \frac{1}{n} \tag{5.24}$$

This formula cannot be computed exactly because c_n is not known. But we can bound the error by looking at the largest possible value for $|f''(c_n)|$. Bound $|f''(x)|$ on $[a, b] = [0, 1]$:

$$\max_{0 \le x \le 1} \frac{2}{(1+x)^3} = 2$$

Then

$$|E_n^T(f)| \le \frac{h^2}{12}(2) = \frac{h^2}{6} \tag{5.25}$$

For $n = 1$ and $n = 2$, we have

$$\left|E_1^T(f)\right| \le \frac{1}{6} \doteq 0.167, \qquad \left|E_2^T(f)\right| \le \frac{(1/2)^2}{6} \doteq 0.0417$$

Comparing these results with the true errors given in (5.7) and (5.9), we see that these bounds are two to three times the actual errors. ■

A possible weakness in the trapezoidal rule can be inferred from the assumptions of Theorem 5.2.1. If $f(x)$ does not have two continuous derivatives on $[a, b]$, then does $T_n(f)$ converge more slowly? The answer is yes for some functions, especially if the first derivative is not continuous. This is explored experimentally in Problem 14.

5.2.1 An Asymptotic Estimate of the Trapezoidal Error

The error formula (5.23) can only be used to bound the error, because $f''(c_n)$ is unknown. This will be improved on by a more careful consideration of the error formula.

A central element of our proof of (5.23) lies in being able to demonstrate the $n = 1$ case for an interval $[\alpha, \alpha + h]$:

$$\int_{\alpha}^{\alpha+h} f(x)\,dx - h\left[\frac{f(\alpha) + f(\alpha + h)}{2}\right] = -\frac{h^3}{12} f''(c) \tag{5.26}$$

for some c in $[\alpha, \alpha + h]$. A short proof of this can be based on the error formula (4.53) for linear interpolation; and another approach that makes use of Taylor polynomial approximations is taken up in Problem 11. Here, we show only how to use (5.26) to obtain the general formula (5.23) in Theorem 5.2.1.

Recall the derivation of the trapezoidal rule $T_n(f)$ as given in and following (5.10) in Section 5.1. Then

$$\begin{aligned} E_n^T(f) &= \int_a^b f(x)\,dx - T_n(f) = \int_{x_0}^{x_n} f(x)\,dx - T_n(f) \\ &= \int_{x_0}^{x_1} f(x)\,dx - h\left[\frac{f(x_0) + f(x_1)}{2}\right] + \int_{x_1}^{x_2} f(x)\,dx - h\left[\frac{f(x_1) + f(x_2)}{2}\right] \\ &\qquad + \cdots + \int_{x_{n-1}}^{x_n} f(x)\,dx - h\left[\frac{f(x_{n-1}) + f(x_n)}{2}\right] \end{aligned} \tag{5.27}$$

Apply (5.26) to each of the terms on the right side of (5.27), to obtain

$$E_n^T(f) = -\frac{h^3}{12} f''(\gamma_1) - \frac{h^3}{12} f''(\gamma_2) - \cdots - \frac{h^3}{12} f''(\gamma_n) \tag{5.28}$$

The unknown constants $\gamma_1, \ldots, \gamma_n$ are located in the respective subintervals

$$[x_0, x_1], [x_1, x_2], \ldots, [x_{n-1}, x_n] \tag{5.29}$$

By factoring (5.28), we obtain

$$E_n^T(f) = -\frac{h^2}{12}[hf''(\gamma_1) + \cdots + hf''(\gamma_n)] \tag{5.30}$$

It is left to Problem 12 to show that the quantity in brackets equals $(b-a)f''(c_n)$ for some c_n in $[a, b]$, thus obtaining the general case of (5.23).

To *estimate* the trapezoidal error, observe that the term in brackets in (5.30) is a Riemann sum for the integral

$$\int_a^b f''(x)\,dx = f'(b) - f'(a) \tag{5.31}$$

The Riemann sum is based on the partition (5.29) of $[a, b]$; as $n \to \infty$, this sum will approach the integral (5.31). Using (5.31) to estimate the right side of (5.30), we find that

$$E_n^T(f) \approx \frac{-h^2}{12}[f'(b) - f'(a)] \tag{5.32}$$

This error estimate will be denoted by $\widetilde{E}_n^T(f)$. It is called an *asymptotic estimate* of the error because it improves as n increases. As long as $f'(x)$ is computable, $\widetilde{E}_n^T(f)$ will be very easy to compute.

Example 5.2.3 Again consider the case $I = \int_0^1 \frac{dx}{1+x}$. Then $f'(x) = -1/(1+x)^2$, and (5.32) yields the estimate

$$\widetilde{E}_n^T(f) = \frac{-h^2}{12}\left[\frac{-1}{(1+1)^2} - \frac{-1}{(1+0)^2}\right] = \frac{-h^2}{16}, \qquad h = \frac{1}{n} \tag{5.33}$$

For $n = 1$ and $n = 2$,

$$\widetilde{E}_1^T(f) = -\frac{1}{16} = -0.0625, \qquad \widetilde{E}_2^T(f) \doteq -0.0156$$

These compare quite closely to the true errors given in (5.7) and (5.9). ■

The estimate $\widetilde{E}_n^T(f)$ has several practical advantages over the earlier error formula (5.23). First, it confirms that when n is doubled (or h is halved), the error decreases by a factor of about 4, *provided* that $f'(b) - f'(a) \neq 0$. This agrees with the results for $I^{(1)}$ and $I^{(2)}$ in Table 5.1. Second, (5.32) implies that the convergence of $T_n(f)$ will be more rapid when $f'(b) - f'(a) = 0$. This is a partial explanation of the very rapid convergence observed with $I^{(3)}$ in Table 5.1; a further discussion is given at the end of

Table 5.3. Example of $CT_n(f)$ and $\tilde{E}_n(f)$

n	$I - T_n(f)$	$\widetilde{E}_n(f)$	$CT_n(f)$	$I - CT_n(f)$	Ratio
2	1.545E − 2	1.533E − 2	0.746698561877	1.26E − 4	
4	3.840E − 3	3.832E − 3	0.746816175313	7.96E − 6	15.8
8	9.585E − 4	9.580E − 4	0.746823634224	4.99E − 7	16.0
16	2.395E − 4	2.395E − 4	0.746824101633	3.12E − 8	16.0
32	5.988E − 5	5.988E − 5	0.746824130863	1.95E − 9	16.0
64	1.497E − 5	1.497E − 5	0.746824132690	2.22E − 10	16.0

this section. Finally, (5.32) leads to a more accurate numerical integration formula by taking $\widetilde{E}_n^T(f)$ into account:

$$I(f) - T_n(f) \approx \frac{-h^2}{12}[f'(b) - f'(a)]$$

$$I(f) \approx T_n(f) - \frac{h^2}{12}[f'(b) - f'(a)] \tag{5.34}$$

This is called the *corrected trapezoidal rule*, and it will be denoted by $CT_n(f)$.

Example 5.2.4 Recall the integral $I^{(1)}$ used in Table 5.1,

$$I = \int_0^1 e^{-x^2}\,dx \doteq 0.74682413281243$$

In Table 5.3, we give the results of using $T_n(f)$ and $CT_n(f)$, including their errors and the estimate

$$\widetilde{E}_n^T(f) = \frac{h^2 e^{-1}}{6}, \qquad h = \frac{1}{n}$$

Note that $\widetilde{E}_n^T(f)$ is a very accurate estimator of the true error. Also, the error in $CT_n(f)$ converges to zero at a more rapid rate than does the error for $T_n(f)$. When n is doubled, the error in $CT_n(f)$ decreases by a factor of about 16. ■

5.2.2 Error Formulas for Simpson's Rule

The type of analysis used in the preceding discussion can also be used to derive corresponding error formulas for Simpson's rule. These are stated in the following theorem, with the proof omitted.

Theorem 5.2.5 Assume $f(x)$ has four continuous derivatives on $[a, b]$, and let n be an even positive integer. Then the error in using Simpson's rule is given by

$$E_n^S(f) = I(f) - S_n(f) = -\frac{h^4(b-a)}{180} f^{(4)}(c_n) \tag{5.35}$$

with c_n an unknown point in $[a, b]$ and $h = (b-a)/n$. Moreover, this error can be estimated with the asymptotic error formula

$$\widetilde{E}_n^S(f) = -\frac{h^4}{180}\left[f'''(b) - f'''(a)\right] \tag{5.36}$$

Note that (5.35) says that Simpson's rule is exact for all $f(x)$ that are polynomials of degree ≤ 3, whereas the quadratic interpolation on which Simpson's rule is based is exact only for $f(x)$ a polynomial of degree ≤ 2. The degree of precision being 3 leads to the power h^4 in the error, rather than the power h^3, which would have been produced on the basis of the error in quadratic interpolation. It is this higher power of h^4 in the error and the simple form of the method that historically have caused Simpson's rule to be the most popular numerical integration rule.

Example 5.2.6 Recall (5.19) where $S_2(f)$ was applied to $I = \displaystyle\int_0^1 \frac{dx}{1+x}$. Then

$$f(x) = \frac{1}{1+x}, \qquad f^{(3)}(x) = \frac{-6}{(1+x)^4}, \qquad f^{(4)}(x) = \frac{24}{(1+x)^5}$$

The exact error is given by

$$E_n^S(f) = -\frac{h^4}{180} f^{(4)}(c_n), \qquad h = \frac{1}{n}$$

for some $0 \le c_n \le 1$. We can bound it by

$$\left|E_n^S(f)\right| \le \frac{h^4}{180}(24) = \frac{2h^4}{15}$$

The asymptotic error is given by

$$\widetilde{E}_n^S(f) = -\frac{h^4}{180}\left[\frac{-6}{(1+1)^4} - \frac{-6}{(1+0)^4}\right] = -\frac{h^4}{32}$$

For $n = 2$, $\widetilde{E}^S(f) \doteq -0.00195$; for comparison from (5.20), the actual error is -0.00130. ■

The behavior in $I(f) - S_n(f)$ can be derived from (5.36). When n is doubled, h is halved, and h^4 decreases by a factor of 16. Thus, the error $E_n^S(f)$ should decrease by the same factor, provided that $f'''(b) \neq f'''(a)$. This is the error behavior observed in Table 5.2 with integrals $I^{(1)}$ and $I^{(2)}$. When $f'''(b) = f'''(a)$, the error will decrease more rapidly, which is a partial explanation of the rapid convergence for $I^{(3)}$ in Table 5.2.

The theory of asymptotic error formulas

$$E_n(f) \approx \widetilde{E}_n(f) \tag{5.37}$$

such as for $E_n^T(f)$ and $E_n^S(f)$, says that (5.37) is valid, provided that

$$\lim_{n\to\infty} \frac{\widetilde{E}_n(f)}{E_n(f)} = 1$$

The needed size of n in (5.37) will vary with the integrand f, which is illustrated with the two cases $I^{(1)}$ and $I^{(2)}$ in Table 5.2. For $I^{(2)}$, the behavior (5.37) is not valid until n becomes larger, $n \geq 64$.

From (5.35) and (5.36), we also are led to infer that Simpson's rule will not perform as well if $f(x)$ is not four times continuously differentiable on $[a, b]$. This is correct for most such functions, and other numerical methods are often necessary for integrating them.

Example 5.2.7 Use Simpson's rule to approximate

$$I = \int_0^1 \sqrt{x}\, dx = \tfrac{2}{3}$$

The results are shown in Table 5.4. The column "Ratio" shows the convergence is much slower. ■

As was done for the trapezoidal rule, a *corrected Simpson's rule* can be defined:

$$CS_n(f) = S_n(f) - \frac{h^4}{180}[f'''(b) - f'''(a)] \tag{5.38}$$

This will usually be a more accurate approximation than $S_n(f)$.

Table 5.4. Simpson's Rule for $\sqrt{x}$

n	Error	Ratio
2	2.860E − 2	
4	1.014E − 2	2.82
8	3.587E − 3	2.83
16	1.268E − 3	2.83
32	4.485E − 4	2.83

5.2.3 Richardson Extrapolation

The error estimates (5.32) and (5.36) are both of the form

$$I - I_n \approx \frac{c}{n^p} \tag{5.39}$$

where I_n denotes the numerical integral and h has been replaced by $(b-a)/n$. The constants c and p vary with the method and the function. With most integrands $f(x)$, $p = 2$ for the trapezoidal rule and $p = 4$ for Simpson's rule. There are other numerical methods that satisfy (5.39), with other values of p and c. We will use (5.39) to obtain a computable estimate of the error $I - I_n$, without needing to know c explicitly.

In (5.39), replace n by $2n$ to obtain

$$I - I_{2n} \approx \frac{c}{2^p n^p} \tag{5.40}$$

Comparing this to (5.39), we see that

$$2^p[I - I_{2n}] \approx \frac{c}{n^p} \approx I - I_n$$

Solving for I gives us

$$(2^p - 1)I \approx 2^p I_{2n} - I_n$$

$$I \approx \frac{1}{2^p - 1}[2^p I_{2n} - I_n] \equiv R_{2n} \tag{5.41}$$

R_{2n} is an improved estimate of I, based on using I_n, I_{2n}, p, and the assumption (5.39). It is called *Richardson's extrapolation formula*, and generally it is a more accurate approximation to I than is I_{2n}. How much more accurate it is depends on the validity of (5.39), (5.40).

To estimate the error in I_{2n}, compare it with the more accurate value R_{2n}.

$$I - I_{2n} \approx R_{2n} - I_{2n} = \frac{1}{2^p - 1}[2^p I_{2n} - I_n] - I_{2n}$$

$$I - I_{2n} \approx \frac{1}{2^p - 1}[I_{2n} - I_n] \tag{5.42}$$

This is *Richardson's error estimate*.

Example 5.2.8 In using the trapezoidal rule to approximate

$$I = \int_0^1 e^{-x^2}\,dx \doteq 0.74682413281243$$

we have

$$T_2 \doteq 0.7313702518, \qquad T_4 \doteq 0.7429840978$$

Using (5.41) with $p = 2$ and $n = 2$, we obtain

$$I \approx R_4 = \tfrac{1}{3}[4I_4 - I_2] = \tfrac{1}{3}[4T_4 - T_2] \doteq 0.7468553797$$

The error in R_4 is -0.0000312; and from Table 5.1, R_4 is more accurate than T_{32}. To estimate the error in T_4, use (5.42) to get

$$I - T_4 \approx \tfrac{1}{3}[T_4 - T_2] \doteq 0.00387$$

The actual error in T_4 is 0.00384; and thus (5.42) is a very accurate error estimate. ■

Richardson's extrapolation and error estimation is not always as accurate as this example might suggest, but it is usually a fairly accurate procedure. The main assumption that must be satisfied is (5.39); and Problem 13(b) gives a way of testing whether this assumption is valid for the actual values of I_n being used.

Most computer program libraries contain one or more programs for *automatic numerical integration*. This means that the user of such a program presents an integral to it along with a desired error tolerance. The program then attempts to find a numerical integral within that error limit. There are many such programs, based on many different numerical integration rules. But most of them use error estimation techniques related to the ideas of Richardson's extrapolation and estimation. We will not consider these programs here, as they are very complex in structure; but they are often the best way to integrate a function if there are not too many such integrals. These programs usually have higher *program overhead* costs, in the form of extensive "bookkeeping" tasks for various quantities used in the integration; but these costs are generally more than offset by the programming time that these programs save the user.

5.2.4 Periodic Integrands

A function $f(x)$ is *periodic* with period τ if

$$f(x) = f(x + \tau), \qquad -\infty < x < \infty \tag{5.43}$$

and this relation should not be true with any smaller value of τ. For example,

$$f(x) = e^{\cos(\pi x)}$$

is periodic with period $\tau = 2$. If $f(x)$ is periodic and differentiable, then its derivatives are also periodic with period τ.

Consider integrating

$$I = \int_a^b f(x)\,dx$$

with the trapezoidal or Simpson's rule, and assume that $b - a$ is an integer multiple of the period τ. Assume $f(x)$ has derivatives of any order. Then for all derivatives of $f(x)$, the periodicity of $f(x)$ implies that

$$f^{(k)}(a) = f^{(k)}(b), \qquad k \geq 0 \tag{5.44}$$

If we now look at the asymptotic error formulas for the trapezoidal and Simpson's rules, they become zero because of (5.44). Thus, the error formulas $E_n^T(f)$ and $E_n^S(f)$ should converge to zero more rapidly when $f(x)$ is a periodic function, provided $b - a$ is an integer multiple of the period of f.

The asymptotic error formulas $\widetilde{E}_n^T(f)$ and $\widetilde{E}_n^S(f)$ can be extended to higher-order terms in h, using what is called the *Euler–MacLaurin expansion* and the higher-order terms are multiples of $f^{(k)}(b) - f^{(k)}(a)$ for all odd integers $k \geq 1$. Using this, we can prove the errors $E_n^T(f)$ and $E_n^S(f)$ converge to zero even more rapidly than was implied by the earlier comments for $f(x)$ periodic. This work is omitted; but note that the trapezoidal rule is the preferred integration rule when we are dealing with smooth periodic integrands. The earlier results for integral $I^{(3)}$ in Tables 5.1 and 5.2 are illustrations of these comments.

Example 5.2.9 The ellipse with boundary

$$\left(\frac{x}{a}\right)^2 + \left(\frac{y}{b}\right)^2 = 1$$

has area πab. For the case in which the area is π (and, thus, $ab = 1$), we study the variation of the perimeter of the ellipse as a and b vary. Some of our results are presented without a detailed discussion; and the reader is expected to fill in these details.

The ellipse has the parametric representation

$$(x, y) = (a\cos\theta, b\sin\theta), \qquad 0 \leq \theta \leq 2\pi \tag{5.45}$$

By using the standard formula for the perimeter, and using the symmetry of the ellipse about the x-axis, we find that the perimeter is given by

$$\begin{aligned} P &= 2\int_0^\pi \sqrt{\left(\frac{dx}{d\theta}\right)^2 + \left(\frac{dy}{d\theta}\right)^2}\,d\theta \\ &= 2\int_0^\pi \sqrt{a^2\sin^2\theta + b^2\cos^2\theta}\,d\theta \end{aligned}$$

Since $ab = 1$, we write this as

$$\begin{aligned} P(b) &= 2\int_0^{\pi} \sqrt{\frac{1}{b^2}\sin^2\theta + b^2\cos^2\theta}\, d\theta \\ &= \frac{2}{b}\int_0^{\pi} \sqrt{(b^4-1)\cos^2\theta + 1}\, d\theta \end{aligned} \tag{5.46}$$

We consider only the case with $1 \le b < \infty$. Since the perimeters for the two ellipses

$$\left(\frac{x}{a}\right)^2 + \left(\frac{y}{b}\right)^2 = 1 \qquad \text{and} \qquad \left(\frac{x}{b}\right)^2 + \left(\frac{y}{a}\right)^2 = 1$$

are equal, we can always consider the case in which the y-axis of the ellipse is larger than or equal to its x-axis; and this also shows

$$P\left(\frac{1}{b}\right) = P(b), \qquad b > 0 \tag{5.47}$$

The integrand of $P(b)$

$$f(\theta) = \frac{2}{b}\left[(b^4-1)\cos^2\theta + 1\right]^{1/2}$$

is periodic with period π. As discussed above, the trapezoidal rule is the natural choice for numerical integration of (5.46). Nonetheless, there is a variation in the behavior of $f(\theta)$ as b varies, and this will affect the accuracy of the numerical integration. In Figure 5.4, we give graphs of $f(\theta)$ for several values of b. In Table 5.5, we give the results of using the trapezoidal rule for these values of b, for increasing values of n. Note that as b increases, the trapezoidal rule converges more slowly. This is due to the integrand $f(\theta)$ changing more rapidly as b increases. For large b, $f(\theta)$ changes very rapidly in the

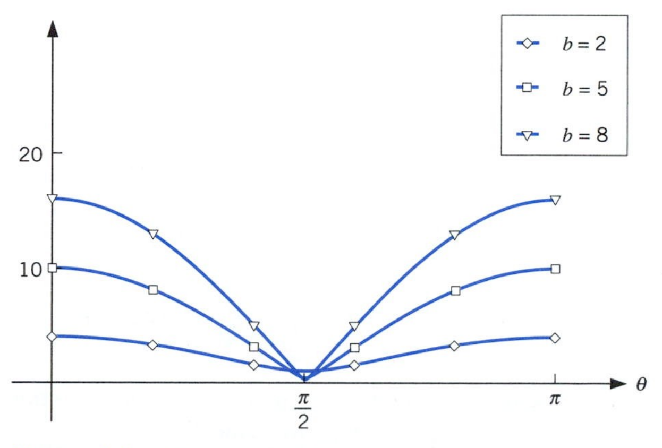

Figure 5.4. The graph of integrand $f(\theta) : b = 2, 5, 8$

Table 5.5. Trapezoidal Rule Approximations of (5.46)

n	$b = 2$	$b = 5$	$b = 8$
8	8.575517	19.918814	31.690628
16	8.578405	20.044483	31.953632
32	8.578422	20.063957	32.008934
64	8.578422	20.065672	32.018564
128	8.578422	20.065716	32.019660
256	8.578422	20.065717	32.019709

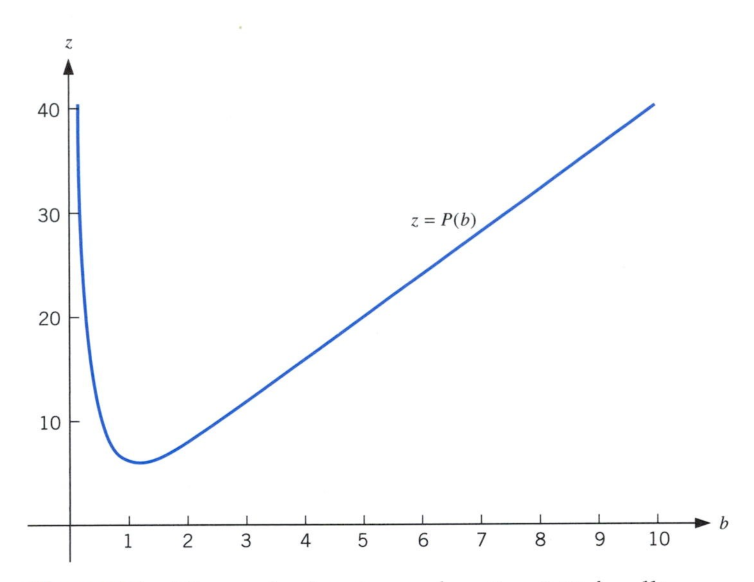

Figure 5.5. The graph of perimeter function $P(b)$ for ellipse

vicinity of $\theta = \frac{1}{2}\pi$; and this causes the trapezoidal rule to be less accurate than when b is smaller, near 1. To obtain a certain accuracy in the perimeter $P(b)$, we must increase n as b increases.

The graph of $P(b)$, given in Figure 5.5, reveals that $P(b) \approx 4b$ for large b. Returning to (5.46), we have for large b,

$$\begin{aligned} P(b) &\approx \frac{2}{b} \int_0^{\pi} \left[b^4 \cos^2\theta\right]^{1/2} d\theta \\ &= \frac{2}{b} b^2 \int_0^{\pi} |\cos\theta| \, d\theta = 4b \end{aligned}$$

We need to estimate the error in the above approximation to know when we can use it to replace $P(b)$; but it provides a way to avoid the integration of (5.46) for the most badly behaved cases. The reader should give a geometric argument to explain the reasonableness of this result. ■

PROBLEMS

1. Using the error formula (5.23), bound the error in $T_n(f)$ applied to the following integrals:

   (a) $\int_0^{\pi/2} \cos(x)\,dx$

   (b) $\int_0^1 e^{-x^2}\,dx$

   (c) $\int_0^{\sqrt{\pi}} \cos(x^2)\,dx$

2. Repeat Problem 1 by using the integrals (a), (b), and (c) of Problem 2 in Section 5.1. Compare your bounds to the actual errors.

3. Apply the trapezoidal error estimate (5.32) to the integrals (a), (b), (c), and (e) of Problem 2 in Section 5.1. Compare the results with the actual errors.

4. Repeat Problem 3 by using the integral $I^{(2)}$ of Table 5.1 in Section 5.1.

5. Using the asymptotic error formula (5.32) for the trapezoidal rule, estimate the number n of subdivisions to evaluate the following integrals to the given accuracy ϵ:

   (a) $\int_1^3 \log(x)\,dx, \qquad \epsilon = 10^{-8}$

   (b) $\int_0^2 \frac{e^x - e^{-x}}{2}\,dx, \qquad \epsilon = 10^{-10}$

   (c) $\int_0^2 e^{-x^2}\,dx, \qquad \epsilon = 10^{-10}$

   (d) $\int_0^4 \frac{dx}{1+x^2}, \qquad \epsilon = 10^{-12}$

6. Repeat Problem 5, but use Simpson's rule and its asymptotic error estimate (5.36).

7. **(a)** Consider using the trapezoidal rule T_n to estimate the integral

$$I = \int_1^3 \log x\,dx$$

   Give both a rigorous error bound for $I - T_n$ and an asymptotic error estimate $I - T_n$. Using the rigorous error bound, determine how large n should be in order that $|I - T_n| \le 5 \times 10^{-8}$.

   (b) Repeat with Simpson's rule.

8. Repeat Problem 7 with

$$I = \int_{-1}^{1} \frac{dx}{2+x}$$

9. Repeat Problem 7 with

$$I = \int_{-1}^{1} \frac{e^x + e^{-x}}{2}\, dx$$

10. The error $E_n(f)$ for both the trapezoidal and Simpson rules has some useful properties that simplify its calculation.

(a) Show that $E_n(f+g) = E_n(f) + E_n(g)$, for all continuous functions $f(x)$ and $g(x)$.

(b) Show that $E_n(cf) = cE_n(f)$, for all continuous functions $f(x)$ and constants c.

(c) If $p(x)$ is a linear polynomial, what is $E_n^T(p)$? If $p(x)$ is a cubic polynomial, what is $E_n^S(p)$?

(d) For a twice continuously differentiable function $f(x)$ on $[a, b]$, use the linear Taylor polynomial

$$p_1(x) = f(a) + (x-a)f'(a)$$

to write $f(x) = p_1(x) + R_1(x)$, with $R_1(x)$ the error in $p_1(x)$. Show that

$$E_n^T(f) = E_n^T(R_1)$$

This can sometimes be used to simplify the calculation of $E_n^T(f)$.

11. The proof of the basic trapezoidal error formula (5.26) is based on using Taylor's formula to expand $f(x)$ about α. To give a heuristic proof of (5.26), write

$$f(x) \approx f(\alpha) + (x-\alpha)f'(\alpha) + \frac{(x-\alpha)^2}{2} f''(\alpha)$$

Substitute this into the left side of (5.26) and obtain something quite close to its right side. Problem 10(d) can be used to simplify these calculations.

12. Recall the formula (5.30), an intermediate step in obtaining the error formula (5.23) for $E_n^T(f)$. To complete the proof, apply the ideas embodied in formula (A.1) and Problem 9 of Appendix A, with $f''(x)$ playing the role of $f(x)$ in those statements.

Hint: Write the term in brackets in (5.30) as

$$(b-a)\left[\frac{f''(\gamma_1) + f''(\gamma_2) + \cdots + f''(\gamma_n)}{n}\right]$$

13. **(a)** From (5.39) derive

$$\frac{I - I_n}{I - I_{2n}} \approx 2^p$$

for all n for which (5.39) is valid.

Hint: Consider (5.39) and (5.40).

(b) From (5.39), derive the computable estimate

$$\frac{I_{2n} - I_n}{I_{4n} - I_{2n}} \approx 2^p$$

This gives a practical means of checking the value of p, using three successive values I_n, I_{2n}, and I_{4n}. Using the log function, we get

$$p = \log\left(\frac{I_{2n} - I_n}{I_{4n} - I_{2n}}\right) / \log 2$$

Hint: Write $I_{2n} - I_n = (I - I_n) - (I - I_{2n})$, and do the same for the denominator. Then apply (5.39).

14. **(a)** Use the formula in Problem 13(a) with the results in Table 5.4 to identify the appropriate p in (5.39) for Simpson's rule applied to $\int_0^1 \sqrt{x}\, dx$.

(b) Apply $T_n(f)$ to $\int_0^1 \sqrt{x}\, dx$ for $n = 2, 4, 8, \ldots, 128$. Compute the value of p in (5.39) for these numerical integrals. Use Problem 13(a) or (b).

15. Following is a table of values of the trapezoidal rule applied to the integral

$$I = \int_0^1 \tan^{-1} x\, dx = \tfrac{1}{4}\pi - \tfrac{1}{2}\ln 2 \doteq 0.43882457311748$$

Using the table, produce the Richardson's error estimate for T_n for $n = 16, 32, 64$. In addition, produce the corrected trapezoidal rule for $n = 64$. Using the true answer, given above, what is the error in your value for the corrected trapezoidal rule?

n	T_n
4	0.4362066157
8	0.4381726803
16	0.4386617597
32	0.4387838797
64	0.4388144004

16. **(a)** Following is a table of numerical integrals I_n for an integral whose true value is $I = 0.3$. Assuming that the error has an asymptotic formula of the form

$$I - I_n \approx \frac{c}{n^p}$$

for some $p > 0$ and c, estimate the order of convergence p. Estimate c. Estimate the size of n in order to have $|I - I_n| \leq 10^{-10}$.

n	I_n	n	I_n
8	0.2993331765	64	0.2999791556
16	0.2997899139	128	0.2999934344
32	0.2999338239	256	0.2999979320

(b) Assuming I is not known (as is usually the case), estimate p.

17. Use Richardson's extrapolation to estimate the errors in Problems 2(a), (b), (c), and (e) of Section 5.1.

18. Use Richardson's extrapolation to estimate the errors in Problems 3(a), (b), (c), and (e) of Section 5.1.

19. In the following table of numerical integrals and their differences, give the likely value of p if we assume the error behaves like $I - I_n \approx c/n^p$. Also, estimate the error in I_{64}.

Hint: Use Problem 13(b).

n	I_n	$I_n - I_{n/2}$
2	0.702877396	
4	0.781978959	0.07910
8	0.804500932	0.02252
16	0.810303086	0.005802
32	0.811764354	0.001461
64	0.812130341	0.0003660

20. **(a)** Apply Simpson's rule to $I = \int_0^1 \sin(\sqrt{x})\, dx$ with $n = 2, 4, 8, \ldots, 128$. Use Problem 13(b) to calculate the rate of convergence.

(b) Transform I by using the change of variable $x = t^2$, obtaining

$$I = 2\int_0^1 t \sin(t)\, dt$$

Apply Simpson's rule to this new integral with $n = 2, 4, 8, \ldots, 128$, and compare the results with those of (a).

21. **(a)** Apply Simpson's rule to $I = \int_0^1 \sin(\sqrt[3]{x})\, dx$ with $n = 2, 4, 8, \ldots, 128$. By using Problem 13(b), calculate the rate of convergence.

(b) Following the ideas in 20(b), find a change of the variable of integration to give a new integrand for which Simpson's rule will have a higher order of convergence.

22. Consider the numerical integration of $I = \int_a^b f(x)\, dx$ when the nodal values $f(x_i)$ are known only approximately. More precisely, let $\hat{f}_i \approx f(x_i)$, $i = 0, 1, \ldots, n$, and suppose

$$\left| f(x_i) - \hat{f}_i \right| \le \epsilon, \qquad i = 0, 1, \ldots, n$$

Let I_n and $\hat{I}_n$ be the numerical integrals computed by using $\{f(x_i)\}$ and $\{\hat{f}_i\}$, respectively.

(a) Show that

$$\left| I_n - \hat{I}_n \right| \le \epsilon(b - a)$$

when I_n is based on either the trapezoidal rule or Simpson's rule.

(b) Generalize this result to any integration formula for which (i) all integration weights are positive, and (ii) the formula has a degree of precision greater than or equal to zero.

5.3. GAUSSIAN NUMERICAL INTEGRATION

The numerical methods studied in the first two sections were based on integrating linear and quadratic interpolating polynomials, and the resulting formulas were applied on subdivisions of ever smaller subintervals. In this section, we consider a numerical method that is based on the exact integration of polynomials of increasing degree; no subdivision of the integration interval is used. To motivate this approach, recall from Section 4.4 of Chapter 4 the material on approximation of functions.

Let $f(x)$ be continuous on $[a, b]$. Then $\rho_n(f)$ denotes the smallest error bound that can be attained in approximating $f(x)$ with a polynomial $p(x)$ of degree $\le n$ on the given interval $a \le x \le b$. The polynomial $m_n(x)$ that yields this approximation is called the minimax approximation of degree n for $f(x)$,

$$\max_{a \le x \le b} |f(x) - m_n(x)| = \rho_n(f) \tag{5.48}$$

Table 5.6. Minimax Errors for e^{-x^2}, $0 \le x \le 1$

n	$\rho_n(f)$	n	$\rho_n(f)$
1	5.30E − 2	6	7.82E − 6
2	1.79E − 2	7	4.62E − 7
3	6.63E − 4	8	9.64E − 8
4	4.63E − 4	9	8.05E − 9
5	1.62E − 5	10	9.16E − 10

and $\rho_n(f)$ is called the minimax error. From formula (4.82), it can be seen that $\rho_n(f)$ will often converge to zero quite rapidly.

Example 5.3.1 Let $f(x) = e^{-x^2}$ for $0 \le x \le 1$. Table 5.6 contains the minimax errors $\rho_n(f)$ for $n = 1, 2, \dots, 10$. They converge to zero rapidly, although not at a uniform rate. ■

If we have a numerical integration formula to integrate low- to moderate-degree polynomials exactly, then the hope is that the same formula will integrate other functions $f(x)$ almost exactly, if $f(x)$ is well approximable by such polynomials. To illustrate the derivation of such integration formulas, we restrict our attention to the integral

$$I(f) = \int_{-1}^{1} f(x)\,dx \tag{5.49}$$

Its relation to integrals over other intervals $[a, b]$ will be discussed later.

The integration formula is to have the general form

$$I_n(f) = \sum_{j=1}^{n} w_j f(x_j) \tag{5.50}$$

and we require that the nodes $\{x_1, \dots, x_n\}$ and weights $\{w_1, \dots, w_n\}$ be so chosen that $I_n(f) = I(f)$ for all polynomials $f(x)$ of as large a degree as possible.

Case $n = 1$ The integration formula has the form

$$\int_{-1}^{1} f(x)\,dx \approx w_1 f(x_1) \tag{5.51}$$

It is to be exact for polynomials of as large a degree as possible.

Using $f(x) \equiv 1$ and forcing equality in (5.51) give us

$$2 = w_1$$

Now use $f(x) = x$ and again force equality in (5.51). Then

$$0 = w_1 x_1$$

which implies $x_1 = 0$. Thus, (5.51) becomes

$$\int_{-1}^{1} f(x)\,dx \approx 2f(0) \equiv I_1(f) \tag{5.52}$$

This is the midpoint formula of Problem 8(a) in Section 5.1. The formula (5.52) is exact for all linear polynomials; the proof is left as Problem 7 for the reader.

To see that (5.52) is not exact for quadratics, let $f(x) = x^2$. Then the error in (5.52) is given by

$$\int_{-1}^{1} x^2\,dx - 2(0)^2 = \tfrac{2}{3} \neq 0$$

Following Problem 11 of Section 5.1, the formula (5.52) has degree of precision 1.

Case $n = 2$ The integration formula is

$$\int_{-1}^{1} f(x)\,dx \approx w_1 f(x_1) + w_2 f(x_2) \tag{5.53}$$

and it has four unspecified quantities: x_1, x_2, w_1, and w_2. To determine these, we require it to be exact for the four monomials

$$f(x) = 1, x, x^2, x^3 \tag{5.54}$$

This leads to the four equations

$$\begin{aligned} 2 &= w_1 + w_2 \\ 0 &= w_1 x_1 + w_2 x_2 \\ \tfrac{2}{3} &= w_1 x_1^2 + w_2 x_2^2 \\ 0 &= w_1 x_1^3 + w_2 x_2^3 \end{aligned} \tag{5.55}$$

This is a nonlinear system in four unknowns; its solution can be shown to be

$$w_1 = w_2 = 1, \qquad x_1 = -\frac{\sqrt{3}}{3}, \qquad x_2 = \frac{\sqrt{3}}{3} \tag{5.56}$$

along with one based on reversing the signs of x_1 and x_2.

This yields the integration formula

$$\int_{-1}^{1} f(x)\,dx \approx f\left(-\frac{\sqrt{3}}{3}\right) + f\left(\frac{\sqrt{3}}{3}\right) \equiv I_2(f) \tag{5.57}$$

From being exact for the monomials in (5.54), one can show this formula will be exact for all polynomials of degree ≤ 3 (cf. Problem 6). It also can be shown by direct calculation to not be exact for the degree 4 polynomial $f(x) = x^4$. Thus, $I_2(f)$ has degree of precision 3.

Example 5.3.2 Approximate

$$I = \int_{-1}^{1} e^x\,dx = e - e^{-1} \doteq 2.3504024$$

Using (5.57), we get

$$I_2 = e^{-\sqrt{3}/3} + e^{\sqrt{3}/3} \doteq 2.3426961$$
$$I - I_2 \doteq 0.00771$$

The error is quite small for using such a small number of node points. ■

Case $n > 2$ We seek the formula (5.50), which has $2n$ unspecified parameters, $x_1, \ldots, x_n, w_1, \ldots, w_n$, by forcing the integration formula to be exact for the $2n$ monomials

$$f(x) = 1, x, x^2, \ldots, x^{2n-1} \tag{5.58}$$

In turn, this forces $I_n(f) = I(f)$ for all polynomials f of degree $\leq 2n - 1$. This leads to the following system of $2n$ nonlinear equations in $2n$ unknowns:

$$\begin{aligned}
2 &= w_1 + w_2 + \cdots + w_n \\
0 &= w_1 x_1 + w_2 x_2 + \cdots + w_n x_n \\
\tfrac{2}{3} &= w_1 x_1^2 + w_2 x_2^2 + \cdots + w_n x_n^2 \\
0 &= w_1 x_1^3 + w_2 x_2^3 + \cdots + w_n x_n^3 \\
&\vdots \\
\frac{2}{2n-1} &= w_1 x_1^{2n-2} + \cdots + w_n x_n^{2n-2} \\
0 &= w_2 x_1^{2n-1} + \cdots + w_n x_n^{2n-1}
\end{aligned} \tag{5.59}$$

The resulting formula $I_n(f)$ has degree of precision $2n - 1$.

Table 5.7. Nodes and Weights of Gaussian Quadrature Formulas

n	x_i	w_i
2	± 0.5773502692	1.0
3	± 0.7745966692	0.5555555556
	0.0	0.8888888889
4	± 0.8611363116	0.3478548451
	± 0.3399810436	0.6521451549
5	± 0.9061798459	0.2369268851
	± 0.5384693101	0.4786286705
	0.0	0.5688888889
6	± 0.9324695142	0.1713244924
	± 0.6612093865	0.3607615730
	± 0.2386191861	0.4679139346
7	± 0.9491079123	0.1294849662
	± 0.7415311856	0.2797053915
	± 0.4058451514	0.3818300505
	0.0	0.4179591837
8	± 0.9602898565	0.1012285363
	± 0.7966664774	0.2223810345
	± 0.5255324099	0.3137066459
	± 0.1834346425	0.3626837834

Solving this system is a formidable problem. Thankfully, the nodes $\{x_i\}$ and weights $\{w_i\}$ have been calculated and collected in tables for the most commonly used values of n. Table 5.7 contains the solutions for $n = 2, 3, \ldots, 8$. For more complete tables, see A. Stroud and D. Secrest (1966). Most computer centers will have programs to produce these nodes and weights or to directly perform the numerical integration.

There is also another approach to the development of the numerical integration formula (5.50), using the *theory of orthogonal polynomials*. From that theory, it can be shown that the nodes $\{x_1, \ldots, x_n\}$ are the zeros of the *Legendre polynomial* of degree n on the interval $[-1, 1]$. Recall that these polynomials were introduced in Section 4.7 of Chapter 4. For example,

$$P_2(x) = \tfrac{1}{2}\left(3x^2 - 1\right)$$

and its roots are the nodes given in (5.56). Since the Legendre polynomials are well known, the nodes $\{x_j\}$ can be found without any recourse to the nonlinear system (5.59). For an introduction to this theory, see Atkinson (1989, p. 270).

The sequence of formulas (5.50) is called the *Gaussian numerical integration* method. From its definition, $I_n(f)$ uses n nodes, and it is exact for all polynomials of degree $\leq 2n-1$. $I_n(f)$ is limited to (5.49), an integral over $[-1, 1]$; but this limitation is easily removed. Given an integral over $[a, b]$

$$I(f) = \int_a^b f(x)\,dx \tag{5.60}$$

introduce the linear change of variable

$$x = \frac{b+a+t(b-a)}{2}, \qquad -1 \leq t \leq 1 \tag{5.61}$$

transforming the integral to

$$I(f) = \frac{b-a}{2}\int_{-1}^{1} \tilde{f}(t)\,dt \tag{5.62}$$

with

$$\tilde{f}(t) = f\left(\frac{b+a+t(b-a)}{2}\right) \tag{5.63}$$

Now apply $I_n(\tilde{f})$ to this new integral.

Example 5.3.3 Apply Gaussian numerical integration to the three integrals $I^{(1)}$, $I^{(2)}$, and $I^{(3)}$ of (5.14) to (5.16), which were used as examples for the trapezoidal and Simpson rules in Section 5.1. All are reformulated as integrals over $[-1, 1]$, in the manner described above. The error results are shown in Table 5.8. The entry $(*)$ means the error was zero relative to the accuracy possible on the computer being used (around 16 decimal digits).

If these results are compared to those in Tables 5.1 and 5.2, then Gaussian integration of $I^{(1)}$ and $I^{(2)}$ is much more efficient than are the trapezoidal and Simpson rules. But

Table 5.8. Gaussian Numerical Integration Examples

n	Error in $I^{(1)}$	Error in $I^{(2)}$	Error in $I^{(3)}$
2	2.29E − 4	−2.33E − 2	8.23E − 1
3	9.55E − 6	−3.49E − 2	−4.30E − 1
4	−3.35E − 7	−1.90E − 3	1.77E − 1
5	6.05E − 9	1.70E − 3	−8.12E − 2
6	−7.77E − 11	2.74E − 4	3.55E − 2
7	7.89E − 13	−6.45E − 5	−1.58E − 2
10	*	1.27E − 6	1.37E − 3
15	*	7.40E − 10	−2.33E − 5
20	*	*	3.96E − 7

Table 5.9. Gaussian Integration of (5.64)

n	$I - I_n$	Ratio
2	$-7.22E-3$	
4	$-1.16E-3$	6.2
8	$-1.69E-4$	6.9
16	$-2.30E-5$	7.4
32	$-3.00E-6$	7.6
64	$-3.84E-7$	7.8

the integration of the periodic integrand of $I^{(3)}$ is not as efficient as with the trapezoidal rule. These results are also true for most other integrals. Except for periodic integrands, Gaussian numerical integration is usually much more accurate than the trapezoidal and Simpson rules. This is even true with many integrals in which the integrand does not have a continuous derivative. ■

Example 5.3.4 Use Gaussian integration on

$$I = \int_0^1 \sqrt{x}\, dx = \tfrac{2}{3} \tag{5.64}$$

The results are shown in Table 5.9, with n the number of node points. The ratio column is defined as

$$\frac{I - I_{\frac{1}{2}n}}{I - I_n}$$

and it shows that the error behaves like

$$I - I_n \approx \frac{c}{n^3} \tag{5.65}$$

for some c. Compare this with Table 5.4, which gives the results of Simpson's rule for (5.64). There the empirical rate of convergence is proportional to only $1/n^{1.5}$, a much slower rate than in (5.65). ■

We give an additional result that relates the minimax error to the Gaussian numerical integration error.

Theorem 5.3.5 Let $f(x)$ be continuous for $a \le x \le b$, and let $n \ge 1$. Then, if we apply Gaussian numerical integration to $I = \int_a^b f(x)\, dx$, the error in I_n satisfies

$$|I(f) - I_n(f)| \le 2(b-a)\rho_{2n-1}(f) \tag{5.66}$$

where $\rho_{2n-1}(f)$ is the minimax error of degree $2n-1$ for $f(x)$ on $[a, b]$.

Example 5.3.6 Using Table 5.6, apply (5.66) to

$$I = \int_0^1 e^{-x^2}\,dx \tag{5.67}$$

For $n = 3$, the above bound implies

$$|I - I_3| \le 2\rho_5(e^{-x^2}) \doteq 3.24 \times 10^{-5}$$

The actual error is 9.55E − 6 from Table 5.8. ■

Gaussian numerical integration is not as simple to use as are the trapezoidal and Simpson rules, partly because the Gaussian nodes and weights do not have simple formulas and also because the error is harder to predict. Nonetheless, the increase in the speed of convergence is so rapid and dramatic in most instances that the method should always be considered seriously when one is doing many integrations. Estimating the error is quite difficult, and most people satisfy themselves by looking at two or more successive values. If n is doubled, then repeatedly comparing two successive values, I_n and I_{2n}, is almost always adequate for estimating the error in I_n

$$I - I_n \approx I_{2n} - I_n \tag{5.68}$$

This is somewhat inefficient, but the speed of convergence in I_n is so rapid that this will still not diminish its advantage over most other methods.

5.3.1 Weighted Gaussian Quadrature

A common problem is the evaluation of integrals of the form

$$I(f) = \int_a^b w(x)f(x)\,dx \tag{5.69}$$

with $f(x)$ a "well-behaved" function and $w(x)$ a possibly (and often) ill-behaved function. Gaussian quadrature has been generalized to handle such integrals for many functions $w(x)$. Examples include

$$\int_{-1}^1 \frac{f(x)}{\sqrt{1-x^2}}\,dx, \qquad \int_0^1 \sqrt{x}\,f(x)\,dx, \qquad \int_0^1 f(x)\log\left(\frac{1}{x}\right)dx$$

The function $w(x)$ is called a *weight function*.

We begin by imitating the development given earlier in this section, and we do so for the special case of

$$I(f) = \int_0^1 \frac{f(x)}{\sqrt{x}}\,dx \tag{5.70}$$

in which $w(x) = 1/\sqrt{x}$. As before, we seek numerical integration formulas of the form

$$I_n(f) = \sum_{j=1}^{n} w_j f(x_j) \tag{5.71}$$

and we require that the nodes $\{x_1, \ldots, x_n\}$ and weights $\{w_1, \ldots, w_n\}$ be so chosen that $I_n(f) = I(f)$ for polynomials $f(x)$ of as large a degree as possible.

Case $n = 1$ The integration formula has the form

$$\int_0^1 \frac{f(x)}{\sqrt{x}}\,dx \approx w_1 f(x_1)$$

We force equality for $f(x) = 1$ and $f(x) = x$. This leads to the equations

$$w_1 = \int_0^1 \frac{1}{\sqrt{x}}\,dx = 2$$

$$w_1 x_1 = \int_0^1 \frac{x}{\sqrt{x}}\,dx = \tfrac{2}{3}$$

Solving for w_1 and x_1, we obtain the formula

$$\int_0^1 \frac{f(x)}{\sqrt{x}}\,dx \approx 2f\left(\tfrac{1}{3}\right) \tag{5.72}$$

and it has degree of precision 1.

Case $n = 2$ The integration formula has the form

$$\int_0^1 \frac{f(x)}{\sqrt{x}}\,dx \approx w_1 f(x_1) + w_2 f(x_2) \tag{5.73}$$

We force equality for $f(x) = 1,\ x,\ x^2,\ x^3$. This leads to the equations

$$w_1 + w_2 = \int_0^1 \frac{1}{\sqrt{x}}\,dx = 2$$

$$w_1 x_1 + w_2 x_2 = \int_0^1 \frac{x}{\sqrt{x}}\,dx = \tfrac{2}{3}$$

$$w_1x_1^2 + w_2x_2^2 = \int_0^1 \frac{x^2}{\sqrt{x}}\,dx = \tfrac{2}{5}$$

$$w_1x_1^3 + w_2x_2^3 = \int_0^1 \frac{x^3}{\sqrt{x}}\,dx = \tfrac{2}{7}$$

This has the solution

$$x_1 = \tfrac{3}{7} - \tfrac{2}{35}\sqrt{30} \doteq 0.11559, \qquad x_2 = \tfrac{3}{7} + \tfrac{2}{35}\sqrt{30} \doteq 0.74156$$

$$w_1 = 1 + \tfrac{1}{18}\sqrt{30} \doteq 1.30429, \qquad w_2 = 1 - \tfrac{1}{18}\sqrt{30} \doteq 0.69571$$

The resulting formula (5.73) has degree of precision 3.

Case $n > 2$ We seek the formula (5.71), which has $2n$ unspecified parameters, $x_1, \ldots, x_n, w_1, \ldots, w_n$, by forcing the integration formula to be exact for the $2n$ monomials

$$f(x) = 1, x, x^2, \ldots, x^{2n-1}$$

In turn, this forces $I_n(f) = I(f)$ for all polynomials f of degree $\leq 2n - 1$. This leads to the following system of $2n$ nonlinear equations in $2n$ unknowns:

$$\begin{aligned} w_1 + w_2 + \cdots + w_n &= 2 \\ w_1x_1 + w_2x_2 + \cdots + w_nx_n &= \tfrac{2}{3} \\ w_1x_1^2 + w_2x_2^2 + \cdots + w_nx_n^2 &= \tfrac{2}{5} \\ &\vdots \\ w_2x_1^{2n-1} + \cdots + w_nx_n^{2n-1} &= \frac{2}{4n-1} \end{aligned} \tag{5.74}$$

The resulting formula $I_n(f)$ has degree of precision $2n - 1$. As before, this system is very difficult to solve directly, but there are alternative methods of deriving $\{x_i\}$ and $\{w_i\}$. It is based on looking at the polynomials that are orthogonal with respect to the weight function

$$w(x) = \frac{1}{\sqrt{x}}$$

on the interval [0, 1].

Example 5.3.7 We evaluate

$$I = \int_0^1 \frac{\cos(\pi x)}{\sqrt{x}}\,dx \doteq 0.74796566683146$$

using (5.72) and (5.73):

$$I_1 = 1.0$$
$$I_2 \doteq 0.740519$$

I_2 is a reasonable estimate of I, with $I - I_2 \doteq 0.00745$. ■

A general theory can be developed for weighted Gaussian quadrature

$$I(f) = \int_a^b w(x) f(x)\, dx \approx \sum_{j=1}^{n} w_j f(x_j) = I_n(f) \tag{5.75}$$

It requires the following assumptions for the weight function $w(x)$:

A1. $w(x) > 0$ for $a < x < b$;

A2. For all integers $n \geq 0$,

$$\int_a^b w(x)\, |x|^n\, dx < \infty$$

These hypotheses are the same as were assumed for the generalized least squares approximation theory following (4.124) in Section 4.7 of Chapter 4. This is not accidental since both Gaussian quadrature and least squares approximation theory are dependent on the subject of orthogonal polynomials. The node points $\{x_i\}$ solving the system (5.74) are the zeros of the degree n orthogonal polynomial on $[a, b]$ with respect to the weight function $w(x) = 1/\sqrt{x}$. For the generalization (5.75), the nodes $\{x_i\}$ are the zeros of the degree n orthogonal polynomial on $[a, b]$ with respect to the weight function $w(x)$. For more on general Gaussian quadrature formulas (5.75), see Atkinson (1989, Section 5.3).

PROBLEMS

1. Recalling the example following (5.57), apply I_3 and I_4 to $\int_{-1}^{1} e^x\, dx$. Use the nodes and weights given in Table 5.7.

2. Apply I_2, I_3, and I_4 to the integrals from Problems 2(a) to (f) of Section 5.1. Calculate the errors and compare them with the earlier results for the trapezoidal and Simpson rules. If your computer center has a Gaussian numerical integration program, then do these same calculations for $I_5, \ldots, I_{10}$.

3. Repeat the example following (5.67) with $n = 2$, 4, and 5.

4. Repeat Problem 4 of Section 5.1, but use the Gaussian numerical integration formula with $n = 1, 2, \ldots, 8$. Begin by transforming the integrals to the standard interval $[-1, 1]$, as in (5.60–5.63).

5. Repeat Problem 5 of Section 5.1, but use the Gaussian numerical integration formula with $n = 1, 2, \ldots, 8$. Begin by transforming the integrals to the standard interval $[-1, 1]$, as in (5.60–5.63).

6. Show that if an integration formula of the form (5.50) is exact when integrating $1, x, x^2, \ldots, x^m$, then it is exact for all polynomials of degree $\le m$.

Hint: Recall Problem 10 in Section 5.2.

7. Show that the one-point formula (5.52) is exact whenever $f(x)$ is a linear polynomial.

8. Determine constants c_1 and c_2 in the formula

$$\int_0^1 f(x) \approx c_1 f(0) + c_2 f(1)$$

so that it is exact for all polynomials of as large a degree as possible. What is the degree of precision of the formula?

9. For the formula

$$\int_0^1 f(x) \approx w_1 f(0) + w_2 f(x_2)$$

determine the weights w_1, w_2 and the node x_2 so that the formula is exact for all polynomials of as large a degree as possible. What is the degree of precision of the formula?

10. Consider integrals

$$I(f) = \int_0^1 f(x) \log\left(\frac{1}{x}\right) dx$$

with $f(x)$ a function with several continuous derivatives on $0 \le x \le 1$. Repeat the ideas of (5.69–5.75) to develop the following formulas:

(a) Find a formula

$$\int_0^1 f(x) \log\left(\frac{1}{x}\right) dx \approx w_1 f(x_1) \equiv I_1(f)$$

which is exact if $f(x)$ is any linear polynomial.

Hint: $\log\left(\frac{1}{x}\right) = -\log x$, and

$$\int_0^1 x^m \log(x)\, dx = \frac{-1}{(m+1)^2}, \qquad m \ge 0$$

(b) To find a formula

$$\int_0^1 f(x) \log\left(\frac{1}{x}\right) dx \approx w_1 f(x_1) + w_2 f(x_2) \equiv I_2(f)$$

which is exact for all polynomials of degree ≤ 3, set up a system of four equations with unknowns w_1, w_2, x_1, x_2. Do not attempt to solve the system, although the methods of Section 5.3 can be used. Instead, show that

$$x_1 = \frac{15 - \sqrt{106}}{42}, \qquad x_2 = \frac{15 + \sqrt{106}}{42},$$

$$w_1 = \frac{21}{\sqrt{106}}\left(x_2 - \frac{1}{4}\right), \qquad w_2 = 1 - w_1$$

is a solution of the system.

(c) Use I_1 of (a) and I_2 of (b) to approximate

$$I = \int_0^1 \cos(x) \log\left(\frac{1}{x}\right) dx$$

whose true value is $I \doteq 0.9460831$.

(d) Apply the midpoint rule [cf. Problem 8(b) of Section 5.1] with $n = 2, 4, 8$ to the integral I. Compare the results with that from (c).

11. Consider approximating integrals of the form

$$I(f) = \int_0^1 \sqrt{x} f(x)\, dx$$

in which $f(x)$ has several continuous derivatives on [0, 1].

(a) Find a formula

$$\int_0^1 \sqrt{x} f(x)\, dx \approx w_1 f(x_1) \equiv I_1(f)$$

which is exact if $f(x)$ is any linear polynomial.

(b) To find a formula

$$\int_0^1 \sqrt{x} f(x)\, dx \approx w_1 f(x_1) + w_2 f(x_2) \equiv I_2(f)$$

which is exact for all polynomials of degree ≤ 3, set up a system of four equations with unknowns w_1, w_2, x_1, x_2. Verify that

$$x_1 = \frac{1}{9}\left(5 + 2\sqrt{\frac{10}{7}}\right), \qquad x_2 = \frac{1}{9}\left(5 - 2\sqrt{\frac{10}{7}}\right),$$

$$w_1 = \frac{1}{15}\left(5 + \sqrt{\frac{7}{10}}\right), \qquad w_2 = \frac{2}{3} - w_1$$

is a solution of the system. The verification can be numerical (say using *Matlab*) or symbolic (say using *Maple* or *Mathematica*).

(c) Apply I_1 and I_2 to the evaluation of

$$I = \int_0^1 \sqrt{x}e^{-x}\,dx \doteq 0.37894469164$$

(d) Apply the midpoint rule [cf. Problem 8(b)] with $n = 2, 4, 8$ to the integral I. Compare the results with that from (c).

12. Consider the proof of Theorem 5.3.5. For a given continuous function $f(x)$ on $[a, b]$ and for $n \geq 0$, let $q_{2n-1}(x)$ be the minimax approximation of degree $2n - 1$ to $f(x)$:

$$\max_{a\leq x\leq b} |f(x) - q_{2n-1}(x)| = \rho_{2n-1}(f)$$

(a) Show $I_n(q_{2n-1}) = I(q_{2n-1})$

(b) Show

$$I(f) - I_n(f) = I(f - q_{2n-1}) - I_n(f - q_{2n-1})$$

(c) It can be shown that for any $n \geq 1$, all weights in the Gaussian quadrature formula $I_n(f)$ satisfy $w_i > 0$. Show that

$$|I(f) - I(q_{2n-1})| \leq (b - a)\rho_{2n-1}(f)$$
$$|I_n(f) - I_n(q_{2n-1})| \leq (b - a)\rho_{2n-1}(f)$$

Then complete the proof of Theorem 5.3.5.

5.4. NUMERICAL DIFFERENTIATION

To numerically calculate the derivative of $f(x)$, begin by recalling the definition of derivative:

$$f'(x) = \lim_{h\to 0}\frac{f(x + h) - f(x)}{h}$$

Table 5.10. Numerical Differentiation of $f(x) = \cos(x)$ Using (5.76)

h	$D_h f$	Error	Ratio
0.1	−0.54243	0.04243	
0.05	−0.52144	0.02144	1.98
0.025	−0.51077	0.01077	1.99
0.0125	−0.50540	0.00540	1.99
0.00625	−0.50270	0.00270	2.00
0.003125	−0.50135	0.00135	2.00

This justifies using

$$f'(x) \approx \frac{f(x+h) - f(x)}{h} \equiv D_h f(x) \tag{5.76}$$

for small values of h. $D_h f(x)$ is called a *numerical derivative* of $f(x)$ with stepsize h.

Example 5.4.1 Use $D_h f(x)$ to approximate the derivative of $f(x) = \cos(x)$ at $x = \pi/6$. Table 5.10 contains the results for various values of h. Looking at the error column, we see the error is nearly proportional to h; when h is halved, the error is almost halved. ■

To explain the behavior in this example, Taylor's theorem can be used to find an error formula. Expanding $f(x+h)$ about x, we get

$$f(x+h) = f(x) + hf'(x) + \frac{h^2}{2} f''(c)$$

for some c between x and $x + h$. Substituting on the right side of (5.76), we obtain

$$\begin{aligned} D_h f(x) &= \frac{1}{h}\left\{\left[f(x) + hf'(x) + \frac{h^2}{2} f''(c)\right] - f(x)\right\} \\ &= f'(x) + \frac{h}{2} f''(c) \\ f'(x) - D_h f(x) &= -\frac{h}{2} f''(c) \end{aligned} \tag{5.77}$$

The error is proportional to h, agreeing with the results in Table 5.10 above. For that example,

$$f'\left(\frac{\pi}{6}\right) - D_h f\left(\frac{\pi}{6}\right) = \frac{h}{2}\cos(c) \tag{5.78}$$

where c is between $\frac{1}{6}\pi$ and $\frac{1}{6}\pi + h$. The reader should check that if c is replaced by $\frac{1}{6}\pi$, then the right side of (5.78) agrees with the error column in Table 5.10.

As seen from Example 5.4.1, we use the formula (5.76) with a positive stepsize $h > 0$. The formula (5.76) is commonly known as the *forward difference formula* for the first derivative. We can formally replace h by $-h$ in (5.76) to obtain the formula

$$f'(x) \approx \frac{f(x) - f(x-h)}{h}, \qquad h > 0 \tag{5.79}$$

This is the *backward difference formula* for the first derivative. A derivation similar to that leading to (5.77) shows that

$$f'(x) - \frac{f(x) - f(x-h)}{h} = \frac{h}{2} f''(c) \tag{5.80}$$

for some c between x and $x - h$. Thus, we expect the accuracy of the backward difference formula to be almost the same as that of the forward difference formula.

5.4.1 Differentiation Using Interpolation

Let $P_n(x)$ denote the degree n polynomial that interpolates $f(x)$ at $n + 1$ node points $x_0, x_1, \ldots, x_n$. To calculate $f'(x)$ at some point $x = t$, use

$$f'(t) \approx P_n'(t) \tag{5.81}$$

Many different formulas can be obtained by varying n and by varying the placement of the nodes $x_0, \ldots, x_n$ relative to the point t of interest.

As an especially useful example of (5.81), take $n = 2$, $t = x_1$, $x_0 = x_1 - h$, $x_2 = x_1 + h$. Then

$$\begin{aligned} P_2(x) &= \frac{(x - x_1)(x - x_2)}{2h^2} f(x_0) + \frac{(x - x_0)(x - x_2)}{-h^2} f(x_1) \\ &\quad + \frac{(x - x_0)(x - x_1)}{2h^2} f(x_2) \\ P_2'(x) &= \left(\frac{2x - x_1 - x_2}{2h^2}\right) f(x_0) + \left(\frac{2x - x_0 - x_2}{-h^2}\right) f(x_1) \\ &\quad + \left(\frac{2x - x_0 - x_1}{2h^2}\right) f(x_2) \\ P_2'(x_1) &= \left(\frac{x_1 - x_2}{2h^2}\right) f(x_0) + \left(\frac{2x_1 - x_0 - x_2}{-h^2}\right) f(x_1) + \left(\frac{x_1 - x_0}{2h^2}\right) f(x_2) \\ &= \frac{f(x_2) - f(x_0)}{2h} \end{aligned} \tag{5.82}$$

Replacing x_0 and x_2 by $x_1 - h$ and $x_1 + h$, from (5.81) and (5.82) we obtain

$$f'(x_1) \approx \frac{f(x_1 + h) - f(x_1 - h)}{2h} \equiv D_h f(x_1) \tag{5.83}$$

another approximation to the derivative of $f(x)$, called the *central difference formula*. It will be shown below that this is a more accurate approximation to $f'(x)$ than is the $D_h f(x)$ of (5.76).

The error for the general procedure (5.81) can be obtained from the interpolation error formula (4.53) in Chapter 4. The main result is given in the following theorem; the proof is omitted.

Theorem 5.4.2 Assume $f(x)$ has $n + 2$ continuous derivatives on an interval $[a, b]$. Let $x_0, x_1, \ldots, x_n$ be $n + 1$ distinct interpolation nodes in $[a, b]$, and let t be an arbitrary given point in $[a, b]$. Then

$$f'(t) - P_n'(t) = \Psi_n(t)\frac{f^{(n+2)}(c_1)}{(n+2)!} + \Psi_n'(t)\frac{f^{(n+1)}(c_2)}{(n+1)!} \tag{5.84}$$

with

$$\Psi_n(t) = (t - x_0)(t - x_1)\cdots(t - x_n)$$

The numbers c_1 and c_2 are unknown points located between the maximum and minimum of $x_0, x_1, \ldots, x_n$ and t.

To illustrate this result, an error formula can be derived for (5.83). Since $t = x_1$ in deriving (5.83), we find that the first term on the right side of (5.84) is zero. Also $n = 2$ and

$$\begin{aligned} \Psi_2(x) &= (x - x_0)(x - x_1)(x - x_2) \\ \Psi_2'(x) &= (x - x_1)(x - x_2) + (x - x_0)(x - x_2) + (x - x_0)(x - x_1) \\ \Psi_2'(x_1) &= (x_1 - x_0)(x_1 - x_2) = -h^2 \end{aligned}$$

Using this in (5.84), we get

$$f'(x_1) - \frac{f(x_1 + h) - f(x_1 - h)}{2h} = -\frac{h^2}{6} f'''(c_2) \tag{5.85}$$

with $x_1 - h \le c_2 \le x_1 + h$. This says that for small values of h, the formula (5.83) should be more accurate than the earlier approximation (5.76), because the error term of (5.83) decreases more rapidly with h.

Example 5.4.3 The earlier example in Table 5.10 is repeated using (5.83). Recall that $f(x) = \cos(x)$ and $x_1 = \frac{1}{6}\pi$. The results are shown in Table 5.11, with (5.83) given in the column

Table 5.11. Numerical Differentiation Using (5.83)

h	$D_h f$	Error	Ratio
0.1	−0.49916708	−0.0008329	
0.05	−0.49979169	−0.0002083	4.00
0.025	−0.49994792	−0.00005208	4.00
0.0125	−0.49998698	−0.00001302	4.00
0.00625	−0.49999674	−0.000003255	4.00

labeled $D_h f$. The results confirm the rate of convergence given in (5.85), and they illustrate that (5.83) will usually be superior to the earlier approximation (5.76). ■

5.4.2 The Method of Undetermined Coefficients

The *method of undetermined coefficients* is a procedure used in deriving formulas for numerical differentiation, interpolation, and integration. We will explain the method by using it to derive an approximation for $f''(x)$.

To approximate $f''(x)$ at some point $x = t$, write

$$f''(t) \approx D_h^{(2)} f(t) \equiv Af(t+h) + Bf(t) + Cf(t-h) \tag{5.86}$$

with A, B, and C unspecified constants. Replace $f(t-h)$ and $f(t+h)$ by the Taylor polynomial approximations

$$\begin{aligned} f(t-h) &\approx f(t) - hf'(t) + \frac{h^2}{2} f''(t) - \frac{h^3}{6} f'''(t) + \frac{h^4}{24} f^{(4)}(t) \\ f(t+h) &\approx f(t) + hf'(t) + \frac{h^2}{2} f''(t) + \frac{h^3}{6} f'''(t) + \frac{h^4}{24} f^{(4)}(t) \end{aligned} \tag{5.87}$$

Including more terms would give higher powers of h; and for small values of h, these additional terms should be much smaller than the terms included in (5.87). Substituting these approximations into the formula for $D_h^{(2)} f(t)$ and collecting together common powers of h give us

$$\begin{aligned} D_h^{(2)} f(t) \approx (A+B+C)f(t) &+ h(A-C)f'(t) + \frac{h^2}{2}(A+C)f''(t) \\ &+ \frac{h^3}{6}(A-C)f'''(t) + \frac{h^4}{24}(A+C)f^{(4)}(t) \end{aligned} \tag{5.88}$$

To have

$$D_h^{(2)} f(t) \approx f''(t) \tag{5.89}$$

for arbitrary functions $f(x)$, it is necessary to require

$$\begin{aligned} A + B + C &= 0: \quad \text{coefficient of } f(t) \\ h(A - C) &= 0: \quad \text{coefficient of } f'(t) \\ \frac{h^2}{2}(A + C) &= 1: \quad \text{coefficient of } f''(t) \end{aligned}$$

This system has the solution

$$A = C = \frac{1}{h^2}, \qquad B = -\frac{2}{h^2} \tag{5.90}$$

This determines

$$D_h^{(2)} f(t) = \frac{f(t+h) - 2f(t) + f(t-h)}{h^2} \tag{5.91}$$

To determine an error formula for $D_h^{(2)} f(t)$, substitute (5.90) into (5.88) to obtain

$$D_h^{(2)} f(t) \approx f''(t) + \frac{h^2}{12} f^{(4)}(t)$$

The approximation in this arises from not including terms in the Taylor polynomials (5.87) corresponding to higher powers of h. Thus,

$$f''(t) - \frac{f(t+h) - 2f(t) + f(t-h)}{h^2} \approx \frac{-h^2}{12} f^{(4)}(t) \tag{5.92}$$

This is an accurate estimate of the error for small values of h. Of course, in a practical situation we would not know $f^{(4)}(t)$. But the error formula shows that the error decreases by a factor of about 4 when h is halved. This can be used to justify Richardson's extrapolation to obtain an even more accurate estimate of the error and of $f''(t)$; see Problem 6.

Example 5.4.4 Let $f(x) = \cos(x)$, $t = \frac{1}{6}\pi$, and use (5.91) to calculate $f''(t) = -\cos\left(\frac{1}{6}\pi\right)$. The results are shown in Table 5.12. Note that the ratio column is consistent with the error formula (5.92). ■

In the derivation of (5.91), the form (5.86) was assumed for the approximate derivative. We could equally well have chosen to evaluate $f(x)$ at points other than those used there, for example,

$$f''(t) \approx Af(t + 2h) + Bf(t + h) + Cf(t) \tag{5.93}$$

Or, we could have chosen more evaluation points, as in

$$f''(t) \approx Af(t + 3h) + Bf(t + 2h) + Cf(t + h) + Df(t) \tag{5.94}$$

Table 5.12. Numerical Differentiation Using $D_h^{(2)} f$

h	$D_h^{(2)} f$	Error	Ratio
0.5	−0.84813289	−1.789E − 2	
0.25	−0.86152424	−4.501E − 3	3.97
0.125	−0.86489835	−1.127E − 3	3.99
0.0625	−0.86574353	−2.819E − 4	4.00
0.03125	−0.86595493	−7.048E − 5	4.00

The extra degree of freedom could have been used to obtain a more accurate approximation to $f''(t)$, by forcing the error term to be proportional to a higher power of h. Both of the possibilities (5.93) and (5.94) are explored in Problem 10.

Many of the formulas derived by the method of undetermined coefficients can also be derived by differentiating and evaluating a suitably chosen interpolation polynomial. But often, it is easier to visualize the desired formula as a combination of certain function values and to then derive the proper combination, as was done above for (5.91).

5.4.3 Effects of Error in Function Values

The formulas derived above are useful for differentiating functions that are known analytically and for setting up numerical methods for solving differential equations. Nonetheless, they are very sensitive to errors in the function values, especially if these errors are not sufficiently small compared with the stepsize h used in the differentiation formula. To explore this, we analyze the effect of such errors in the formula $D_h^{(2)} f(t)$ approximating $f''(t)$.

Rewrite (5.91) as

$$D_h^{(2)} f(x_1) = \frac{f(x_2) - 2f(x_1) + f(x_0)}{h^2} \approx f''(x_1) \tag{5.95}$$

where $x_2 = x_1 + h, x_0 = x_1 - h$. Let the actual function values used in the computation be denoted by $\widehat{f_0}$, $\widehat{f_1}$, and $\widehat{f_2}$ with

$$f(x_i) - \widehat{f_i} = \epsilon_i, \qquad i = 0, 1, 2 \tag{5.96}$$

the errors in the function values. Thus, the actual quantity calculated is

$$\widehat{D}_h^{(2)} f(x_1) = \frac{\widehat{f_2} - 2\widehat{f_1} + \widehat{f_2}}{h^2} \tag{5.97}$$

For the error in this quantity, replace $\widehat{f}_j$ by $f(x_j) - \epsilon_j$, $j = 0, 1, 2$, to obtain

$$\begin{aligned} f''(x_1) - \widehat{D}_h^{(2)} f(x_1) &= f''(x_1) - \frac{[f(x_2) - \epsilon_2] - 2[f(x_1) - \epsilon_1] + [f(x_0) - \epsilon_0]}{h^2} \\ &= \left[f''(x_1) - \frac{f(x_2) - 2f(x_1) + f(x_0)}{h^2} \right] + \frac{\epsilon_2 - 2\epsilon_1 + \epsilon_0}{h^2} \\ &\approx -\frac{h^2}{12} f^{(4)}(x_1) + \frac{\epsilon_2 - 2\epsilon_1 + \epsilon_0}{h^2} \end{aligned} \tag{5.98}$$

The last step used (5.92).

The errors $\epsilon_0, \epsilon_1, \epsilon_2$ are generally random in some interval $[-\delta, \delta]$. If the values $\widehat{f}_0$, $\widehat{f}_1$, $\widehat{f}_2$ are experimental data, then δ is a bound on the experimental error. Also, if these function values $\widehat{f}_j$ are obtained from computing $f(x)$ in a computer, then the errors ϵ_j are the combination of rounding or chopping errors and δ is a bound on these errors. In either case, (5.98) yields the approximate inequality

$$\left| f''(x_1) - \widehat{D}_h^{(2)} f(x_1) \right| \le \frac{h^2}{12} \left| f^{(4)}(x_1) \right| + \frac{4\delta}{h^2} \tag{5.99}$$

This error bound suggests that as $h \to 0$, the error will eventually increase, because of the final term $4\delta / h^2$.

As a contrast to this behavior in the error, recall Problem 22 from Section 5.2. In that problem, it was shown that the change in the numerical integral due to error in the function values was bounded independently of the integration parameter n (or equivalently, of the mesh size h). This is not true of numerical differentiation, as indicated in (5.98) or (5.99).

Example 5.4.5 Calculate $\widehat{D}_h^{(2)}(x_1)$ for $f(x) = \cos(x)$ at $x_1 = \frac{1}{6}\pi$. To show the effect of rounding errors, the values $\widehat{f}_i$ are obtained by rounding $f(x_i)$ to six significant digits; and the errors satisfy

$$|\epsilon_i| \le 5.0 \times 10^{-7} = \delta, \qquad i = 0, 1, 2$$

Other than these rounding errors, the formula $\widehat{D}_h^{(2)} f(x_1)$ is calculated exactly. The results are shown in Table 5.13. In this example, the bound (5.99) becomes

$$\begin{aligned} \left| f''(x_1) - \widehat{D}_h^{(2)} f(x_1) \right| &\le \frac{h^2}{12} \cos\left(\frac{1}{6}\pi\right) + \left(\frac{4}{h^2}\right)(5 \times 10^{-7}) \\ &\doteq 0.0722h^2 + \frac{2 \times 10^{-6}}{h^2} \equiv E(h) \end{aligned} \tag{5.100}$$

For $h = 0.125$, the bound $E(h) \doteq 0.00126$, which is not too far off from the actual error given in the table.

The bound $E(h)$ indicates that there is a smallest value of h, call it $h*$, below which the error bound will begin to increase. To find it, let $E'(h) = 0$, with its root being $h*$.

Table 5.13. Calculation of $\tilde{D}_h^{(2)} f(x_1)$

h	$\widehat{D}_h^{(2)} f(x_1)$	Error
0.5	−0.848128	−0.017897
0.25	−0.861504	−0.004521
0.125	−0.864832	−0.001193
0.0625	−0.865536	−0.000489
0.03125	−0.865280	−0.000745
0.015625	−0.860160	−0.005865
0.0078125	−0.851968	−0.014057
0.00390625	−0.786432	−0.079593

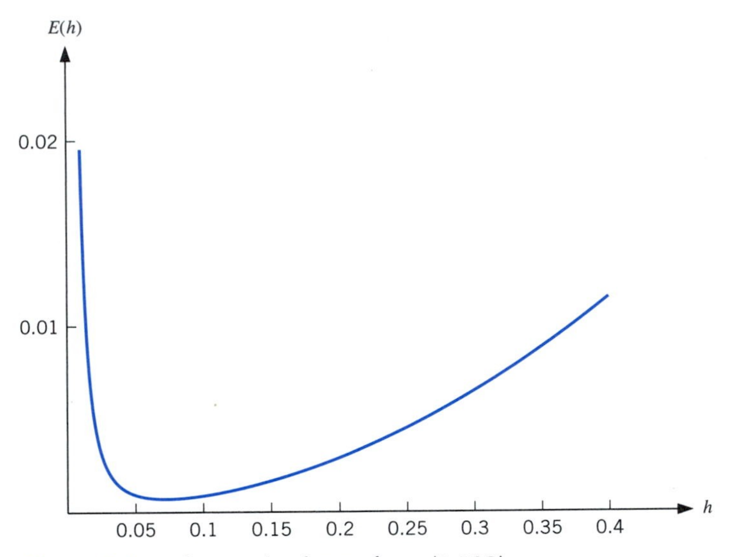

Figure 5.6. The graph of $E(h)$ from (5.100)

This leads to $h* \doteq 0.0726$, which is consistent with the behavior of the errors in Table 5.13. Also, see the graph of $E(h)$ in Figure 5.6. ■

One must be very cautious in using numerical differentiation, because of the sensitivity to errors in the function values. This is especially true if the function values are obtained empirically with relatively large experimental errors, as is common in practice. In this latter case, one should probably use a carefully prepared package program for numerical differentiation. Such programs take into account the error in the data, attempting to find numerical derivatives that are as accurate as can be justified by the data. In the absence of such a program, one should consider producing a cubic spline function that approximates the data, and then use its derivative as a numerical derivative for the data. The cubic spline function could be based on interpolation; or better for data with relatively large errors, construct a cubic spline that is a *least squares approximation* to the data. The concept of least squares approximation is introduced in Section 7.1 of Chapter 7.

PROBLEMS

1. In the following instances, find the numerical derivative $D_h f(x)$ at the indicated point, using formula (5.76). Use $h = 0.1, 0.05, 0.025, 0.0125, 0.00625$. As in Table 5.10, calculate the error and the ratio with which the error decreases. Also, estimate the error by using (5.77) with c replaced by x.

(a) $f(x) = e^x$ at $x = 0$.

(b) $f(x) = \tan^{-1}(x^2 - x + 1)$ at $x = 1$.

(c) $f(x) = \tan^{-1}(100x^2 - 199x + 100)$ at $x = 1$.

2. Repeat Problem 1, but use the backward difference formula (5.79) and estimate the error using (5.80), replacing c by x. Compare the errors of the backward difference formula with those of the forward difference formula calculated in Problem 1.

3. Repeat Problem 1, but use the numerical derivative $D_h f(x_1)$ from (5.83) and estimate the error using (5.85) with c_2 replaced by x_1.

4. Let $h > 0$, $x_j = x_0 + jh$ for $j = 0, 1, 2, \ldots, n$, and let $P_n(x)$ be the degree n polynomial interpolating $f(x)$ at $x_0, \ldots, x_n$. Use this polynomial to estimate $f'(x_0)$. Produce the actual formulas involving $f(x_0), \ldots, f(x_n)$ for $n = 1, 2, 3, 4$. Also produce their error formulas.

5. Use the degree 4 polynomial $P_4(x)$, defined in Problem 4, to find an approximation to $f'(x_2)$. Also find its error formula.

6. The error formulas (5.77), (5.85), and (5.92) all are nearly proportional to a power of h. These justify the use of Richardson's extrapolation, as in Section 5.2. For (5.76), derive the extrapolation formula

$$f'(x) \approx 2D_h f(x) - D_{2h} f(x)$$

and show its error converges to zero more rapidly than does the error for (5.76). Derive corresponding extrapolation formulas for (5.83) and (5.91), based on the error formulas (5.85) and (5.92).

7. Use the extrapolation formula for (5.76), given in Problem 6, to improve the answers given in Table 5.10. Produce a table of extrapolated values from Table 5.10. Include the errors in the extrapolated values and the ratios by which they decrease.

8. Use the polynomial $P_2(x)$ preceding (5.82) to obtain an estimate for $f''(x_1)$. Comment on the resulting formula.

9. Use $D_h^{(2)} f(x)$ from (5.91) to estimate $f''(x)$ for the functions and points in Problem 1. Also calculate the errors and the ratios by which they decrease. Use $h = 0.5$, 0.25, 0.125, 0.0625, 0.03125.

10. **(a)** Use the method of undetermined coefficients to derive the formula (5.93), with the error as small as possible.

(b) Use the method of undetermined coefficients to derive the formula (5.94), with the error as small as possible.

11. Repeat the example summarized in Table 5.13, but use $f(x) = e^x$ and $x_1 = 0$.

12. Repeat the rounding error analysis that led to (5.98), but do it for $D_h f(t)$ in (5.83).

13. On your computer with single precision arithmetic, perform the numerical differentiation $D_h^{(2)} f(t)$ of (5.91). Successively halve h, calculating $D_h^{(2)} f(t)$ and its error; continue this until the error begins to increase.

14. Using the following table of rounded values of $f(x)$, estimate $f''(0.5)$ numerically with stepsizes $h = 0.2, 0.1$. Also estimate the possible size of that part of the error in your answer that is due to the rounding errors in the table entries. Is this a serious source of error in this case?

x	$f(x)$	x	$f(x)$
0.3	7.3891	0.6	7.6141
0.4	7.4633	0.7	7.6906
0.5	7.5383		

CHAPTER 6

SOLUTION OF SYSTEMS OF LINEAR EQUATIONS

Systems of simultaneous linear equations occur in solving problems in a wide variety of disciplines, including mathematics, statistics, the physical, biological, and social sciences, engineering, and business. They arise directly in solving real-world problems, and they also occur as part of the solution process for other problems, for example, solving systems of simultaneous nonlinear equations. Numerical solutions of boundary value problems and initial boundary value problems for differential equations are a rich source of linear systems, especially large-size ones. In this chapter, we will examine some classical methods for solving linear systems, including direct methods such as the Gaussian elimination method, and iterative methods such as the Jacobi method and Gauss–Seidel method.

The notation and theory for linear systems are given in the first section, and a general method of solution is given in Section 6.3. Most work with systems of linear equations is simpler to express and understand when matrix algebra is used, and this is introduced in Section 6.2. By using matrix algebra, in Sections 6.4 and 6.5 we extend the ideas of Section 6.3 for solving linear systems and consider special types of linear systems and the effects of rounding errors. Section 6.6 introduces the idea of an iteration method for solving linear systems, giving some of the more common and simple iteration procedures.

6.1. SYSTEMS OF LINEAR EQUATIONS

One of the topics studied in elementary algebra is the solution of pairs of linear equations, such as

$$\begin{aligned} ax + by &= c \\ dx + ey &= f \end{aligned} \tag{6.1}$$

The coefficients $a, b, \ldots, f$ are given constants, and the task is to find the unknown values x, y. In this chapter, we examine the problem of finding solutions to larger systems of linear equations, containing more equations and unknowns.

To write the most general system of linear equations that we will study, we must change the notation used in (6.1) to something more convenient. Let n be a positive integer. The general form for a system of n linear equations in the n unknown components $x_1, x_2, \ldots, x_n$ is

$$\begin{aligned} a_{11}x_1 + a_{12}x_2 + a_{13}x_3 + \cdots + a_{1n}x_n &= b_1 \\ a_{21}x_1 + a_{22}x_2 + a_{23}x_3 + \cdots + a_{2n}x_n &= b_2 \\ &\vdots \\ a_{n1}x_1 + a_{n2}x_2 + a_{n3}x_3 + \cdots + a_{nn}x_n &= b_n \end{aligned} \tag{6.2}$$

The coefficients are given symbolically by a_{ij}, with i the number of the equation and j the number of the associated unknown component. On some occasions, to avoid possible confusion, we also use the symbol $a_{i,j}$. The right-hand sides $b_1, \ldots, b_n$ are given numbers; and the problem is to calculate the unknowns $x_1, \ldots, x_n$. The linear system is said to be of *order n*.

Example 6.1.1 Define the coefficients of (6.2) by

$$a_{ij} = \max\{i, j\} \tag{6.3}$$

for $1 \le i, j \le n$. Also define the right sides by

$$b_1 = b_2 = \cdots = b_n = 1 \tag{6.4}$$

The solution of this linear system, (6.2) to (6.4), is

$$x_1 = x_2 = \cdots = x_{n-1} = 0, \qquad x_n = \frac{1}{n} \tag{6.5}$$

This can also be shown to be the only solution to this linear system. To make the example more concrete, let $n = 3$. Then the linear system is

$$\begin{aligned} x_1 + 2x_2 + 3x_3 &= 1 \\ 2x_1 + 2x_2 + 3x_3 &= 1 \\ 3x_1 + 3x_2 + 3x_3 &= 1 \end{aligned} \tag{6.6}$$

and its solution is

$$x_1 = x_2 = 0, \qquad x_3 = \tfrac{1}{3} \quad \blacksquare$$

For linear systems of small orders, such as the systems (6.1) and (6.6), it is possible to solve them by hand calculations or with the help of a calculator, with methods learned in elementary algebra. For linear systems arising in most applications, however, it is common to have much larger orders, from several dozens to millions. Evidently, there is no hope to solve such large systems by hand. We need to employ numerical methods for their solutions. For this purpose, it is most convenient to use the matrix/vector notation to represent linear systems and to use the corresponding matrix/vector arithmetic for their numerical treatment.

The linear system of equations (6.2) is completely specified by knowing the coefficients a_{ij} and the right-hand constants b_i. These coefficients are arranged as the elements of a *matrix*:

$$A = \begin{bmatrix} a_{11} & a_{12} & \cdots & a_{1n} \\ a_{21} & a_{22} & \cdots & a_{2n} \\ \vdots & \vdots & \ddots & \vdots \\ a_{n1} & a_{n2} & \cdots & a_{nn} \end{bmatrix} \tag{6.7}$$

We say a_{ij} is the (i, j) element or (i, j) entry of the matrix A. Similarly, the right-hand constants b_i are arranged in the form of a vector:

$$b = \begin{bmatrix} b_1 \\ b_2 \\ \vdots \\ b_n \end{bmatrix} \tag{6.8}$$

The letters A and b are the names given to the matrix and the vector. The indices of a_{ij} now give the numbers of the row and column of A that contain a_{ij}. The solution $x_1, \ldots, x_n$ is written similarly

$$x = \begin{bmatrix} x_1 \\ x_2 \\ \vdots \\ x_n \end{bmatrix} \tag{6.9}$$

Example 6.1.2 For the linear system (6.6),

$$A = \begin{bmatrix} 1 & 2 & 3 \\ 2 & 2 & 3 \\ 3 & 3 & 3 \end{bmatrix}, \qquad b = \begin{bmatrix} 1 \\ 1 \\ 1 \end{bmatrix}, \qquad x = \begin{bmatrix} 0 \\ 0 \\ \frac{1}{3} \end{bmatrix} \tag{6.10}$$

In MATLAB, the matrix A can be created by any of the following:

```
A =[1 2 3; 2 2 3; 3 3 3];
A =[1, 2, 3; 2, 2, 3; 3, 3, 3];
A =[1 2 3
    2 2 3
    3 3 3];
```

and commas can also be used in the last case to replace the spaces as delimiters. The vector b can be created by

```
b = ones(3, 1);
```

More generally, for the system (6.2) to (6.4), we create the matrix as follows:

```
A = zeros(n,n);
for i = 1:n
  A(i,1:i) = i;
  A(i,i+1:n) = i+1:n;
end
```

and set up the vector b by

```
b = ones(n,1);
```
■

With the notations introduced in (6.7), (6.8), and (6.9), the linear system (6.2) is then written in a compact form

$$Ax = b \tag{6.11}$$

A reader with some knowledge of linear algebra will immediately recognize that the left-hand side of (6.11) is the matrix A multiplied by the vector x, and (6.11) expresses the equality between the two vectors Ax and b. The system (6.11) consists of n relations, and for $i = 1, \ldots, n$, the ith relation is precisely the ith equation in the system (6.2).

Since a knowledge of linear algebra is not assumed of the reader, in the next section we review the arithmetic and properties of matrices and vectors.

PROBLEMS

1. **(a)** Consider the problem of cubic polynomial interpolation

$$p(x_i) = y_i, \qquad i = 0, 1, 2, 3$$

with $\deg(p) \leq 3$ and x_0, x_1, x_2, x_3 distinct. Convert the problem of finding $p(x)$ to another problem involving the solution of a system of linear equations.

Hint: Write

$$p(x) = a_0 + a_1 x + a_2 x^2 + a_3 x^3$$

and determine a_0, a_1, a_2, and a_3. Use the interpolation conditions to obtain equations involving $a_0, \ldots, a_3$.

(b) Expressing the system from (a) in the form (6.11), identify the matrix A and the vectors b and x.

2. Repeat Problem 1, but with the following Hermite interpolation conditions:

$$p(x_i) = y_i, \qquad p'(x_i) = y_i', \qquad i = 1, 2$$

where $\{x_i, y_i, y_i'\}$ are given and $x_1 \neq x_2$.

3. Recall the equations (4.64) to (4.65) for the interpolating cubic spline function. Assuming the interpolation node points are evenly spaced, with a spacing of h, set up the coefficient matrix and the right-hand side vector for this linear system.

4. Consider the matrix

$$A = \begin{bmatrix} 3 & 1 & 0 & \cdots & 0 \\ 1 & 3 & 1 & \ddots & \vdots \\ 0 & \ddots & \ddots & \ddots & 0 \\ \vdots & \ddots & 1 & 3 & 1 \\ 0 & \cdots & 0 & 1 & 3 \end{bmatrix}$$

and the vectors

$$b = \begin{bmatrix} 4 \\ 5 \\ 5 \\ \vdots \\ 5 \\ 4 \end{bmatrix}, \qquad x = \begin{bmatrix} 1 \\ 1 \\ 1 \\ \vdots \\ 1 \\ 1 \end{bmatrix}$$

Set up the linear system (6.2) of order n associated with A and b. Verify that the given x solves this linear system.

5. Set up the following matrix A and vector b in MATLAB:

$$a_{ij} = \begin{cases} \dfrac{i}{j}, & 1 \le i \le j \le n \\[2ex] \dfrac{j}{i}, & 1 \le j \le i \le n \end{cases}, \qquad b_i = \begin{cases} -1, & i \text{ even} \\ 1, & i \text{ odd} \end{cases}$$

6. Repeat Problem 5 with the following matrices A and b:

$$\begin{aligned} a_{1j} = a_{j1} &= \beta, & j &= 1, 2, \ldots, n \\ a_{ij} &= a_{i-1,j} + a_{i,j-1}, & i, j &= 2, 3, \ldots, n \\ b_i &= (-1)^{i-1}/i, & i &= 1, \ldots, n \end{aligned}$$

Let β be an input parameter to the program, $\beta \ne 0$. Write out a specific example when $\beta = 1$.

7. Set up in MATLAB the coefficient matrix H_{n+1} of the linear system (4.113). Note that

$$(H_{n+1})_{i,j} = \frac{1}{i + j - 1}, \qquad 1 \le i, j \le n + 1$$

6.2. MATRIX ARITHMETIC

In this section, we formally define matrices and vectors, their arithmetic operations, and related properties that are useful for the numerical solution of linear systems.

A matrix is a rectangular array of numbers

$$B = \begin{bmatrix} b_{11} & \cdots & b_{1n} \\ \vdots & \ddots & \vdots \\ b_{m1} & \cdots & b_{mn} \end{bmatrix} \tag{6.12}$$

It is said to have *order* $m \times n$ if m is the number of rows and n is the number of columns. Square matrices with n rows and n columns are said to have order n. Matrices consisting of a single row or column are called *row* and *column matrices*, respectively. They are more often called row or column vectors, or simply, *vectors*. This is because they can be identified with geometric vectors drawn from the origin of a space to the point with coordinates given by the row or column matrix. This adds an important geometric perspective to the study of matrices, but we will omit it here. Since vectors are special types of rectangular matrices, the operations and properties for general rectangular matrices are valid for vectors also.

As notation, capital letters will be used to denote matrices; and the corresponding lowercase letters with double indices will usually be used to denote matrix elements.

Vectors will be denoted by lowercase letters and its elements are denoted by the same lowercase letters with one index; see (6.7) to (6.9) for examples. Two matrices are said to be equal if (1) they have the same order, and (2) corresponding elements are equal. Finally, the *transpose* of the general matrix in (6.12) is

$$B^T = \begin{bmatrix} b_{11} & \cdots & b_{m1} \\ \vdots & \ddots & \vdots \\ b_{1n} & \cdots & b_{mn} \end{bmatrix} \tag{6.13}$$

It is obtained by interchanging the rows and columns of B; the order of B^T is $n \times m$. In particular, the transpose of a row vector is a column vector, and vice versa. In MATLAB, once a matrix `A` is defined, its transpose can be obtained by `A'`.

Example 6.2.1

$$A \equiv \begin{bmatrix} a & b & 1 \\ 2 & c & d \end{bmatrix} = \begin{bmatrix} 5 & 6 & 1 \\ 2 & 2 & a+1 \end{bmatrix}$$

implies

$$a = 5, \qquad b = 6, \qquad c = 2, \qquad d = a + 1 = 6$$

Also

$$A^T = \begin{bmatrix} a & 2 \\ b & c \\ 1 & d \end{bmatrix}$$

A has order 2×3, and A^T has order 3×2. ■

6.2.1 Arithmetic Operations

There are three basic arithmetic operations related to matrices. The first arithmetic operation is the multiplication of a number and a matrix. Let B be the $m \times n$ matrix given in (6.12) and let a be an arbitrary real number. Then aB is a matrix of order $m \times n$, defined by

$$aB = \begin{bmatrix} ab_{11} & \cdots & ab_{1n} \\ \vdots & \ddots & \vdots \\ ab_{m1} & \cdots & ab_{mn} \end{bmatrix} \tag{6.14}$$

Each element of B is multiplied by a to get the corresponding element in aB. Particularly, with $a = -1$,

$$(-1)\,B = \begin{bmatrix} -b_{11} & \cdots & -b_{1n} \\ \vdots & \ddots & \vdots \\ -b_{m1} & \cdots & -b_{mn} \end{bmatrix}$$

The new matrix is called the negative of B and is denoted by $-B$, that is,

$$-B = (-1)\,B$$

The second arithmetic operation is the summation for matrices of the same order. Let A and B be matrices of order $m \times n$. Then $A + B$ is a new matrix of order $m \times n$, defined by

$$A + B = \begin{bmatrix} a_{11} + b_{11} & \cdots & a_{1n} + b_{1n} \\ \vdots & \ddots & \vdots \\ a_{m1} + b_{m1} & \cdots & a_{mn} + b_{mn} \end{bmatrix}$$

The (i, j) element of $A + B$ is $a_{ij} + b_{ij}$. For matrix addition, there is a zero matrix. Define the *zero matrix* of order $m \times n$ as having all zero entries. It is denoted by $O_{m\times n}$ or, more commonly, O. It has the property that

$$A + O = O + A = A$$

for any matrix A. The order of the zero matrix O is implicitly determined from the context.

With the summation and scalar multiplication at our disposal, we can define the matrix subtraction for matrices of the same order:

$$A - B = A + (-B)$$

For any matrix A, we have

$$A - A = A + (-A) = O$$

Example 6.2.2

$$-5\begin{bmatrix} 2 & 3 \\ 6 & -1 \end{bmatrix} = \begin{bmatrix} -10 & -15 \\ -30 & 5 \end{bmatrix}$$

$$\begin{bmatrix} 3 & 2 & 1 \\ 1 & 2 & 3 \end{bmatrix} + \begin{bmatrix} 1 & 2 & 3 \\ 3 & 2 & 1 \end{bmatrix} = \begin{bmatrix} 4 & 4 & 4 \\ 4 & 4 & 4 \end{bmatrix} = 4\begin{bmatrix} 1 & 1 & 1 \\ 1 & 1 & 1 \end{bmatrix}$$

$$\begin{bmatrix} 3 & 2 & 1 \\ 1 & 2 & 3 \end{bmatrix} - \begin{bmatrix} 1 & 2 & 3 \\ 3 & 2 & 1 \end{bmatrix} = \begin{bmatrix} 2 & 0 & -2 \\ -2 & 0 & 2 \end{bmatrix} = 2\begin{bmatrix} 1 & 0 & -1 \\ -1 & 0 & 1 \end{bmatrix}$$

$$\begin{bmatrix} 2 & 3 \\ -1 & 5 \end{bmatrix} + \begin{bmatrix} -2 & -3 \\ 1 & -5 \end{bmatrix} = \begin{bmatrix} 0 & 0 \\ 0 & 0 \end{bmatrix} = O \quad \blacksquare$$

The third arithmetic operation, matrix multiplication, is more complicated to define and to calculate. Let A have order $m \times n$ and B have order $n \times p$. Then $C = AB$, the product of A and B, is a matrix of order $m \times p$, and its general element c_{ij} is defined by

$$c_{ij} = \sum_{k=1}^{n} a_{ik}b_{kj}, \qquad 1 \le i \le m, \qquad 1 \le j \le p \tag{6.15}$$

This is the sum of the products of corresponding elements from row i of A and column j of B. If A and B are square matrices, then AB is a square matrix of the same order. The product of an $m \times n$ matrix A and an n column vector b is an m column vector Ab.

The definition of matrix multiplication is associated with the composition of linear functions. A linear function mapping a vector of order n to a vector of order m

$$f(x) = \begin{bmatrix} a_{11}x_1 + \cdots + a_{1n}x_n \\ \vdots \\ a_{m1}x_1 + \cdots + a_{mn}x_n \end{bmatrix} \qquad \text{for} \qquad x = \begin{bmatrix} x_1 \\ \vdots \\ x_n \end{bmatrix}$$

can be expressed as the product of a matrix and a vector

$$f(x) = Ax$$

with

$$A = \begin{bmatrix} a_{11} & \cdots & a_{1n} \\ \vdots & \ddots & \vdots \\ a_{m1} & \cdots & a_{mn} \end{bmatrix}$$

Let linear functions f and g be defined by

$$f(x) = Ax \qquad \text{for} \qquad x = \begin{bmatrix} x_1 \\ \vdots \\ x_n \end{bmatrix}$$

$$g(y) = By \qquad \text{for} \qquad y = \begin{bmatrix} y_1 \\ \vdots \\ y_p \end{bmatrix}$$

Then the matrix associated with the composite function $h(y) \equiv f(g(y))$ is $C = AB$, as defined in (6.15).

Example 6.2.3 (a)

$$\begin{bmatrix} 2 & 3 & 2 \\ -1 & 2 & -1 \end{bmatrix} \begin{bmatrix} 1 & 0 \\ 2 & -1 \\ 3 & 1 \end{bmatrix} = \begin{bmatrix} 14 & -1 \\ 0 & -3 \end{bmatrix}$$

$$\begin{bmatrix} 1 & 0 \\ 2 & -1 \\ 3 & 1 \end{bmatrix} \begin{bmatrix} 2 & 3 & 2 \\ -1 & 2 & -1 \end{bmatrix} = \begin{bmatrix} 2 & 3 & 2 \\ 5 & 4 & 5 \\ 5 & 11 & 5 \end{bmatrix}$$

(b)

$$\begin{bmatrix} 3 & 2 & 1 \\ 1 & 2 & 3 \\ 1 & -1 & 1 \end{bmatrix} \begin{bmatrix} 1 \\ 1 \\ 1 \end{bmatrix} = \begin{bmatrix} 6 \\ 6 \\ 1 \end{bmatrix}$$

(c)

$$\begin{bmatrix} 3 & 2 & 1 \\ 1 & 2 & 3 \\ 1 & -1 & 1 \end{bmatrix} \begin{bmatrix} x_1 \\ x_2 \\ x_3 \end{bmatrix} = \begin{bmatrix} 3x_1 + 2x_2 + x_3 \\ x_1 + 2x_2 + 3x_3 \\ x_1 - x_2 + x_3 \end{bmatrix}$$ ■

As suggested by the last example, a system of linear equations can be written as a matrix equation by using matrix multiplication. For example, the system

$$\begin{aligned} x_1 + 2x_2 + x_3 &= 0 \\ 2x_1 + 2x_2 + 3x_3 &= 3 \\ -x_1 - 3x_2 &= 2 \end{aligned}$$

can be written as

$$\begin{bmatrix} 1 & 2 & 1 \\ 2 & 2 & 3 \\ -1 & -3 & 0 \end{bmatrix} \begin{bmatrix} x_1 \\ x_2 \\ x_3 \end{bmatrix} = \begin{bmatrix} 0 \\ 3 \\ 2 \end{bmatrix}$$

The reader should check that the system will follow after performing the matrix multiplication on the left side and matching the corresponding elements on the two sides of the equation.

As was noted at the end of the previous section, the general linear system

$$\begin{aligned} a_{11}x_1 + \cdots + a_{1n}x_n &= b_1 \\ &\vdots \\ a_{n1}x_1 + \cdots + a_{nn}x_n &= b_n \end{aligned} \tag{6.16}$$

can be concisely written as

$$Ax = b \tag{6.17}$$

When written in this way, many results become conceptually clearer. Also, results from the theory of matrices and linear algebra can be applied to (6.17), to obtain both theoretical results and computational methods for solving the linear system (6.16).

6.2.2 Elementary Row Operations

The Gaussian elimination method for solving linear systems, to be discussed in the next section, will be most conveniently described in matrix form. To prepare for this, we introduce three elementary row operations on general rectangular matrices. They are:

(i) Interchange of two rows.

(ii) Multiplication of a row by a nonzero scalar.

(iii) Addition of a nonzero multiple of one row to another row.

Example 6.2.4 Consider the rectangular matrix

$$A = \begin{bmatrix} 3 & 3 & 3 & 1 \\ 2 & 2 & 3 & 1 \\ 1 & 2 & 3 & 1 \end{bmatrix}$$

We add row 2 times (-1) to row 1, and then add row 3 times (-1) to row 2 to obtain the matrix:

$$\begin{bmatrix} 1 & 1 & 0 & 0 \\ 1 & 0 & 0 & 0 \\ 1 & 2 & 3 & 1 \end{bmatrix}$$

Add row 2 times (-1) to row 1, and to row 3 as well:

$$\begin{bmatrix} 0 & 1 & 0 & 0 \\ 1 & 0 & 0 & 0 \\ 0 & 2 & 3 & 1 \end{bmatrix}$$

Add row 1 times (-2) to row 3:

$$\begin{bmatrix} 0 & 1 & 0 & 0 \\ 1 & 0 & 0 & 0 \\ 0 & 0 & 3 & 1 \end{bmatrix}$$

Interchange row 1 and row 2:

$$\begin{bmatrix} 1 & 0 & 0 & 0 \\ 0 & 1 & 0 & 0 \\ 0 & 0 & 3 & 1 \end{bmatrix}$$

Finally, we multiply row 3 by $\frac{1}{3}$:

$$\begin{bmatrix} 1 & 0 & 0 & 0 \\ 0 & 1 & 0 & 0 \\ 0 & 0 & 1 & \frac{1}{3} \end{bmatrix}$$

This is a matrix obtained from the original matrix A through elementary row operations. ■

6.2.3 The Matrix Inverse

The number 1 satisfies

$$1 \cdot x = x \cdot 1 = x$$

for all real numbers x. With matrix multiplication, the analog of 1 is the *identity matrix* I, also called the *unit matrix* in some references. Let $n \geq 1$, and define I_n to be the square matrix of order n whose (i, j) element is

$$\delta_{ij} = \begin{cases} 1, & i = j \\ 0, & i \neq j \end{cases}$$

With $n = 3$,

$$I_3 = \begin{bmatrix} 1 & 0 & 0 \\ 0 & 1 & 0 \\ 0 & 0 & 1 \end{bmatrix}$$

The most important property of the identity matrix is

$$AI_n = A, \qquad I_m A = A \tag{6.18}$$

where A has order $m \times n$. This is the analog of the above identity with the real number 1.

There is an identity matrix of each order $n \geq 1$. Instead of writing I_n, the subscript n is usually omitted and I denotes any possible identity, with the order chosen suitably for the given situation. For example, (6.18) would be written as

$$AI = IA = A$$

Again, in analogy with the real numbers, we can generalize the concept of the reciprocal of a number. Let A be a square matrix. If there is a square matrix B such that

$$AB = I, \qquad BA = I \tag{6.19}$$

then B is called the *inverse* of A. It is shown later that if A has an inverse, then it has exactly one inverse. The inverse of A is denoted by A^{-1}, in analogy with the reciprocal of a real number. Moreover, only one of the above statements in (6.19) about B need be verified, for each of the statements can be deduced from the other.

Example 6.2.5 (a)

$$A = \begin{bmatrix} 2 & 1 & 0 \\ 1 & 2 & 1 \\ 0 & 1 & 2 \end{bmatrix}, \qquad A^{-1} = \begin{bmatrix} \frac{3}{4} & -\frac{1}{2} & \frac{1}{4} \\ -\frac{1}{2} & 1 & -\frac{1}{2} \\ \frac{1}{4} & -\frac{1}{2} & \frac{3}{4} \end{bmatrix}$$

(b) Let

$$A = \begin{bmatrix} a & b \\ c & d \end{bmatrix}$$

Then it can be shown that A has an inverse if and only if $ad - bc \neq 0$, and in that case

$$A^{-1} = \frac{1}{ad - bc} \begin{bmatrix} d & -b \\ -c & a \end{bmatrix} \tag{6.20}$$

In both (a) and (b), check the formula for A^{-1} by multiplying with A and seeing whether the product is the identity matrix. ■

With real numbers, the number b has a reciprocal (or inverse) if and only if $b \neq 0$. For matrices, this generalizes as follows.

Theorem 6.2.6 Let A be a square matrix. Then A has an inverse if and only if $\det(A) \neq 0$.

In this theorem, $\det(A)$ denotes the *determinant* of A. Many people have studied the determinant in beginning algebra, in association with solving (6.16), so we omit any theoretical discussion of it. Matrices A for which $\det(A) = 0$ are called *singular*, and those with $\det(A) \neq 0$ are called *nonsingular*.

For A of order $n = 2$,

$$A = \begin{bmatrix} a_{11} & a_{12} \\ a_{21} & a_{22} \end{bmatrix}$$

its determinant is

$$\det(A) = a_{11}a_{22} - a_{12}a_{21} \tag{6.21}$$

6.2.4 Matrix Algebra Rules

Many of the rules of arithmetic with real numbers are still valid for arithmetic with matrices, but there are some important differences. We simply list these rules, illustrating and showing a few of them. The matrices in these statements are not restricted to be square, except when referring to inverses and determinants. The orders of the matrices in a relation are assumed compatible so that the listed arithmetic expressions make sense. For example, in (6.24), A and B are rectangular matrices of the same order.

$$(A + B) + C = A + (B + C) \tag{6.22}$$

$$(AB)C = A(BC) \tag{6.23}$$

$$A + B = B + A \tag{6.24}$$

$$A(B + C) = AB + AC \tag{6.25}$$

$$(A + B)C = AC + BC \tag{6.26}$$

$$(AB)^T = B^T A^T \tag{6.27}$$

$$(A + B)^T = A^T + B^T \tag{6.28}$$

$$(cA)^{-1} = \frac{1}{c} A^{-1}, \qquad c = \text{nonzero constant} \tag{6.29}$$

$$(AB)^{-1} = B^{-1}A^{-1} \tag{6.30}$$

$$\det(AB) = \det(A)\det(B) \tag{6.31}$$

$$\det(A^T) = \det(A) \tag{6.32}$$

$$\det(cA) = c^n \det(A), \qquad \text{order } (A) = n, \qquad c = \text{constant} \tag{6.33}$$

Properties (6.22) and (6.23) are the associative laws, (6.24) is the commutative law, (6.25) and (6.26) are the distributive laws. Implicit in the relations (6.29) and (6.30) are the assumptions that A and B are square matrices of the same order, both invertible. Then (6.29) states that for any constant $c \neq 0$, cA is also invertible and its inverse is given by $\frac{1}{c}A^{-1}$. Similarly, (6.30) states that AB is invertible with $(AB)^{-1}$ given by $B^{-1}A^{-1}$. The properties (6.31), (6.32), and (6.33) give results for the matrix determinant.

Let us prove (6.23) as an example. Let A, B, and C have orders $m \times n$, $n \times p$, and $p \times q$, respectively. Then AB has order $m \times p$, and BC has order $n \times q$. In the following, D_{ij} will denote the (i, j) element in a matrix D. Then

$$[(AB)C]_{ij} = \sum_{l=1}^{p} (AB)_{il} C_{lj} = \sum_{l=1}^{p} \left[\sum_{k=1}^{n} A_{ik} B_{kl} \right] C_{lj}$$

$$[A(BC)]_{ij} = \sum_{k=1}^{n} A_{ik} (BC)_{kj} = \sum_{k=1}^{n} A_{ik} \left[\sum_{l=1}^{p} B_{kl} C_{lj} \right]$$

These two right sides are the same, being rearrangements of the sum of all products $A_{ik} B_{kl} C_{lj}$ for $1 \le k \le n$, $1 \le l \le p$. This proves (6.23).

Example 6.2.7 Show that if A has an inverse, then it is unique. To show this, suppose there are two such inverses B_1 and B_2, with

$$B_1 A = A B_1 = I, \qquad B_2 A = A B_2 = I$$

Let us examine the product $B_1 A B_2$. First, by the associative law,

$$(B_1 A) B_2 = B_1 (A B_2)$$

Then from the assumptions on B_1 and B_2,

$$(B_1 A) B_2 = I B_2 = B_2$$
$$B_1 (A B_2) = B_1 I = B_1$$

Putting these together proves $B_1 = B_2$. Thus, A can have only one inverse. ■

Example 6.2.8 Let

$$A = \begin{bmatrix} 1 & 2 \\ 3 & 4 \end{bmatrix}, \qquad B = \begin{bmatrix} 1 & -1 \\ 1 & 1 \end{bmatrix}$$

Then

$$\det(A) = -2, \qquad \det(B) = 2$$

$$AB = \begin{bmatrix} 3 & 1 \\ 7 & 1 \end{bmatrix}$$

$$\det(AB) = -4 = \det(A)\det(B)$$

This illustrates (6.31). ■

Missing from the preceding list of rules is the commutative law for multiplication

$$AB = BA \tag{6.34}$$

because this is *not true* for most square matrices A and B. Two matrices that do satisfy (6.34) are said to *commute* with each other.

Example 6.2.9 Use the matrices A and B from the last example. Then

$$AB = \begin{bmatrix} 3 & 1 \\ 7 & 1 \end{bmatrix}, \qquad BA = \begin{bmatrix} -2 & -2 \\ 4 & 6 \end{bmatrix}$$

The matrices A and B do not commute. ■

6.2.5 Solvability Theory of Linear Systems

Consider the linear system (6.17), $Ax = b$, where A is a square matrix of order n. In the theory of the solvability of the linear system (6.17), an important role is played by the special system obtained by letting all of the constants $b_i = 0$. With this definition of b, the system is called *homogeneous*; all other systems are called *nonhomogeneous*. The following theorem summarizes some important theoretical results about linear systems.

Theorem 6.2.10 Let n be a positive integer, and let A be a square matrix of order n. Then the following are equivalent statements about the linear system (6.17).

S1. For each right side b, the system (6.17) has exactly one solution x.

S2. For each right side b, the system (6.17) has at least one solution x.

S3. The homogeneous form of (6.17) has exactly one solution

$$x_1 = x_2 = \cdots = x_n = 0$$

S4. $\det(A) \neq 0$.

This chapter considers the solution of nonsingular systems of linear equations. From (S1) in the above theorem, such systems have a unique solution $x_1, \ldots, x_n$ for each set of right-hand constants $b_1, \ldots, b_n$. The remaining parts of the theorem, (S2) to (S4), are equivalent to (S1); but it is often easier to show that (S1) is true for your system by instead showing (S2), (S3), or (S4). For example, questions involving the existence and uniqueness of interpolating polynomials are often treated by first reformulating them as questions involving linear systems of the form (6.17); then, these systems are often treated using (S3). We omit such examples here, but they occur commonly in higher-level numerical analysis.

From (S4), the linear system of order 2 associated with this A

$$\begin{aligned} a_{11}x_1 + a_{12}x_2 &= b_1 \\ a_{21}x_1 + a_{22}x_2 &= b_2 \end{aligned}$$

is uniquely solvable for all right-hand sides b if and only if

$$a_{11}a_{22} - a_{12}a_{21} \neq 0$$

which is an easy condition to check. For systems of larger order, (S4) is also used in some cases to determine the solvability of the linear system associated with A.

Referring back to Theorem 6.2.6, we see that the existence of A^{-1} is equivalent to any of the statements (S1) to (S4) of Theorem 6.2.10. Thus, A has an inverse if and only if the linear system $Ax = b$ has a unique solution x for any right-hand column vector b.

PROBLEMS

1. Simplify the following matrix expressions to obtain a single matrix for each case:

(a)

$$2\begin{bmatrix} 1 & 0 \\ -1 & 3 \end{bmatrix} + \begin{bmatrix} 1 & 1 \\ 1 & -1 \end{bmatrix}\begin{bmatrix} 2 & 3 \\ 1 & 3 \end{bmatrix}$$

(b)

$$\begin{bmatrix} 1 & 2 \\ 0 & 3 \end{bmatrix}\begin{bmatrix} a & b \\ 0 & c \end{bmatrix}$$

(c)

$$A = I_3 - 2ww^T, \qquad w^T = \left[\tfrac{1}{3}, \tfrac{2}{3}, \tfrac{2}{3}\right], \qquad B = A^2$$

(d)

$$A = \begin{bmatrix} \cos(\theta) & \sin(\theta) \\ -\sin(\theta) & \cos(\theta) \end{bmatrix}, \qquad B = AA^T$$

(e)

$$A = \begin{bmatrix} 2 & 1 & 0 & \cdots & 0 \\ 1 & 2 & 1 & \ddots & \vdots \\ 0 & \ddots & \ddots & \ddots & 0 \\ \vdots & \ddots & 1 & 2 & 1 \\ 0 & \cdots & 0 & 1 & 2 \end{bmatrix}, \qquad B = A^2$$

2. Let A be an arbitrary square matrix of order 3, and let

$$D = \begin{bmatrix} \lambda_1 & 0 & 0 \\ 0 & \lambda_2 & 0 \\ 0 & 0 & \lambda_3 \end{bmatrix}$$

calculate AD and DA. The matrix D is called a *diagonal matrix*. Give a simple rule for multiplication of a general matrix by a diagonal matrix.

3. Let A be a square matrix of order n, and let it satisfy

$$a_{ij} = 0 \qquad \text{for} \qquad i > j$$

Such a matrix is called *upper triangular*. Show that the sum and products of such matrices are also upper triangular. Is the inverse upper triangular?

4. A square matrix A is called *lower triangular* if its transpose A^T is upper triangular. Use this definition, results of the previous problem, and properties of matrix transposition to show that the sum and product of lower triangular matrices are also lower triangular.

5. Let w be a column vector for which $w^T w = 1$. The product $A = ww^T$ is a square matrix. Show that $A^2 = A$.

6. A matrix B is called *symmetric* if $B^T = B$. Write a general formula for all symmetric matrices of order 2×2.

7. Combining ideas from Problems 5 and 6, define a new matrix

$$B = I - 2ww^T$$

where $w^T w = 1$. Show that B is symmetric and that $B^2 = I$. What is B^{-1}?

8. Let A be $m \times n$ and let B be $n \times p$. Do an operations count for calculating AB. Consider, in particular, the cases $m = n = p$ and $m = n$, $p = 1$.

9. Let u, v, and w be column vectors of length n. Define $x = u^{\mathrm{T}} v w^{\mathrm{T}}$ by the associative law for matrix multiplication, $\left(u^{\mathrm{T}} v\right) w^{\mathrm{T}} = u^{\mathrm{T}} \left(v w^{\mathrm{T}}\right)$. Do an operations count on these two ways of computing x. Which way is preferable from the perspective of cost, or does it make any difference?

10. Let A, B, and C have orders $m \times n, n \times p$, and $p \times q$, respectively. Do operation counts for the multiplications involved in calculating $(AB)C$ and $A(BC)$. Show with particular values of m, n, p, and q that these can be very different.

11. Define $B = ww^T$, with w a column vector of length n.

(a) Give an operations count for forming B. Be as efficient as possible.

(b) Let A be an arbitrary matrix of order $n \times n$. Give the additional number of operations needed to form the product A and B, using the matrix B formed in (a).

(c) Give an alternative and less costly way, in operations, to form the product $AB = A(ww^T)$.

12. Let u and v be column vectors of length n. Define $A = uv^{\mathrm{T}}$ and note that A is a square matrix of order $n \times n$.

(a) Give the general element $A_{i,j}$. What is the cost in number of operations to create A from u and v in the computer?

(b) What is the operations cost of evaluating Ax, where x is a general column vector of length n?

(c) Give an alternative and less costly way, in operations, to evaluate $Ax = (uv^{\mathrm{T}})x$.

13. Use elementary row operations to transform the following matrices to an identity matrix:

(a) $\begin{bmatrix} 1 & 2 \\ 3 & 4 \end{bmatrix}$ **(b)** $\begin{bmatrix} 2 & 3 \\ 5 & 6 \end{bmatrix}$

14. Use elementary row operations to transform the following matrices to the form:

$$\begin{bmatrix} 1 & 0 & \mathrm{x} \\ 0 & 1 & \mathrm{x} \end{bmatrix}$$

where x represents some arbitrary number.

(a) $\begin{bmatrix} 2 & 1 & 1 \\ 1 & 2 & 0 \end{bmatrix}$ **(b)** $\begin{bmatrix} 1 & 4 & 1 \\ 9 & 35 & 7 \end{bmatrix}$

15. Produce two square matrices A and B of order 2 for which $AB = O$, but A and B have all nonzero elements. Can either of the two matrices be nonsingular?

16. Let A and B be two square matrices such that AB is singular. Show that either A or B is singular.

17. In order that

$$A = \begin{bmatrix} a & b \\ c & d \end{bmatrix}$$

will commute with

$$B = \begin{bmatrix} 1 & 2 \\ 2 & 1 \end{bmatrix}$$

what conditions must be satisfied by a, b, c, and d?

18. Let A be a square matrix. Show that

$$\left(I + 2A + 3A^2\right)(2I - A) = (2I - A)\left(I + 2A + 3A^2\right)$$

thus showing that the matrices $I + 2A + 3A^2$ and $2I - A$ commute.

Remark: Let

$$p(t) = a_0 + a_1 t + \cdots + a_k t^k$$
$$q(t) = b_0 + b_1 t + \cdots + b_\ell t^\ell$$

be polynomials of degrees k and ℓ. Define matrices

$$p(A) = a_0 I + a_1 A + \cdots + a_k A^k$$
$$q(A) = b_0 I + b_1 A + \cdots + b_\ell A^\ell$$

It can be shown that $p(A)q(A) = q(A)p(A)$.

19. Using the polynomial matrix notation introduced in Problem 18, let

$$p(t) = (t - r_1)(t - r_2) = t^2 - t(r_1 + r_2) + r_1 r_2$$

For an arbitrary square matrix A, show

$$p(A) = (A - r_1 I)(A - r_2 I) = A^2 - A(r_1 + r_2) + r_1 r_2 I$$

20. Derive the formula (6.20) for the inverse of an order 2 matrix.

21. Check that the inverse of the $n \times n$ matrix

$$A = \begin{bmatrix} 1 & -1 & 0 & \cdots & 0 \\ -1 & 2 & -1 & \ddots & \vdots \\ 0 & \ddots & \ddots & \ddots & 0 \\ \vdots & \ddots & -1 & 2 & -1 \\ 0 & \cdots & 0 & -1 & 2 \end{bmatrix}$$

is given by

$$A^{-1} = \begin{bmatrix} n & n-1 & n-2 & n-3 & \cdots & 1 \\ n-1 & n-1 & n-2 & n-3 & \cdots & 1 \\ n-2 & n-2 & n-2 & n-3 & \cdots & 1 \\ \vdots & \vdots & \vdots & \vdots & & \vdots \\ 1 & 1 & 1 & 1 & \cdots & 1 \end{bmatrix}$$

22. Let A be $m \times n$ and B be $n \times p$. Show $(AB)^T = B^T A^T$.

Hint: Follow the type of proof used for showing $(AB)C = A(BC)$, following (6.33).

23. Let A and B be nonsingular matrices of the same order. Show that AB is nonsingular and $(AB)^{-1} = B^{-1}A^{-1}$.

Hint: Multiply by AB.

24. Show $(A^T)^{-1} = (A^{-1})^T$.

25. For linear systems of order 2, verify that (S1) is equivalent to (S4) in Theorem 6.2.10.

26. Recall the reformulation of cubic polynomial interpolation given in Problem 1, Section 6.1. Show that the linear system for a_0, a_1, a_2, a_3 always has a unique solution by proving the equivalent statement (S3) in Theorem 6.2.10.

Hint: Statement (S3) can be interpreted as a statement about zeros of a cubic polynomial. How many zeros can a nonzero cubic polynomial possess?

27. Consider showing the system of Problem 4 of Section 6.1 is nonsingular, by showing (S3) in Theorem 6.2.10. Let x be the solution of the homogeneous system, and let

$$\alpha = \max_{1 \le i \le n} |x_i|$$

This maximum will be attained for at least one of the indices, say, $\alpha = |x_k|$. Look at equation k. By using it, one can show that $\alpha = 0$ is the only possibility for α, thus showing $x_i = 0$ for all i. Do this. Consider separately the possibilities $k = 1$ or n, and $1 < k < n$.

28. By using Theorem 6.2.6 and assuming $\det(A) = 0$, then there is at least one vector $x \neq 0$ for which $Ax = 0$. For the following singular matrix, find such an x:

$$\begin{bmatrix} 0 & 4 & 1 & 1 \\ 4 & 0 & 1 & 1 \\ 1 & 1 & -1 & 2 \\ 1 & 1 & 2 & -1 \end{bmatrix}$$

29. **(a)** Find the values of λ for which

$$\det[\lambda I - B] = 0, \qquad B = \begin{bmatrix} 2 & 1 \\ 1 & 2 \end{bmatrix}$$

These λ values are called the eigenvalues of the matrix B, cf. Section 7.2.

(b) Using the λ values found in (a), and using the first sentence of Problem 28 with $A = \lambda I - B$, find an $x \neq 0$ for which $(\lambda I - B)x = 0$ or, equivalently, $Bx = \lambda x$. Such a vector x is called an eigenvector associated with the eigenvalue λ.

30. Consider the following matrices, called elementary matrices. Let I be the identity matrix of order n, and modify its column #k below the diagonal, for some $1 \leq k \leq n-1$. Define

$$E_k = \begin{bmatrix} 1 & 0 & \cdots & 0 & & \cdots & & & 0 \\ 0 & 1 & & 0 & & & & & \\ \vdots & & \ddots & \vdots & & & & & \\ & & & 1 & 0 & & & & \\ & & & 0 & 1 & \ddots & & & \vdots \\ & & & 0 & a_{k+1} & 1 & & & \\ & & & 0 & a_{k+2} & 0 & \ddots & & \\ \vdots & & & \vdots & \vdots & \vdots & & 1 & 0 \\ 0 & 0 & \cdots & 0 & a_n & 0 & \cdots & 0 & 1 \end{bmatrix}$$

(a) Show that the products of elementary matrices are lower triangular matrices.

(b) If A is a matrix of order $n \times m$, show that $E_k A$ is formed from A as follows: Add a_i times row #k of A to row #i of A, forming the new row #i of $E_k A$, for $k+1 \leq i \leq n$. Rows $1, 2, \ldots, k$ of A are left unchanged.

(c) Calculate the inverse of E_k.

Hint: Use the result of part (b), reversing the operations used in forming $E_k A$.

6.3. GAUSSIAN ELIMINATION

To solve large systems of linear equations, we use a structured form of the method taught in introductory algebra courses. It is introduced here by first using it to solve a particular system of order 3. Consider the system

$$\begin{aligned} x_1 + 2x_2 + x_3 &= 0 \qquad E(1) \\ 2x_1 + 2x_2 + 3x_3 &= 3 \qquad E(2) \\ -x_1 - 3x_2 \qquad &= 2 \qquad E(3) \end{aligned} \tag{6.35}$$

The individual equations have been labeled for easier reference.

Step 1 Eliminate x_1 from $E(2)$ and $E(3)$. Subtract 2 times $E(1)$ from $E(2)$; and subtract (-1) times $E(1)$ from $E(3)$. This results in the new system:

$$\begin{aligned} x_1 + 2x_2 + x_3 &= 0 \qquad E(1) \\ -2x_2 + x_3 &= 3 \qquad E(2) \\ -x_2 + x_3 &= 2 \qquad E(3) \end{aligned}$$

Step 2 Eliminate x_2 from $E(3)$. Subtract $\frac{1}{2}$ times $E(2)$ from $E(3)$. This yields

$$\begin{aligned} x_1 + 2x_2 + x_3 &= 0 \qquad E(1) \\ -2x_2 + x_3 &= 3 \qquad E(2) \\ \tfrac{1}{2}x_3 &= \tfrac{1}{2} \qquad E(3) \end{aligned} \tag{6.36}$$

Step 3 In succession, solve for x_3, x_2, and x_1.

$$\begin{aligned} x_3 &= 1 \\ -2x_2 + 1 &= 3 \\ x_2 &= -1 \\ x_1 + 2(-1) + 1 &= 0 \\ x_1 &= 1 \end{aligned}$$

Steps 1 and 2 are the *elimination* steps, resulting in (6.36), which is called an *upper triangular system* of linear equations. This system has exactly the same solutions as (6.35), but (6.36) is in a form that is easier to solve. Step 3 is called solution by *back substitution*. The entire process is called *Gaussian elimination*, and it is generally the most efficient means of solving a linear system of small to medium size, say, of an order up to several hundreds.

The elimination steps are more conveniently carried out using the matrix notation and operations. For this purpose, we form the augmented matrix for the system (6.35):

$$[\,A \mid b\,] = \left[\begin{array}{rrr|r} 1 & 2 & 1 & 0 \\ 2 & 2 & 3 & 3 \\ -1 & -3 & 0 & 2 \end{array}\right] \tag{6.37}$$

Notice that eliminating a variable from an equation is equivalent to making the corresponding coefficient zero through elementary row operations. For example, to eliminate x_1 from $E(2)$ and $E(3)$, we multiply row 1 by (-2) and add it to row 2, and then add row 1 to row 3 to obtain

$$\left[\begin{array}{rrr|r} 1 & 2 & 1 & 0 \\ 0 & -2 & 1 & 3 \\ 0 & -1 & 1 & 2 \end{array}\right]$$

Then, to eliminate x_2 from $E(3)$, row 2 is multiplied by $(-1/2)$ and is added to row 3:

$$\left[\begin{array}{ccc|c} 1 & 2 & 1 & 0 \\ 0 & -2 & 1 & 3 \\ 0 & 0 & \frac{1}{2} & \frac{1}{2} \end{array}\right]$$

This matrix is obtained from the matrix (6.37) through elementary row operations and we have reduced the original system (6.35) to the upper triangular system

$$\begin{bmatrix} 1 & 2 & 1 \\ 0 & -2 & 1 \\ 0 & 0 & \frac{1}{2} \end{bmatrix} \begin{bmatrix} x_1 \\ x_2 \\ x_3 \end{bmatrix} = \begin{bmatrix} 0 \\ 3 \\ \frac{1}{2} \end{bmatrix} \tag{6.38}$$

which is the same as (6.36). Then the backward substitution procedure can be applied to solve (6.38).

In elementary algebra courses, variables are often eliminated in a somewhat ad hoc manner. One is taught to look for simple integer coefficients to use in the elimination, in order to minimize calculations with fractions. But on a computer, dealing with fractions is no more difficult than using integers. To write a computer program, the algorithm must have a precisely stated set of directions, those that do not depend on whether or not a coefficient happens to be an integer.

To see the formal structure of *Gaussian elimination*, we again look at the $n = 3$ case, but with general coefficients. The system to be solved is

$$\begin{aligned} a_{11}x_1 + a_{12}x_2 + a_{13}x_3 &= b_1 \qquad E(1) \\ a_{21}x_1 + a_{22}x_2 + a_{23}x_3 &= b_2 \qquad E(2) \\ a_{31}x_1 + a_{32}x_2 + a_{33}x_3 &= b_3 \qquad E(3) \end{aligned} \tag{6.39}$$

Step 1 Eliminate x_1 from $E(2)$ and $E(3)$. To simplify the presentation, assume $a_{11} \neq 0$; this assumption will be removed later in the section. Define

$$m_{21} = \frac{a_{21}}{a_{11}}, \qquad m_{31} = \frac{a_{31}}{a_{11}}$$

Subtract m_{21} times $E(1)$ from $E(2)$, and subtract m_{31} times $E(1)$ from $E(3)$. This changes (6.39) to the equivalent system:

$$\begin{aligned} a_{11}x_1 + a_{12}x_2 + a_{13}x_3 &= b_1 \qquad E(1) \\ a_{22}^{(2)}x_2 + a_{23}^{(2)}x_3 &= b_2^{(2)} \qquad E(2) \\ a_{32}^{(2)}x_2 + a_{33}^{(2)}x_3 &= b_3^{(2)} \qquad E(3) \end{aligned} \tag{6.40}$$

The coefficients $a_{ij}^{(2)}$ are defined by

$$a_{ij}^{(2)} = a_{ij} - m_{i1}a_{1j}, \qquad i, j = 2, 3$$
$$b_i^{(2)} = b_i - m_{i1}b_1, \qquad i = 2, 3$$

Step 2 Eliminate x_2 from $E(3)$. Again assume temporarily that $a_{22}^{(2)} \neq 0$. Define

$$m_{32} = \frac{a_{32}^{(2)}}{a_{22}^{(2)}}$$

Subtract m_{32} times $E(2)$ from $E(3)$. This yields

$$\begin{aligned} a_{11}x_1 + a_{12}x_2 + a_{13}x_3 &= b_1 & E(1) \\ a_{22}^{(2)}x_2 + a_{23}^{(2)}x_3 &= b_2^{(2)} & E(2) \\ a_{33}^{(3)}x_3 &= b_3^{(3)} & 3E(3) \end{aligned}$$

The new coefficients are defined by

$$a_{33}^{(3)} = a_{33}^{(2)} - m_{32}a_{23}^{(2)}, \qquad b_3^{(3)} = b_3^{(2)} - m_{32}b_2^{(2)}$$

Step 3 Use back substitution to solve successively for x_3, x_2, and x_1.

$$\begin{aligned} x_3 &= \frac{b_3^{(3)}}{a_{33}^{(3)}} \\ x_2 &= \frac{b_2^{(2)} - a_{23}^{(2)}x_3}{a_{22}^{(2)}} \\ x_1 &= \frac{b_1 - a_{12}x_2 - a_{13}x_3}{a_{11}} \end{aligned} \tag{6.41}$$

This algorithm for $n = 3$ is easily extended to one for a general nonsingular system of n linear equations. We will again assume that certain elements will be nonzero, but this assumption will be removed later in the section. Also, for uniformity in the presentation, denote the original system by

$$\begin{aligned} a_{11}^{(1)}x_1 + \cdots + a_{1n}^{(1)}x_n &= b_1^{(1)} & E(1) \\ &\vdots \\ a_{n1}^{(1)}x_1 + \cdots + a_{nn}^{(1)}x_n &= b_n^{(1)} & E(n) \end{aligned} \tag{6.42}$$

For $k = 1, 2, \ldots, n-1$, carry out the following elimination step.

Step k Eliminate x_k from $E(k+1)$ through $E(n)$. The results of the preceding steps $1, \ldots, k-1$ will have yielded a system of the form

$$
\begin{aligned}
a_{11}^{(1)} x_1 + a_{12}^{(1)} x_2 + \cdots + a_{1n}^{(1)} x_n &= b_1^{(1)} && E(1)\\
a_{22}^{(2)} x_2 + \cdots + a_{2n}^{(2)} x_n &= b_2^{(2)} && E(2)\\
\ddots \qquad &\vdots\\
a_{kk}^{(k)} x_k + \cdots + a_{kn}^{(k)} x_n &= b_k^{(k)} && E(k)\\
&\vdots\\
a_{nk}^{(k)} x_k + \cdots + a_{nn}^{(k)} x_n &= b_n^{(k)} && E(n)
\end{aligned}
\tag{6.43}
$$

Assume $a_{kk}^{(k)} \neq 0$, and define the *multipliers*

$$
m_{ik} = \frac{a_{ik}^{(k)}}{a_{kk}^{(k)}}, \qquad i = k+1, \ldots, n \tag{6.44}
$$

For equations $i = k+1, \ldots, n$, subtract m_{ik} times $E(k)$ from $E(i)$, eliminating x_k from $E(i)$. The new coefficients and the right-hand side numbers in $E(k+1)$ through $E(n)$ are defined by

$$
\begin{aligned}
a_{ij}^{(k+1)} &= a_{ij}^{(k)} - m_{ik} a_{kj}^{(k)}, && i, j = k+1, \ldots, n\\
b_i^{(k+1)} &= b_i^{(k)} - m_{ik} b_k^{(k)}, && i = k+1, \ldots, n
\end{aligned}
\tag{6.45}
$$

When step $n-1$ is completed, the linear system will be in upper triangular form. We will denote it by

$$
\begin{aligned}
u_{11} x_1 + \cdots + u_{1n} x_n &= g_1\\
&\vdots\\
u_{nn} x_n &= g_n
\end{aligned}
\tag{6.46}
$$

These coefficients are related to the earlier ones by

$$
u_{ij} = a_{ij}^{(i)}, \qquad g_i = b_i^{(i)}
$$

Step n Solve (6.46) using back substitution. Solve successively for $x_n, x_{n-1}, \ldots, x_1$ in (6.46).

$$
\begin{aligned}
x_n &= \frac{g_n}{u_{nn}}\\
x_i &= \frac{g_i - \sum_{j=i+1}^{n} u_{ij} x_j}{u_{ii}}, \qquad i = n-1, \ldots, 1
\end{aligned}
\tag{6.47}
$$

This completes the definition of Gaussian elimination.

Note that every step of the elimination procedure can be achieved through elementary row operations on certain rows of the augmented matrix

$$[\ A \mid b\ ]$$

just as that shown in the example at the beginning of the section.

Example 6.3.1 Solve the system

$$\begin{aligned} 4x_1 + 3x_2 + 2x_3 + x_4 &= 1 \\ 3x_1 + 4x_2 + 3x_3 + 2x_4 &= 1 \\ 2x_1 + 3x_2 + 4x_3 + 3x_4 &= -1 \\ x_1 + 2x_2 + 3x_3 + 4x_4 &= -1 \end{aligned} \tag{6.48}$$

We first set up the augmented matrix and give the results of steps 1, 2, 3 of Gaussian elimination for this system.

$$\left[\begin{array}{cccc|c} 4 & 3 & 2 & 1 & 1 \\ 3 & 4 & 3 & 2 & 1 \\ 2 & 3 & 4 & 3 & -1 \\ 1 & 2 & 3 & 4 & -1 \end{array}\right] \quad \begin{array}{c} m_{21} = \frac{3}{4} \\ \longrightarrow \\ m_{31} = \frac{1}{2} \\ m_{41} = \frac{1}{4} \end{array} \quad \left[\begin{array}{cccc|c} 4 & 3 & 2 & 1 & 1 \\ 0 & \frac{7}{4} & \frac{3}{2} & \frac{5}{4} & \frac{1}{4} \\ 0 & \frac{3}{2} & 3 & \frac{5}{2} & -\frac{3}{2} \\ 0 & \frac{5}{4} & \frac{5}{2} & \frac{15}{4} & -\frac{5}{4} \end{array}\right]$$

$$m_{32} = \frac{6}{7} \downarrow m_{42} = \frac{5}{7}$$

$$\left[\begin{array}{cccc|c} 4 & 3 & 2 & 1 & 1 \\ 0 & \frac{7}{4} & \frac{3}{2} & \frac{5}{4} & \frac{1}{4} \\ 0 & 0 & \frac{12}{7} & \frac{10}{7} & -\frac{12}{7} \\ 0 & 0 & 0 & \frac{5}{3} & 0 \end{array}\right] \quad \begin{array}{c} \longleftarrow \\ m_{43} = \frac{5}{6} \end{array} \quad \left[\begin{array}{cccc|c} 4 & 3 & 2 & 1 & 1 \\ 0 & \frac{7}{4} & \frac{3}{2} & \frac{5}{4} & \frac{1}{4} \\ 0 & 0 & \frac{12}{7} & \frac{10}{7} & -\frac{12}{7} \\ 0 & 0 & \frac{10}{7} & \frac{20}{7} & -\frac{10}{7} \end{array}\right]$$

Using back substitution to solve successively for x_4, x_3, x_2, x_1, we obtain the solution

$$x_1 = 0, \qquad x_2 = 1, \qquad x_3 = -1, \qquad x_4 = 0 \quad \blacksquare$$

6.3.1 Partial Pivoting

In the elimination, it was assumed that $a_{kk}^{(k)} \neq 0$ at each of the steps $k = 1, 2, \ldots, n-1$. To remove this assumption when $a_{kk}^{(k)}$ equals zero, examine the equations following $E(k)$ in (6.43). From the assumption that the original system (6.42) is nonsingular, it can be shown that one of the equations following $E(k)$ must contain a term involving x_k with a nonzero coefficient (see Problem 4). This equation can be interchanged with $E(k)$ to obtain a new, equivalent system in which $a_{kk}^{(k)} \neq 0$ is satisfied. But in numerical computations, having $a_{kk}^{(k)}$ be nonzero is not sufficient, as the following example demonstrates.

Example 6.3.2 Solve

$$\begin{aligned} 6x_1 + 2x_2 + 2x_3 &= -2 \\ 2x_1 + \tfrac{2}{3}x_2 + \tfrac{1}{3}x_3 &= 1 \\ x_1 + 2x_2 - x_3 &= 0 \end{aligned} \tag{6.49}$$

This system has the solution

$$x_1 = 2.6, \qquad x_2 = -3.8, \qquad x_3 = -5.0 \tag{6.50}$$

We use a decimal computer to solve this system. The computer uses a floating-point representation with four digits in the significand, and all operations will be rounded. The augmented matrix will be used to display the steps of the elimination, beginning with the machine representation of (6.49).

$$\left[\begin{array}{ccc|c} 6.000 & 2.000 & 2.000 & -2.000 \\ 2.000 & 0.6667 & 0.3333 & 1.000 \\ 1.000 & 2.000 & -1.000 & 0.0 \end{array}\right]$$

$$\downarrow m_{21} = 0.3333, \qquad m_{31} = 0.1667$$

$$\left[\begin{array}{ccc|c} 6.000 & 2.000 & 2.000 & -2.000 \\ 0.0 & 0.0001000 & -0.3333 & 1.667 \\ 0.0 & 1.667 & -1.333 & 0.3334 \end{array}\right] \tag{6.51}$$

$$\downarrow m_{32} = 16670$$

$$\left[\begin{array}{ccc|c} 6.000 & 2.000 & 2.000 & -2.000 \\ 0.0 & 0.0001000 & -0.3333 & 1.667 \\ 0.0 & 0.0 & 5555 & -27790 \end{array}\right]$$

Using back substitution, we obtain the following approximate solution:

$$x_1 = 1.335, \qquad x_2 = 0.0, \qquad x_3 = -5.003$$

Compare it with the true solution in (6.50).

The difficulty with this elimination process is that in (6.51), the element in row 2, column 2 should have been zero, but rounding error prevented it. This means that the coefficient in this (2,2) position had essentially infinite relative error, and this was carried through into computations involving this coefficient. To avoid this, interchange rows 2 and 3 in (6.51) and then continue the elimination. The final matrix now equals

$$\left[\begin{array}{ccc|c} 6.000 & 2.000 & 2.000 & -2.000 \\ 0.0 & 1.667 & -1.333 & 0.3334 \\ 0.0 & 0.0 & -0.3332 & 1.667 \end{array}\right]$$

and the final multiplier was $m_{32} = 0.00005999$. With back substitutions, we then obtain the approximate solution

$$x_1 = 2.602, \qquad x_2 = -3.801, \qquad x_3 = -5.003$$

This compares well with the true solution in (6.50). ■

To avoid the problem presented by this example, we use the following strategy. At step k [see (6.43)], calculate

$$c = \max_{k \le i \le n} \left| a_{ik}^{(k)} \right|$$

This is the maximum size of the elements in column k of the coefficient matrix of step k, beginning at row k and going downward. If the element $|a_{kk}^{(k)}| < c$, then interchange $E(k)$ with one of the following equations, to obtain a new equation $E(k)$ in which $|a_{kk}^{(k)}| = c$. This strategy makes $a_{kk}^{(k)}$ as far away from zero as possible. The element $a_{kk}^{(k)}$ is called the *pivot element* for step k of the elimination, and the process described in this paragraph is called *partial pivoting* or, more simply, pivoting. There are other forms of pivoting, but partial pivoting is the most popular one in practice.

Partial pivoting has been introduced here to avoid using coefficients that are nearly zero as pivot elements. But there is another equally important reason to use pivoting: In most instances, it decreases the propagated effects of rounding errors. With partial pivoting, the multipliers m_{ik} in (6.44) will satisfy

$$|m_{ik}| \le 1, \qquad 1 \le k < i \le n$$

This will help reduce loss-of-significance errors, because multiplications by m_{ik} will not lead to much larger numbers.

Example 6.3.3 Using the four-decimal-place computer described in the last example, solve

$$\begin{aligned} 0.729x_1 + 0.81x_2 + 0.9x_3 &= 0.6867 \\ x_1 + \quad x_2 + \quad x_3 &= 0.8338 \\ 1.331x_1 + 1.21x_2 + 1.1x_3 &= 1.000 \end{aligned}$$

Its exact solution, rounded to four places, is

$$x_1 = 0.2245, \qquad x_2 = 0.2814, \qquad x_3 = 0.3279 \tag{6.52}$$

Solution without Pivoting

$$\left[\begin{array}{ccc|c} 0.7290 & 0.8100 & 0.9000 & 0.6867 \\ 1.000 & 1.000 & 1.000 & 0.8338 \\ 1.331 & 1.210 & 1.100 & 1.000 \end{array}\right]$$

$$\downarrow \begin{array}{l} m_{21} = 1.372 \\ m_{31} = 1.826 \end{array}$$

$$\left[\begin{array}{ccc|c} 0.7290 & 0.8100 & 0.9000 & 0.6867 \\ 0.0 & -0.1110 & -0.2350 & -0.1084 \\ 0.0 & -0.2690 & -0.5430 & -0.2540 \end{array}\right]$$

$$\downarrow \quad m_{32} = 2.423$$

$$\left[\begin{array}{ccc|c} 0.7290 & 0.8100 & 0.9000 & 0.6867 \\ 0.0 & -0.1110 & -0.2350 & -0.1084 \\ 0.0 & 0.0 & 0.02640 & 0.008700 \end{array}\right]$$

The solution is

$$x_1 = 0.2251, \qquad x_2 = 0.2790, \qquad x_3 = 0.3295 \tag{6.53}$$

Solution with Pivoting

To indicate the interchange of rows i and j, we will use the notation $r_i \longleftrightarrow r_j$.

$$\left[\begin{array}{ccc|c} 0.7290 & 0.8100 & 0.9000 & 0.6867 \\ 1.000 & 1.000 & 1.000 & 0.8338 \\ 1.331 & 1.210 & 1.100 & 1.000 \end{array}\right]$$

$$r_1 \longleftrightarrow r_3 \downarrow \begin{array}{l} m_{21} = 0.7513 \\ m_{31} = 0.5477 \end{array}$$

$$\left[\begin{array}{ccc|c} 1.331 & 1.210 & 1.100 & 1.000 \\ 0.0 & 0.09090 & 0.1736 & 0.08250 \\ 0.0 & 0.1473 & 0.2975 & 0.1390 \end{array}\right]$$

$$r_2 \longleftrightarrow r_3 \downarrow m_{32} = 0.6171$$

$$\left[\begin{array}{ccc|c} 1.331 & 1.210 & 1.100 & 1.000 \\ 0.0 & 0.1473 & 0.2975 & 0.1390 \\ 0.0 & 0.0 & -0.01000 & -0.003280 \end{array}\right]$$

The solution is

$$x_1 = 0.2246, \qquad x_2 = 0.2812, \qquad x_3 = 0.3280 \tag{6.54}$$

Comparing this to (6.52), we observe that this is a much more accurate solution than (6.53). ■

6.3.2 Calculation of Inverse Matrices

To calculate A^{-1}, we can use the Gaussian elimination method. To explain what is involved, consider matrices of order $n = 3$. Let X denote the unknown inverse matrix, $X = A^{-1}$, and let the three columns of X be denoted by x_{*1}, x_{*2}, x_{*3}. Also, let the columns of I_3 be denoted by e_1, e_2, and e_3

$$e_1 = \begin{bmatrix} 1 \\ 0 \\ 0 \end{bmatrix}, \qquad e_2 = \begin{bmatrix} 0 \\ 1 \\ 0 \end{bmatrix}, \qquad e_3 = \begin{bmatrix} 0 \\ 0 \\ 1 \end{bmatrix}$$

The statement

$$AX = I$$

becomes

$$\begin{bmatrix} a_{11} & a_{12} & a_{13} \\ a_{21} & a_{22} & a_{23} \\ a_{31} & a_{32} & a_{33} \end{bmatrix} \begin{bmatrix} x_{11} & x_{12} & x_{13} \\ x_{21} & x_{22} & x_{23} \\ x_{31} & x_{32} & x_{33} \end{bmatrix} = \begin{bmatrix} 1 & 0 & 0 \\ 0 & 1 & 0 \\ 0 & 0 & 1 \end{bmatrix} \tag{6.55}$$

Consider only the first column of the product. It is

$$\begin{bmatrix} a_{11}x_{11} + a_{12}x_{21} + a_{13}x_{31} \\ a_{21}x_{11} + a_{22}x_{21} + a_{23}x_{31} \\ a_{31}x_{11} + a_{32}x_{21} + a_{33}x_{31} \end{bmatrix} = Ax_{*1}$$

Similarly, the second and third columns of AX are Ax_{*2} and Ax_{*3}. Thus, (6.55) can be written as

$$AX = [Ax_{*1}, Ax_{*2}, Ax_{*3}] = [e_1, e_2, e_3]$$

By matching corresponding columns, we get

$$Ax_{*1} = e_1, \qquad Ax_{*2} = e_2, \qquad Ax_{*3} = e_3 \tag{6.56}$$

Thus, the columns of X are the solutions of three simultaneous linear systems, all with the same matrix of coefficients A. Gaussian elimination can be used to solve these systems.

Example 6.3.4 Find the inverse of

$$A = \begin{bmatrix} 1 & 1 & -1 \\ 1 & 2 & -2 \\ -2 & 1 & 1 \end{bmatrix}$$

Perform Gaussian elimination on the augmented matrix

$$\left[\begin{array}{rrr|rrr} 1 & 1 & -1 & 1 & 0 & 0 \\ 1 & 2 & -2 & 0 & 1 & 0 \\ -2 & 1 & 1 & 0 & 0 & 1 \end{array}\right]$$

This simultaneously takes account of the three systems in (6.56). The steps of the elimination are

$$\left[\begin{array}{rrr|rrr} 1 & 1 & -1 & 1 & 0 & 0 \\ 0 & 1 & -1 & -1 & 1 & 0 \\ 0 & 3 & -1 & 2 & 0 & 1 \end{array}\right] \qquad \begin{array}{l} m_{21} = 1 \\ m_{31} = -2 \end{array}$$

$$\downarrow$$

$$\left[\begin{array}{rrr|rrr} 1 & 1 & -1 & 1 & 0 & 0 \\ 0 & 1 & -1 & -1 & 1 & 0 \\ 0 & 0 & 2 & 5 & -3 & 1 \end{array}\right] \qquad m_{32} = 3$$

Using back substitution, we solve the system with x_{*1} as its solution. This is

$$\begin{aligned} x_{11} + x_{21} - x_{31} &= 1 \\ x_{21} - x_{31} &= -1 \\ 2x_{31} &= 5 \end{aligned}$$

The solution is

$$x_{11} = 2, \qquad x_{21} = \tfrac{3}{2}, \qquad x_{31} = \tfrac{5}{2}$$

The second and third columns can be found similarly. And then

$$X = A^{-1} = \begin{bmatrix} 2 & -1 & 0 \\ \frac{3}{2} & -\frac{1}{2} & \frac{1}{2} \\ \frac{5}{2} & -\frac{3}{2} & \frac{1}{2} \end{bmatrix} \tag{6.57}$$

This can be checked by multiplication by A, to obtain the identity.

Alternatively, for the triangular systems, instead of using the backward substitution to solve, we may use the same kind of calculations to eliminate x_3 from the first two equations, and to eliminate x_2 from the first equation. In other words, we continue to apply elementary row operations until the left half of the augmented matrix is transformed to an identity matrix.

$$\left[\begin{array}{ccc|ccc} 1 & 1 & -1 & 1 & 0 & 0 \\ 0 & 1 & -1 & -1 & 1 & 0 \\ 0 & 0 & 2 & 5 & -3 & 1 \end{array}\right]$$

$\downarrow$ divide row 3 by 2

$$\left[\begin{array}{ccc|ccc} 1 & 1 & -1 & 1 & 0 & 0 \\ 0 & 1 & -1 & -1 & 1 & 0 \\ 0 & 0 & 1 & \frac{5}{2} & -\frac{3}{2} & \frac{1}{2} \end{array}\right]$$

$\downarrow$ add row 3 to row 1 and row 2

$$\left[\begin{array}{ccc|ccc} 1 & 1 & 0 & \frac{7}{2} & -\frac{3}{2} & \frac{1}{2} \\ 0 & 1 & 0 & \frac{3}{2} & -\frac{1}{2} & \frac{1}{2} \\ 0 & 0 & 1 & \frac{5}{2} & -\frac{3}{2} & \frac{1}{2} \end{array}\right]$$

$\downarrow$ subtract row 2 from row 1

$$\left[\begin{array}{ccc|ccc} 1 & 0 & 0 & 2 & -1 & 0 \\ 0 & 1 & 0 & \frac{3}{2} & -\frac{1}{2} & \frac{1}{2} \\ 0 & 0 & 1 & \frac{5}{2} & -\frac{3}{2} & \frac{1}{2} \end{array}\right]$$

Thus, the original systems

$$AX = I$$

are transformed to

$$I\,X = \begin{bmatrix} 2 & -1 & 0 \\ \frac{3}{2} & \frac{1}{2} & \frac{1}{2} \\ \frac{5}{2} & -\frac{3}{2} & \frac{1}{2} \end{bmatrix}$$

Therefore, the inverse matrix is given by (6.57). ■

The above procedure generalizes to matrices A of arbitrary order $n \geq 1$. Augment A by the identity of the same order and then apply Gaussian elimination in the manner illustrated above.

6.3.3 Operations Count

It is important to know the length of a computation and, for that reason, we count the number of arithmetic operations involved in Gaussian elimination. For reasons that will be apparent in later sections, the count will be divided into three parts. Also, as a notational convenience, let U and g denote the coefficient matrix and the right-hand side vector in (6.46).

Table 6.1. Operations Count for $A \to U$

Step	Additions	Multiplications	Divisions
1	$(n-1)^2$	$(n-1)^2$	$n-1$
2	$(n-2)^2$	$(n-2)^2$	$n-2$
$\vdots$	$\vdots$	$\vdots$	$\vdots$
$n-1$	1	1	1
Total	$\frac{n(n-1)(2n-1)}{6}$	$\frac{n(n-1)(2n-1)}{6}$	$\frac{n(n-1)}{2}$

1. *The elimination step.* We count the additions/subtractions, multiplications, and divisions in going from the system (6.42) to the triangular system (6.46). We consider only the operations for the coefficients of A and not for the right-hand side b in (6.42). Table 6.1 contains the number of these operations for each of the steps in the elimination. The totals in the last row of the table are obtained by using the formulas

$$\sum_{j=1}^{p} j = \frac{p(p+1)}{2}, \qquad \sum_{j=1}^{p} j^2 = \frac{p(p+1)(2p+1)}{6}, \qquad p \geq 1$$

Generally, the divisions and multiplications are counted together, since they are about the same in operation time. Doing this gives us

$$AS(A \to U) = \frac{n(n-1)(2n-1)}{6}$$

$$MD(A \to U) = \frac{n(n-1)(2n-1)}{6} + \frac{n(n-1)}{2} = \frac{n(n^2-1)}{3}$$

$AS(\cdot)$ denotes the number of additions and subtractions for the enclosed operation, and $MD(\cdot)$ denotes that of multiplications and divisions. The notation $A \to U$ denotes the elimination process that converts A in (6.42) into U in (6.46).

2. *Modification of the right side b to g.* Proceeding as before, we get

$$AS(b \to g) = (n-1) + (n-2) + \cdots + 1 = \frac{n(n-1)}{2}$$

$$MD(b \to g) = (n-1) + (n-2) + \cdots + 1 = \frac{n(n-1)}{2}$$

3. *The back substitution step, find x from* (6.46). As before,

$$AS(g \to x) = 0 + 1 + \cdots + (n-1) = \frac{n(n-1)}{2}$$

$$MD(g \to x) = 1 + 2 + \cdots + n = \frac{n(n+1)}{2}$$

Combining these results, we observe that the total number of operations to obtain x is

$$\begin{aligned} AS(x) &= AS(A \to U) + AS(b \to g) + AS(g \to x) \\ &= \frac{n(n-1)(2n-1)}{6} + \frac{n(n-1)}{2} + \frac{n(n-1)}{2} \\ &= \frac{n(n-1)(2n+5)}{6} \\ MD(x) &= \frac{n(n^2+3n-1)}{3} \end{aligned}$$

Since AS and MD are almost the same in all of these counts, only MD is discussed. These operations are also slightly more expensive in running time. For larger values of n, the operation count for Gaussian elimination is about $\frac{1}{3}n^3$. This means that as n is doubled, the cost of solving the linear system goes up by a factor of 8. In addition, most of the cost of Gaussian elimination is in the elimination step, $A \to U$, since for the remaining steps

$$MD(b \to g) + MD(g \to x) = \frac{n(n-1)}{2} + \frac{n(n+1)}{2} = n^2 \tag{6.58}$$

Thus, once the $A \to U$ step has been completed, it is much less expensive to solve the linear system. We return to this later in Section 6.5, where we give a simple and inexpensive way to estimate the error in the answer obtained by Gaussian elimination.

Consider solving the linear system of equations

$$Ax = b$$

where A has order n and is nonsingular. Then A^{-1} exists and

$$\begin{aligned} A^{-1}(Ax) &= A^{-1}b \\ x &= A^{-1}b \end{aligned} \tag{6.59}$$

Thus, if A^{-1} is known, then x can be found by matrix multiplication.

From this, it might at first seem reasonable to find A^{-1} and to solve for x using (6.59). But this is not an efficient procedure, because of the greater cost needed to find

A^{-1}. The operations cost for finding A^{-1} can be shown to be

$$MD(A \to A^{-1}) = n^3$$

This is about three times the cost of finding x by the Gaussian elimination method. Even if we wish to solve $Ax = b$ with several different right sides b, it is no more efficient to use (6.59) than to use Gaussian elimination, doing the elimination step, $A \to U$ in (6.42) to (6.46), only once. From (6.58), the cost of elimination (when $A \to U$ has already been completed) is only n^2, and this is exactly the same as the number of multiplications to calculate $x = A^{-1}b$. Thus, there is no savings in using A^{-1}.

The chief value of A^{-1} is as a theoretical tool for examining the solution of nonsingular systems of linear equations. With a few exceptions, one seldom needs to calculate A^{-1} explicitly.

MATLAB PROGRAM. The following MATLAB program implements the Gaussian elimination method with partial pivoting. The user provides the coefficient matrix A and the right-hand side vector b. The program computes the solution of the corresponding linear system $Ax = b$. It also produces a matrix that stores the upper triangular part of the upper triangular matrix resulting from the Gaussian elimination with partial pivoting, as well as the multipliers used in the elimination. The third output is a vector that records the row permutations made in the partial pivoting.

```
function [x,lu,piv] = GEpivot(A,b)
%
% function [x,lu,piv] = GEpivot(A,b)
%
% This program employs the Gaussian elimination method with
% partial pivoting to solve the linear system Ax=b.
%
% Input
% A: coefficient square matrix
% b:  right side column vector
%
% Output
% x:  solution vector
% lu:  a matrix whose upper triangular part is the upper
% triangular matrix resulting from the Gaussian
% elimination with partial pivoting, and whose strictly
% lower triangular part stores the multipliers.  In
% other words, it stores the LU factorization of the
% matrix A with row permutations determined by the
% pivoting vector piv.
% piv:  a pivoting vector recording the row permutations.
```

```
% check the order of the matrix and the size of the vector
[m,n] = size(A);
if m ~= n
  error('The matrix is not square.')
end

m = length(b);
if m ~= n
  error('The matrix and the vector do not match in size.')
end

% initialization of the pivoting vector
piv = (1:n)';

% elimination step
for k = 1:n-1
% Find the maximal element in the pivot column, below the
% pivot position, along with the index of that maximal element.

  [col_max index] = max(abs(A(k:n,k)));
  index = index + k-1;
  if index ~= k
%   Switch rows k and index, in columns k through n.
%   Do similarly for the right-hand side b.

    tempA = A(k,k:n);
    A(k,k:n) = A(index,k:n);
    A(index,k:n) = tempA;

    tempb = b(k);
    b(k) = b(index);
    b(index) = tempb;

    temp = piv(k);
    piv(k) = piv(index);
    piv(index) = temp;
  end

% Form the needed multipliers and store them into the pivot
% column, below the diagonal.
  A(k+1:n,k) = A(k+1:n,k)/A(k,k);

% Carry out the elimination step, first modifying the matrix,
% and then modifying the right-hand side.
  for i = k+1:n
```

```
    A(i,k+1:n) = A(i,k+1:n) - A(i,k)*A(k,k+1:n);
  end
  b(k+1:n) = b(k+1:n) - A(k+1:n,k)*b(k);
end

% Solve the upper triangular linear system.
x = zeros(n,1);
x(n) = b(n)/A(n,n);
for i = n-1:-1:1
  x(i) = (b(i)-A(i,i+1:n)*x(i+1:n))/A(i,i);
end

% Record the LU factorization with row permutation.
lu = A;
```

PROBLEMS

1. Modify the program `GEpivot` so that no pivoting is used. Then use the modified program to solve the following linear systems by using Gaussian elimination without pivoting. The systems are specified by A and b, with the system then written as in (6.42).

(a)

$$A = \begin{bmatrix} 2 & 1 & -1 \\ 4 & 0 & -1 \\ -8 & 2 & 2 \end{bmatrix}, \qquad b = \begin{bmatrix} 6 \\ 6 \\ -8 \end{bmatrix}$$

(b)

$$A = \begin{bmatrix} 2 & 1 & -1 & -2 \\ 4 & 4 & 1 & 3 \\ -6 & -1 & 10 & 10 \\ -2 & 1 & 8 & 4 \end{bmatrix}, \qquad b = \begin{bmatrix} 2 \\ 4 \\ -5 \\ 1 \end{bmatrix}$$

(c)

$$A = \begin{bmatrix} 1 & -1 & 2 \\ -1 & 5 & 4 \\ 2 & 4 & 29 \end{bmatrix}, \qquad b = \begin{bmatrix} 1 \\ -3 \\ 15 \end{bmatrix}$$

2. Use the program `GEpivot` given in this section to solve the linear systems of order n whose coefficients are given in (6.3) to (6.4) of Section 6.1. Do so for various values of n, say, $n = 2, 5, 10, 20$.

3. Repeat Problem 2, but use the arrays A and b given by

$$a_{ij} = \min\{i, j\}, \qquad b_i = 1, \qquad 1 \le i, j \le n$$

4. In step 2 for the $n = 3$ case (6.39), suppose that both $a_{22}^{(2)}$ and $a_{32}^{(2)}$ are zero. Then the system (6.40) becomes

$$\begin{aligned} a_{11}x_1 + a_{12}x_2 + a_{13}x_3 &= b_1 \\ a_{23}^{(2)}x_3 &= b_2^{(2)} \\ a_{33}^{(2)}x_3 &= b_3^{(2)} \end{aligned}$$

Show that this system is not uniquely solvable but, instead, has either no solution or an infinity of solutions. The same type of argument can be used with larger systems to show that there must be a nonzero pivot element for a nonsingular system.

5. Repeat Problem 2 for the systems with the following coefficients:

(a)

$$A = \begin{bmatrix} 5 & 7 & 6 & 5 \\ 7 & 10 & 8 & 7 \\ 6 & 8 & 10 & 9 \\ 5 & 7 & 9 & 10 \end{bmatrix}, \quad b = \begin{bmatrix} 1 \\ -1 \\ -1 \\ 1 \end{bmatrix}, \quad x = \begin{bmatrix} 136 \\ -82 \\ -35 \\ 21 \end{bmatrix}$$

(b)

$$A = \begin{bmatrix} 1 & \frac{1}{2} & \frac{1}{3} & \frac{1}{4} \\ \frac{1}{2} & \frac{1}{3} & \frac{1}{4} & \frac{1}{5} \\ \frac{1}{3} & \frac{1}{4} & \frac{1}{5} & \frac{1}{6} \\ \frac{1}{4} & \frac{1}{5} & \frac{1}{6} & \frac{1}{7} \end{bmatrix}, \quad b = \begin{bmatrix} 1 \\ -1 \\ 1 \\ -1 \end{bmatrix}, \quad x = \begin{bmatrix} 516 \\ -5700 \\ 13620 \\ -8820 \end{bmatrix}$$

True answers are given for comparison with your calculated answers. Print out the full number of digits justified by your computer arithmetic. Note any anomalies in the computed values.

6. Compute the inverses of the following matrices by hand calculation:

(a) $\begin{bmatrix} 3 & 1 & 1 \\ 1 & 3 & 1 \\ 1 & 1 & 3 \end{bmatrix}$ **(b)** $\begin{bmatrix} 0 & 1 & 2 \\ 1 & 0 & 1 \\ 2 & 1 & 0 \end{bmatrix}$ **(c)** $\begin{bmatrix} 1 & 2 & 4 \\ 1 & 3 & 9 \\ 1 & 4 & 16 \end{bmatrix}$

(d) $\begin{bmatrix} -3 & 1 & 0 & 0 \\ 1 & -2 & 1 & 0 \\ 0 & 0 & -2 & 1 \\ 0 & 0 & 1 & -1 \end{bmatrix}$ **(e)** $\begin{bmatrix} 1 & 1 & 1 & 1 \\ 1 & 2 & 3 & 4 \\ 1 & 3 & 6 & 10 \\ 1 & 4 & 10 & 20 \end{bmatrix}$

7. In the system (6.42), let $a_{ij} = 0$ whenever $i - j \geq 2$. Write out the general form of this system. Use Gaussian elimination to solve it, taking advantage of the elements that are known to be zero. Do an operations count for $A \to U$ in this case.

6.4. THE *LU* FACTORIZATION

When we use matrix multiplication, another meaning can be given to the Gaussian elimination method of Section 6.3. The matrix of coefficients of the linear system being solved can be factored into the product of two triangular matrices. We begin this section with that result, and then we discuss the implications and applications of it. Let $Ax = b$ denote the system to be solved, as before, with A the $n \times n$ coefficient matrix. In the elimination steps (6.42) to (6.45) of Section 6.3, the linear system was reduced to the upper triangular system $Ux = g$ of (6.46), with

$$U = \begin{bmatrix} u_{11} & u_{12} & \cdots & u_{1n} \\ 0 & u_{22} & \cdots & u_{2n} \\ \vdots & \ddots & \ddots & \vdots \\ 0 & \cdots & 0 & u_{nn} \end{bmatrix} \tag{6.60}$$

and $u_{ij} = a_{ij}^{(i)}$. Introduce an auxiliary lower triangular matrix L based on the multipliers m_{ij} of (6.44)

$$L = \begin{bmatrix} 1 & 0 & \cdots & 0 \\ m_{21} & 1 & \ddots & \vdots \\ \vdots & \ddots & \ddots & 0 \\ m_{n1} & \cdots & m_{n,n-1} & 1 \end{bmatrix} \tag{6.61}$$

The relationship of the matrices L and U to the original matrix A is given by the following theorem.

Theorem 6.4.1 Let A be a nonsingular matrix, and let L and U be defined by (6.60) and (6.61). If U is produced without pivoting, as in (6.42) to (6.45), then

$$LU = A \tag{6.62}$$

This is called the *LU factorization of* A.

We omit the proof, but the $n = 3$ case is considered in Problems 3 and 4. Also, there is a result analogous to (6.62) when pivoting is used, but it will not be needed here.

Example 6.4.2 Recall the system (6.48). Then

$$L = \begin{bmatrix} 1 & 0 & 0 & 0 \\ \frac{3}{4} & 1 & 0 & 0 \\ \frac{1}{2} & \frac{6}{7} & 1 & 0 \\ \frac{1}{4} & \frac{5}{7} & \frac{5}{6} & 1 \end{bmatrix}, \qquad U = \begin{bmatrix} 4 & 3 & 2 & 1 \\ 0 & \frac{7}{4} & \frac{3}{2} & \frac{5}{4} \\ 0 & 0 & \frac{12}{7} & \frac{10}{7} \\ 0 & 0 & 0 & \frac{5}{3} \end{bmatrix} \tag{6.63}$$

and

$$LU = \begin{bmatrix} 4 & 3 & 2 & 1 \\ 3 & 4 & 3 & 2 \\ 2 & 3 & 4 & 3 \\ 1 & 2 & 3 & 4 \end{bmatrix} = A \qquad \blacksquare$$

The LU factorization leads to another perspective on Gaussian elimination. From (6.62), solving $Ax = b$ is equivalent to solving $LUx = b$. And this is equivalent to solving the two systems

$$Lg = b, \qquad Ux = g \tag{6.64}$$

$Ux = g$ is the upper triangular system whose solution was given in (6.47), based on back substitution. $Lg = b$ is the lower triangular system

$$\begin{aligned} g_1 \qquad\qquad\qquad\qquad\qquad\qquad &= b_1 \\ m_{21}g_1 + \quad g_2 \qquad\qquad\qquad\qquad &= b_2 \\ \vdots \qquad\qquad\qquad\qquad\qquad\qquad & \;\;\vdots \\ m_{n1}g_1 + m_{n2}g_2 + \cdots + m_{n,n-1}g_{n-1} + g_n &= b_n \end{aligned}$$

It is solved by forward substitution, yielding the column vector g that appears in $Ux = g$. Thus, once the factorization $A = LU$ is known, the solution of $Ax = b$ is reduced to solving two triangular systems, a relatively simple and computationally inexpensive task.

Example 6.4.3 Again recall the system (6.48), with

$$b = [1, 1, -1, -1]^T$$

Use the factorization $A = LU$ given in (6.63). Then $Lg = b$ has the solution

$$g = \left[1, \tfrac{1}{4}, -\tfrac{12}{7}, 0\right]^T$$

and $Ux = g$ has the solution

$$x = [0, 1, -1, 0]^T$$

The reader should check these results. ■

6.4.1 Compact Variants of Gaussian Elimination

Rather than constructing L and U by using the elimination steps (6.42) to (6.45) of Section 6.3, it is possible to solve directly for these matrices. The number of operations will be the same as with Gaussian elimination, but there are some advantages to such *compact variants* of Gaussian elimination.

To illustrate the direct computation of L and U, consider the $n = 3$ case. Write $A = LU$ as

$$\begin{bmatrix} a_{11} & a_{12} & a_{13} \\ a_{21} & a_{22} & a_{23} \\ a_{31} & a_{32} & a_{33} \end{bmatrix} = \begin{bmatrix} 1 & 0 & 0 \\ m_{21} & 1 & 0 \\ m_{31} & m_{32} & 1 \end{bmatrix} \begin{bmatrix} u_{11} & u_{12} & u_{13} \\ 0 & u_{22} & u_{23} \\ 0 & 0 & u_{33} \end{bmatrix}$$

Multiply L and U, and match the elements of the product with the corresponding elements in A. Doing this for the first row of the product yields

$$a_{11} = u_{11}, \qquad a_{12} = u_{12}, \qquad a_{13} = u_{13} \tag{6.65}$$

This gives the first row of U. Next, multiply row 2 of L with U, to obtain

$$a_{21} = m_{21}u_{11}, \qquad a_{22} = m_{21}u_{12} + u_{22}, \qquad a_{23} = m_{21}u_{13} + u_{23} \tag{6.66}$$

These can be solved for m_{21}, u_{22}, and u_{23}, that is, the unknowns in the second rows of L and U. Finally, multiply row 3 of L with U to obtain

$$\begin{aligned} a_{31} &= m_{31}u_{11} \\ a_{32} &= m_{31}u_{12} + m_{32}u_{22} \\ a_{33} &= m_{31}u_{13} + m_{32}u_{23} + u_{33} \end{aligned} \tag{6.67}$$

These equations yield values for m_{31}, m_{32}, and u_{33}, completing the construction of L and U. In this process, we must have $u_{11} \neq 0$, $u_{22} \neq 0$ in order to solve for L. There are modifications of the method to avoid this assumption, using pivoting, but we will simply consider those cases where the necessary elements will be nonzero.

Example 6.4.4 Let

$$A = \begin{bmatrix} 1 & 1 & -1 \\ 1 & 2 & -2 \\ -2 & 1 & 1 \end{bmatrix}$$

Then using (6.65), we get

$$u_{11} = a_{11} = 1, \qquad u_{12} = a_{12} = 1, \qquad u_{13} = a_{13} = -1$$

From (6.66) and (6.67),

$$\begin{aligned}
m_{21} &= \frac{a_{21}}{u_{11}} = 1 \\
u_{22} &= a_{22} - m_{21}u_{12} = 2 - 1 \cdot 1 = 1 \\
u_{23} &= a_{23} - m_{21}u_{13} = -2 - 1 \cdot (-1) = -1 \\
m_{31} &= \frac{a_{31}}{u_{11}} = -2 \\
m_{32} &= (a_{32} - m_{31}u_{12})/u_{22} = (1 - (-2) \cdot 1)/1 = 3 \\
u_{33} &= a_{33} - m_{31}u_{13} - m_{32}u_{23} = 1 - (-2)(-1) - 3(-1) = 2
\end{aligned}$$

Thus,

$$A = \begin{bmatrix} 1 & 0 & 0 \\ 1 & 1 & 0 \\ -2 & 3 & 1 \end{bmatrix} \cdot \begin{bmatrix} 1 & 1 & -1 \\ 0 & 1 & -1 \\ 0 & 0 & 2 \end{bmatrix} \quad \blacksquare$$

The above procedure (6.65) to (6.67) can be extended to any $n \geq 2$; it is called *Doolittle's method.* It can also be modified to allow for the pivoting procedure discussed in Section 6.3. The detailed definition of Doolittle's method will not be given here, but it does have an important computational advantage over Gaussian elimination.

The two methods for computing L and U, Gaussian elimination and Doolittle's method, are mathematically equivalent and have equal operation counts. But Doolittle's method can be used to greatly reduce the number of rounding errors, with only a minimal increase in cost. The formulas for L and U, with larger values of n, will involve many sums of products, as suggested in (6.67). If we assume that the numbers are all stored in single precision, then the limited use of double precision can be employed to decrease greatly the number of single precision rounding errors. Multiply all of the products in double precision, sum them in double precision, and then round back to single precision. Thus, the computation of a sum of products of single precision numbers will involve only one single precision rounding error, rather than many. The number of such rounding errors can be reduced from about $\frac{2}{3}n^3$ for regular Gaussian elimination to a number proportional to n^2 for Doolittle's method, a significant reduction for larger values of n. This limited use of double precision arithmetic was discussed earlier at the end of Chapter 2. Well-written programs for solving linear systems use Doolittle's method or a close relation to it, and they use the limited double precision arithmetic described above.

One topic not yet discussed is the storage of the matrix L. Returning to the elimination steps (6.42) to (6.45) in Section 6.3, we see that the multipliers m_{ik} can be stored in those memory positions of the machine array A that have been zeroed in the elimination step. Thus, m_{ik} is stored in $A(i, k)$ in the machine, for $i > k$. And as discussed previously, u_{ij} is stored in $A(i, j)$ for $j \geq i$. The diagonal elements of L are all 1 and thus they do not need to be stored. Hence, after Gaussian elimination or Doolittle's method is applied to $Ax = b$, the elements of L and U will have replaced the original elements of A in the computer, as in the program `GEpivot` in Section 6.3.

6.4.2 Tridiagonal Systems

When the coefficient matrix A has a special form, it is often possible to simplify Gaussian elimination. We will do this for tridiagonal systems of linear equations.

A system $Ax = f$ is called *tridiagonal* if its matrix has the form

$$A = \begin{bmatrix} b_1 & c_1 & 0 & 0 & \cdots & 0 \\ a_2 & b_2 & c_2 & 0 & \cdots & 0 \\ 0 & a_3 & b_3 & c_3 & \ddots & \vdots \\ \vdots & \ddots & \ddots & \ddots & \ddots & 0 \\ 0 & \cdots & 0 & a_{n-1} & b_{n-1} & c_{n-1} \\ 0 & \cdots & \cdots & 0 & a_n & b_n \end{bmatrix} \tag{6.68}$$

This is called a *tridiagonal matrix.* Tridiagonal systems occur commonly in solving problems in many different areas of numerical analysis. In Chapter 5, spline function interpolation led to the tridiagonal system in (4.64). Finite difference solution of two-point boundary value problems of second-order ordinary differential equations also leads to tridiagonal systems, cf. Section 8.8.

When Gaussian elimination is applied to $Ax = f$, most of the multipliers $m_{ik} = 0$ and most of the elements of U are also zero. With this in mind, it can be shown that the LU factorization will have the following general form, provided that pivoting is not used.

$$LU = \begin{bmatrix} 1 & 0 & 0 & \cdots & 0 \\ \alpha_2 & 1 & 0 & \cdots & 0 \\ 0 & \alpha_3 & 1 & \ddots & \vdots \\ \vdots & \ddots & \ddots & \ddots & 0 \\ 0 & \cdots & 0 & \alpha_n & 1 \end{bmatrix} \cdot \begin{bmatrix} \beta_1 & c_1 & 0 & \cdots & 0 \\ 0 & \beta_2 & c_2 & \ddots & \vdots \\ \vdots & \ddots & \ddots & \ddots & 0 \\ 0 & \cdots & 0 & \beta_{n-1} & c_{n-1} \\ 0 & \cdots & \cdots & 0 & \beta_n \end{bmatrix} \tag{6.69}$$

Multiply in succession each row of L with various columns of U, and then equate the results to the corresponding elements of A. This yields

$$\begin{aligned} \beta_1 &= b_1 & & : \text{ row 1 of } LU \\ \alpha_2\beta_1 &= a_2, \quad & \alpha_2 c_1 + \beta_2 = b_2 & : \text{ row 2 of } LU \\ \alpha_j\beta_{j-1} &= a_j, \quad & \alpha_j c_{j-1} + \beta_j = b_j & : \text{ row } j \text{ of } LU \end{aligned} \tag{6.70}$$

for $j = 3, \ldots, n$. The reader should check these equations for the first few rows of A.

The equations (6.70) are easily solved for the elements $\{\alpha_i\}$ and $\{\beta_i\}$.

$$\begin{aligned}&\beta_1 = b_1\\ &\alpha_j = a_j/\beta_{j-1}, \qquad \beta_j = b_j - \alpha_j c_{j-1}, \qquad j = 2, \dots, n\end{aligned} \tag{6.71}$$

The system $Ax = f$ is now converted to the pair of triangular systems

$$Lg = f, \qquad Ux = g$$

Forward substitution in $Lg = f$ gives

$$g_1 = f_1, \qquad g_j = f_j - \alpha_j g_{j-1}, \qquad j = 2, 3, \dots, n \tag{6.72}$$

Back substitution in $Ux = g$ yields

$$x_n = \frac{g_n}{\beta_n}, \qquad x_j = \frac{g_j - c_j x_{j+1}}{\beta_j}, \qquad j = n-1, \dots, 1 \tag{6.73}$$

The entire procedure (6.71) to (6.73) is extremely rapid, with an operations count of only $(5n - 4)$ multiplications and divisions. Showing this is left as Problem 11. Also, to store the matrix A, only three one-dimensional arrays for the three diagonals of A are needed. This means that very large systems can be solved, and systems of order over $n = 10{,}000$ are not unusual in some applications, for example, in solving boundary value problems of differential equations.

Example 6.4.5 Let

$$A = \begin{bmatrix} 2 & 1 & 0 & 0 & 0 \\ 1 & 2 & 1 & 0 & 0 \\ 0 & 1 & 2 & 1 & 0 \\ 0 & 0 & 1 & 2 & 1 \\ 0 & 0 & 0 & 1 & 2 \end{bmatrix} \tag{6.74}$$

In the notation of (6.68),

$$a_j = 1, \qquad b_j = 2, \qquad c_j = 1, \qquad \text{all } j$$

From (6.71), $\beta_1 = 2$ and

$$\alpha_j = \frac{1}{\beta_{j-1}}, \qquad \beta_j = 2 - \alpha_j, \qquad j = 2, 3, 4, 5$$

This leads to

$$A = LU = \begin{bmatrix} 1 & 0 & 0 & 0 & 0 \\ \frac{1}{2} & 1 & 0 & 0 & 0 \\ 0 & \frac{2}{3} & 1 & 0 & 0 \\ 0 & 0 & \frac{3}{4} & 1 & 0 \\ 0 & 0 & 0 & \frac{4}{5} & 1 \end{bmatrix} \begin{bmatrix} 2 & 1 & 0 & 0 & 0 \\ 0 & \frac{3}{2} & 1 & 0 & 0 \\ 0 & 0 & \frac{4}{3} & 1 & 0 \\ 0 & 0 & 0 & \frac{5}{4} & 1 \\ 0 & 0 & 0 & 0 & \frac{6}{5} \end{bmatrix} \quad \blacksquare$$

The procedure (6.69) to (6.73) assumes that pivoting is not necessary. This can be shown to be true if the following conditions are satisfied:

$$\begin{aligned} &|b_1| > |c_1| \\ &|b_j| \geq |a_j| + |c_j|, \qquad j = 2, 3, \ldots, n-1 \\ &|b_n| > |a_n| \end{aligned} \tag{6.75}$$

These conditions are satisfied in the most important cases that occur in applications. The case of interpolating cubic splines is considered in Problem 12.

Sparse linear systems are systems for which the percentage of nonzero elements is very small for the coefficient matrix. Tridiagonal linear systems are an example of such sparse systems. As with tridiagonal systems, there are also efficient extensions of Gaussian elimination to solve many other forms of sparse linear systems; but we do not consider them here.

MATLAB PROGRAM. The following MATLAB program solves tridiagonal systems by means of the method (6.69) to (6.73). The coefficients $\{a_j, b_j, c_j\}$ are given by the user in the vectors A, B, and C, and the right-hand constants are supplied in F. On exit, the vectors A and B contain the constants $\{\alpha_j, \beta_j\}$ of (6.71), and F contains the solution x. In many applications, a tridiagonal system is solved with many right sides f. To allow for this option, the variable `iflag` indicates whether or not the LU factorization has already been computed and stored in the arrays A, B, and C. See the comment statements in the program for additional information. This program will be used later in solving finite difference systems for two-point boundary value problems of ordinary differential equations in Section 8.8.

```
function [x, ier, alpha, beta] = tridiag(a,b,c,f,n,iflag)
%
% function [x, ier, alpha, beta] = tridiag(a,b,c,f,n,iflag)
%
% Solve a tridiagonal linear system M*x=f
%
```

```
% INPUT:
% The order of the linear system is given as n.
% The subdiagonal, diagonal, and superdiagonal of M are given
% by the arrays a,b,c, respectively.  More precisely,
% M(i,i-1) = a(i), i=2,...,n
% M(i,i) = b(i), i=1,...,n
% M(i,i+1) = c(i), i=1,...,n-1
% iflag=0 means that the original matrix M is given as
% specified above.
% iflag=1 means that the LU factorization of M is already known
% and is stored in a,b,c.  This will have been accomplished by
% a previous call to this routine.  In that case, the vectors
% alpha and beta should have been substituted for a and b
% in the calling sequence.
%
% OUTPUT:
% Upon exit, the LU factorization of M is already known and
% is stored in alpha,beta,c.  The solution x is given as well.
% ier=0 means the program was completed satisfactorily.
% ier=1 means that a zero pivot element was encountered and the
% solution process was abandoned.

a(1) = 0;
if iflag == 0

% Compute LU factorization of matrix M.
  for j = 2:n
    if b(j-1) == 0
      ier = 1;
      return
    end
    a(j) = a(j)/b(j-1);
    b(j) = b(j) - a(j)*c(j-1);
  end
  if b(n) == 0
    ier = 1;
    return
  end
end

% Compute solution x to M*x = f.
% Do forward substitution to solve lower triangular system.
for j = 2:n
  f(j) = f(j) - a(j)*f(j-1);
end
```

```
% Do backward substitution to solve upper triangular system.
f(n) = f(n)/b(n);
for j = n-1:-1:1
  f(j) = (f(j) - c(j)*f(j+1))/b(j);
end

% Set output variables.
ier = 0;
x = f;
alpha = a; beta = b;
```

6.4.3 MATLAB Built-in Functions for Solving Linear Systems

For a linear system $Ax = b$ with a nonsingular square matrix A, the solution x can be found through the backslash operation:

$$\mathtt{x} = \mathtt{A}\backslash\mathtt{b}$$

When A is a triangular matrix, or is obtained from a triangular matrix through row interchanges, the solution is computed by substitution. If A is symmetric and positive definite, then the Cholesky factorization of A is calculated (see Problem 6), and the resulting two triangular systems are solved by substitutions. In the case of a general nonsingular matrix A, the solution is based on the *LU* factorization of A followed by substitutions. When the matrix is detected to be close to singular, a warning message is issued for a possible inaccurate result.

The *LU* factorization (with partial pivoting) is done within MATLAB by the function `lu`. For a square matrix A, the call

$$[\mathtt{L}, \mathtt{U}, \mathtt{P}] = \mathtt{lu(A)}$$

provides a lower triangular matrix L with unity diagonal elements, an upper triangular matrix U, and a permutation matrix P that records the row interchanges used during the factorization. The output matrices satisfy the relation $PA = LU$. Another form of calling the `lu` function is

$$[\mathtt{L}, \mathtt{U}] = \mathtt{lu(A)}$$

Then $LU = A$ and L is a permutation of a lower triangular matrix with unity diagonal elements, U an upper triangular matrix.

When the inverse of a nonsingular matrix is needed, a call to the `inv` function does the work. As is discussed in Subsection 6.3.3, it is not advisable to use the inverse matrix to solve a given linear system. Typically, doing

$$\mathtt{x} = \mathtt{A}\backslash\mathtt{b}$$

is two to three times faster than

```
x = inv(A) * b
```

The reader is encouraged to experiment with these and other MATLAB built-in functions related to the solution of linear systems.

PROBLEMS

1. Recall the solution of the system (6.35) in Section 6.3. Using that computation, give the LU factorization of the matrix of coefficients of that system.

2. Calculate the LU factorization of the matrices in Problem 1 of Section 6.3.

3. Consider the proof of Theorem 6.4.1 when $n = 3$. To do so, recall the general derivation (6.39) to (6.41) for Gaussian elimination with three equations. Form the product

$$LU = \begin{bmatrix} 1 & 0 & 0 \\ m_{21} & 1 & 0 \\ m_{31} & m_{32} & 1 \end{bmatrix} \begin{bmatrix} a_{11} & a_{12} & a_{13} \\ 0 & a_{22}^{(2)} & a_{23}^{(2)} \\ 0 & 0 & a_{33}^{(3)} \end{bmatrix}$$

Use the definitions of m_{ij} and $a_{ij}^{(i)}$ to simplify the results.

4. Recall the definition of elementary matrices from Problem 30 of Section 6.2. Define

$$E_1 = \begin{bmatrix} 1 & 0 & 0 \\ -m_{21} & 1 & 0 \\ -m_{31} & 0 & 1 \end{bmatrix}, \qquad E_2 = \begin{bmatrix} 1 & 0 & 0 \\ 0 & 1 & 0 \\ 0 & -m_{32} & 1 \end{bmatrix}$$

Define $A_2 = E_1 A$, $A_3 = E_2 A_2$. Show $A_3 = U$, the upper triangular matrix obtained by Gaussian elimination. Thus, $E_2 E_1 A = U$. Use this to show $A = LU$, with L derived from E_2 and E_1. How would this proof generalize to A, a matrix of order n?

5. The LU factorization of A is not unique if one only requires that L be lower triangular and U be upper triangular.

(a) Given an LU factorization of A, define

$$L_1 = LD, \qquad U_1 = D^{-1}U$$

for some nonsingular diagonal matrix D. Show that $L_1 U_1$ is another such factorization of A.

(b) Let $L_1 U_1 = L_2 U_2 = A$, with L_1, L_2 lower triangular, U_1, U_2 upper triangular. Also let A be nonsingular. Show that $L_1 = L_2 D$, $U_1 = D^{-1} U_2$ for some diagonal matrix D.

Hint: Derive $L_2^{-1}L_1 = U_2U_1^{-1}$. Show that these matrices must be diagonal. To make matters simpler, initially let $n = 2$ be the order of A. Also recall Problem 3 of Section 6.2.

6. When A is a symmetric, positive definite matrix, it can be factored in the form

$$A = LL^T$$

for some lower triangular matrix L that has nonzero diagonal elements, usually taken to be positive. This is called the Cholesky factorization. Recall that a symmetric matrix A is said to be positive definite if for any $x \neq 0$, $x^T Ax > 0$.

(a) For

$$A = \begin{bmatrix} 1 & -1 \\ -1 & 5 \end{bmatrix}$$

let

$$L = \begin{bmatrix} l_{11} & 0 \\ l_{21} & l_{22} \end{bmatrix}$$

Find L such that $LL^T = A$.

(b) Repeat this process for

$$A = \begin{bmatrix} 2.25 & -3 & 4.5 \\ -3 & 5 & -10 \\ 4.5 & -10 & 34 \end{bmatrix}$$

finding a lower triangular matrix L for which $LL^T = A$.

7. Write a MATLAB code to solve the linear systems (6.64). Assume the storage conventions for L and U described in the paragraph preceding the discussion of tridiagonal systems.

8. For the matrix A in (6.74), solve the linear system with

$$f = [1, 0, 0, 0, 0]^T$$

9. Solve the tridiagonal system

$$\begin{aligned}
2x_1 + x_2 \qquad\qquad\qquad\qquad &= 3 \\
-x_1 + 2x_2 + x_3 \qquad\qquad\quad &= 2 \\
-x_2 + 2x_3 + x_4 \qquad &= 2 \\
-x_3 + 2x_4 + x_5 &= 2 \\
-x_4 + 2x_5 &= 1
\end{aligned}$$

10. Solve the tridiagonal system $Ax = f$ with

$$A_{ii} = 4, \qquad A_{i,i-1} = A_{i,i+1} = 1$$

for all i. Let the order of the system be $n = 100$, and let

$$f = [1, 1, \ldots, 1]^T$$

11. Let $Ax = f$ be a tridiagonal system of order n.

(a) Give an operation count for forming $A = LU$, using (6.71).

(b) Assuming LU is known, give an operations count for solving $Ax = f$ by using (6.72) to (6.73).

12. Recall the tridiagonal system (4.64) to (4.65) for cubic spline interpolation. Show that the conditions (6.75) are satisfied by this system.

6.5. ERROR IN SOLVING LINEAR SYSTEMS

On a computer or calculator, Gaussian elimination involves rounding errors through its arithmetic operations, and these errors lead to error in the computed solution of the linear system being solved. In this section, we look at a method for predicting and correcting the error in the computed solution. Then, we examine the sensitivity of the solution to small changes such as rounding errors.

Let $Ax = b$ denote the system being solved, and let $\hat{x}$ denote the computed solution. Define

$$r = b - A\hat{x} \tag{6.76}$$

This is called the *residual* in the approximation of b by $A\hat{x}$. If $\hat{x}$ is the exact solution, then r equals zero. To relate r to the error in x, substitute $b = Ax$ in (6.76), yielding

$$r = Ax - A\hat{x} = A(x - \hat{x})$$

Let $e = x - \hat{x}$, the error in $\hat{x}$. Then

$$Ae = r \tag{6.77}$$

The error e satisfies a linear system with the same coefficient matrix A as in the original system $Ax = b$.

To find e, solve (6.77). In the original solution of $Ax = b$, the matrices L and U should have been saved, along with a knowledge of any possible row interchanges, cf. the program `GEpivot` in Section 6.3. This information can be used to solve $Ae = r$ at a relatively small computational cost. Recalling (6.58), we see the operations count for finding e from (6.77) is

$$MD(r \to e) = n^2$$

In addition, the residual r must be computed from (6.76), and for it,

$$MD(r) = n^2$$

Thus, the operations count for calculating e from $\hat{x}$ is only $2n^2$, and this is relatively small compared with the original operations count of about $\frac{1}{3}n^3$ for solving $Ax = b$.

In the practical implementation of this procedure for computing the error e, there are two possible sources of difficulty. First, the computation of the residual r in (6.76) will involve loss-of-significance errors. Generally, each component

$$r_i = b_i - \sum_{j=1}^{n} a_{ij}\hat{x}_j \tag{6.78}$$

will be very close to zero, and this occurs by the subtraction of the nearly equal quantities b_i and $\sum a_{ij}\hat{x}_j$. To avoid obtaining a very inaccurate value for r_i, the calculation of (6.78) should be carried out in a higher-precision arithmetic. If the solution of $Ax = b$ was done in single precision arithmetic, then r should be computed by using double precision arithmetic and then rounded back to single precision. If the solution was done in double precision arithmetic, then r should be computed in some form of extended precision arithmetic. The IEEE floating-point arithmetic standard contains such extended arithmetic.

The second source of difficulty in finding e is that solving $Ae = r$ will involve the same rounding errors that led to the error in the calculated value $\hat{x}$. Thus, we will not obtain e, but only an approximation $\hat{e}$ of e. How closely $\hat{e}$ approximates e will depend on the sensitivity or conditioning of the matrix, a topic considered later in this section. As a general rule, the approximation $\hat{e}$ will have at least one more digit of accuracy relative to e, except for the most ill-behaved of linear systems $Ax = b$.

Example 6.5.1 Recall the system preceding (6.52) in Section 6.3

$$\begin{aligned} 0.729x_1 + 0.81x_2 + 0.9x_3 &= 0.6867 \\ x_1 + \quad x_2 + \quad x_3 &= 0.8338 \\ 1.331x_1 + 1.21x_2 + 1.1x_3 &= 1.000 \end{aligned} \tag{6.79}$$

As before, we use a four-digit decimal machine with rounding. The true solution of this system is

$$x_1 = 0.2245, \qquad x_2 = 0.2814, \qquad x_3 = 0.3279 \tag{6.80}$$

correctly rounded to four digits.

To illustrate the estimation of the error e, we consider the solution of the system by Gaussian elimination without pivoting. This leads to the answers given earlier in (6.53)

$$\hat{x}_1 = 0.2251, \qquad \hat{x}_2 = 0.2790, \qquad \hat{x}_3 = 0.3295 \tag{6.81}$$

Calculating the residual r in the manner described above, using eight-digit floating-point decimal arithmetic, and rounding the resultant components of the residual to four significant digits, we obtain

$$r = [0.00006210,\ 0.0002000,\ 0.0003519]^T$$

Solving the linear system in (6.79) with this new right-hand side yields the approximation

$$\hat{e} = [-0.0004471,\ 0.002150,\ -0.001504]^T$$

Compare this to the true error of

$$e = x - \hat{x} = [-0.0007,\ 0.0024,\ -0.0016]^T$$

obtained from (6.80) and (6.81). This $\hat{e}$ gives a fairly good idea of the size of the error e in the computed solution $\hat{x}$. ■

6.5.1 The Residual Correction Method

A further use of this error estimation procedure is to define an iterative method for improving the computed value x. Let $x^{(0)} = \hat{x}$, the initial computed value for x, generally obtained by using Gaussian elimination. Define

$$r^{(0)} = b - Ax^{(0)} \tag{6.82}$$

Then as before in (6.77),

$$Ae^{(0)} = r^{(0)}, \qquad e^{(0)} = x - x^{(0)}$$

Solving by Gaussian elimination, we obtain an approximate value $\hat{e}^{(0)} \approx e^{(0)}$. Using it, we define an improved approximation

$$x^{(1)} = x^{(0)} + \hat{e}^{(0)} \tag{6.83}$$

Now we repeat the entire process, calculating

$$r^{(1)} = b - Ax^{(1)}$$
$$x^{(2)} = x^{(1)} + \hat{e}^{(1)}$$

where $\hat{e}^{(1)}$ is the approximate solution of

$$Ae^{(1)} = r^{(1)}, \qquad e^{(1)} = x - x^{(1)} \tag{6.84}$$

Continue this process until there is no further decrease in the size of $\hat{e}^{(k)}$, $k \geq 0$.

This method usually gives a way to obtain answers that are as accurate as possible relative to the number of digits in the arithmetic being used. The procedure is called the *residual correction method* or *iterative improvement.*

Example 6.5.2 Use a computer with four-digit floating-point decimal arithmetic with rounding, and use Gaussian elimination with pivoting. The system to be solved is

$$\begin{aligned} x_1 + \quad 0.5x_2 + 0.3333x_3 &= 1 \\ 0.5x_1 + 0.3333x_2 + \quad 0.25x_3 &= 0 \\ 0.3333x_1 + \quad 0.25x_2 + \quad 0.2x_3 &= 0 \end{aligned} \tag{6.85}$$

Then

$$\begin{aligned} x^{(0)} &= [8.968, \ -35.77, \ 29.77]^T \\ r^{(0)} &= [-0.005341, \ -0.004359, \ -0.0005344]^T \\ \hat{e}^{(0)} &= [0.09216, \ -0.5442, \ 0.5239]^T \\ x^{(1)} &= [9.060, \ -36.31, \ 30.29]^T \\ r^{(1)} &= [-0.0006570, \ -0.0003770, \ -0.0001980]^T \\ \hat{e}^{(2)} &= [0.001707, \ -0.01300, \ 0.01241]^T \\ x^{(2)} &= [9.062, \ -36.32, \ 30.30]^T \end{aligned}$$

The iterate $x^{(2)}$ is the correctly rounded solution of the system (6.85). This illustrates the usefulness of the residual correction method. ■

6.5.2 Stability in Solving Linear Systems

Rounding errors in Gaussian elimination will lead to errors in the computed solution of $Ax = b$, as has been illustrated above. We would now like to examine the degree of sensitivity of the solution x to these errors. The complexity of Gaussian elimination with its many arithmetic operations makes it difficult to do a direct analysis of the effects of these rounding errors. Instead, we look for the effect on the solution x of making a small change in the right side b of the linear system.

Let $\hat{b}$ be an approximation to b, and consider the solution $\hat{x}$ of

$$A\hat{x} = \hat{b}$$

To examine the sensitivity of x to small changes, we ask what is the size of $x - \hat{x}$ in comparison to $b - \hat{b}$. The residual equation (6.76) fits into this framework, with $\hat{b} = b - r$.

Example 6.5.3 The linear system

$$\begin{aligned} 5x_1 + \;\; 7x_2 &= 0.7 \\ 7x_1 + 10x_2 &= 1 \end{aligned} \tag{6.86}$$

has the solution

$$x_1 = 0, \qquad x_2 = 0.1$$

The perturbed system

$$\begin{aligned} 5\hat{x}_1 + \;\; 7\hat{x}_2 &= 0.69 \\ 7\hat{x}_1 + 10\hat{x}_2 &= 1.01 \end{aligned}$$

has the solution

$$\hat{x}_1 = -0.17, \qquad \hat{x}_2 = 0.22$$

A relatively small change in the right side of (6.86) has led to a relatively large change in the solution. ■

To more precisely express the relationship of $x - \hat{x}$ to $b - \hat{b}$, we introduce the following measures for the size of a vector and a matrix. Let us define

$$\begin{aligned} \|z\| &= \max_{1 \le i \le n} |z_i|, \qquad \text{order}(z) = n \times 1 \\ \|A\| &= \max_{1 \le i \le n} \sum_{j=1}^{n} |a_{ij}|, \qquad \text{order}(A) = n \times n \end{aligned} \tag{6.87}$$

These quantities are called *vector and matrix norms*, respectively, and they can be shown to satisfy the following inequalities:

$$\begin{array}{ll} \textbf{P1.} & \|y + z\| \le \|y\| + \|z\| \\ \textbf{P2.} & \|A + B\| \le \|A\| + \|B\| \\ \textbf{P3.} & \|AB\| \le \|A\| \, \|B\| \\ \textbf{P4.} & \|Az\| \le \|A\| \, \|z\| \end{array} \tag{6.88}$$

To examine the error $x - \hat{x}$, we look at

$$\|x - \hat{x}\| = \max_{1 \le i \le n} |x_i - \hat{x}_i|$$

the maximum of the errors in the components of $\hat{x}$. There are many other definitions of vector and matrix norms [see Atkinson (1989, §7.3)], but the above (6.87) is sufficient for our needs.

Theorem 6.5.4 Let A be nonsingular. Then, the solutions of $Ax = b$ and $A\hat{x} = \hat{b}$ satisfy

$$\frac{\|x - \hat{x}\|}{\|x\|} \le \|A\| \; \|A^{-1}\| \frac{\|b - \hat{b}\|}{\|b\|} \tag{6.89}$$

And for suitable choices of b and $\hat{b}$, the inequality becomes an equality.

Proof. Subtracting $A\hat{x} = \hat{b}$ from $Ax = b$, we get

$$\begin{aligned} A(x - \hat{x}) &= b - \hat{b} \\ x - \hat{x} &= A^{-1}(b - \hat{b}) \end{aligned}$$

Using (P4) of (6.88) yields

$$\|x - \hat{x}\| = \|A^{-1}(b - \hat{b})\| \le \|A^{-1}\| \; \|b - \hat{b}\|$$

Dividing by $\|x\|$, we obtain

$$\frac{\|x - \hat{x}\|}{\|x\|} \le \frac{\|A^{-1}\| \; \|b - \hat{b}\|}{\|x\|} = \|A\| \; \|A^{-1}\| \frac{\|b - \hat{b}\|}{\|A\| \; \|x\|} \tag{6.90}$$

Using (P4) again, we obtain

$$\|b\| = \|Ax\| \le \|A\| \; \|x\|$$

and substitute this on the right side of (6.90) to obtain (6.89). We omit the proof of the final statement in the theorem. ■

The result (6.89) compares the relative error in $\hat{x}$ to the relative error in $\hat{b}$. It says that if $\|A\| \; \|A^{-1}\|$ is very large, then the relative error in $\hat{x}$ may be much larger than that for $\hat{b}$; this is guaranteed to happen for certain choices of b and $\hat{b}$, although the actual choices are unknown in practice. The number

$$\text{cond}(A) = \|A\| \; \|A^{-1}\| \tag{6.91}$$

is called a *condition number*. When it is very large, the solution of $Ax = b$ will be very sensitive to relatively small changes in b. Or in the earlier notation of the residual in (6.76), a relatively small residual will quite possibly lead to a relatively large error in $\hat{x}$ as compared with x. Such linear systems are called *ill-conditioned*, whereas systems with a small condition number are called *well-conditioned*.

These comments are also valid when the changes are made to A rather than to b. In particular, if we compare the solutions of $Ax = b$ and $\hat{A}\hat{x} = b$, then it can be shown that

$$\frac{\|x - \hat{x}\|}{\|x\|} \leq \frac{\text{cond}(A)}{1 - \text{cond}(A)\dfrac{\|A - \hat{A}\|}{\|A\|}} \cdot \frac{\|A - \hat{A}\|}{\|A\|}$$

provided that

$$\|A - \hat{A}\| < 1/\|A^{-1}\|$$

Again, small changes in A will lead to possibly large changes in x, if $\text{cond}(A)$ is a large number. The condition number can be computed with the MATLAB `cond` function. The call

```
condA = cond(A, inf)
```

returns the condition number of the matrix A. In most situations, only the size of the condition number matters; its accurate value is not needed. Much of the current computer software for solving linear systems contains techniques for estimating $\text{cond}(A)$, without calculating A^{-1}. In MATLAB, for a different matrix norm [the so-called 1-norm, which is also the norm defined in (6.87) when it is applied to A^{T}], two functions are available for this purpose. The call

```
condA1 = condest(A)
```

returns an estimate of the condition number in the 1-norm, while

```
rcondA1 = rcond(A)
```

returns the reciprocal of an estimate of the condition number. Then the approximate condition number is used in estimating the error in the computed solution.

Example 6.5.5 A well-known example of an ill-conditioned matrix is the *Hilbert matrix*:

$$H_n = \begin{bmatrix} 1 & \frac{1}{2} & \frac{1}{3} & \cdots & \frac{1}{n} \\ \frac{1}{2} & \frac{1}{3} & \frac{1}{4} & \cdots & \frac{1}{n+1} \\ \vdots & \vdots & \vdots & & \vdots \\ \frac{1}{n} & \frac{1}{n+1} & \frac{1}{n+2} & \cdots & \frac{1}{2n-1} \end{bmatrix} \tag{6.92}$$

As n increases, H_n becomes increasingly ill-conditioned. To illustrate the ill-conditioning, let $n = 4$ and define $\hat{H}_4$ by rounding the elements of H_4 to five significant digits.

Then

$$\hat{H}_4 = \begin{bmatrix} 1.0000 & 0.50000 & 0.33333 & 0.25000 \\ 0.50000 & 0.33333 & 0.25000 & 0.20000 \\ 0.33333 & 0.25000 & 0.20000 & 0.16667 \\ 0.25000 & 0.20000 & 0.16667 & 0.14286 \end{bmatrix}$$

To see that H_4 is ill-conditioned, compare H_4^{-1} and $\hat{H}_4^{-1}$. These are the solutions of $H_4 X = I$ and $\hat{H}_4 \hat{X} = I$. Then

$$H_4^{-1} = \begin{bmatrix} 16 & -120 & 240 & -140 \\ -120 & 1200 & -2700 & 1680 \\ 240 & -2700 & 6480 & -4200 \\ -140 & 1680 & -4200 & 2800 \end{bmatrix}$$

$$\hat{H}_4^{-1} = \begin{bmatrix} 16.248 & -122.72 & 246.49 & -144.20 \\ -122.72 & 1229.9 & -2771.3 & 1726.1 \\ 246.49 & -2771.3 & 6650.1 & -4310.0 \\ -144.20 & 1726.1 & -4310.0 & 2871.1 \end{bmatrix}$$

The errors in $\hat{H}_4$ are in the sixth decimal place, but some of the errors in $\hat{H}_4^{-1}$ are in the second decimal place. This is a significant change in $\hat{H}_4^{-1}$ as compared to that in $\hat{H}_4$. Also, by direct computation,

$$\text{cond}(H_4) = \|H_4\| \, \|H_4^{-1}\| = \left(\tfrac{25}{12}\right)(13{,}620) \doteq 28{,}000$$

and this is consistent with the loss of four significant digits in $\hat{H}_4^{-1}$ when compared to H_4^{-1}. ■

Ill-conditioned matrices do not occur in many applications, but they do occur. The Hilbert matrix arises naturally out of the application presented in the fitting of functions and data in the sense of least squares (cf. Sections 4.7, 7.1); and the discretization of many partial differential equations leads to moderately as well as severely ill-conditioned linear systems. There are other ill-conditioned matrices that arise as naturally. For that reason, it is best to use linear equation solvers that have some way to detect ill-conditioning, if possible; or else, the residual correction technique described previously should be used to estimate the accuracy in the computed solution $\hat{x}$ to your system $Ax = b$.

PROBLEMS

1. Write a program to solve a linear system $Ax = b$, using the program `GEpivot` given in Section 6.3. Have the program contain the following features. (i) Create or read A and b. (ii) Solve $Ax = b$ using `GEpivot`, calling the computed solution $\hat{x}$. (iii) Compute the residual $r = b - A\hat{x}$, using the procedure described following

(6.78). (iv) Solve for the error in $Ae = r$, calling the solution $\hat{e}$. (v) Print $\hat{e}$ and $\hat{x} + \hat{e}$. Also print $\|\hat{e}\| / \|\hat{x}\|$ to approximate $\|e\| / \|x\|$. Apply this program to solving the following linear systems:

(a)
$$\begin{aligned} 4x_1 - 6x_2 + 4x_3 - x_4 &= 0.43 \\ -6x_1 + 14x_2 - 11x_3 + 3x_4 &= -1 \\ 4x_1 - 11x_2 + 10x_3 - 3x_4 &= 0.82 \\ -x_1 + 3x_2 - 3x_3 + x_4 &= -0.23 \end{aligned}$$

(b)
$$\begin{aligned} 5x_1 + 7x_2 + 6x_3 + 5x_4 &= 23 \\ 7x_1 + 10x_2 + 8x_3 + 7x_4 &= 32 \\ 6x_1 + 8x_2 + 10x_3 + 9x_4 &= 33 \\ 5x_1 + 7x_2 + 9x_3 + 10x_4 &= 31 \end{aligned}$$

Its solution is $x_1 = x_2 = x_3 = x_4 = 1$.

(c) $H_5 x = [1, 0.6, 0.4, 0.3, 0.3]^T$, where H_5 is defined in (6.92).

2. **(a)** For the vector norm of (6.87), show (i) $\|x\| = 0$ if and only if $x = 0$, and (ii) $\|\alpha x\| = |\alpha| \, \|x\|$, for all numbers α. The same properties also hold for the matrix norm in (6.87).

(b) Verify the norm properties (P1) to (P4) of (6.88).

3. **(a)** Calculate $\text{cond}(A)$ for the system (6.86).

(b) Graph the equations in (6.86). What is the effect geometrically of changing the right-hand constants in the equations? Interpret the moderate ill-conditioning of this system geometrically.

4. Consider the linear system

$$\begin{aligned} 19x_1 + 20x_2 &= b_1 \\ 20x_1 + 21x_2 &= b_2 \end{aligned}$$

Compute the condition number of the coefficient matrix. Is the system well-conditioned with respect to perturbations of the right-hand side constants $\{b_1, b_2\}$?

5. Calculate $\text{cond}(A)$ for

$$A = \begin{bmatrix} 1 & c \\ c & 1 \end{bmatrix}, \qquad |c| \neq 1$$

When does A become ill-conditioned? What does this say about the linear system $Ax = b$? How is $\text{cond}(A)$ related to det(A)?

6. Prove that $\text{cond}(A) \geq 1$ for any A.

Hint: First show $\|I\| = 1$.

7. Define the matrix A_n of order n by

$$A_n = \begin{bmatrix} 1 & -1 & -1 & \cdots & -1 \\ 0 & 1 & -1 & \cdots & -1 \\ 0 & 0 & 1 & \ddots & \vdots \\ \vdots & \vdots & \ddots & \ddots & -1 \\ 0 & 0 & \cdots & 0 & 1 \end{bmatrix}$$

(a) Find the inverse of A_n, explicitly.

Hint: Find the inverse of A_6, by solving $BA_6 = I$, with B an unknown upper triangular matrix. Then use this result to "guess" the inverse of A_n in general.

(b) Calculate $\text{cond}(A_n)$.

(c) With $b = [-n+2, -n+3, \ldots, -1, 0, 1]^T$, the solution of $A_n x = b$ is $x = [1, 1, \ldots, 1]^T$. Perturb b to $\hat{b} = b + [0, \ldots, 0, \epsilon]^T$. Solve for $\hat{x}$ in $A_n \hat{x} = \hat{b}$.

Hint: Use $\hat{x} = x + A_n^{-1}(\hat{b} - b)$. Show that these values of b, $\hat{b}$, x, $\hat{x}$ satisfy Theorem 6.5.4.

6.6. ITERATION METHODS

The linear systems $Ax = b$ that occur in many applications can have a very large order. For such systems, the Gaussian elimination method of Section 6.3 is often too expensive in either computation time or computer memory requirements, or possibly both. Moreover, the accumulation of round-off errors can sometimes prevent the numerical solution from being accurate. As an alternative, such linear systems are usually solved with iteration methods, and that is the subject of this section.

In an iterative method, a sequence of progressively accurate iterates is produced to approximate the solution. Thus, in general, we do not expect to get the exact solution in a finite number of iteration steps, even if the round-off error effect is not taken into account. In contrast, if round-off errors are ignored, the Gaussian elimination method produces the exact solution after $(n-1)$ steps of elimination and backward substitution for the resulting upper triangular system. Gaussian elimination method and its variants are thus usually called *direct methods.*

In the study of iteration methods, a most important issue is the convergence property. We provide a framework for the convergence analysis of a general iteration method. For the two classical iteration methods, the Jacobi and Gauss–Seidel methods, studied in this section, a sufficient condition for convergence is stated.

6.6.1 Jacobi Method and Gauss–Seidel Method

We begin with some numerical examples that illustrate two popular iteration methods. Following that, we give a more general discussion of iteration methods.

Consider the linear system

$$\begin{aligned} 9x_1 + x_2 + x_3 &= b_1 \\ 2x_1 + 10x_2 + 3x_3 &= b_2 \\ 3x_1 + 4x_2 + 11x_3 &= b_3 \end{aligned} \tag{6.93}$$

One class of iteration methods for solving (6.93) proceeds as follows. In the equation numbered k, solve for x_k in terms of the remaining unknowns. In the above case,

$$\begin{aligned} x_1 &= \tfrac{1}{9}[b_1 - x_2 - x_3] \\ x_2 &= \tfrac{1}{10}[b_2 - 2x_1 - 3x_3] \\ x_3 &= \tfrac{1}{11}[b_3 - 3x_1 - 4x_2] \end{aligned} \tag{6.94}$$

Let $x^{(0)} = [x_1^{(0)}, x_2^{(0)}, x_3^{(0)}]^T$ be an initial guess of the true solution x. Then define an iteration sequence:

$$\begin{aligned} x_1^{(k+1)} &= \tfrac{1}{9}\left[b_1 - x_2^{(k)} - x_3^{(k)}\right] \\ x_2^{(k+1)} &= \tfrac{1}{10}\left[b_2 - 2x_1^{(k)} - 3x_3^{(k)}\right] \\ x_3^{(k+1)} &= \tfrac{1}{11}\left[b_3 - 3x_1^{(k)} - 4x_2^{(k)}\right] \end{aligned} \tag{6.95}$$

for $k = 0, 1, 2, \ldots$. This is called the *Jacobi iteration method* or the *method of simultaneous replacements*.

In Table 6.2, we give a number of the iterations for the case that

$$b = [10,\ 19,\ 0]^T$$

which yields

$$x = [1,\ 2,\ -1]^T$$

In the table,

$$\text{Error} = \|x - x^{(k)}\|$$

using the norm of (6.87); and *Ratio* denotes the ratio of successive values of the error. The floating-point arithmetic contained approximately 16 decimal digits of precision, and thus the results for *Error* are correct to the number of digits shown in the table.

Table 6.2. Jacobi Iteration (6.95) for Solving (6.93)

k	$x_1^{(k)}$	$x_2^{(k)}$	$x_3^{(k)}$	Error	Ratio
0	0	0	0	2.00E + 0	
1	1.1111	1.9000	0	1.00E + 0	0.500
2	0.9000	1.6778	−0.9939	3.22E − 1	0.322
3	1.0351	2.0182	−0.8556	1.44E − 1	0.448
4	0.9819	1.9496	−1.0162	5.06E − 2	0.349
5	1.0074	2.0085	−0.9768	2.32E − 2	0.462
6	0.9965	1.9915	−1.0051	8.45E − 3	0.364
7	1.0015	2.0022	−0.9960	4.03E − 3	0.477
8	0.9993	1.9985	−1.0012	1.51E − 3	0.375
9	1.0003	2.0005	−0.9993	7.40E − 4	0.489
10	0.9999	1.9997	−1.0003	2.83E − 4	0.382
30	1.0000	2.0000	−1.0000	3.01E − 11	0.447
31	1.0000	2.0000	−1.0000	1.35E − 11	0.447

The errors decrease as k increases; and the values of *Ratio* eventually approach a limiting constant of approximately 0.447 as k becomes much larger. Note that although *Error* is decreasing, the errors in individual components may increase. As an example, compare the errors in $x_3^{(k)}$ for $k = 4$ and 5. With many linear systems and iteration methods for their solution, the values of *Ratio* approach a limiting value; but there are also many cases where *Ratio* does not have such a limit.

As another approach to the iterative solution of (6.93) through the use of (6.94), we use all the information we obtain in the calculation of each new component. Specifically, let us define

$$\begin{aligned} x_1^{(k+1)} &= \tfrac{1}{9}\left[b_1 - x_2^{(k)} - x_3^{(k)}\right] \\ x_2^{(k+1)} &= \tfrac{1}{10}\left[b_2 - 2x_1^{(k+1)} - 3x_3^{(k)}\right] \\ x_3^{(k+1)} &= \tfrac{1}{11}\left[b_3 - 3x_1^{(k+1)} - 4x_2^{(k+1)}\right] \end{aligned} \tag{6.96}$$

for $k = 0, 1, 2, \ldots$. This is called the *Gauss–Seidel method* or the *method of successive replacements*. This method is usually more rapidly convergent than the Jacobi method.

In Table 6.3, we give a number of iterations for solving the system (6.93). Compare these results with those in Table 6.2. The speed of convergence is much faster than with the Jacobi method (6.95). The values of *Ratio* do not appear to approach a limiting value, even when looking at values of k larger than those in the table.

Table 6.3. Gauss–Seidel Iteration (6.96) for Solving (6.93)

k	$x_1^{(k)}$	$x_2^{(k)}$	$x_3^{(k)}$	Error	Ratio
0	0	0	0	2.00E + 0	
1	1.1111	1.6778	−0.9131	3.22E − 1	0.161
2	1.0262	1.9687	−0.9958	3.13E − 2	0.097
3	1.0030	1.9981	−1.0001	3.00E − 3	0.096
4	1.0002	2.0000	−1.0001	2.24E − 4	0.074
5	1.0000	2.0000	−1.0000	1.65E − 5	0.074
6	1.0000	2.0000	−1.0000	2.58E − 6	0.155

6.6.2 General Schema

To understand the behavior of iteration methods, it is best to put them into a vector-matrix format. Rewrite the linear system $Ax = b$ as

$$Nx = b + Px \tag{6.97}$$

where $A = N - P$ is a *splitting* of A. The matrix N must be nonsingular; and usually it is chosen so that the linear systems

$$Nz = f$$

are relatively easy to solve for general vectors f. For example, N could be diagonal, triangular, or tridiagonal. The iteration method is defined by

$$Nx^{(k+1)} = b + Px^{(k)}, \qquad k = 0, 1, 2, \ldots \tag{6.98}$$

Example 6.6.1 (a) The Jacobi method (6.95) corresponds to

$$N = \begin{bmatrix} 9 & 0 & 0 \\ 0 & 10 & 0 \\ 0 & 0 & 11 \end{bmatrix}, \qquad P = \begin{bmatrix} 0 & -1 & -1 \\ -2 & 0 & -3 \\ -3 & -4 & 0 \end{bmatrix} \tag{6.99}$$

(b) The Gauss–Seidel method (6.96) corresponds to

$$N = \begin{bmatrix} 9 & 0 & 0 \\ 2 & 10 & 0 \\ 3 & 4 & 11 \end{bmatrix}, \qquad P = \begin{bmatrix} 0 & -1 & -1 \\ 0 & 0 & -3 \\ 0 & 0 & 0 \end{bmatrix} \tag{6.100}$$

For a general matrix $A = [a_{ij}]$ of order n, the Jacobi method is defined with

$$N = \begin{bmatrix} a_{11} & 0 & \cdots & 0 \\ 0 & a_{22} & \ddots & \vdots \\ \vdots & \ddots & \ddots & 0 \\ 0 & \cdots & 0 & a_{nn} \end{bmatrix} \tag{6.101}$$

and $P = N - A$. For the Gauss–Seidel method, let

$$N = \begin{bmatrix} a_{11} & 0 & \cdots & 0 \\ a_{21} & a_{22} & \ddots & \vdots \\ \vdots & \vdots & \ddots & 0 \\ a_{n1} & a_{n2} & \cdots & a_{nn} \end{bmatrix} \tag{6.102}$$

and $P = N - A$. The linear system $Nz = f$ is easily solvable because N is diagonal for the Jacobi iteration methods and lower triangular for the Gauss–Seidel method. For systems $Ax = b$ to which the Jacobi and Gauss–Seidel methods are often applied, the above matrices N of (6.101) and (6.102) are nonsingular. [For example, see (6.107) and the following.] ■

To analyze the convergence of the iteration, subtract (6.98) from (6.97) and let $e^{(k)} = x - x^{(k)}$, obtaining

$$\begin{aligned} Ne^{(k+1)} &= Pe^{(k)} \\ e^{(k+1)} &= Me^{(k)} \end{aligned} \tag{6.103}$$

where $M = N^{-1}P$. Using the vector and matrix norms of (6.87), we get

$$\|e^{(k+1)}\| \le \|M\| \, \|e^{(k)}\| \tag{6.104}$$

By induction on k, this implies

$$\|e^{(k)}\| \le \|M\|^k \|e^{(0)}\| \tag{6.105}$$

Thus, the errors $e^{(k)}$ converge to zero if

$$\|M\| < 1 \tag{6.106}$$

We attempt to choose the splitting $A = N - P$ so that this will be true, while also having the systems $Nz = f$ be easily solvable. The result (6.104) also says the error will decrease at each step by at least a factor of $\|M\|$; and (6.105) along with (6.106) says that convergence will occur for any initial guess $x^{(0)}$.

Example 6.6.2 (a) Consider the Jacobi method for (6.93). From (6.99),

$$M = \begin{bmatrix} 0 & -\frac{1}{9} & -\frac{1}{9} \\ -\frac{2}{10} & 0 & -\frac{3}{10} \\ -\frac{3}{11} & -\frac{4}{11} & 0 \end{bmatrix}$$

$$\|M\| = \tfrac{7}{11} \doteq 0.636$$

This is consistent with the column *Ratio* in Table 6.2, although the actual speed of convergence is better than that predicted by (6.104).

(b) Consider the Gauss–Seidel method for (6.93). From (6.100),

$$M = \begin{bmatrix} 0 & -\frac{1}{9} & -\frac{1}{9} \\ 0 & \frac{1}{45} & -\frac{5}{18} \\ 0 & \frac{1}{45} & \frac{13}{99} \end{bmatrix}$$

$$\|M\| = 0.3$$

Again, this is consistent with the results in Table 6.3, with the actual speed of convergence being better than that predicted by (6.104). ■

For the Jacobi method for a general matrix A, with N defined as in (6.101), we have

$$\|M\| = \max_{1 \le i \le n} \sum_{\substack{j=1 \\ j \ne i}}^{n} \left| \frac{a_{ij}}{a_{ii}} \right|$$

The condition (6.106) is equivalent to requiring

$$\sum_{\substack{j=1 \\ j \ne i}}^{n} \left| a_{ij} \right| < |a_{ii}|, \qquad 1 \le i \le n \tag{6.107}$$

A matrix satisfying this condition is said to be *diagonally dominant*, and the Jacobi method converges for all such linear systems $Ax = b$.

With the Gauss–Seidel method, using (6.102), the matrix M cannot be easily calculated. But by other methods, it can be shown that if (6.107) is satisfied, then the Gauss–Seidel method will converge at least as rapidly as the Jacobi method; and usually the Gauss–Seidel method will be faster. See Atkinson (1989, p. 548).

We emphasize that (6.107) is only a sufficient condition for the convergence of the Jacobi and Gauss–Seidel methods. Sharper convergence conditions for the two methods are possible, but we will not get into the details of such conditions in this text. There are many linear systems for which both methods converge and the condition (6.107) is violated. For example, the $n \times n$ matrix

$$A_n = \begin{bmatrix} 2 & -1 & 0 & \cdots & 0 \\ -1 & 2 & -1 & \ddots & \vdots \\ 0 & \ddots & \ddots & \ddots & 0 \\ \vdots & \ddots & -1 & 2 & -1 \\ 0 & \cdots & 0 & -1 & 2 \end{bmatrix} \tag{6.108}$$

does not satisfy (6.107), but it is a classical result that both the Jacobi and Gauss–Seidel methods converge on the corresponding linear systems, for any n. Matrices of this type arise in the finite difference solution of certain differential equation problems, cf. Section 8.8.

Returning to the reformulated system (6.97), how might N and P be chosen? A helpful perspective is to consider this as a problem in approximating A by a "simpler" matrix N. Since

$$\|M\| = \|N^{-1}P\| \le \|N^{-1}\| \, \|P\|$$

the requirement that $\|M\| < 1$ is satisfied if N is chosen so that

$$\|N^{-1}\| \, \|P\| < 1$$

If we use $P = N - A$, this can be rewritten as

$$\|A - N\| < \frac{1}{\|N^{-1}\|} \tag{6.109}$$

Often, it is possible to approximate A by a matrix N such that this bound is true.

A more general convergence result for the iteration (6.98) can be given, providing the necessary and sufficient conditions for convergence. In this result, we need the concept of matrix eigenvalues, a topic reviewed in Section 7.2.

Theorem 6.6.3 The iteration (6.98) will converge, for all right sides b and all initial guesses $x^{(0)}$, if and only if all eigenvalues λ of $M = N^{-1}P$ satisfy

$$|\lambda| < 1 \tag{6.110}$$

We take up a partial proof of this theorem in Problem 10. Note from Problem 9 of Section 7.2 that $\|M\| < 1$ implies $|\lambda| < 1$ for all eigenvalues λ of M. Thus, the condition (6.110) is satisfied when $\|M\| < 1$.

We should note here that one of the main applications of iteration methods is to the solution of discretizations of partial differential equations. The partial derivatives of the unknown function, say, $u(x, y)$, in a partial differential equation are approximated using the techniques of Section 5.4 from Chapter 5. This is done on a rectangular grid of points in the xy-plane, in such a way that we obtain a large sparse system of linear equations for approximate values of the unknown function $u(x, y)$. For details, see Section 9.1 of Chapter 9. With such problems, the Jacobi and Gauss–Seidel methods have been supplanted by better and more efficient iteration methods; but many of these methods are based on ideas similar to those introduced here.

6.6.3 The Residual Correction Method

We generalize the method (6.82) to (6.84) of Section 6.5, which was used in correcting the solution obtained by Gaussian elimination. This will also give another perspective on the above general schema (6.97) to (6.106) for iteration methods.

Let N be an invertible approximation of the matrix A, and let $x^{(0)}$ be an initial guess for the solution of $Ax = b$. Define

$$r^{(0)} = b - Ax^{(0)}$$

Since $Ax = b$ for the true solution x, we have

$$r^{(0)} = Ax - Ax^{(0)} = A(x - x^{(0)}) = Ae^{(0)}$$

with $e^{(0)} = x - x^{(0)}$, the error in $x^{(0)}$. Let $\hat{e}^{(0)}$ be the solution of

$$N\hat{e}^{(0)} = r^{(0)}$$

and then define

$$x^{(1)} = x^{(0)} + \hat{e}^{(0)}$$

Repeat this process inductively:

$$r^{(k)} = b - Ax^{(k)} \tag{6.111}$$

$$N\hat{e}^{(k)} = r^{(k)} \tag{6.112}$$

$$x^{(k+1)} = x^{(k)} + \hat{e}^{(k)} \tag{6.113}$$

This is the general *residual correction method.*

To see that this method is exactly the same as our earlier method (6.98), proceed as follows:

$$x^{(k+1)} = x^{(k)} + \hat{e}^{(k)} = x^{(k)} + N^{-1}r^{(k)}$$
$$= x^{(k)} + N^{-1}(b - Ax^{(k)})$$

Thus,

$$Nx^{(k+1)} = Nx^{(k)} + b - Ax^{(k)}$$
$$= b + (N - A)x^{(k)} = b + Px^{(k)}$$

Since this derivation is reversible, this shows the equivalence of (6.111–6.113) with (6.98).

Sometimes, one form of the iteration may be preferable over the other. For example, with the residual correction method of Section 6.5, it would be costly in time and memory to actually produce the matrix P. In other cases, the choice of form is a matter of personal preference. Both perspectives on iteration methods are common in the literature on the subject.

MATLAB PROGRAM. We include two sample programs for the Jacobi method and the Gauss–Seidel method. In general, for small-size systems, the iterative methods cannot compete with the Gaussian elimination method with pivoting. The purpose of having the programs here is to test the performance of the methods. Also, in these programs, possible special structures of the coefficient matrix are not taken into account. Therefore, for particular linear systems, the part of the program related to the matrix/vector multiplication should be modified for efficiency. For instance, for linear systems with coefficient matrices such as that in Problem 4 of Section 6.1 and that in Problem 21 of Section 6.2, the coefficient matrix should not be stored, and the matrix/vector multiplication can be carried out with only a few scalar multiplications and additions.

In the Jacobi method to solve the linear system $Ax = b$, denote D the diagonal part of A. Then the iteration formula is

$$x^{(k+1)} = D^{-1}\left(b - (A - D)x^{(k)}\right)$$

We implement the Jacobi method based on this form. We stop the iteration when the latest iterates $x^{(k+1)}$ and $x^{(k)}$ satisfy

$$\frac{\|x^{(k+1)} - x^{(k)}\|}{\|x^{(k+1)}\| + \texttt{eps}} < \delta$$

with a given error tolerance δ, or when the maximal number of iterations is reached. Here, eps is the machine epsilon, and it is used to avoid possible division by 0 when the relative error is computed. The quantity $\| \cdot \|$ is the vector norm defined in (6.87). This norm is calculated by MATLAB with

```
norm(x, inf)
```

for a vector x.

```
function [x, iflag, itnum] = Jacobi(A,b,x0,delta,max_it)
%
% function [x, iflag, itnum] = Jacobi(A,b,x0,delta,max_it)
%
% A program implementing the Jacobi iteration method to
% solve the linear system Ax=b.
%
% Input
% A: square coefficient matrix.  It is modified on output
% b:  right side vector
% x0:  initial guess
% delta:  error tolerance for the relative difference
%          between two consecutive iterates
% max_it:  maximum number of iterations to be allowed
%
% Output
% x:  numerical solution vector
% iflag:  1 if a numerical solution satisfying the error
%          tolerance is found within max_it iterations
%       -1 if the program fails to produce a numerical
%          solution in max_it iterations
% itnum:  the number of iterations used to compute x

% initialization
n = length(b);
iflag = 1;
k = 0;

% create a vector with diagonal elements of A
diagA = diag(A);
% modify A to make its diagonal elements zero
A = A-diag(diagA);

% iteration
while k < max_it
  k = k+1;
  x = (b-A*x0)./diagA;
  relerr = norm(x-x0,inf)/(norm(x,inf)+eps);
```

```
  x0 = x;
  if relerr < delta
    break
  end
end

itnum = k;
if itnum == max_it
  iflag = -1
end
```

For the Gauss–Seidel method, the iteration formula is

$$x_i^{(k+1)} = \frac{1}{a_{ii}}\left(b_i - \sum_{j=1}^{i-1} a_{ij}x_j^{(k+1)} - \sum_{j=i+1}^{n} a_{ij}x_j^{(k)}\right), \qquad 1 \le i \le n$$

Notice that the two summations can be evaluated as products of row vectors and column vectors.

```
function [x, iflag, itnum] = GS(A,b,x0,delta,max_it)
%
% function [x, iflag, itnum] = GS(A,b,x0,delta,max_it)
%
% A program implementing the Gauss-Seidel iteration
% method to solve the linear system Ax=b.
%
% Input
% A: square coefficient matrix
% b:  right side vector
% x0:  initial guess
% delta:  error tolerance for the relative difference
% between two consecutive iterates
% max_it:  maximum number of iterations to be allowed
%
% Output
% x:  numerical solution vector
% iflag:  1 if a numerical solution satisfying the error
%          tolerance is found within max_it iterations
%        -1 if the program fails to produce a numerical
%           solution in max_it iterations
% itnum:  the number of iterations used to compute x

% initialization
n = length(b);
```

```
iflag = 1;
k = 0;
x = x0;

% iteration
while k < max_it
  k = k+1;
  x(1) = (b(1)-A(1,2:n)*x0(2:n))/A(1,1);
  for i = 2:n-1
    x(i) = (b(i)-A(i,1:i-1)*x(1:i-1) ...
              -A(i,i+1:n)*x0(i+1:n))/A(i,i);
  end
  x(n) = (b(n)-A(n,1:n-1)*x(1:n-1))/A(n,n);
  relerr = norm(x-x0,inf)/(norm(x,inf)+eps);
  x0 = x
  if relerr < delta
    break
  end
end

itnum = k;
if itnum == max_it
  iflag = -1
end
```

PROBLEMS

1. Repeat the examples of the iterative solution of (6.93) by the Jacobi and Gauss–Seidel iteration methods, but do it for the following choices of right side b and true solution x. Use $x^{(0)} = [0\ 0\ 0]^T$. Iterate until $\|x - x^{(k)}\| \le 0.00005$. Look at the ratios of successive errors, as in Tables 6.2 and 6.3.

(a) $b = [16\ \ 44\ \ 59]^T, \qquad x = [1\ \ 3\ \ 4]^T$

(b) $b = [0\ \ -7\ \ 7]^T, \qquad x = [0\ \ -1\ \ 1]^T$

2. Repeat Problem 1 for the following system:

$$A = \begin{bmatrix} 9 & 1 & 1 & 1 \\ 1 & 8 & 1 & 1 \\ 1 & 1 & 7 & 1 \\ 1 & 1 & 1 & 6 \end{bmatrix}, \qquad b = \begin{bmatrix} 75 \\ 54 \\ 43 \\ 34 \end{bmatrix}, \qquad x = \begin{bmatrix} 7 \\ 5 \\ 4 \\ 3 \end{bmatrix}$$

3. Consider the linear system

$$\begin{aligned} x_1 + 4x_2 &= 1 \\ 4x_1 + x_2 &= 0 \end{aligned}$$

The true solution is $x_1 = -1/15$, $x_2 = 4/15$. Apply the Jacobi and Gauss–Seidel methods with $x^{(0)} = [0, 0]^T$ to the system and find out which methods diverge more rapidly. Next, interchange the two equations to write the system as

$$\begin{aligned} 4x_1 + x_2 &= 0 \\ x_1 + 4x_2 &= 1 \end{aligned}$$

and apply both methods with $x^{(0)} = [0, 0]^T$. Iterate until $\|x - x^{(k)}\| \le 10^{-5}$. Which method converges faster?

4. Consider the iterative solution of $Ax = b$ with

$$A = \begin{bmatrix} 4 & -1 & -1 & 0 \\ -1 & 4 & 0 & -1 \\ -1 & 0 & 4 & -1 \\ 0 & -1 & -1 & 4 \end{bmatrix}, \qquad b = \begin{bmatrix} 5 \\ -3 \\ -7 \\ 9 \end{bmatrix}, \qquad x = \begin{bmatrix} 1 \\ 0 \\ -1 \\ 2 \end{bmatrix}$$

(a) Use the Jacobi method with $x^{(0)} = 0$. Iterate until $\|x - x^{(k)}\| \le 0.00005$. Observe the ratios of the successive errors. Calculate $\|M\|$ for this case.

(b) Repeat part (a) with the Gauss–Seidel method. Although it is more difficult than in (a), also compute $\|M\|$ in this case.

5. Consider the linear system

$$\begin{bmatrix} 4 & 1 & 1 \\ 1 & 2 & 0 \\ 0 & 2 & 3 \end{bmatrix} \begin{bmatrix} x_1 \\ x_2 \\ x_3 \end{bmatrix} = \begin{bmatrix} b_1 \\ b_2 \\ b_3 \end{bmatrix}$$

(a) Show that for any right-hand side vector $b = [b_1, b_2, b_3]^{\mathrm{T}}$, the system has a unique solution $x = [x_1, x_2, x_3]^{\mathrm{T}}$.

(b) When the Jacobi or Gauss–Seidel method is applied to the system, do you expect convergence? Explain!

6. Consider a linear system with the coefficient matrix A_n given in (6.108). Explore the convergence property of the Jacobi method and the Gauss–Seidel method by applying Theorem 6.6.3 and using MATLAB to compute the eigenvalues of the iteration matrices M. Do so for $n = 5, 10, 20$.

Note: In MATLAB, use `eig(B)` to find all the eigenvalues of a matrix `B`.

7. Consider solving the system

$$A_n x = b_n \tag{6.114}$$

with A_n given in (6.108) and

$$b_n = \frac{1}{(n+1)^2}[1, \ldots, 1]^T$$

Use the Jacobi and Gauss–Seidel methods with $x^{(0)} = 0$. Iterate until the difference between two consecutive iterates is smaller than ϵ, that is, until $\|x^{(k)} - x^{(k-1)}\| < \epsilon$. Do the computation with $n = 10$, 100, 1000, and $\epsilon = 10^{-6}, 10^{-8}, 10^{-10}$. Produce a table of the integer k with which the computation is completed. Comment on the behavior of the two methods based on the results reported in the table.

8. Continuing Problem 7, for a fixed n, write the solution as

$$x = [x_1, \ldots, x_n]^T$$

It can be shown that x_i is close to $t_i(1 - t_i)/2$, $t_i = i/(n+1)$, $1 \le i \le n$. For each n and each ϵ, denoting $x^{(k)}$ the numerical solution computed in Problem 7, graph the error vector

$$e = [e_1, \ldots, e_n]^T$$
$$e_i = x_i^{(k)} - \tfrac{1}{2} t_i(1 - t_i), \quad 1 \le i \le n$$

against the vector $t = [t_1, \ldots, t_n]^T$. Compute $\|e\|$ and observe how $\|e\|$ decreases when n increases.

9. Repeat Problem 4 for the system given below. To calculate $M = N^{-1}P$ for the Gauss–Seidel method will require the use of a program, to calculate N^{-1} and to then multiply it by P. As another means of checking for convergence, use Theorem 6.6.3. Thus, you must calculate the maximum eigenvalue of M in magnitude.

$$A = \begin{bmatrix} 4 & -1 & 0 & -1 & 0 & 0 & 0 & 0 & 0 \\ -1 & 4 & -1 & 0 & -1 & 0 & 0 & 0 & 0 \\ 0 & -1 & 4 & 0 & 0 & -1 & 0 & 0 & 0 \\ -1 & 0 & 0 & 4 & -1 & 0 & -1 & 0 & 0 \\ 0 & -1 & 0 & -1 & 4 & -1 & 0 & -1 & 0 \\ 0 & 0 & -1 & 0 & -1 & 4 & 0 & 0 & -1 \\ 0 & 0 & 0 & -1 & 0 & 0 & 4 & -1 & 0 \\ 0 & 0 & 0 & 0 & -1 & 0 & -1 & 4 & -1 \\ 0 & 0 & 0 & 0 & 0 & -1 & 0 & -1 & 4 \end{bmatrix}$$

$$b = \begin{bmatrix} 4 \\ -1 \\ -5 \\ -2 \\ 2 \\ 2 \\ -1 \\ 1 \\ 6 \end{bmatrix}, \qquad x = \begin{bmatrix} 1 \\ 0 \\ -1 \\ 0 \\ 1 \\ 1 \\ 0 \\ 1 \\ 2 \end{bmatrix}$$

This system arises from solving the *partial differential equation*

$$\frac{\partial^2 x(s,t)}{\partial s^2} + \frac{\partial^2 x(s,t)}{\partial t^2} = f(s,t), \qquad 0 \le s, t \le 1$$

10. **(a)** For the iteration (6.98), show

$$e^{(k)} = M^k e^{(0)}, \qquad k \ge 0, \qquad M = N^{-1}P$$

To have the error $e^{(k)}$ go to zero as $k \to \infty$, for all choices of $e^{(0)}$, the above shows we want to have $M^k \to 0$, the zero matrix, as $k \to \infty$.

(b) Show that if convergence of (6.98) takes place for all initial guesses $x^{(0)}$, or equivalently all $e^{(0)}$, then all eigenvalues λ of the matrix $M = N^{-1}P$ satisfy $|\lambda| < 1$.

Hint: Show the contrapositive of this statement. Assume some eigenvalue λ of M satisfies $|\lambda| \ge 1$. Show this implies $e^{(k)}$ does not go to zero for some choices of $e^{(0)}$, by considering a choice of $e^{(0)}$ based on the eigenvector associated with λ.

11. Many linear systems $Ax = b$ are directly written in the form $x = b + Mx$ with $A = I - M$. Do a convergence analysis for

$$x^{(k+1)} = b + Mx^{(k)}, \qquad k \ge 0$$

using the ideas of this section.

12. Consider the iteration

$$x^{(k+1)} = b + \alpha \begin{bmatrix} 2 & 1 \\ 1 & 2 \end{bmatrix} x^{(k)}, \qquad k = 0, 1, 2, \ldots$$

where α is a real constant. For some values of α, the iteration method converges for any choice of initial guess $x^{(0)}$; and for some other values of α, the method diverges. Find the values of α for which the method converges.

13. By using the iteration method of Problem 11, solve the linear system $x = b + Mx$ with

$$M_{ij} = \frac{1}{2n}\left[\frac{t_i^3}{1+t_j} + 1\right], \qquad b_i = \frac{1}{4} + t_i - \frac{1}{2}t_i^3$$

and $t_i = (2i-1)/2n$, $1 \le i \le n$. The true solution is

$$x_i = 1 + t_i, \qquad 1 \le i \le n$$

Solve this for several values of n, say, $n = 3, 6, 12, 24$ and perhaps larger. Calculate the errors $\|x - x^{(k)}\|$ and the ratios with which they decrease. This arises from solving numerically the following *integral equation:*

$$x(s) - \frac{1}{2}\int_0^1 \left[\frac{s^3}{1+t} + 1\right] x(t)\,dt = \frac{1}{4} + s - \frac{1}{2}s^3, \qquad 0 \le s \le 1$$

14. Consider solving the linear system $A_\epsilon x = b$, where $A_\epsilon = A_0 + \epsilon B$

$$A_0 = \begin{bmatrix} 2 & 1 & 0 \\ 1 & 2 & 1 \\ 0 & 1 & 2 \end{bmatrix}, \qquad B = \begin{bmatrix} 0 & 1 & 1 \\ -1 & 0 & 1 \\ -1 & -1 & 0 \end{bmatrix}$$

Use the residual correction method (6.111–6.113) to solve $A_\epsilon x = b$ for various values of ϵ, with $N = A_0$. Give a condition of the form $|\epsilon| \le \alpha$, for some α, that will ensure convergence of the residual correction method.

CHAPTER 7

NUMERICAL LINEAR ALGEBRA: ADVANCED TOPICS

Section 7.1 defines the least squares data fitting method, which leads to some well-known linear systems. Section 7.2 introduces the matrix eigenvalue problem, along with one numerical method for its solution. Systems of nonlinear equations are discussed in Section 7.3, and Newton's method is generalized to solve these systems.

7.1. LEAST SQUARES DATA FITTING

In scientific and engineering experiments, the measurement of physical quantities will generally be somewhat inaccurate. This may be due to human error but, more often, it is due to limitations inherent in the equipment being used to make the measurements. In this section, we present a method for obtaining relationships between empirically obtained variables, with a minimization of the effects of the measurement errors present in these variables.

Suppose an experiment involves measuring two related variables x and y to obtain an unknown relationship

$$y = f(x) \tag{7.1}$$

In the experiment, one chooses various values of x, say, $x_1, x_2, \ldots, x_n$, and then one measures a corresponding set of values for y. Let the actual measurements be denoted by $y_1, \ldots, y_n$, and let

$$\epsilon_i = f(x_i) - y_i$$

denote the measurement errors. The experimenter wants to use the points $(x_1, y_1), \ldots, (x_n, y_n)$ to determine the analytic relationship (7.1) as accurately as possible.

An analytic relation (7.1) can always be obtained by using interpolation with polynomials, spline functions, or some other form of interpolating function. But this ignores the presence of the errors $\{\epsilon_i\}$, and it often leads to a poor estimate of the true function $f(x)$. Often, experimenters will know or will suspect that the unknown function $f(x)$ lies within some known class of functions, for example, polynomials. Then they will want to choose the member of that class of functions that will best approximate their unknown function $f(x)$, taking into account the experimental errors $\{\epsilon_i\}$. As an example of such a situation, consider the data in Table 7.1 and the plot of it in Figure 7.1. From this plot, it is reasonable to expect $f(x)$ to be close to a linear polynomial

$$f(x) = mx + b \tag{7.2}$$

If we assume this to be the true form, the problem of determining $f(x)$ is now reduced to that of determining the constants m and b.

To determine a reasonable approximation to $f(x)$, we must assume something more is known about the errors $\{\epsilon_i\}$. A standard assumption is that each of these errors is a

Table 7.1. Empirical Data

x_i	y_i	x_i	y_i
1.0	−1.945	3.2	0.764
1.2	−1.253	3.4	0.532
1.4	−1.140	3.6	1.073
1.6	−1.087	3.8	1.286
1.8	−0.760	4.0	1.502
2.0	−0.682	4.2	1.582
2.2	−0.424	4.4	1.993
2.4	−0.012	4.6	2.473
2.6	−0.190	4.8	2.503
2.8	0.452	5.0	2.322
3.0	0.337		

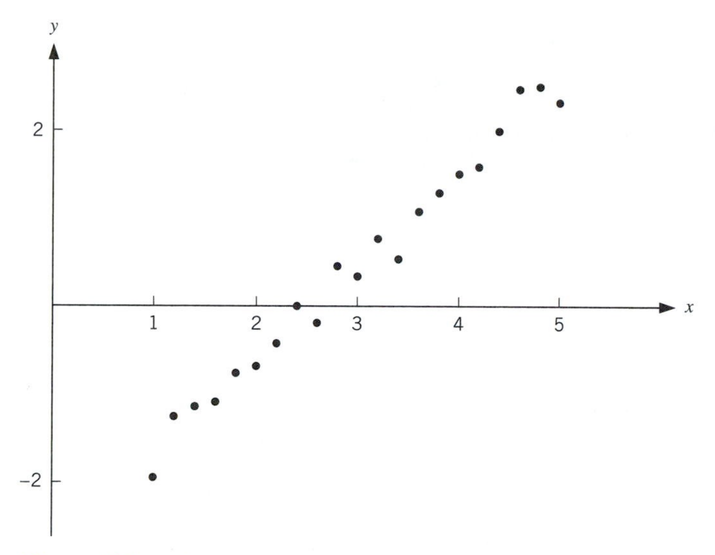

Figure 7.1. The plot of empirical data from Table 7.1

random variable chosen from a *normal probability distribution*. This is a topic developed in probability theory and statistics courses, and it is too involved to consider here. But errors satisfying this assumption have the following general characteristics. (1) If the experiment is repeated many times for the same $x = x_i$, then the associated unknown errors ϵ_i in the empirical values y_i will be likely to have an average of zero; and (2) for this same experimental case with $x = x_i$, as the size of ϵ_i increases, the likelihood of its occurring will decrease rapidly. There are very good reasons for accepting this assumption on the behavior of the errors $\{\epsilon_i\}$, and it will be denoted as the *normal error assumption* in what follows. Also, to further simplify the following presentation, we will assume that the individual errors ϵ_i, $1 \le i \le n$, are all random variables from the same normal probability distribution function, meaning that the size of ϵ_i is unrelated to the size of x_i or y_i.

Let the true function $f(x)$ be assumed to lie in a known class of functions, call it C. An example is the assumption that $f(x)$ is linear for the data in Table 7.1. Then among all functions $\widehat{f}(x)$ in C, it can be shown that the function $\widehat{f}^*$ that is most likely to equal f will also minimize the expression

$$E = \sqrt{\frac{1}{n}\sum_{i=1}^{n}\left[\widehat{f}(x_i) - y_i\right]^2} \tag{7.3}$$

among all functions $\widehat{f}$ in C. This is called the *root-mean-square-error* in the approximation of the data $\{y_i\}$ by $\widehat{f}(x)$. The function $\widehat{f}^*(x)$ that minimizes E relative to all $\widehat{f}$ in C is called the *least squares approximation* to the data $\{(x_i, y_i)\}$.

7.1.1 The Linear Least Squares Approximation

We consider the use of $\widehat{f}(x) = mx + b$ in E. Before attempting to minimize (7.3) relative to such functions $\widehat{f}(x)$, note that minimizing E is equivalent to minimizing the sum in (7.3), although the minimum values will be different. We seek to minimize

$$G(b, m) = \sum_{i=1}^{n} [mx_i + b - y_i]^2 \tag{7.4}$$

as b and m are allowed to vary arbitrarily, and the resulting choice of b and m will also minimize E.

From multivariable calculus, the values of (b, m) that minimize $G(b, m)$ will satisfy

$$\frac{\partial G(b, m)}{\partial b} = 0, \qquad \frac{\partial G(b, m)}{\partial m} = 0 \tag{7.5}$$

using the partial derivatives of $G(b, m)$. From (7.4),

$$\frac{\partial G}{\partial b} = \sum_{i=1}^{n} 2\,[mx_i + b - y_i]$$

$$\frac{\partial G}{\partial m} = \sum_{i=1}^{n} 2\,[mx_i + b - y_i]\,x_i = \sum_{i=1}^{n} 2\left[mx_i^2 + bx_i - x_i y_i\right]$$

Combining these with (7.5), we obtain the linear system of order 2

$$\begin{aligned} nb + \left(\sum_{i=1}^{n} x_i\right) m &= \sum_{i=1}^{n} y_i \\ \left(\sum_{i=1}^{n} x_i\right) b + \left(\sum_{i=1}^{n} x_i^2\right) m &= \sum_{i=1}^{n} x_i y_i \end{aligned} \tag{7.6}$$

This can be solved for a unique solution (b, m), provided that the determinant is nonzero

$$n \sum_{i=1}^{n} x_i^2 - \left(\sum_{i=1}^{n} x_i\right)^2 \neq 0 \tag{7.7}$$

This condition will be satisfied except in the one trivial case

$$x_1 = x_2 = \cdots = x_n = \text{constant}$$

and it cannot happen here, since the numbers $\{x_j\}$ are assumed to be distinct.

Example 7.1.1 Determine the least squares fit for the data in Table 7.1. We have $n = 21$ and

$$\sum_{i=1}^{n} x_i = 63.0, \qquad \sum_{i=1}^{n} x_i^2 = 219.8$$

$$\sum_{i=1}^{n} y_i = 9.326, \qquad \sum_{i=1}^{n} x_i y_i = 60.7302$$

Using this in (7.6) and then solving for b and m give the solution

$$b \doteq -2.74605, \qquad m \doteq 1.06338$$

The linear least squares fit to the data in Table 7.1 is

$$\widehat{f}^*(x) = 1.06338x - 2.74605 \tag{7.8}$$

This is optimum in the sense of minimizing the root-mean-square-error E in (7.3), relative to all linear functions $\widehat{f}(x) = mx + b$. The value of E for (7.8) is

$$E \doteq 0.171$$

It can be considered as an average error in the approximation of the data in Table 7.1 by the least squares fit (7.8). A graph of (7.8) is given in Figure 7.2, along with the data points from Table 7.1. ■

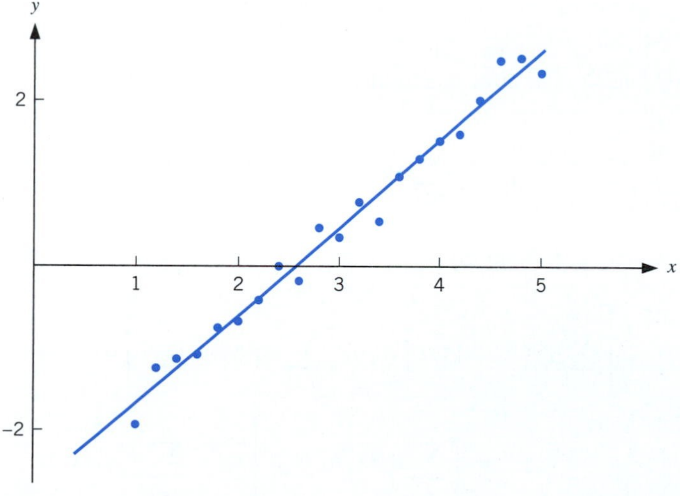

Figure 7.2. The least squares fit $\widehat{f}^*(x)$ of (7.8)

7.1.2 Polynomial Least Squares Approximation

Most data will not correspond to a linear function, but to something more complicated. Let the functions $\widehat{f}(x)$ that are being used to represent the data $\{(x_i, y_i)\}$ be given by

$$\widehat{f}(x) = a_1\varphi_1(x) + a_2\varphi_2(x) + \cdots + a_m\varphi_m(x) \tag{7.9}$$

where $a_1, a_2, \ldots, a_m$ are arbitrary numbers and $\varphi_1(x), \ldots, \varphi_m(x)$ are given functions. For example, if $\widehat{f}(x)$ was to be a quadratic polynomial, we could write

$$\widehat{f}(x) = a_1 + a_2 x + a_3 x^2 \tag{7.10}$$

where

$$\varphi_1(x) \equiv 1, \qquad \varphi_2(x) = x, \qquad \varphi_3(x) = x^2$$

Under the normal error assumption, the function $\widehat{f}(x)$ is to be chosen to minimize the root-mean-square-error of (7.3). To show what is involved, we will consider the specific case $m = 3$. Then

$$\widehat{f}(x) = a_1\varphi_1(x) + a_2\varphi_2(x) + a_3\varphi_3(x)$$

and the constants a_1, a_2, a_3 are to be chosen so as to minimize the expression

$$G(a_1, a_2, a_3) = \sum_{j=1}^{n}[a_1\varphi_1(x_j) + a_2\varphi_2(x_j) + a_3\varphi_3(x_j) - y_j]^2$$

The points (a_1, a_2, a_3) that minimize $G(a_1, a_2, a_3)$ will satisfy

$$\frac{\partial G}{\partial a_1} = 0, \qquad \frac{\partial G}{\partial a_2} = 0, \qquad \frac{\partial G}{\partial a_3} = 0$$

This leads to the three equations

$$0 = \frac{\partial G}{\partial a_i} = \sum_{j=1}^{n} 2[a_1\varphi_1(x_j) + a_2\varphi_2(x_j) + a_3\varphi_3(x_j) - y_j]\varphi_i(x_j)$$

for $i = 1, 2, 3$. Simplifying, we get

$$\begin{aligned}&\left[\sum_{j=1}^{n}\varphi_1(x_j)\varphi_i(x_j)\right]a_1 + \left[\sum_{j=1}^{n}\varphi_2(x_j)\varphi_i(x_j)\right]a_2 \\ &\quad + \left[\sum_{j=1}^{n}\varphi_3(x_j)\varphi_i(x_j)\right]a_3 = \sum_{j=1}^{n} y_j\varphi_i(x_j), \qquad i = 1, 2, 3\end{aligned} \tag{7.11}$$

These are three linear equations for the three unknowns a_1, a_2, a_3.

Example 7.1.2 Apply (7.11) to the quadratic formula for $f(x)$ in (7.10). Then the three equations are

$$
\begin{aligned}
na_1 + \left[\sum_{j=1}^{n} x_j\right] a_2 + \left[\sum_{j=1}^{n} x_j^2\right] a_3 &= \sum_{j=1}^{n} y_j \\
\left[\sum_{j=1}^{n} x_j\right] a_1 + \left[\sum_{j=1}^{n} x_j^2\right] a_2 + \left[\sum_{j=1}^{n} x_j^3\right] a_3 &= \sum_{j=1}^{n} y_j x_j \\
\left[\sum_{j=1}^{n} x_j^2\right] a_1 + \left[\sum_{j=1}^{n} x_j^3\right] a_2 + \left[\sum_{j=1}^{n} x_j^4\right] a_3 &= \sum_{j=1}^{n} y_j x_j^2
\end{aligned}
\tag{7.12}
$$

This can be shown to be a nonsingular system due to the assumption that the points $\{x_j\}$ are distinct. ■

When $\widehat{f}(x)$ takes the more general form given in (7.9), the choice for which the root-mean-square-error E in (7.3) is minimized has the coefficients $a_1, \ldots, a_m$ determined by the linear system

$$\sum_{k=1}^{m} a_k \left[\sum_{j=1}^{n} \varphi_k(x_j)\varphi_i(x_j)\right] = \sum_{j=1}^{n} y_j \varphi_i(x_j), \qquad i = 1, \ldots, m \tag{7.13}$$

The derivation of this is left as Problem 9. The special case in which $\widehat{f}(x)$ is a polynomial of degree $(m-1)$ is very important, and we consider it in greater detail.

Write $\widehat{f}(x)$ in the standard form for a polynomial of degree $(m-1)$

$$\widehat{f}(x) = a_1 + a_2 x + a_3 x^2 + \cdots + a_m x^{m-1}$$

Then this corresponds to (7.9) with

$$\varphi_1(x) = 1, \quad \varphi_2(x) = x, \quad \varphi_3(x) = x^2, \quad \ldots, \quad \varphi_m(x) = x^{m-1} \tag{7.14}$$

The system (7.13) for determining $\{a_j\}$ becomes

$$\sum_{k=1}^{m} a_k \left[\sum_{j=1}^{n} x_j^{i+k-2}\right] = \sum_{j=1}^{n} y_j x_j^{i-1}, \qquad i = 1, 2, \ldots, m \tag{7.15}$$

When $m = 3$, this yields the system (7.12) obtained earlier.

The system (7.15) is nonsingular (for $m < n$), but unfortunately it becomes increasingly ill-conditioned as the degree of $\widehat{f}(x)$ increases; furthermore, the size of the condition number for the matrix of coefficients can be very large for fairly small values of m, say, $m = 4$. For this reason, it is seldom advisable to use (7.14) to do a least squares polynomial fit, except for degree ≤ 2.

To do a least squares fit to data $\{(x_i, y_i)\}$ with a higher-degree polynomial $\widehat{f}(x)$, it is necessary to write it in a form

$$\widehat{f}(x) = a_1\varphi_1(x) + \cdots + a_m\varphi_m(x)$$

where $\varphi_1(x), \ldots, \varphi_m(x)$ are so chosen that the matrix of coefficients in (7.13) is not ill-conditioned. There are optimal choices of these functions $\varphi_j(x)$, with $\deg(\varphi_j) = j - 1$ and with the coefficient matrix for (7.13) becoming diagonal. A nonoptimal but still satisfactory choice in general can be based on the Chebyshev polynomials $\{T_k(x)\}$ of Section 4.5, and a somewhat better choice is the Legendre polynomials of Section 4.7.

Suppose that the nodes $\{x_i\}$ are chosen from an interval $[\alpha, \beta]$. Introduce modified Chebyshev polynomials

$$\varphi_k(x) = T_{k-1}\left(\frac{2x - \alpha - \beta}{\beta - \alpha}\right), \qquad \alpha \le x \le \beta, \qquad k \ge 1 \tag{7.16}$$

Then degree $(\varphi_k) = k - 1$; and any polynomial $\widehat{f}(x)$ of degree $(m - 1)$ can be written as a combination of $\varphi_1(x), \ldots, \varphi_m(x)$.

Example 7.1.3 Consider the data in Table 7.2 and the plot of it in Figure 7.3. We first find the least squares cubic polynomial fit $\widehat{f}(x)$ by using the normal representation:

$$\widehat{f}(x) = a_1 + a_2x + a_3x^2 + a_4x^3 \tag{7.17}$$

Table 7.2. Data for a Cubic Least Squares Fit

x_i	y_i	x_i	y_i
0.00	0.486	0.55	1.102
0.05	0.866	0.60	1.099
0.10	0.944	0.65	1.017
0.15	1.144	0.70	1.111
0.20	1.103	0.75	1.117
0.25	1.202	0.80	1.152
0.30	1.166	0.85	1.265
0.35	1.191	0.90	1.380
0.40	1.124	0.95	1.575
0.45	1.095	1.00	1.857
0.50	1.122		

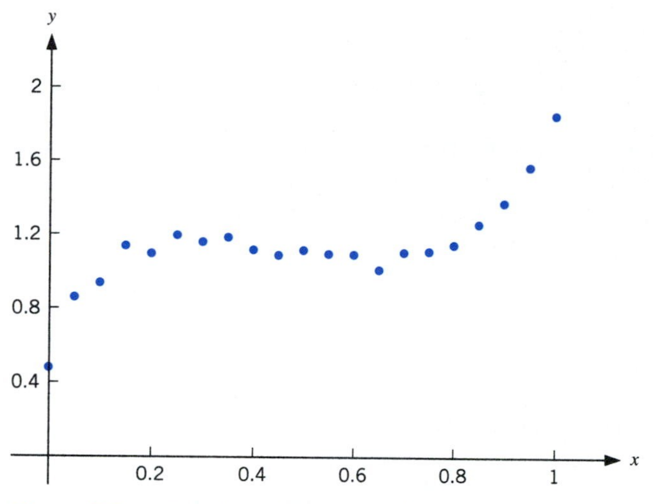

Figure 7.3. The plot of data of Table 7.2

The resulting linear system (7.15), denoted here by $La = b$, is given by

$$L = \begin{bmatrix} 21 & 10.5 & 7.175 & 5.5125 \\ 10.5 & 7.175 & 5.5125 & 4.51666 \\ 7.175 & 5.5125 & 4.51666 & 3.85416 \\ 5.5125 & 4.51666 & 3.85416 & 3.38212 \end{bmatrix}$$

$$a = [a_1, a_2, a_3, a_4]^T$$

$$b = [24.1180, 13.2345, 9.46836, 7.55944]^T$$

The solution is

$$a = [0.5747,\ 4.7259,\ -11.1282,\ 7.6687]^T$$

The condition number, defined in (6.91), can be calculated to be

$$\text{cond}(L) = \|L\|\,\|L^{-1}\| \doteq 22{,}000 \tag{7.18}$$

This is very large, and it indicates possible difficulty in finding an accurate answer for $La = b$. To verify this, perturb b above by adding to it the perturbation

$$[0.01,\ -0.01,\ 0.01,\ -0.01]^T$$

This will change b in its second place to the right of the decimal point, within the range of possible perturbations due to errors in the data. The solution of the new perturbed system is

$$a = [0.7408,\ 2.6825,\ -6.1538,\ 4.4550]^T$$

which is dramatically different from the earlier result for a. The main point here is that the use of (7.17) leads to a rather ill-conditioned system of linear equations. To find a cubic least squares approximation for the data in Table 7.2, with a better-behaved linear system, use the functions in (7.16) on [0, 1] and write

$$f(x) = a_1\varphi_1(x) + a_2\varphi_2(x) + a_3\varphi_3(x) + a_4\varphi_4(x) \tag{7.19}$$

where

$$\begin{aligned}
\varphi_1(x) &= T_0(2x-1) \equiv 1\\
\varphi_2(x) &= T_1(2x-1) = 2x-1\\
\varphi_3(x) &= T_2(2x-1) = 2(2x-1)^2 - 1 = 8x^2 - 8x + 1\\
\varphi_4(x) &= T_3(2x-1) = 4(2x-1)^3 - 3(2x-1) = 32x^3 - 48x^2 + 18x - 1
\end{aligned}$$

Because of the different functions used in the representation of $\widehat{f}(x)$ of (7.19), the values a_1, a_2, a_3, a_4 will be completely different than in the representation (7.17). The linear system (7.13) will again be denoted by $La = b$, now

$$L = \begin{bmatrix} 21 & 0 & -5.6 & 0\\ 0 & 7.7 & 0 & -2.8336\\ -5.6 & 0 & 10.4664 & 0\\ 0 & -2.8336 & 0 & 11.01056 \end{bmatrix}$$
$$b = [24.118,\ 2.351,\ -6.01108,\ 1.523576]^T$$

The solution is

$$a = [1.160969,\ 0.393514,\ 0.046850,\ 0.239646]^T$$

The linear system is very stable with respect to the type of perturbation made in b with the earlier approach to the cubic least squares fit, using (7.17). This also is implied by the small condition number of L

$$\text{cond}(L) = \|L\|\,\|L^{-1}\| \doteq (26.6)(0.1804) \doteq 4.8 \tag{7.20}$$

From the result (6.89) in Section 6.5, relatively small perturbations in b will lead to correspondingly small changes in the solution a. Also, compare (7.20) to the condition number (7.18) for the earlier system. To give some idea of the accuracy of $\widehat{f}(x)$ in approximating the data in Table 7.2, we easily compute the root-mean-square-error from (7.3) to be

$$E \doteq 0.0421$$

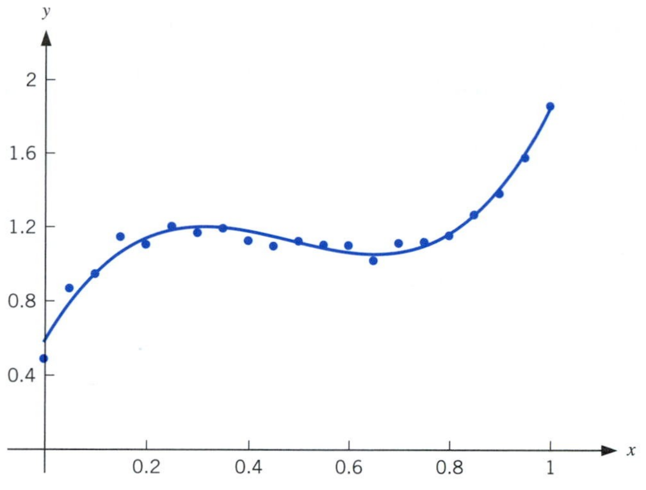

Figure 7.4. The cubic least squares fit for Table 7.2

a fairly small value when compared with the function values of $\widehat{f}(x)$. The graph of $\widehat{f}(x)$ is shown in Figure 7.4. ■

The MATLAB polynomial data fitting function is `polyfit`. Given vectors `x` and `y` of the same length, with `x` having distinct components, the call `polyfit(x,y,n)` returns the least squares polynomial fitting of a degree less than or equal to n. As an example, we store the data points from Example 7.1.3 in vectors `xx` and `yy`. Then Figure 7.5 is

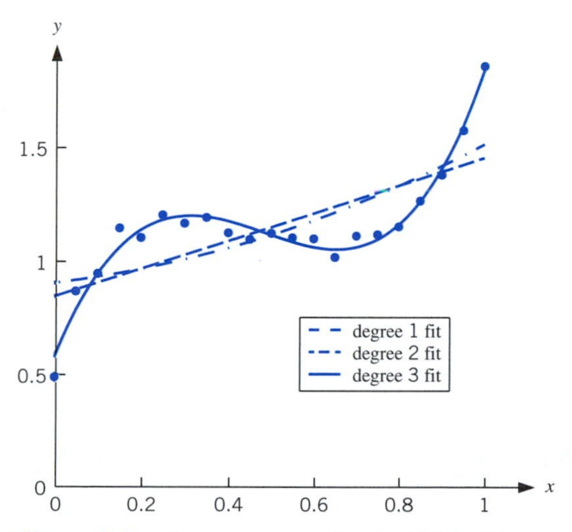

Figure 7.5. Least squares fits for Table 7.2

produced by using the following MATLAB commands:

```
p1 = polyfit(xx,yy,1);
p2 = polyfit(xx,yy,2);
p3 = polyfit(xx,yy,3);
plot(xx,yy,'ko',xx,polyval(p1,xx),'--',xx,...
   polyval(p2,xx),'-.',xx,polyval(p3,xx),'LineWidth',2)
legend('data points','degree 1 fit',...
   'degree 2 fit','degree 3 fit')
```

PROBLEMS

1. Calculate the linear least squares fit for the following data. Graph the data and the least squares fit. Also, find the root-mean-square-error in the least squares fit.

x_i	y_i	x_i	y_i	x_i	y_i
0	−1.466	1.2	1.068	2.4	4.148
0.3	−0.062	1.5	1.944	2.7	4.464
0.6	0.492	1.8	2.583	3.0	5.185
0.9	0.822	2.1	3.239		

2. Repeat Problem 1 with the following data:

x_i	y_i	x_i	y_i	x_i	y_i
−1.0	1.032	−0.3	1.139	0.4	−0.415
−0.9	1.563	−0.2	0.646	0.5	−0.112
−0.8	1.614	−0.1	0.474	0.6	−0.817
−0.7	1.377	0	0.418	0.7	−0.234
−0.6	1.179	0.1	0.067	0.8	−0.623
−0.5	1.189	0.2	0.371	0.9	−0.536
−0.4	0.910	0.3	0.183	1.0	−1.173

3. Do a quadratic least squares polynomial fit to the following data. Use the standard form

$$\widehat{f}(x) = a_1 + a_2 x + a_3 x^2$$

with a_1, a_2, a_3 the solution of the system (7.12). Also, calculate the root-mean-square-error, plot the data, and plot the least squares fit.

x_i	y_i	x_i	y_i	x_i	y_i
−1.0	7.904	−0.3	0.335	0.4	−0.711
−0.9	7.452	−0.2	−0.271	0.5	0.224
−0.8	5.827	−0.1	−0.963	0.6	0.689
−0.7	4.400	0	−0.847	0.7	0.861
−0.6	2.908	0.1	−1.278	0.8	1.358
−0.5	2.144	0.2	−1.335	0.9	2.613
−0.4	0.581	0.3	−0.656	1.0	4.599

4. The process described in this section is called the least squares approximation method for discrete data. There is a corresponding method for approximating data given as a continuous function, although the reason for doing this is different than in the discrete case. To keep matters somewhat simple, consider approximating $f(x)$ by a quadratic polynomial

$$\widehat{f}(x) = a_1 + a_2 x + a_3 x^2$$

on the interval $0 \le x \le 1$. Do so by choosing a_1, a_2, and a_3 to minimize the root-mean-square-error

$$E = \left[\int_0^1 [a_1 + a_2 x + a_3 x^2 - f(x)]^2 \, dx \right]^{1/2}$$

Derive the linear system satisfied by the optimum choices of a_1, a_2, a_3. What is its relation to the Hilbert matrix of Section 6.5?

5. In this problem, we explore a statistical point of view of the linear least squares approximation. Let (x_j, y_j), $1 \le j \le n$, be the given data points. We are interested in a linear curve fit. Define the mean values

$$\overline{x} = \frac{1}{n} \sum_{j=1}^{n} x_j, \qquad \overline{y} = \frac{1}{n} \sum_{j=1}^{n} y_j$$

It is natural to require the point $(\overline{x}, \overline{y})$ to lie on the line. Thus, the linear function is of the form

$$y = \overline{y} + a\,(x - \overline{x})$$

To determine a, we minimize the standard deviation

$$\sqrt{\frac{1}{n}\sum_{j=1}^{n}[y_j - (\overline{y} + a\,(x_j - \overline{x}))]^2}$$

that is, we choose a that minimizes the quadratic function

$$\sum_{j=1}^{n}[y_j - (\overline{y} + a\,(x_j - \overline{x}))]^2 = \sum_{j=1}^{n}(y_j - \overline{y})^2 - 2\,a\sum_{j=1}^{n}(x_j - \overline{x})(y_j - \overline{y}) + a^2\sum_{j=1}^{n}(x_j - \overline{x})^2$$

Easily,

$$a = \frac{\sum_{j=1}^{n}(x_j - \overline{x})(y_j - \overline{y})}{\sum_{j=1}^{n}(x_j - \overline{x})^2}$$

Show that the linear function determined in this way is identical to the function $\widehat{f}^*(x) = mx + b$ with b and m determined by the linear system (7.6).

6. Let (x_j, y_j), $x_j \geq 0$, $1 \leq j \leq n$, be n points. Let $\mu > 0$ be given. Show that for the least squares power curve $y = a\,x^{\mu}$, we have

$$a = \frac{\sum_{j=1}^{n} x_j^{\mu}\, y_j}{\sum_{j=1}^{n} x_j^{2}\mu}$$

7. For $n = 2$ and 3, verify that the inequality (7.7) holds except when $x_1 = \cdots = x_n =$ constant.

8. Use the MATLAB built-in function `cond(A,inf)` to explore the conditioning of the system (7.15). Let $n = 11$, $x_j = (j-1)h$, $1 \leq j \leq n$, with $h = 1/(n-1)$. Compute the condition numbers of the system for $m = 1$ to 5. Observe how rapidly the condition number increases with m.

9. Derive the linear system (7.13) by imitating the derivation of the system (7.11).

10. Replace the assumption of $\widehat{f}(x) = mx + b$ with

$$\widehat{f}(x) = a + be^{-x}$$

Find a and b by minimizing the root-mean-square-error

$$E = \left[\frac{1}{n}\sum_{j=1}^{n}[a + be^{-x_j} - y_j]^2\right]^{1/2}$$

Find a linear system satisfied by the optimum choice of a and b.

Hint: To simplify the work, minimize the sum in E, and then follow the derivation of (7.4) to (7.6).

11. Some more difficult curve fitting problems can be converted to linear least squares approximation problems. One example is the fitting function $y = a\,x^{\mu}$, where both a and μ are to be determined. Suppose (x_j, y_j), $x_j > 0$, $y_j > 0$, $1 \le j \le n$, are given points. We apply natural log to rewrite the fitting function as

$$\ln y = \ln a + \mu \ln x$$

Demonstrate how to determine a and μ.

12. Do the same as in the Problem 11 for the fitting function

$$y = \frac{1}{(a\,x + b)^3}$$

7.2. THE EIGENVALUE PROBLEM

For a square matrix A, we define the concept of an eigenvalue of A as follows. The number λ is called an *eigenvalue* of A and the column vector v is a corresponding *eigenvector* if

$$Av = \lambda v \tag{7.21}$$

and $v \neq 0$, the zero column vector. The eigenvalue-eigenvector pairs $\{\lambda, v\}$ of a matrix A can be used in understanding what happens when a general column vector x is multiplied by A and when solving a linear system $Ax = b$. The eigenvalues and eigenvectors also arise naturally in many physical problems, often leading to certain "basic solutions" that are of intrinsic physical interest, as in the natural frequencies of vibration of a taut string or a musical instrument. In this section, we give a brief introduction to the theory of eigenvalues and eigenvectors, and then we present a numerical method for calculating the largest eigenvalue of a matrix.

Example 7.2.1 For the matrix

$$A = \begin{bmatrix} 1.25 & 0.75 \\ 0.75 & 1.25 \end{bmatrix} \tag{7.22}$$

eigenvalue-eigenvector pairs are

$$\begin{aligned} \lambda_1 &= 2, & v^{(1)} &= \begin{bmatrix} 1 \\ 1 \end{bmatrix} \\ \lambda_2 &= 0.5, & v^{(2)} &= \begin{bmatrix} -1 \\ 1 \end{bmatrix} \end{aligned} \tag{7.23}$$

The reader should verify that (7.21) is satisfied with both of these pairs. We now illustrate the use of the eigenvalues and eigenvectors in further understanding the multiplication of a general column vector x by A. For $x = [x_1, x_2]^T$, it can be verified directly that

$$x = c_1 v^{(1)} + c_2 v^{(2)} \tag{7.24}$$

with

$$c_1 = \frac{x_1 + x_2}{2}, \qquad c_2 = \frac{x_2 - x_1}{2}$$

Consider x as a vector from the origin to the point with coordinates (x_1, x_2) in the x_1x_2-plane. Then (7.24) says that x can be written as the sum of the two vectors $c_1 v^{(1)}$ and $c_2 v^{(2)}$, which is illustrated in Figure 7.2.1(a). Now compute Ax:

$$\begin{aligned} Ax &= A[c_1 v^{(1)} + c_2 v^{(2)}] \\ &= c_1 A v^{(1)} + c_2 A v^{(2)} \\ &= c_1 \lambda_1 v^{(1)} + c_2 \lambda_2 v^{(2)} \\ Ax &= 2c_1 v^{(1)} + 0.5 c_2 v^{(2)} \end{aligned}$$

Thus, Ax is based on doubling the length of one part of x, the component $c_1 v^{(1)}$, and halving the other part, the component $c_2 v^{(2)}$. The resulting value of Ax is shown in Figure 7.2.1(b). The figure suggests that the problem which led to calculating Ax might

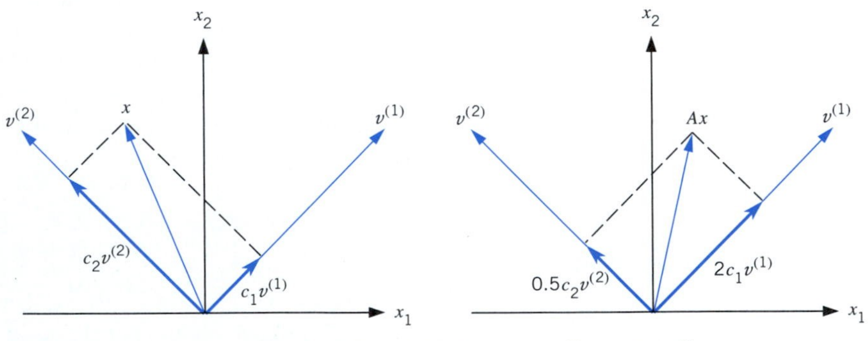

Figure 7.6. (a) $x = c_1 v^{(1)} + c_2 v^{(2)}$ and (b) $Ax = 2c_1 v^{(1)} + 0.5 c_2 v^{(2)}$

have been better formulated in the coordinate system determined by axes containing the eigenvectors $v^{(1)}$ and $v^{(2)}$. These same ideas extend to most other matrices of order 2 and to most other square matrices of higher order. ■

7.2.1 The Characteristic Polynomial

Rewrite the definition (7.21) as

$$(\lambda I - A)v = 0, \qquad v \neq 0 \tag{7.25}$$

This is a homogeneous system of linear equations with the coefficient matrix $\lambda I - A$ and the nonzero solution v. From Theorem 6.2.10 of Section 6.3, this can be true if and only if

$$\det(\lambda I - A) = 0$$

The function

$$f(\lambda) = \det(\lambda I - A)$$

is called the *characteristic polynomial* of A, and its roots are the eigenvalues of A. By using the properties of determinants and assuming A has order n, it can be shown that

$$f(\lambda) = \lambda^n + \alpha_{n-1}\lambda^{n-1} + \cdots + \alpha_1\lambda + \alpha_0 \tag{7.26}$$

a polynomial of degree n. For the case $n = 2$,

$$\begin{aligned} f(\lambda) &= \det \begin{bmatrix} \lambda - a_{11} & -a_{12} \\ -a_{21} & \lambda - a_{22} \end{bmatrix} \\ &= (\lambda - a_{11})(\lambda - a_{22}) - a_{21}a_{12} \\ &= \lambda^2 - (a_{11} + a_{22})\lambda + a_{11}a_{22} - a_{21}a_{12} \end{aligned}$$

The formula (7.26) shows that a matrix A of order n can have at most n distinct eigenvalues.

Example 7.2.2 For the earlier example A of (7.22),

$$\begin{aligned} f(\lambda) &= \det \begin{bmatrix} \lambda - 1.25 & -0.75 \\ -0.75 & \lambda - 1.25 \end{bmatrix} \\ &= \lambda^2 - 2.5\lambda + 1 \end{aligned}$$

This polynomial has the roots $\lambda = 0.5, 2$, as given earlier in (7.23). ■

Once the eigenvalues are known, then the eigenvectors can be found by solving the linear system (7.25), with λ replaced by the actual eigenvalue. Note that the solution v is not unique, because any constant multiple is also a solution. The solution process for v is nontrivial, since there are places where rounding errors can propagate and expand rapidly in size. To illustrate the general solution process with a simple case, where rounding error is not a problem, we include the following example.

Example 7.2.3 Let

$$A = \begin{bmatrix} -7 & 13 & -16 \\ 13 & -10 & 13 \\ -16 & 13 & -7 \end{bmatrix} \tag{7.27}$$

Then

$$\begin{aligned} f(\lambda) &= \det \begin{bmatrix} \lambda + 7 & -13 & 16 \\ -13 & \lambda + 10 & -13 \\ 16 & -13 & \lambda + 7 \end{bmatrix} \\ &= \lambda^3 + 24\lambda^2 - 405\lambda + 972 \end{aligned}$$

The roots are

$$\lambda_1 = -36, \qquad \lambda_2 = 9, \qquad \lambda_3 = 3 \tag{7.28}$$

For $\lambda = \lambda_1 = -36$, let us find an associated eigenvector $v = v^{(1)}$. If we substitute the value $\lambda = \lambda_1$ into (7.25), v satisfies

$$(-36I - A)v = 0$$

$$\begin{bmatrix} -29 & -13 & 16 \\ -13 & -26 & -13 \\ 16 & -13 & -29 \end{bmatrix} \begin{bmatrix} v_1 \\ v_2 \\ v_3 \end{bmatrix} = \begin{bmatrix} 0 \\ 0 \\ 0 \end{bmatrix}$$

If $v_1 = 0$ in this system, then it is straightforward to show that $v_2 = v_3 = 0$ as well. Thus, $v_1 \neq 0$, for $v \neq 0$. Since any multiple of v is also a solution, we can choose $v_1 = 1$. This leads to the system

$$\begin{aligned} -13v_2 + 16v_3 &= 29 \\ -26v_2 - 13v_3 &= 13 \\ -13v_2 - 29v_3 &= -16 \end{aligned}$$

The solution is $v_2 = -1$, $v_3 = 1$. Thus,

$$v^{(1)} = [1, -1, 1]^T \tag{7.29}$$

or any nonzero multiple of this vector. We leave the construction of $v^{(2)}$ and $v^{(3)}$ to Problem 2. ■

In general, eigenvalues are not computed by finding the roots of the characteristic polynomial. Later in the section, we discuss the power method that computes the dominant eigenvalue (i.e., the largest-size eigenvalue). More efficient techniques usually first transform A into a matrix with the same eigenvalues and whose eigenvalues can be computed more easily. Discussion of these methods is beyond the scope of this textbook, though.

7.2.2 Eigenvalues for Symmetric Matrices

The matrix $A = [a_{ij}]$ is symmetric if

$$A^T = A$$

or, equivalently,

$$a_{ij} = a_{ji}, \qquad 1 \le i, j \le n$$

This means that A is visually symmetric about the diagonal of A, the line between the elements a_{11} and a_{nn}. The matrix

$$A = \begin{bmatrix} a & b & c \\ b & d & e \\ c & e & f \end{bmatrix}$$

is a general symmetric matrix of order 3. Symmetric matrices occur commonly in applications, and both the theory and the numerical analysis of the eigenvalue problem are simpler in this case. For symmetric matrices we have the following important result.

Theorem 7.2.4 Let A be symmetric and have order n. Then, there is a set of n eigenvalue-eigenvector pairs $\{\lambda_i, v^{(i)}\}$, $1 \le i \le n$, that satisfy the following properties:

i. The numbers $\lambda_1, \lambda_2, \ldots, \lambda_n$ are all of the roots of the characteristic polynomial $f(\lambda)$ of A, repeated according to their multiplicity. Moreover, all the λ_i are real numbers.

ii. If the column matrices $v^{(i)}$, $1 \le i \le n$ are regarded as vectors in n-dimensional space, then they are mutually perpendicular and of length 1.

iii. For each column vector $x = [x_1, x_2, \ldots, x_n]^T$, there is a unique choice of constants $c_1, \ldots, c_n$ for which

$$x = c_1 v^{(1)} + \cdots + c_n v^{(n)} \tag{7.30}$$

The constants are given by

$$c_i = \sum_{j=1}^{n} x_j v_j^{(i)} = x^T v^{(i)}, \qquad 1 \le i \le n \tag{7.31}$$

where $v^{(i)} = [v_1^{(i)}, \ldots, v_n^{(i)}]^T$.

iv. Define the matrix U of order n by

$$U = [v^{(1)}, v^{(2)}, \ldots, v^{(n)}] \tag{7.32}$$

Then

$$U^T A U = D \equiv \begin{bmatrix} \lambda_1 & 0 & \cdots & 0 \\ 0 & \lambda_2 & \ddots & \vdots \\ \vdots & \ddots & \ddots & 0 \\ 0 & \cdots & 0 & \lambda_n \end{bmatrix} \tag{7.33}$$

and

$$UU^T = U^T U = I \tag{7.34}$$

We omit the proof of this theorem, as it is quite lengthy and requires concepts of linear algebra that have not been included in this chapter. Matrices satisfying (7.34) are called *orthogonal matrices* and, as a class, they are important in the numerical analysis of matrix algebra problems. The property (7.34) also says the vectors $\{v^{(1)}, \ldots, v^{(n)}\}$ form an *orthogonal basis* for the vector space of all column vectors of length n. For examples of such orthogonal matrices, see Problem 15, and also see Problem 7 of Section 6.2.

Example 7.2.5 Recall the matrix A in (7.22) and its eigenvalue-eigenvector pairs in (7.23). Figure 7.2.1 illustrates that the eigenvectors are perpendicular. To have them be of length 1, replace (7.23) by

$$\lambda_1 = 2, \qquad v^{(1)} = \begin{bmatrix} 1/\sqrt{2} \\ 1/\sqrt{2} \end{bmatrix}$$

$$\lambda_2 = 0.5, \qquad v^{(2)} = \begin{bmatrix} -1/\sqrt{2} \\ 1/\sqrt{2} \end{bmatrix}$$

The matrix U of (7.32) is given by

$$U = \begin{bmatrix} 1/\sqrt{2} & -1/\sqrt{2} \\ 1/\sqrt{2} & 1/\sqrt{2} \end{bmatrix}$$

Easily, $U^T U = I$. Also,

$$U^T AU = \begin{bmatrix} 1/\sqrt{2} & 1/\sqrt{2} \\ -1/\sqrt{2} & 1/\sqrt{2} \end{bmatrix} \begin{bmatrix} 1.25 & 0.75 \\ 0.75 & 1.25 \end{bmatrix} \begin{bmatrix} 1/\sqrt{2} & -1/\sqrt{2} \\ 1/\sqrt{2} & 1/\sqrt{2} \end{bmatrix}$$
$$= \begin{bmatrix} 2 & 0 \\ 0 & 0.5 \end{bmatrix}$$

as specified in (7.33). ■

The result (7.33) is often rewritten as

$$A = UDU^T$$

This decomposition of A is useful in many applications, both inside and outside of mathematics.

7.2.3 The Nonsymmetric Eigenvalue Problem

With nonsymmetric matrices, there are many more possibilities for the eigenvalues and eigenvectors. The roots of the characteristic polynomial $f(\lambda)$ may be complex numbers, and these lead to eigenvectors with complex numbers as components. Also, for multiple roots of $f(\lambda)$, it may not be possible to write an arbitrary column vector x as a combination of eigenvectors, as was done in (7.30) for symmetric matrices; and there is no longer a simple formula such as (7.31) for the coefficients. The nonsymmetric eigenvalue problem is important, but it is too sophisticated in its general theory to be discussed here.

Example 7.2.6 We illustrate the existence of complex eigenvalues. Let

$$A = \begin{bmatrix} 0 & 1 \\ -1 & 0 \end{bmatrix}$$

The characteristic polynomial is

$$f(\lambda) = \det \begin{bmatrix} \lambda & -1 \\ 1 & \lambda \end{bmatrix} = \lambda^2 + 1$$

The roots of $f(\lambda)$ are complex

$$\lambda_1 = i \equiv \sqrt{-1}, \qquad \lambda_2 = -i$$

and corresponding eigenvectors are

$$v^{(1)} = \begin{bmatrix} i \\ -1 \end{bmatrix}, \qquad v^{(2)} = \begin{bmatrix} i \\ 1 \end{bmatrix}$$ ■

We know from Theorem 7.2.4 that for a symmetric matrix of order n, there are n eigenvalue-eigenvector pairs such that the eigenvectors are orthogonal. In the next example, we show that for a nonsymmetric matrix, multiple eigenvalues may have fewer number of associated eigenvectors.

Example 7.2.7 Consider the matrix

$$A = \begin{bmatrix} a & 1 & 0 \\ 0 & a & 1 \\ 0 & 0 & a \end{bmatrix}$$

where a is a constant. It is left as a problem for the reader to verify that $\lambda = a$ is the eigenvalue of A with multiplicity 3, and any associated eigenvector must be of the form

$$v = c\,[1,\ 0,\ 0]^T$$

for some $c \neq 0$. Thus, up to a nonzero multiplicative constant, we have only one eigenvector

$$v = [1,\ 0,\ 0]^T$$

for the three equal eigenvalues $\lambda_1 = \lambda_2 = \lambda_3 = a$. ■

7.2.4 The Power Method

We now describe a numerical method that calculates the eigenvalue of A that is largest in size. It has some limitations as a general procedure, but it is simple to program and is a satisfactory method for many matrices of interest. The matrix A need not be symmetric; but we will need to assume that the eigenvalues λ_i of A satisfy

$$|\lambda_1| > |\lambda_2| \geq \cdots \geq |\lambda_n| \tag{7.35}$$

where these eigenvalues are repeated according to their multiplicity as a root of the characteristic polynomial.

We define an iteration process for computing λ_1 and its associated eigenvector $v^{(1)}$. Choose an initial guess to $v^{(1)}$, calling it $z^{(0)}$. Generally, it is chosen randomly, by choosing each component using a random number generator; and this is usually sufficient. Define

$$w^{(1)} = Az^{(0)}$$

and let α_1 be the maximum component of $w^{(1)}$, in size. If there is more than one such component, choose the first such component as α_1. Then define

$$z^{(1)} = \frac{1}{\alpha_1} w^{(1)}$$

Repeat the process iteratively. Define

$$w^{(m)} = Az^{(m-1)} \tag{7.36}$$

and let α_m be the maximum component of $w^{(m)}$, in size. Define

$$z^{(m)} = \frac{1}{\alpha_m} w^{(m)} \tag{7.37}$$

Then, roughly speaking, the vectors $z^{(m)}$ will converge to some multiple of $v^{(1)}$.

To find λ_1 by this process, also pick some nonzero component of the vectors $z^{(m)}$ and $w^{(m)}$, say, component k, and fix k. Often this is picked as the maximal component of $z^{(l)}$, for some sufficiently large l. Define

$$\lambda_1^{(m)} = \frac{w_k^{(m)}}{z_k^{(m-1)}} \tag{7.38}$$

where $z_k^{(m-1)}$ denotes component k of $z^{(m-1)}$. It can be shown that $\lambda_1^{(m)}$ converges to λ_1 as $m \to \infty$.

Example 7.2.8 We apply the power method to the earlier example (7.22), where

$$A = \begin{bmatrix} 1.25 & 0.75 \\ 0.75 & 1.25 \end{bmatrix} \tag{7.39}$$

The results are shown in Table 7.3. The column of successive differences $\lambda_1^{(m)} - \lambda_1^{(m-1)}$ and their successive ratios are included to show that there is a regular pattern to the convergence. The reason for this will be explained below. Note that in this example, $\lambda_1 = 2$ and $v^{(1)} = [1, 1]^T$, and the numerical results are converging to these values. ■

Table 7.3. Power Method Example for (7.40)

m	$z_1^{(m)}$	$z_2^{(m)}$	$\lambda_1^{(m)}$	$\lambda_1^{(m)} - \lambda_1^{(m-1)}$	Ratio
0	1.0	0.5			
1	1.0	0.8461538	1.6250000		
2	1.0	0.9591837	1.8846154	2.596E − 1	
3	1.0	0.9896373	1.9693878	8.477E − 2	0.327
4	1.0	0.9973992	1.9922280	2.284E − 2	0.269
5	1.0	0.9993492	1.9980494	5.821E − 3	0.255
6	1.0	0.9998373	1.9995119	1.462E − 3	0.251
7	1.0	0.9999593	1.9998779	3.660E − 4	0.250

7.2.5 Convergence of the Power Method

To analyze the convergence of the power method, we consider only the case when A is a symmetric matrix. Also assume that λ_1 is dominant, as in (7.35). We begin the analysis by showing that

$$z^{(m)} = \sigma_m \cdot \frac{A^m z^{(0)}}{\|A^m z^{(0)}\|} \tag{7.40}$$

where $\sigma_m = \pm 1$. For $m = 1$,

$$\alpha_1 = \pm\|w^{(1)}\| = \pm\|Az^{(0)}\|$$

with α_1 the maximal component of $w^{(1)}$. Then

$$z^{(1)} = \frac{1}{\alpha_1} w^{(1)} = \pm \frac{Az^{(0)}}{\|Az^{(0)}\|}$$

The maximum component of $z^{(1)}$, or the first such component if there are several maximal components in size, is forced to equal 1 in this definition. Thus, the sign on the right side is to be chosen in accordance with that requirement on the maximal component. The proof of the general case of (7.40) can be done inductively, using the same ideas as for the $m = 1$ case. The sign σ_m is to be chosen so that the maximum component on the right side of (7.40) is 1, since that is true for $z^{(m)}$ by its definition (7.37).

Using (7.30), we write

$$z^{(0)} = c_1 v^{(1)} + \cdots + c_n v^{(n)} \tag{7.41}$$

where the c_j's are given in (7.31), with x replaced by $z^{(0)}$. We assume $c_1 \neq 0$. If $z^{(0)}$ is chosen randomly, then it is extremely likely that $c_1 \neq 0$; even if $c_1 = 0$, the power method will usually introduce a modified $c_1 \neq 0$, because of rounding errors. Apply A to $z^{(0)}$ in (7.41) to get

$$\begin{aligned} Az^{(0)} &= c_1 Av^{(1)} + \cdots + c_n Av^{(n)} \\ &= \lambda_1 c_1 v^{(1)} + \cdots + \lambda_n c_n v^{(n)} \end{aligned}$$

Apply A repeatedly to obtain

$$\begin{aligned} A^m z^{(0)} &= \lambda_1^m c_1 v^{(1)} + \cdots + \lambda_n^m c_n v^{(n)} \\ &= \lambda_1^m \left[c_1 v^{(1)} + \left(\frac{\lambda_2}{\lambda_1}\right)^m c_2 v^{(2)} + \cdots + \left(\frac{\lambda_n}{\lambda_1}\right)^m c_n v^{(n)} \right] \end{aligned}$$

From (7.40),

$$z^{(m)} = \sigma_m \left(\frac{\lambda_1}{|\lambda_1|}\right)^m \cdot \frac{c_1 v^{(1)} + \left(\frac{\lambda_2}{\lambda_1}\right)^m c_2 v^{(2)} + \cdots + \left(\frac{\lambda_n}{\lambda_1}\right)^m c_n v^{(n)}}{\left\| c_1 v^{(1)} + \left(\frac{\lambda_2}{\lambda_1}\right)^m c_2 v^{(2)} + \cdots + \left(\frac{\lambda_n}{\lambda_1}\right)^m c_n v^{(n)} \right\|} \tag{7.42}$$

As $m \to \infty$, the terms $(\lambda_j/\lambda_1)^m \to 0$, for $2 \le j \le n$, with $(\lambda_2/\lambda_1)^m$ the largest. Also,

$$\sigma_m \left(\frac{\lambda_1}{|\lambda_1|}\right)^m = \pm 1$$

Thus as $m \to \infty$, most terms in (7.42) are converging to zero, and we can cancel c_1 from the numerator and denominator to see that $z^{(m)}$ looks like

$$z^{(m)} \approx \hat{\sigma}_m \frac{v^{(1)}}{\|v^{(1)}\|}$$

where $\hat{\sigma}_m = \pm 1$. If the normalization of $z^{(m)}$ is modified, to always have some particular component be positive, then

$$z^{(m)} \to \pm \frac{v^{(1)}}{\|v^{(1)}\|} \equiv \hat{v}^{(1)} \tag{7.43}$$

with a fixed sign independent of m. Our earlier normalization of sign, dividing by α_m, will usually accomplish this, but not always; see Problem 18. The error in $z^{(m)}$ will satisfy

$$\|z^{(m)} - \hat{v}^{(1)}\| \le c \left|\frac{\lambda_2}{\lambda_1}\right|^m, \qquad m \ge 0 \tag{7.44}$$

for some constant $c > 0$.

An analysis similar to the above can be given for the convergence of $\lambda_1^{(m)}$ to λ_1, with the same kind of error bound. Moreover, if we also assume

$$|\lambda_1| > |\lambda_2| > |\lambda_3| \ge |\lambda_4| \ge \cdots \ge |\lambda_n| \ge 0 \tag{7.45}$$

then we can show that

$$\lambda_1 - \lambda_1^{(m)} \approx c \left(\frac{\lambda_2}{\lambda_1}\right)^m \tag{7.46}$$

for some constant c, as $m \to \infty$. For example, recall the calculations in Table 7.2 for the matrix A in (7.39). There $\lambda_1 = 2$, $\lambda_2 = 0.5$, and $\lambda_2/\lambda_1 = 0.25$. Note that the ratios in the table are approaching 0.25, agreeing with (7.46). Note, however, that the assumption

(7.35) is not true if λ_2 and λ_3 are conjugate complex eigenvalues; and it is also not true if λ_2 and λ_3 are real, but of opposite sign. The power method can be quite powerful, but it is difficult to program as an automatic method.

Example 7.2.9 Apply the power method to the symmetric matrix

$$A = \begin{bmatrix} -7 & 13 & -16 \\ 13 & -10 & 13 \\ -16 & 13 & -7 \end{bmatrix} \tag{7.47}$$

From earlier, in (7.28), the eigenvalues are

$$\lambda_1 = -36, \qquad \lambda_2 = 9, \qquad \lambda_3 = 3$$

and the eigenvector $v^{(1)}$ associated with λ_1 is

$$v^{(1)} = [1,\ -1,\ 1]^T$$

The results of using the power method are shown in Table 7.4. Note that the ratios of the successive differences of $\lambda_1^{(m)}$ are approaching

$$\frac{\lambda_2}{\lambda_1} = -0.25$$

Also note that the maximal component of $z^{(m)}$ changes from one iteration to the next. The initial guess $z^{(0)}$ was chosen closer to $v^{(1)}$ than would be usual in actual practice. It was so chosen for purposes of illustration. ■

The error result (7.46) for the iterates $\{\lambda_1^{(m)}\}$ can be used to produce an acceleration method. From (7.46),

$$\lambda_1 - \lambda_1^{(m+1)} \approx r(\lambda_1 - \lambda_1^{(m)}), \qquad r = \lambda_2/\lambda_1 \tag{7.48}$$

Table 7.4. The Power Method for (7.47)

m	$z_1^{(m)}$	$z_2^{(m)}$	$z_3^{(m)}$	$\lambda_1^{(m)}$	$\lambda_1^{(m)} - \lambda_1^{(m-1)}$	Ratio
0	1.000000	−0.800000	0.900000			
1	−0.972477	1.000000	−1.000000	−31.80000		
2	1.000000	−0.995388	0.993082	−36.82075	−5.03E + 0	
3	0.998265	−0.999229	1.000000	−35.82936	9.91E − 1	−0.197
4	1.000000	−0.999775	0.999566	−36.04035	−2.11E − 1	−0.213
5	0.999891	−0.999946	1.000000	−35.99013	5.02E − 2	−0.238
6	1.000000	−0.999986	0.999973	−36.00245	−1.23E − 2	−0.245
7	0.999993	−0.999997	1.000000	−35.99939	3.06E − 3	−0.249

for all sufficiently large values of m. Choose r using

$$r \approx \frac{\lambda_1^{(m+1)} - \lambda_1^{(m)}}{\lambda_1^{(m)} - \lambda_1^{(m-1)}} \tag{7.49}$$

The justification of this is left as Problem 19(d). Using this r, solve for λ_1 in (7.48)

$$\begin{aligned} \lambda_1 &\approx \frac{1}{1-r}\left[\lambda_1^{(m+1)} - r\lambda_1^{(m)}\right] \\ &= \lambda_1^{(m+1)} + \frac{r}{1-r}\left[\lambda_1^{(m+1)} - \lambda_1^{(m)}\right] \end{aligned} \tag{7.50}$$

This is called *Aitken's extrapolation formula*. It also gives us the error estimate

$$\lambda_1 - \lambda_1^{(m+1)} \approx \frac{r}{1-r}\left[\lambda_1^{(m+1)} - \lambda_1^{(m)}\right] \tag{7.51}$$

which is called *Aitken's error estimation formula*. Under the same assumption (7.45), similar results can also be shown for the iterates $\{z^{(m)}\}$. Related material on Aitken extrapolation was given in Section 3.4, when discussing fixed point iteration methods for rootfinding.

Example 7.2.10 In Table 7.4, take $m + 1 = 7$. Then (7.51) yields

$$\begin{aligned} \lambda_1 - \lambda_1^{(7)} &\approx \frac{r}{1-r}\left[\lambda_1^{(7)} - \lambda_1^{(6)}\right] \\ &= \frac{-0.249}{1+0.249}[0.00306] \doteq -0.00061 \end{aligned}$$

which is the actual error. Also the Aitken formula (7.50) will give the exact answer for λ_1, to seven significant digits. ■

7.2.6 MATLAB Eigenvalue Calculations

The MATLAB built-in function `eig` computes the eigenvalues and associated eigenvectors of a given square matrix. Let a square matrix A of order n be defined. Then the call

```
lambda = eig(A)
```

returns an n-vector for the n eigenvalues. The call

```
[V, D] = eig(A)
```

returns two $n \times n$ matrices V and D, with D diagonal. The diagonal elements of D are the eigenvalues of A, and columns of V are associated eigenvectors. The matrices satisfy the relation $AV = VD$. For A symmetric, the column vectors of V are made orthogonal, and then V is always invertible. When A is not symmetric and has multiple eigenvalues, the matrix V may happen to be singular or nearly singular.

MATLAB PROGRAM. We give a MATLAB program for the power method.

```
function [lambda, eigenvec, diff, ratio] = ...
            power_method(A,x_0,index,err_tol,max_iterates)
%
% The calling sequence for power_method is
%
% [lambda, eigenvec, diff, ratio]
%          = power_method(A,x_0,index,err_tol,max_iterates)
%
% It implements the power method for finding the eigenvalue
% of A which is largest in magnitude.
%
% The vector x_0 is the user's initial guess at the
% eigenvector corresponding to the eigenvalue of
% largest magnitude.  If the user has no idea of how to
% choose x_0, then we suggest using
%    x_0 = rand(n,1)
% with n the order of A. DO NOT use x_0 = 0, the zero vector.
%
% The input parameter index is used to specify the method of
% approximating the eigenvalue.  With index=0, we generate a
% random vector, call it v, and we use
%    lambda approximately (v'*w)/(v'*z)
% with Az=w.  For index in [1,n], we use
% lambda approximately w(index)/z(index).
%
% The parameter err_tol is used in the error test.  However,
% note this test needs to be made more sophisticated.  It uses
% error approx lambda_present - lambda_previous
% at each step of the iteration, and this is not an accurate
% estimate for slowly convergent iterations.
%
% The output variables are:
% lambda:  A vector of the power method iterations for the
%          approximate eigenvalue of largest magnitude.
% eigenvec:  The corresponding normalized eigenvector.
% diff:  The differences of the successive values in lambda.
% ratio:  The ratios of the successive values in diff.
%         The values of diff and ratio are of use in
```

```
%         examining the rate of convergence of the
%         power method.

% Initialization.
% Find the order of A.
n = size(A,1);
x_init = zeros(n,1);
lambda = zeros(max_iterates,1);

% Define x_init and require it to be a column vector.
for i=1:n
   x_init(i) = x_0(i);
end

% Find the max element in size.
[max_x,i_max] = max(abs(x_init));
% Normalize the initial vector to have an
% infinity norm of 1.
z = x_init/x_init(i_max);
% If index=0, generate a special vector for the
% approximate eigenvalue computation.
if index == 0
   spec_vec = rand(1,n);
end

format long
error = inf;
it_count = 0;
% Main loop.
while abs(error) > err_tol & it_count < max_iterates
   it_count = it_count + 1;
   w = A*z;
   [max_w,i_max] = max(abs(w));
% Calculate approximate eigenvalue.
   if index == 0
      lambda(it_count) = (spec_vec*w)/(spec_vec*z);
   else
      lambda(it_count) = w(index)/z(index);
   end

   if it_count > 1
      error = lambda(it_count) - lambda(it_count-1);
   end

   z = w/w(i_max); % Save approximate eigenvector.
```

```
end

if it_count == max_iterates
   disp(' ')
   disp('*** Your answers may possibly not...
%                satisfy your error tolerance.')
end

% Setting of output variables.
eigenvec = z;
lambda = lambda(1:it_count);
diff = zeros(it_count,1);
ratio = zeros(it_count,1);
diff(2:it_count) = lambda(2:it_count)-lambda(1:it_count-1);
ratio(3:it_count) = diff(3:it_count)./diff(2:it_count-1);
```

PROBLEMS

1. For the following matrices, find the characteristic polynomial, the eigenvalues, and the eigenvectors. Do it by hand computation.

(a) $\begin{bmatrix} 1 & 2 \\ 5 & 4 \end{bmatrix}$ **(b)** $\begin{bmatrix} 2 & 36 \\ 36 & 23 \end{bmatrix}$ **(c)** $\begin{bmatrix} -9 & 6 \\ 3 & 8 \end{bmatrix}$

2. Find the eigenvectors $v^{(2)}$ and $v^{(3)}$ for (7.27), corresponding to the eigenvalues λ_2 and λ_3.

3. Find the characteristic polynomial, eigenvalues, and eigenvectors of

$$A = \begin{bmatrix} 2 & 1 & 0 \\ 1 & 3 & 1 \\ 0 & 1 & 2 \end{bmatrix}$$

4. Let $A = \begin{bmatrix} 5 & -2 \\ -1 & 4 \end{bmatrix}$. Apply the power method with $z^{(0)} = [1, 1]^T$. Determine the sequences $\lambda^{(m)}, z^{(m)}, m \geq 1$. Is the method convergent to the dominant eigenvalue and corresponding eigenvector? If not, what do you suggest to do?

5. Verify the claims made in Example 7.2.7.

6. Consider the matrix

$$A_1 = \begin{bmatrix} a & 1 & 0 \\ 0 & a & 0 \\ 0 & 0 & a \end{bmatrix}$$

where a is a constant. Determine all the eigenvalues and associated eigenvectors. Compare the results for A_1 with that for A in Example 7.2.7.

7. Let A be a square matrix of order n, and let $f(\lambda)$ be its characteristic polynomial, as in (7.26).

(a) Show that the constant term α_0 of $f(\lambda)$ is

$$\alpha_0 = (-1)^n \det(A)$$

(b) For readers knowledgeable about determinants, expand $f(\lambda)$ in powers of λ to show that the coefficient α_{n-1} of λ^{n-1} is given by

$$\alpha_{n-1} = -[a_{11} + a_{22} + \cdots + a_{nn}]$$

The term in brackets is called the *trace* of A, and it is the sum of the diagonal terms of A.

8. Continuing Problem 7, show that

$$\det(A) = \lambda_1 \lambda_2 \cdots \lambda_n$$
$$\text{trace}(A) = \lambda_1 + \lambda_2 + \cdots + \lambda_n$$

where $\lambda_1, \ldots, \lambda_n$ are the roots of $f(\lambda)$, each repeated as many times as its multiplicity.

Hint: Expand $f(\lambda) = (\lambda - \lambda_1) \cdots (\lambda - \lambda_n)$.

9. Let $Ax = \lambda x$ with $x \neq 0$. Show $|\lambda| \leq \|A\|$ with the matrix norm of (6.87).

Hint: Use the property P4 of (6.88). This gives a simple way of bounding all eigenvalues of a matrix A.

10. Let A be a square matrix, c_0 be a constant. Show that λ is an eigenvalue of A if and only if $\lambda + c_0$ is an eigenvalue of $A + c_0 I$.

11. By a direct calculation, show that the $n \times n$ matrix

$$A = [a_1, a_2, \ldots, a_n]^T [1, 1, \ldots, 1]$$

has one eigenvalue $\lambda = a_1 + a_2 + \cdots + a_n$ with the remaining eigenvalues all equal to zero.

12. Show that the eigenvectors of (7.27) satisfy Theorem 7.2.4, provided that they are properly normalized to represent vectors of length 1. For example, change $v^{(1)}$ in (7.29) to

$$v^{(1)} = \left[\frac{1}{\sqrt{3}}, \frac{-1}{\sqrt{3}}, \frac{1}{\sqrt{3}}\right]^T$$

(The length of a vector is the square root of the sum of the squares of its components. For a general vector

$$x = [x_1, \ldots, x_n]^T$$

the length of x is $\sqrt{x_1^2 + \cdots + x_n^2}$). Produce the matrix U of (7.32), and verify that it satisfies $U^T U = I$.

13. Repeat Problem 12 for the matrix in Problem 1(b).

14. Calculate the characteristic polynomial, eigenvalues, and eigenvectors of

(a) $\begin{bmatrix} 1 & 0 \\ 0 & 1 \end{bmatrix}$ **(b)** $\begin{bmatrix} 1 & 0 \\ 1 & 1 \end{bmatrix}$

Do the results contradict Theorem 7.2.4?

15. Show that for any angle θ,

$$U = \begin{bmatrix} \cos\theta & \sin\theta \\ -\sin\theta & \cos\theta \end{bmatrix}$$

is an orthogonal matrix. These are called *rotation matrices*, because of their relationship to changes of rectangular coordinates by rotation of the coordinate axes.

16. Apply the power method to the matrices of Problem 1. In each case, also calculate the successive differences $\lambda_1^{(m)} - \lambda_1^{(m-1)}$ and their successive ratios. By using (7.46) and these ratios, try to estimate λ_2.

17. Repeat Problem 16 for the following matrices:

(a) $\begin{bmatrix} 1 & \frac{1}{2} \\ \frac{1}{2} & \frac{1}{3} \end{bmatrix}$ **(b)** $\begin{bmatrix} 1 & 5 & -8 \\ 5 & -2 & 5 \\ -8 & 5 & 1 \end{bmatrix}$

(c) $\begin{bmatrix} 1 & 2 & 3 \\ 2 & 3 & 4 \\ 3 & 4 & 5 \end{bmatrix}$ **(d)** $\begin{bmatrix} 33 & 16 & 72 \\ -24 & -10 & -57 \\ -8 & -4 & -17 \end{bmatrix}$

(e) $\begin{bmatrix} 1 & \frac{1}{2} & \frac{1}{3} \\ \frac{1}{2} & \frac{1}{3} & \frac{1}{4} \\ \frac{1}{3} & \frac{1}{4} & \frac{1}{5} \end{bmatrix}$ **(f)** $\begin{bmatrix} 6 & 4 & 4 & 1 \\ 4 & 6 & 1 & 4 \\ 4 & 1 & 6 & 4 \\ 1 & 4 & 4 & 6 \end{bmatrix}$

18. Use the power method on the example of (7.47), and choose $z^{(0)} = [1, 0, 1]^T$. Describe the pattern on convergence of the approximate $z^{(m)}$.

19. **(a)** Using the definition (7.38) of $\lambda_1^{(m)}$ and the assumption (7.35), show

$$\lim_{m\to\infty} \lambda_1^{(m)} = \lambda_1$$

(b) Show

$$\left|\lambda_1 - \lambda_1^{(m)}\right| \le c\,|\lambda_2/\lambda_1|^m, \qquad m \ge 0$$

for some constant c.

(c) Under the assumption (7.45), prove formula

$$\lambda_1 - \lambda_1^{(m)} \approx c(\lambda_2/\lambda_1)^m, \qquad c = \text{constant}$$

with the approximation becoming better as $m \to \infty$. More precisely,

$$\lim_{m\to\infty} \frac{\lambda_1 - \lambda_1^{(m)}}{(\lambda_2/\lambda_1)^m} = c$$

a constant that is usually nonzero.

(d) Use (7.46) to show that

$$\lim_{m\to\infty} \frac{\lambda_1^{(m+1)} - \lambda_1^{(m)}}{\lambda_1^{(m)} - \lambda_1^{(m-1)}} = \frac{\lambda_2}{\lambda_1}$$

thus justifying (7.49).

20. For symmetric matrices A, define the following variant of the power method, for computing λ_1:

$$\lambda_1^{(m)} = \frac{(z^{(m)})^T A(z^{(m)})}{(z^{(m)})^T(z^{(m)})} = \frac{(z^{(m)})^T(w^{(m+1)})}{(z^{(m)})^T(z^{(m)})}$$

Using Theorem 7.2.4, verify that

$$\left|\lambda_1 - \lambda_1^{(m)}\right| \le c\left(\frac{\lambda_2}{\lambda_1}\right)^{2m}, \qquad m \ge 0$$

This is a faster rate of linear convergence, with the error decreasing by $(\lambda_2/\lambda_1)^2$ rather than (λ_2/λ_1). Illustrate the method by applying it to the matrix (7.47).

21. Compare the eigenvalues of

$$A = \begin{bmatrix} 101 & -90 \\ 110 & -98 \end{bmatrix}, \qquad \hat{A} = \begin{bmatrix} 100.999 & -90.001 \\ 110 & -98 \end{bmatrix}$$

How would you describe the sensitivity of the eigenvalues of A?

7.3. NONLINEAR SYSTEMS

Systems of nonlinear equations occur widely in the mathematical modeling of real-world phenomena, often giving a more realistic model than is obtained by using systems of linear equations. In this section, we present an introduction to systems of nonlinear equations and to numerical methods for their solution. To help develop some intuition for the solvability of nonlinear systems and to illustrate some of the main ideas used in their numerical solution, we begin the section with an examination of systems of two nonlinear equations and numerical methods for their solution. Following that, methods are given for systems of an arbitrary order.

Consider the following system of two nonlinear equations:

$$\begin{aligned} f(x, y) &\equiv x^2 + 4y^2 - 9 = 0 \\ g(x, y) &\equiv 18y - 14x^2 + 45 = 0 \end{aligned} \tag{7.52}$$

What are the pairs (x, y) for which these two equations are satisfied simultaneously and how can they be calculated numerically? More generally, consider the general systems of two nonlinear equations:

$$\begin{aligned} f(x, y) &= 0 \\ g(x, y) &= 0 \end{aligned} \tag{7.53}$$

How do we develop some geometric insight concerning these equations and their numerical solution?

If we graph $z = f(x, y)$ in xyz-space, then the solutions (x, y) of $f(x, y) = 0$ lie on the intersection of the xy-plane and the graph of $z = f(x, y)$. We shall call this intersection the "zero curve" of $f(x, y)$. As an example, the zero curves of the functions f and g of (7.52) are given in Figure 7.7. To visualize the points (x, y) that satisfy simultaneously both equations in (7.53), we look at the intersection of the zero curves of the functions f and g. This is illustrated in Figure 7.7 for the equations of (7.52). The zero curves intersect at four points, each of which corresponds to a solution of the system (7.52). For example, there is a solution near $(1, -1)$.

7.3.1 Newton's Method

To derive an iteration method for solving (7.53), we generalize Newton's method from Section 3.2. We begin by reviewing the use of a tangent plane approximation, from multivariable calculus.

Let (x_0, y_0) be an initial guess for a solution $\alpha = (\xi, \eta)$ for the system (7.53). The graph of $z = f(x, y)$ is a surface in xyz-space, and we approximate it with a plane that is tangent to it at $(x_0, y_0, f(x_0, y_0))$. The equation of this tangent plane is $z = p(x, y)$ with

$$p(x, y) \equiv f(x_0, y_0) + (x - x_0) f_x(x_0, y_0) + (y - y_0) f_y(x_0, y_0) \tag{7.54}$$

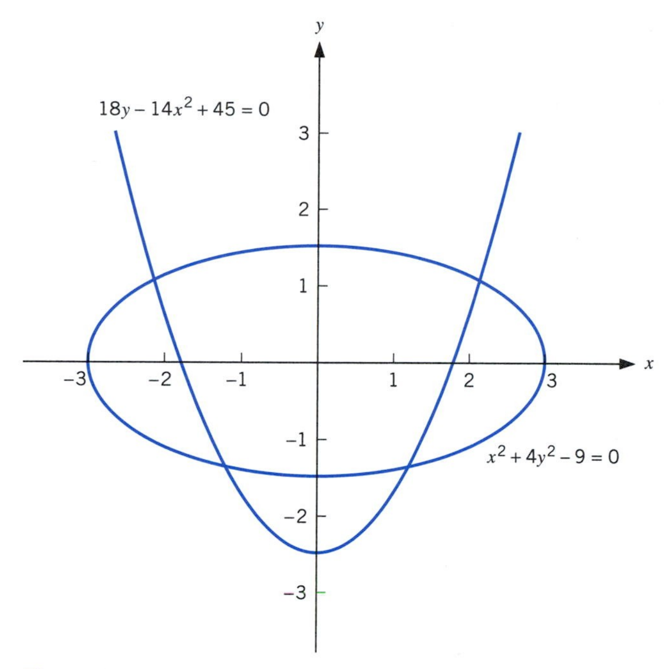

Figure 7.7. The graphs of $f(x, y) = 0$ and $g(x, y) = 0$

using the notation

$$f_x(x, y) = \frac{\partial f(x, y)}{\partial x}, \qquad f_y(x, y) = \frac{\partial f(x, y)}{\partial y}$$

the partial derivatives of f with respect to x and y, respectively. If $f(x_0, y_0)$ is sufficiently close to zero, then the zero curve of $p(x, y)$ will be an approximation of the zero curve of $f(x, y)$ for those points (x, y) near (x_0, y_0). Because the graph of $z = p(x, y)$ is a plane, its zero curve is simply a straight line.

Example 7.3.1 Consider the function $f(x, y) \equiv x^2 + 4y^2 - 9$ of (7.53). Easily,

$$f_x(x, y) = 2x, \qquad f_y(x, y) = 8y$$

At $(x_0, y_0) = (1, -1)$,

$$f(x_0, y_0) = -4, \qquad f_x(x_0, y_0) = 2, \qquad f_y(x_0, y_0) = -8$$

The tangent plane at $(1, -1, -4)$ of the surface $z = f(x, y)$ has the equation

$$z = p(x, y) \equiv -4 + 2(x - 1) - 8(y + 1)$$

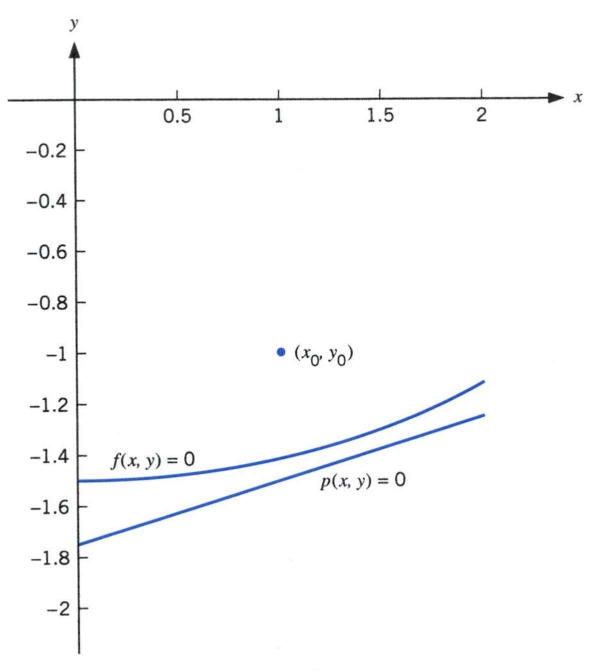

Figure 7.8. $f(x, y) = 0$ and $p(x, y) = 0$

The graphs of the zero curves of $f(x, y)$ and $p(x, y)$ for (x, y) near (x_0, y_0) are given in Figure 7.8. [Note that the graph of $f(x, y) = 0$ in this figure corresponds to only a portion of the graph of $f(x, y) = 0$ in Figure 7.7.] ■

The above construction of a tangent plane can also be applied to the surface $z = g(x, y)$; and we denote the equation of the plane tangent at $(x_0, y_0, g(x_0, y_0))$ by $z = q(x, y)$:

$$q(x, y) \equiv g(x_0, y_0) + (x - x_0)g_x(x_0, y_0) + (y - y_0)g_y(x_0, y_0)$$

Recall that the solution of the system (7.53) is the intersection of the zero curves of $z = f(x, y)$ and $z = g(x, y)$. We approximate these zero curves by those of the tangent planes $z = p(x, y)$ and $z = q(x, y)$, and we use the intersection of these latter zero curves as an approximate solution to the system (7.53). For the equations of (7.52) with $(x_0, y_0) = (1, -1)$, this is illustrated in Figure 7.9.

To find the intersection of the zero curves of the tangent planes, we must solve the linear system

$$\begin{aligned} f(x_0, y_0) + (x - x_0)f_x(x_0, y_0) + (y - y_0)f_y(x_0, y_0) &= 0 \\ g(x_0, y_0) + (x - x_0)g_x(x_0, y_0) + (y - y_0)g_y(x_0, y_0) &= 0 \end{aligned}$$

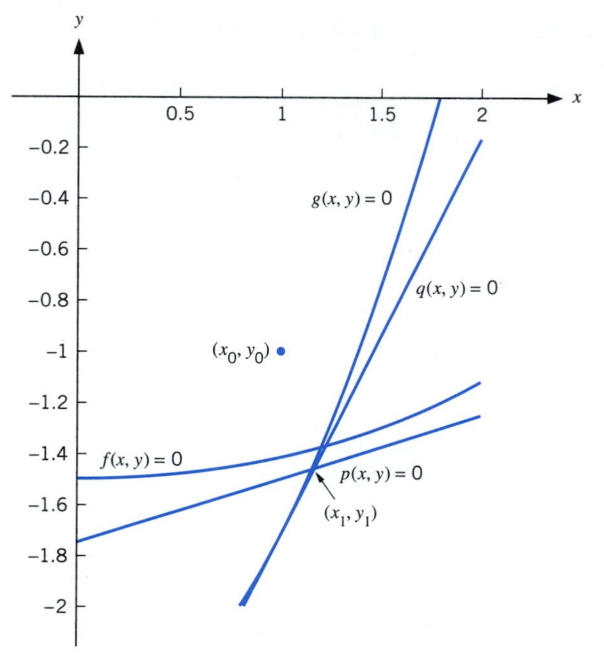

Figure 7.9. $f = g = 0$ and $p = q = 0$

In matrix form, this becomes

$$\begin{bmatrix} f_x(x_0, y_0) & f_y(x_0, y_0) \\ g_x(x_0, y_0) & g_y(x_0, y_0) \end{bmatrix} \begin{bmatrix} x - x_0 \\ y - y_0 \end{bmatrix} = - \begin{bmatrix} f(x_0, y_0) \\ g(x_0, y_0) \end{bmatrix}$$

We denote the solution (x, y) of this system by (x_1, y_1). It is actually computed as follows. Define δ_x and δ_y to be the solution of the linear system

$$\begin{bmatrix} f_x(x_0, y_0) & f_y(x_0, y_0) \\ g_x(x_0, y_0) & g_y(x_0, y_0) \end{bmatrix} \begin{bmatrix} \delta_x \\ \delta_y \end{bmatrix} = - \begin{bmatrix} f(x_0, y_0) \\ g(x_0, y_0) \end{bmatrix}$$

and define the new iterate by

$$\begin{bmatrix} x_1 \\ y_1 \end{bmatrix} = \begin{bmatrix} x_0 \\ y_0 \end{bmatrix} + \begin{bmatrix} \delta_x \\ \delta_y \end{bmatrix}$$

Usually, the point (x_1, y_1) is closer to the solution α than is the original point (x_0, y_0). Using (x_1, y_1) as a new initial guess, we repeat this process to obtain an improved estimate (x_2, y_2); and this iteration process is continued until a solution with sufficient accuracy is obtained. The general iteration is given by

$$\begin{bmatrix} f_x(x_k, y_k) & f_y(x_k, y_k) \\ g_x(x_k, y_k) & g_y(x_k, y_k) \end{bmatrix} \begin{bmatrix} \delta_{x,k} \\ \delta_{y,k} \end{bmatrix} = - \begin{bmatrix} f(x_k, y_k) \\ g(x_k, y_k) \end{bmatrix} \tag{7.55}$$

$$\begin{bmatrix} x_{k+1} \\ y_{k+1} \end{bmatrix} = \begin{bmatrix} x_k \\ y_k \end{bmatrix} + \begin{bmatrix} \delta_{x,k} \\ \delta_{y,k} \end{bmatrix} \tag{7.56}$$

for $k = 0, 1, \ldots$. This is *Newton's method* for solving (7.53).

Example 7.3.2 If we consider again the system (7.52), Newton's method (7.55–7.56) becomes

$$\begin{bmatrix} 2x_k & 8y_k \\ -28x_k & 18 \end{bmatrix} \begin{bmatrix} \delta_{x,k} \\ \delta_{y,k} \end{bmatrix} = - \begin{bmatrix} x_k^2 + 4y_k^2 - 9 \\ 18y_k - 14x_k^2 + 45 \end{bmatrix} \tag{7.57}$$

$$\begin{bmatrix} x_{k+1} \\ y_{k+1} \end{bmatrix} = \begin{bmatrix} x_k \\ y_k \end{bmatrix} + \begin{bmatrix} \delta_{x,k} \\ \delta_{y,k} \end{bmatrix} \tag{7.58}$$

We compute the solution of this with $(x_0, y_0) = (1, -1)$, obtaining the solution (x_1, y_1). The process is repeated to obtain iterates (x_k, y_k), $k = 1, 2, \ldots$. The first few iterates are given in Table 7.5, along with

$$\text{Error } = \|\alpha - (x_k, y_k)\| \equiv \max\{|\xi - x_k|, |\eta - y_k|\}$$

The final iterate is accurate to the precision of the computer arithmetic. ■

7.3.2 The General Newton Method

To help motivate the form of Newton's method for nonlinear systems of order larger than 2, we first change the notation used for solving systems of order 2. Replace (7.53)

Table 7.5. Newton Iterates for the System (7.52)

k	x_k	y_k	Error
0	1.0	−1.0	3.74E − 1
1	1.170212765957447	−1.457446808510638	8.34E − 2
2	1.202158829506705	−1.376760321923060	2.68E − 3
3	1.203165807091535	−1.374083486949713	2.95E − 6
4	1.203166963346410	−1.374080534243534	3.59E − 12
5	1.203166963347774	−1.374080534239942	2.22E − 16

by

$$\begin{aligned} F_1(x_1, x_2) &= 0 \\ F_2(x_1, x_2) &= 0 \end{aligned} \tag{7.59}$$

Introduce the symbols

$$x = \begin{bmatrix} x_1 \\ x_2 \end{bmatrix}, \qquad F(x) = \begin{bmatrix} F_1(x_1, x_2) \\ F_2(x_1, x_2) \end{bmatrix}$$

$$F'(x) = \begin{bmatrix} \dfrac{\partial F_1}{\partial x_1} & \dfrac{\partial F_1}{\partial x_2} \\ \dfrac{\partial F_2}{\partial x_1} & \dfrac{\partial F_2}{\partial x_2} \end{bmatrix}$$

$F'(x)$ is called the *Frechet derivative* of $F(x)$, and it is a generalization to higher dimensions of the ordinary derivative of a function of one variable. The system (7.59) can now be written as

$$F(x) = 0 \tag{7.60}$$

A solution of this equation will be denoted by α.

Newton's method (7.55–7.56) for solving (7.53) becomes

$$\begin{aligned} F'(x^{(k)})\delta^{(k)} &= -F(x^{(k)}) \\ x^{(k+1)} &= x^{(k)} + \delta^{(k)} \end{aligned} \tag{7.61}$$

for $k = 0, 1, \ldots$. This can be rewritten in the shorter and mathematically equivalent form

$$x^{(k+1)} = x^{(k)} - \left[F'(x^{(k)})\right]^{-1} F(x^{(k)}) \tag{7.62}$$

for $k = 0, 1, \ldots$. This formula is often used in discussing and analyzing Newton's method for nonlinear systems; but (7.61) is used for practical computations, since it is usually less expensive to solve a linear system than to find the inverse of the coefficient matrix. Also note the analogy of (7.62) with (3.11) in Section 3.2, Newton's method for a single equation.

Now consider the system of n nonlinear equations

$$\begin{aligned} F_1(x_1, \ldots, x_n) &= 0 \\ &\vdots \\ F_n(x_1, \ldots, x_n) &= 0 \end{aligned} \tag{7.63}$$

Define

$$x = \begin{bmatrix} x_1 \\ \vdots \\ x_n \end{bmatrix}, \qquad F(x) = \begin{bmatrix} F_1(x_1, \dots, x_n) \\ \vdots \\ F_n(x_1, \dots, x_n) \end{bmatrix}$$

$$F'(x) = \begin{bmatrix} \dfrac{\partial F_1}{\partial x_1} & \cdots & \dfrac{\partial F_1}{\partial x_n} \\ \vdots & \ddots & \vdots \\ \dfrac{\partial F_n}{\partial x_1} & \cdots & \dfrac{\partial F_n}{\partial x_n} \end{bmatrix}$$

The nonlinear system (7.63) can be written as $F(x) = 0$, as in (7.60); and *Newton's method* is given by (7.61) or (7.62), exactly the same as for the case of two equations in two unknowns. The derivation of Newton's method for (7.63) can be based on using tangent hyperplane approximations to each of the functions $z = F_i(x)$ in (7.63). Since this derivation is essentially just an extension of the argument used previously for the case of two equations, we omit it here. The notation of (7.62) unifies Newton's method for finite-dimensional systems of any order $n \geq 1$. It also can be extended to "infinite dimensional problems," such as nonlinear differential and integral equations, although we do not consider generalizations of this kind in this text.

The derivation of Newton's method for systems is another illustration of General Observation (1.10) in Section 1.1 and General Observation (3.34) in Section 3.3. We are linearizing the nonlinear system $F(x) = 0$, thereby replacing it by a simpler nearby rootfinding problem.

Example 7.3.3 Consider solving the integral equation

$$z(s) + 4\int_0^1 \cos(st)[z(t)]^2\,dt = 4, \qquad 0 \leq s \leq 1 \tag{7.64}$$

for the unknown function $z(s)$. We approximate this equation by a system of nonlinear equations, and the solution of this system will yield an approximation to $z(s)$ at selected points in $[0, 1]$. For any integer $n > 0$, let $h = 1/n$ and $t_j = (j - \frac{1}{2})h$, $j = 1, 2, \dots, n$. Recall the *midpoint numerical integration rule* [cf. Problem 8, Section 5.1]:

$$\int_0^1 f(t)\,dt \approx h \sum_{j=1}^{n} f(t_j)$$

Apply this to the integral in (7.64), and then evaluate the equation at each of the points $s = t_i$. This yields the approximations

$$z(t_i) + 4h \sum_{j=1}^{n} \cos(t_i t_j)[z(t_j)]^2 \approx 4, \qquad i = 1, 2, \dots, n$$

We solve the system

$$x_i + 4h \sum_{j=1}^{n} \cos(t_i t_j)[x_j]^2 = 4, \qquad i = 1, 2, \ldots, n \tag{7.65}$$

This is of order n, and the unknowns are $x_1, \ldots, x_n$. We hope that

$$z(t_i) \approx x_i, \qquad i = 1, 2, \ldots, n$$

for n small to moderate in size. The system (7.65) is solved with Newton's method, using an initial guess of

$$x_i^{(0)} = 1, \qquad i = 1, 2, \ldots, n \tag{7.66}$$

and again we let α denote the true solution of the nonlinear system. Because of the rapid speed of convergence of Newton's method, we use the error estimate

$$\alpha - x^{(k)} \approx x^{(k+1)} - x^{(k)} = \delta^{(k)} \tag{7.67}$$

with $\delta^{(k)}$ the correction obtained in (7.61). In Table 7.6, we give the estimated iteration errors in solving (7.65) with Newton's method. The iterate $x^{(4)}$ is accurate to the precision of the computer arithmetic. The true solution $z(s)$ of the integral equation (7.64) is shown in Figure 7.10, along with the solution obtained by solving (7.65) for $n = 5$. A larger value of n will give an even better agreement of x_i with $z(t_i)$, $i = 1, 2, \ldots, n$. ■

The convergence analysis of Newton's method for solving nonlinear systems is somewhat complicated and is still being studied, and we will just state the principal result without proof.

Theorem 7.3.4 Let α be a solution of the nonlinear system $F(x) = 0$, given explicitly in (7.63). Assume that the functions $F_i(x_1, \ldots, x_n)$ have continuous first- and second-order partial

Table 7.6. Iteration Errors in Solving (7.65) with $n = 5$

k	$\|\delta^{(k)}\|$
0	3.34E − 1
1	3.39E − 2
2	4.29E − 4
3	8.98E − 8
4	3.53E − 15

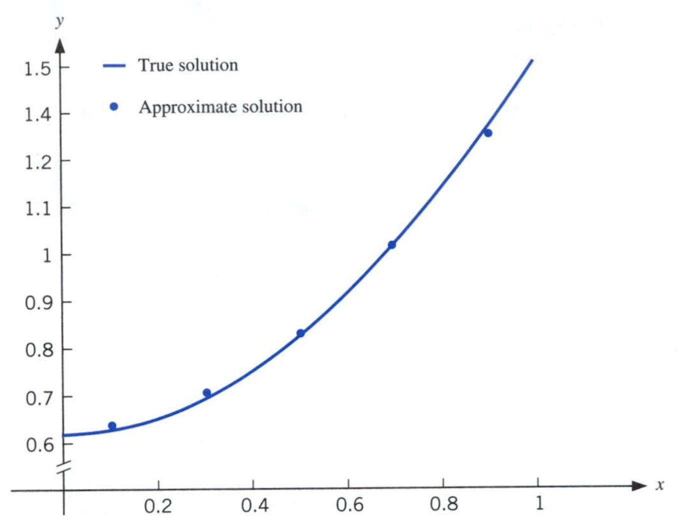

Figure 7.10. The true and approximate solution of (7.64)

derivatives with respect to all of the variables $x_1, \dots, x_n$, for all points x satisfying $\|\alpha - x\| < \epsilon$, for some $\epsilon > 0$, and for all $i = 1, 2, \dots, n$. Further assume that

$$\det F'(\alpha) \neq 0 \tag{7.68}$$

Then if $x^{(0)}$ is chosen sufficiently close to α, the Newton iterates $x^{(k)}$ of (7.62) will converge to α. In addition,

$$\|\alpha - x^{(k+1)}\| \leq c\|\alpha - x^{(k)}\|^2, \qquad k = 0, 1, \dots \tag{7.69}$$

for a suitably chosen constant $c > 0$.

This theorem generalizes the discussion in Section 3.2 for Newton's method for solving a single nonlinear equation [cf. (3.19) and following]. The condition (7.68) generalizes the condition $f'(\alpha) \neq 0$, which was needed for the previous analysis in Section 3.2 of quadratic convergence. The result (7.69) shows that Newton's method is *quadratically convergent*. With (7.69), it can be shown that

$$\lim_{k\to\infty} \frac{\|\alpha - x^{(k)}\|}{\|x^{(k+1)} - x^{(k)}\|} = 1 \tag{7.70}$$

and this is a partial justification for the error estimate (7.67) used earlier.

7.3.3 A Modified Newton's Method

Consider the cost of using Newton's method (7.61) to solve a system $F(x) = 0$ of order n. For each iterate, the derivative matrix $F'(x^{(k)})$ must be evaluated, and this costs n^2 regular function evaluations, since the matrix usually contains n^2 nonzero components. In addition, the linear system in (7.54) must be solved, and this costs approximately $\frac{2}{3}n^3$ arithmetic operations per iterate. Because of these large operation counts, Newton's method (7.61) is usually modified to reduce its cost. We will give one such modification.

The matrix $F'(x^{(k)})$ changes very little when $x^{(k)}$ is close to α. In a computer program, we try to detect when $x^{(k)}$ is sufficiently close to α and then we fix the associated derivative matrix in (7.61). For simplicity of notation, let A denote a matrix that is "close to $F'(\alpha)$." Then define a *modified Newton's method* by

$$\begin{aligned} A\delta^{(k)} &= -F(x^{(k)}) \\ x^{(k+1)} &= x^{(k)} + \delta^{(k)} \end{aligned} \tag{7.71}$$

$k = 0, 1, \ldots$. The matrix A is often chosen as $F'(x^{(0)})$, where $x^{(0)}$ is chosen sufficiently close to α.

What is the cost of using (7.71), and what is the speed of convergence? Once A has been chosen, there are no further costs in obtaining it, in contrast to the evaluation of $F'(x)$ with each iteration of Newton's method. In addition, the LU factorization of A can be done once at a cost of about $\frac{2}{3}n^3$ arithmetic operations. In subsequent iterations, the cost of solving the linear system in (7.71) will be only about $2n^2$ arithmetic operations per iterate. For large n, this can greatly reduce the overall cost of the iteration.

There is a reduction in speed of convergence with (7.71). In particular, it can be proved that if (1) the assumptions of Theorem 7.3.4 are satisfied, (2) A is chosen sufficiently close to $F'(\alpha)$, and (3) $x^{(0)}$ is chosen sufficiently close to α, then the iterates $x^{(k)}$ will converge to α. Moreover,

$$\|\alpha - x^{(k+1)}\| \le c(A)\|\alpha - x^{(k)}\|, \qquad \text{for } k = 0, 1, \ldots$$

with some $c(A) < 1$. This shows *linear convergence* of $x^{(k)}$ to α. The constant $c(A)$ is a bound on the linear rate of convergence; and the constant $c(A)$ can be shown to converge to zero as A approaches $F'(\alpha)$. Since the iteration converges more slowly than does Newton's method, more iterates will need to be computed. But generally, this added cost is very small compared to the extra costs involved in using the original Newton's method (7.61) if n is at all large.

Example 7.3.5 Consider again the integral equation (7.64) and the approximating nonlinear system (7.65). We solve this system (for $n = 5$, as before), and we choose $A = F'(x^{(0)})$, with $x^{(0)}$ as in (7.66). The iteration convergence results are shown in Table 7.7; note that the true errors are given. The column labeled "Ratio" gives the ratio of the successive estimated errors. The values of Ratio show the convergence of the iterates is somewhat irregular in this case, and $c(A) \ge 0.247$. ■

Table 7.7. The Modified Newton's Method for Solving the Nonlinear System (7.65) with $n = 5$

k	$\|\alpha - x^{(k)}\|$	Ratio
0	3.58E − 1	
1	3.42E − 2	0.096
2	7.03E − 3	0.206
3	6.83E − 4	0.097
4	1.54E − 4	0.225
5	2.09E − 5	0.136
6	2.30E − 6	0.110
7	5.68E − 7	0.247
8	5.53E − 8	0.097
9	7.24E − 9	0.131

Newton's method (7.62) and the modified Newton's method (7.71)

$$x^{(k+1)} = x^{(k)} - A^{-1} F(x^{(k)})$$

are examples of fixed point iteration:

$$x^{(k+1)} = g(x^{(k)}), \qquad k \geq 0$$

where $g(x)$ is a vector function

$$g(x) = \begin{bmatrix} g_1(x_1, \ldots, x_n) \\ \vdots \\ g_n(x_1, \ldots, x_n) \end{bmatrix}$$

There is a general theory for such fixed-point iteration methods involving a strengthened form of Theorem 3.4.2; but we do not consider it here [e.g., see Atkinson (1989, §2.10)].

MATLAB PROGRAM. We give a MATLAB program for the Newton method applied to the solution of a systems of two nonlinear equations in two unknowns:

$$F_1(x_1, x_2) = 0$$
$$F_2(x_1, x_2) = 0$$

```
function solution = newton_sys(x_init,err_tol,max_iterates)
%
% The calling sequence for newton_sys.m is
```

```
% solution = newton_sys(x_init,err_tol,max_iterates)
% This solves a pair of two nonlinear equations in two unknowns,
%             f(x) = 0
% with f(x) a column vector of length 2.  The definition of f(x)
% is to be given below in the function named fsys; and you
% also need to give the Jacobian matrix for f(x) in the
% function named deriv_fsys.
%
% x_init is a vector of length 2, and it is an initial guess
% at the solution.
%
% The parameters err_tol and max_iterates are upper limits on
% the desired error in the solution and the maximum number of
% iterates to be computed.

% Initialization.
x0 = zeros(2,1);
for i=1:2
   x0(i) = x_init(i);
end

error = inf;
it_count = 0;

% Begin the main loop.
while error > err_tol & it_count < max_iterates
   it_count = it_count + 1;
   rhs = fsys(x0);
   A = deriv_fsys(x0);
   delta = A\rhs;
   x1 = x0 - delta;
   error = norm(delta,inf);
   % The following statement is an internal print to show
   % the course of the iteration.  It and the pause
   % statement following it can be commented out.
   [it_count x1' error]
   pause
   x0 = x1;
end

% Return with the solution.
solution = x1;
if it_count == max_iterates
   disp(' ')
   disp('*** Your answers may possibly not ...
```

```
%                 satisfy your error tolerance.')
end

%%%%%%%%%%% Definition of functions %%%%%%%%%%%%%%%%%
function f_val = fsys(x)
%
% The equations being solved are
%   x(1)^2 + 4*x(2)^2 - 9 = 0
%   18*x(2) - 14*x(1)^2 + 45 = 0

f_val = [x(1)^2+4*x(2)^2-9, 18*x(2)-14*x(1)^2+45]';

function df_val = deriv_fsys(x)
%
% This defines the Jacobian matrix for the function
% given in fsys

df_val = [2*x(1), 8*x(2); -28*x(1), 18];
```

PROBLEMS

1. Find the remaining three solutions of the system (7.52). Note that symmetry can be used to reduce the number of solutions that must be found computationally.

2. Find all solutions to the following systems, using Newton's method (7.55–7.56).
   (a) $x^2 + y^2 = 4, \quad x^2 - y^2 = 1$
   (b) $x^2 + 4y^2 = 4, \quad y = x^2 - 0.4x - 1.96$
   (c) $x^2 + y^2 = 1, \quad 2y = 2x^3 + x + 1$
   (d) $x + y - 2xy = 0, \quad x^2 + y^2 - 2x + 2y + 1 = 0$

3. Find all solutions to the system
$$x^2 + xy^3 = 9, \qquad 3x^2y - y^3 = 4$$
   Use each of the initial guesses $(x_0, y_0) = (1.2, 2.5)$, $(-2, 2.5)$, $(-1.2, -2.5)$, $(2, -2.5)$. Observe the root to which each of the iterations converges and the number of iterates computed. Comment on your results.

4. Solve the nonlinear system (7.65) for $n = 10, 15, 20$, to the full accuracy of your computer. Compare the needed number of iterates. Graph the solutions x for each n, as in Figure 7.10.

5. Use the modified Newton's method (7.71) to solve for the roots of (7.52), presenting results as in Table 7.7. Use $A = F'(x^{(0)})$.

6. Repeat Problem 5 for the systems given in Problem 2.

7. Reformulate the problem of finding a minimum or maximum for a function $D(x_1, x_2)$ as a rootfinding problem for a system of two equations in two unknowns. We assume the function $D(x_1, x_2)$ is differentiable with respect to both x_1 and x_2.

8. Apply Problem 7 to the minimizing of the function

$$D(x_1, x_2) = x_1^4 + x_1 x_2 + (1 + x_2)^2$$

Use Newton's method to solve the resulting system of equations. Experiment with various choices of initial estimates $x^{(0)}$.

CHAPTER 8

NUMERICAL SOLUTION OF ORDINARY DIFFERENTIAL EQUATIONS

Differential equations are among the most important mathematical tools used in producing models of physical and biological sciences, and engineering. In this chapter, we consider numerical methods for solving ordinary differential equations, that is, those differential equations that have only one independent variable. Numerical methods for partial differential equations, that is, differential equations with more than one independent variable, are discussed in Chapter 9.

The differential equations we consider in most sections of the chapter are of the form

$$Y'(x) = f(x, Y(x)), \qquad x \geq x_0 \tag{8.1}$$

where $Y(x)$ is an unknown function that is being sought. The given function $f(x, z)$ of two variables defines the differential equation; examples are given in Section 8.1. This equation is called a *first-order differential equation* because it contains a first-order derivative of the unknown function, but no higher-order derivative. A higher-order differential equation can be reformulated as a system of first-order equations, and the numerical methods for first-order equations can be extended in a straightforward way to this new system. The reformulation of higher-order equations and the numerical solution of systems of first-order differential equations are discussed in later sections of this chapter.

A discussion of the theory of the initial value problem for ordinary differential equations is given in Section 8.1; the idea of stability of differential equations is also introduced. The simplest of the numerical methods for solving (8.1), *Euler's method*, is given in Sections 8.2 and 8.3. It is not an efficient numerical method, but it is an intuitive way to introduce many important ideas. In Section 8.4, we discuss some numerical methods with better numerical stability for practical computation. Sections 8.5 and 8.6 cover more sophisticated and rapidly convergent methods. Higher-order equations and systems of first-order equations are considered in Section 8.7. In the final section of the chapter, we consider the finite difference solution of boundary value problems for second-order ordinary differential equations.

8.1. THEORY OF DIFFERENTIAL EQUATIONS: AN INTRODUCTION

For simple differential equations, it is possible to find closed-form solutions. For example, given a function g, the general solution of the simplest equation

$$Y'(x) = g(x)$$

is

$$Y(x) = \int g(x)\,dx + c$$

with c an arbitrary integration constant. Here, $\int g(x)\,dx$ denotes any fixed antiderivative of g. The constant c, and thus, a particular solution can be obtained by specifying the value of $Y(x)$ at some given point

$$Y(x_0) = Y_0$$

Example 8.1.1 The general solution of the equation

$$Y'(x) = \sin(x)$$

is

$$Y(x) = -\cos(x) + c$$

If we specify the condition

$$Y\left(\frac{\pi}{3}\right) = 2$$

then it is easy to find $c = 2.5$. Thus, the desired solution is

$$Y(x) = 2.5 - \cos(x) \quad \blacksquare$$

The more general equation

$$Y'(x) = f(x, Y(x))$$

is approached in a similar spirit, in the sense that there is a general solution dependent on a constant. To further illustrate this, we consider some more examples that can be solved analytically. First, and foremost, is the first-order linear equation

$$Y'(x) = a(x)Y(x) + b(x) \tag{8.2}$$

The given functions $a(x)$ and $b(x)$ are assumed continuous. For this equation,

$$f(x, z) = a(x)z + b(x)$$

and the general solution of the equation can be found by the so-called *method of integrating factors*.

We illustrate the method of integrating factors through the particularly useful case

$$Y'(x) = \lambda Y(x) + b(x), \qquad x \geq x_0 \tag{8.3}$$

with λ a given constant. Multiplying the equation (8.3) by the integrating factor $e^{-\lambda x}$, we can reformulate the equation as

$$\frac{d}{dx}\left(e^{-\lambda x}Y(x)\right) = e^{-\lambda x}b(x)$$

Integrating both sides from x_0 to x, we obtain

$$e^{-\lambda x}Y(x) = c + \int_{x_0}^{x} e^{-\lambda t}b(t)\,dt$$

So the general solution of (8.3) is

$$Y(x) = e^{\lambda x}\left[c + \int_{x_0}^{x} e^{-\lambda t}b(t)\,dt\right] = ce^{\lambda x} + \int_{x_0}^{x} e^{\lambda(x-t)}b(t)\,dt \tag{8.4}$$

with c an arbitrary constant. It is easy to see that

$$c = e^{-\lambda x_0}Y(x_0) \tag{8.5}$$

As we have seen from the above discussions, the general solution of the first-order equation (8.1) normally depends on an arbitrary integration constant. To single out a particular solution, we need to specify an additional condition. Usually, such a condition is taken to be of the form

$$Y(x_0) = Y_0 \tag{8.6}$$

In many applications of the ordinary differential equation (8.1), the independent variable x plays the role of time, and x_0 can be interpreted as the initial time. So it is customary to call (8.6) an initial value condition. The differential equation (8.1) and the initial value condition (8.6) together form an *initial value problem*

$$\begin{cases} Y'(x) = f(x, Y(x)), & x \geq x_0 \\ Y(x_0) = Y_0 \end{cases} \tag{8.7}$$

For the initial value problem of the linear equation (8.3), the solution is given by the formulas (8.4) and (8.5). We observe that the solution exists on any open interval where the data function $b(x)$ is continuous. This is a property for linear equations. For the initial value problem of the general linear equation (8.2), its solution exists on any open interval where the functions $a(x)$ and $b(x)$ are continuous. As we will see next through examples, when the equation (8.1) is nonlinear, even if the right side function $f(x, y)$ has derivatives of any order, the solution of the corresponding initial value problem may exist only on a smaller interval.

Example 8.1.2 **1.** By a direct computation, it is easy to verify that the equation

$$Y'(x) = -[Y(x)]^2 + Y(x)$$

has a so-called trivial solution $Y(x) \equiv 0$ and a general solution

$$Y(x) = \frac{1}{1 + c\,e^{-x}} \tag{8.8}$$

with c arbitrary. Alternatively, this equation is a so-called separable equation and its solution can be found by a standard method described in Problem 4. To find the solution of the equation satisfying $Y(0) = 4$, we use the solution formula at $x = 0$:

$$4 = \frac{1}{1 + c}$$

$$c = -0.75$$

So the solution of the initial value problem is

$$Y(x) = \frac{1}{1 - 0.75e^{-x}}, \qquad x \geq 0$$

With a general initial value $Y(0) = Y_0 \neq 0$, the constant c in the solution formula (8.8) is given by $c = Y_0^{-1} - 1$. If $Y_0 > 0$, then $c > -1$, and the solution $Y(x)$ exists for $0 \leq x < \infty$. However, for $Y_0 < 0$, the solution exists only on the finite interval $[0, \log(1 - Y_0^{-1}))$; the value $x = \log(1 - Y_0^{-1})$ is the zero of the denominator in the formula (8.8).

2. Consider the equation

$$Y'(x) = -[Y(x)]^2$$

It has a trivial solution $Y(x) \equiv 0$ and a general solution

$$Y(x) = \frac{1}{x + c} \tag{8.9}$$

with c arbitrary. This can be verified by a direct calculation or by the method described in Problem 4. To find the solution of the equation satisfying the initial value condition $Y(0) = Y_0$, we distinguish several cases according to the value of Y_0.

- If $Y_0 = 0$, then the solution of the initial value problem is $Y(x) \equiv 0$ for any $x \geq 0$.
- If $Y_0 \neq 0$, then the solution of the initial value problem is

$$Y(x) = \frac{1}{x + Y_0^{-1}}$$

- For $Y_0 > 0$, the solution exists for any $x \geq 0$. For $Y_0 < 0$, the solution exists only on the interval $[0, -Y_0^{-1})$.
- As a side note, observe that for $0 < Y_0 < 1$, the solution (8.8) increases for $x \geq 0$, whereas for $Y_0 > 0$, the solution (8.9) decreases for $x \geq 0$.

3. The solution of

$$Y'(x) = \lambda Y(x) + e^{-x}, \qquad Y(0) = 1, \qquad \lambda \neq -1$$

is obtained from (8.4) as

$$\begin{aligned} Y(x) &= e^{\lambda x} + \int_0^x e^{\lambda(x-t)} e^{-t}\, dt \\ &= e^{\lambda x} \left\{ 1 + \frac{1}{\lambda + 1} [1 - e^{-(\lambda+1)x}] \right\} \end{aligned}$$ ■

We remark that for a general right-hand-side function $f(x, z)$, it is usually not possible to solve the initial value problem (8.7) analytically. One such example is for the equation

$$Y' = e^{-x Y^4}$$

In such a case, numerical methods are the only plausible way to compute solutions. Moreover, even when a differential equation can be solved analytically, the solution formula, such as (8.4), usually involves integrations of general functions. The integrals mostly have to be evaluated numerically. As an example, it is easy to verify that the solution of the problem

$$\begin{cases} Y' = 2xY + 1, & x > 0 \\ Y(0) = 1 \end{cases}$$

is

$$Y(x) = e^{x^2} \int_0^x e^{-t^2} dt + e^{x^2}$$

For such a situation, it is usually more efficient to use numerical methods from the outset to solve the differential equation.

8.1.1 General Solvability Theory

Before we consider numerical methods, it is useful to have some discussions on properties of the differential equation (8.1). The following well-known result concerns the existence and unique solvability of the initial value problem (8.7).

Theorem 8.1.3 Let $f(x, z)$ and $\partial f(x, z)/\partial z$ be continuous functions of x and z at all points (x, z) in some neighborhood of the initial point (x_0, Y_0). Then there is a unique function $Y(x)$ defined on some interval $[x_0 - \alpha, x_0 + \alpha]$, satisfying

$$\begin{aligned} Y'(x) &= f(x, Y(x)), \qquad x_0 - \alpha \le x \le x_0 + \alpha \\ Y(x_0) &= Y_0 \end{aligned}$$

The number α in the statement of the theorem depends on the initial value problem (8.7). For some equations, such as the linear equation given in (8.3) with a continuous function $b(x)$, solutions exist for any x, and we can take α to be ∞. For many nonlinear equations, solutions can exist only in bounded intervals. We have seen such examples in Example 8.1.2. Let us look at one more such example.

Example 8.1.4 Consider the initial value problem

$$Y'(x) = 2x[Y(x)]^2, \qquad Y(0) = 1$$

Here,

$$f(x, z) = 2xz^2, \qquad \frac{\partial f(x, z)}{\partial z} = 4xz$$

and both of these functions are continuous for all (x, z). Thus, by Theorem 8.1.3 there is a unique solution to this initial value problem for x in a neighborhood of $x_0 = 0$. This solution is

$$Y(x) = \frac{1}{1 - x^2}, \qquad -1 < x < 1$$

This example illustrates that the continuity of $f(x, z)$ and $\partial f(x, z)/\partial z$ for all (x, z) does not imply the existence of a $Y(x)$ that is continuous for all x. ■

8.1.2 Stability of the Initial Value Problem

When numerically solving the initial value problem (8.7), we will generally assume that the solution $Y(x)$ is being sought on a given finite interval $x_0 \le x \le b$. In that case, it is possible to obtain the following result on stability. Make a small change in the initial value for the initial value problem, changing Y_0 to $Y_0 + \epsilon$. Call the resulting solution $Y_\epsilon(x)$

$$Y_\epsilon'(x) = f(x, Y_\epsilon(x)), \qquad x_0 \le x \le b, \qquad Y_\epsilon(x_0) = Y_0 + \epsilon \tag{8.10}$$

Then under hypotheses similar to those of Theorem 8.1.3, it can be shown that for all small values of ϵ,

$$\max_{x_0 \le x \le b} |Y_\epsilon(x) - Y(x)| \le c\,\epsilon, \qquad \text{some } c > 0 \tag{8.11}$$

Thus, small changes in the initial value Y_0 will lead to small changes in the solution $Y(x)$ of the initial value problem. This is a desirable property for a variety of very practical reasons.

Example 8.1.5 The problem

$$Y'(x) = -Y(x) + 1, \qquad 0 \le x \le b, \qquad Y(0) = 1 \tag{8.12}$$

has the solution $Y(x) \equiv 1$. The perturbed problem

$$Y_\epsilon'(x) = -Y_\epsilon(x) + 1, \qquad 0 \le x \le b, \qquad Y_\epsilon(0) = 1 + \epsilon$$

has the solution $Y_\epsilon(x) = 1 + \epsilon e^{-x}$, $x \ge 0$. Thus,

$$Y(x) - Y_\epsilon(x) = -\epsilon e^{-x}$$
$$|Y(x) - Y_\epsilon(x)| \le |\epsilon|, \qquad x \ge 0$$

The problem (8.12) is said to be stable. ■

Virtually all initial value problems are stable in the sense specified in (8.11); but this is only a partial picture of the effect of small perturbations of the initial value Y_0. If the maximum error in (8.11) is much larger than ϵ [i.e., the minimal possible constant c in the estimate (8.11) is large], then the initial value problem (8.7) is usually considered to be *ill-conditioned*. Attempting to numerically solve such a problem will usually lead to large errors in the computed solution.

Example 8.1.6 The problem

$$Y'(x) = \lambda\,[Y(x) - 1]\,, \qquad 0 \le x \le b, \qquad Y(0) = 1 \tag{8.13}$$

has the solution

$$Y(x) = 1, \qquad \text{all } x$$

The perturbed problem

$$Y_\epsilon'(x) = \lambda[Y_\epsilon(x) - 1], \qquad 0 \le x \le b, \qquad Y_\epsilon(0) = 1 + \epsilon$$

has the solution

$$Y_\epsilon(x) = 1 + \epsilon e^{\lambda x}, \qquad x \ge 0$$

For the error,

$$Y(x) - Y_\epsilon(x) = -\epsilon e^{\lambda x} \tag{8.14}$$

$$\max_{0 \le x \le b} |Y(x) - Y_\epsilon(x)| = \begin{cases} |\epsilon|, & \lambda \le 0 \\ |\epsilon|\, e^{\lambda b}, & \lambda \ge 0 \end{cases}$$

If $\lambda < 0$, the error $Y(x) - Y_\epsilon(x)$ decreases as x increases. We see that (8.13) is well-conditioned when $\lambda \le 0$. In contrast, for $\lambda > 0$, the error $Y(x) - Y_\epsilon(x)$ increases as x increases. And for moderately large λb, say, $\lambda b \ge 10$, the change in $Y(x)$ is quite significant at $x = b$. The problem (8.13) is increasingly ill-conditioned as λ increases. ■

For the more general initial value problem (8.7) and the perturbed problem (8.10), one can show that

$$Y(x) - Y_\epsilon(x) \approx -\epsilon \exp\left(\int_{x_0}^{x} g(t)\, dt\right) \tag{8.15}$$

where

$$g(t) = \left.\frac{\partial f(t, z)}{\partial z}\right|_{z=Y(t)}$$

for x sufficiently close to x_0. Note that this formula correctly predicts (8.14), since in that case

$$\begin{aligned} f(x, z) &= \lambda[z - 1] \\ \frac{\partial f(x, z)}{\partial z} &= \lambda \\ \int_0^x g(t)\, dt &= \lambda x \end{aligned}$$

Then (8.15) yields

$$Y(x) - Y_\epsilon(x) \approx -\epsilon e^{\lambda x}$$

which agrees with the earlier formula (8.14).

Example 8.1.7 The problem

$$Y'(x) = -[Y(x)]^2, \qquad Y(0) = 1 \tag{8.16}$$

has the solution

$$Y(x) = \frac{1}{x + 1}$$

For the perturbed problem,

$$Y_\epsilon'(x) = -[Y_\epsilon(x)]^2, \qquad Y_\epsilon(0) = 1 + \epsilon \tag{8.17}$$

We use (8.15) to estimate $Y(x) - Y_\epsilon(x)$. First,

$$\begin{aligned} f(x, z) &= -z^2 \\ \frac{\partial f(x, z)}{\partial x} &= -2z \\ g(t) &= -2y(t) = -\frac{2}{t + 1} \\ \int_0^x g(t)\, dt &= -2 \int_0^x \frac{dt}{t + 1} = -2 \log(1 + x) = \log(1 + x)^{-2} \\ \exp\left[\int_0^x g(t)\, dt\right] &= e^{\log(x+1)^{-2}} = \frac{1}{(x + 1)^2} \end{aligned}$$

Substituting into (8.15) gives us for sufficiently small $x \geq 0$,

$$Y(x) - Y_\epsilon(x) \approx \frac{-\epsilon}{(1 + x)^2} \tag{8.18}$$

This indicates that (8.16) is a well-conditioned problem. ■

In general, if

$$\frac{\partial f(x, Y(x))}{\partial z} \leq 0, \qquad x_0 \leq x \leq b \tag{8.19}$$

then we say the initial value problem is well-conditioned. Although this test depends on $Y(x)$ on the interval $[x_0, b]$, often one can show (8.19) without knowing $Y(x)$ explicitly.

8.1.3 Direction Fields

Direction fields are a useful tool to understand the behavior of solutions of a differential equation. We notice that the graph of a solution of the equation $Y' = f(x, Y)$ is such that at any point (x, y) on the solution curve, the slope is $f(x, y)$. The slopes can be represented graphically in direction field diagrams. In MATLAB, direction fields can be generated by using the `meshgrid` and `quiver` commands.

Example 8.1.8 Consider the equation $Y' = Y$. The slope of a solution curve at a point (x, y) on the curve is y, which is independent of x. We generate a direction field diagram with the following MATLAB code. First draw the direction field.

```
[x,y] = meshgrid(-2:0.5:2,-2:0.5:2);
dx = ones(9);           %Generates a matrix of 1's.
dy = y; quiver(x,y,dx,dy);
```

Then draw two solution curves.

```
hold on
x = -2:0.01:1;
y1 = exp(x);
y2 = -exp(x);
plot(x,y1,x,y2)
text(1.1,2.8,'\itY=e^x','FontSize',14)
text(1.1,2.8,'\itY=-e^x','FontSize',14)
hold off
```

The result is shown in Figure 8.1. ■

Example 8.1.9 Continuing Example 8.1.4, we use the following MATLAB M-file to generate a direction field diagram and the particular solution $Y = 1/(1 - x^2)$ in Figure 8.2.

```
[x,y] = meshgrid(-1:0.2:1,1:0.5:4);
dx = ones(7,11);
dy = 2*x.*y.^2;
quiver(x,y,dx,dy);
```

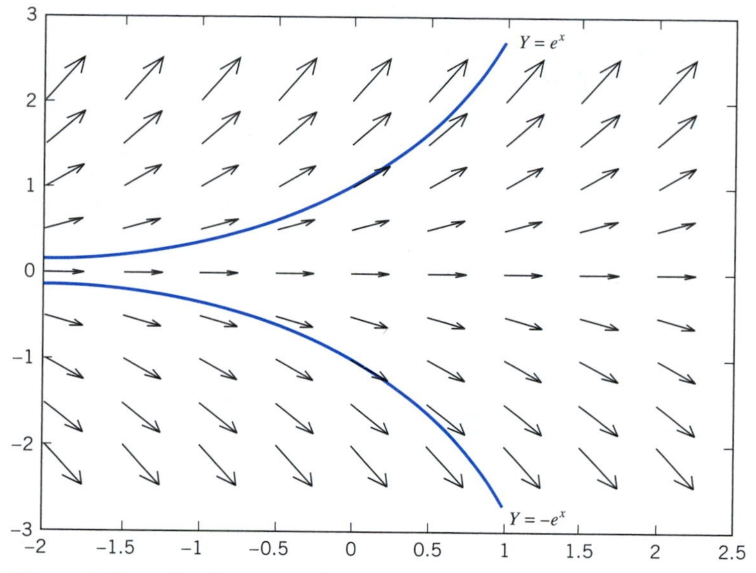

Figure 8.1. The direction field of the equation $Y' = Y$ and solutions $Y = \pm e^x$

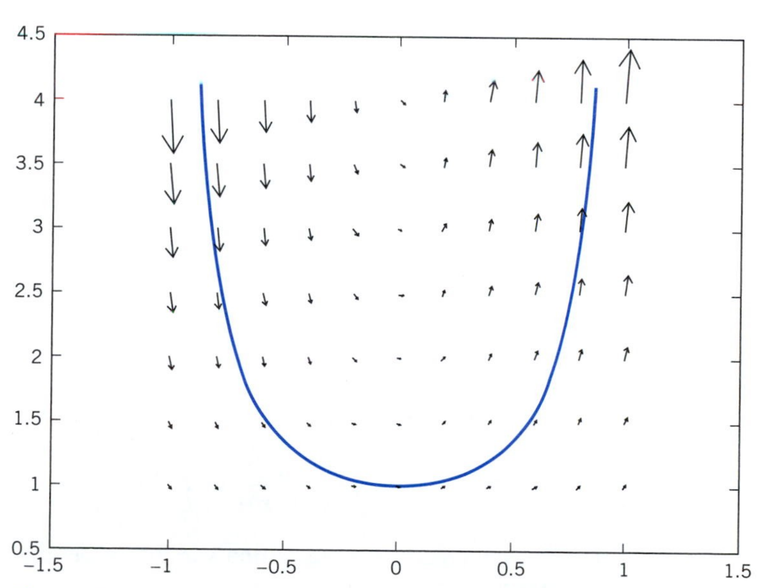

Figure 8.2. The direction field of the equation $Y' = 2xY^2$ and the solution $Y = 1/(1 - x^2)$

```
hold on
xx = -0.87:0.01:0.87;
yy = 1./(1-xx.^2);
plot(xx,yy)
hold off
```

Note that for large y values, the arrows in the direction field diagram point almost vertically. This suggests a solution of the equation may exist only in a bounded interval of the x-axis. ■

PROBLEMS

1. In each of the following cases, show that the given function $Y(x)$ satisfies the associated differential equation. Then determine the value of c required by the initial condition. Finally, with reference to the general format in (8.7), identify $f(x, z)$ for each differential equation.

   (a) $Y'(x) = -Y(x) + \sin(x) + \cos(x), \quad Y(0) = 1; \quad Y(x) = \sin(x) + ce^{-x}$

   (b) $Y'(x) = (Y(x) - Y(x)^2)/x, \quad Y(1) = 2; \quad Y(x) = x/(x + c), \quad x > 0$

   (c) $Y'(x) = \cos^2(Y(x)), \quad Y(0) = \pi/4; \quad Y(x) = \tan^{-1}(x + c)$

   (d) $Y'(x) = Y(x)[Y(x) - 1], \quad Y(0) = \frac{1}{2}; \quad Y(x) = 1/(1 + ce^x)$

2. Use MATLAB to draw direction fields for the differential equations listed in Problem 1.

3. Solve the following problems by using (8.4):

   (a) $Y'(x) = \lambda Y(x) + 1, \quad Y(0) = 1$

   (b) $Y'(x) = \lambda Y(x) + x, \quad Y(0) = 3$

4. Consider the differential equation

$$Y'(x) = f_1(x) f_2(Y(x))$$

for some given functions $f_1(x)$ and $f_2(z)$. This is called a *separable* differential equation, and it can be solved by direct integration. Write the equation as

$$\frac{Y'(x)}{f_2(Y(x))} = f_1(x)$$

and find the antiderivative of each side:

$$\int \frac{Y'(x)\,dx}{f_2(Y(x))} = \int f_1(x)\,dx$$

On the left side, change the integration variable by letting $z = Y(x)$. Then the equation becomes

$$\int \frac{dz}{f_2(z)} = \int f_1(x)\,dx$$

After integrating, replace z by $Y(x)$; then solve for $Y(x)$, if possible. If these integrals can be evaluated, then the differential equation can be solved. Do so for the following problems, finding the general solution and the solution satisfying the given initial condition:

(a) $Y'(x) = x/Y(x), \qquad Y(0) = 2$

(b) $Y'(x) = xe^{-Y(x)}, \qquad Y(1) = 0$

(c) $Y'(x) = Y(x)[a - Y(x)], \qquad Y(0) = a/2, \qquad a > 0$

5. Check the conditioning of the initial value problems in Problem 1 above. Use the test (8.19).

6. Use (8.19) to discuss the conditioning of the problem

$$Y'(x) = Y(x)^2 - 5\sin(x) - 25\cos^2(x), \qquad Y(0) = 6$$

You do not need to know the true solution.

7. Consider the solutions $Y(x)$ of

$$Y'(x) + aY(x) = de^{-bx}$$

with a, b, d constants and $a, b > 0$. Calculate

$$\lim_{x\to\infty} Y(x)$$

Hint: Consider separately the cases $a \neq b$ and $a = b$.

8.2. EULER'S METHOD

The simplest method for solving the initial value problem is called *Euler's method.* We define it in this section and then analyze it in the next section. Euler's method is not an efficient numerical method, but many of the ideas involved in the numerical solution of differential equations are introduced most simply with it.

Before beginning, we establish some notation that will be used in the rest of this chapter. As before, $Y(x)$ denotes the true solution of the initial value problem, with the initial value Y_0:

$$\begin{cases} Y'(x) = f(x, Y(x)), & x_0 \le x \le b \\ Y(x_0) = Y_0 \end{cases} \tag{8.20}$$

Numerical methods for solving (8.20) will yield an approximate solution $y(x)$ at a discrete set of nodes

$$x_0 < x_1 < x_2 < \cdots < x_N \le b \tag{8.21}$$

For simplicity, we will take these nodes to be evenly spaced:

$$x_n = x_0 + nh, \qquad n = 0, 1, \ldots, N$$

The approximate solution will be denoted using $y(x)$, with some variations. The following notations are all used for the approximate solution at the node points:

$$y(x_n) = y_h(x_n) = y_n, \qquad n = 0, 1, \ldots, N$$

To obtain an approximate solution $y(x)$ at points in $[x_0, b]$ other than those in (8.21), some form of interpolation must be used. We will not consider that problem here, although the techniques of Chapter 4 are easily applied.

To derive Euler's method, recall the derivative approximation (5.76) from Chapter 5

$$Y'(x) \approx \frac{1}{h}[Y(x+h) - Y(x)]$$

Applying this to the initial value problem (8.20) at $x = x_n$

$$Y'(x_n) = f(x_n, Y(x_n))$$

we obtain

$$\frac{1}{h}[Y(x_{n+1}) - Y(x_n)] \approx f(x_n, Y(x_n))$$

$$Y(x_{n+1}) \approx Y(x_n) + hf(x_n, Y(x_n)) \tag{8.22}$$

Euler's method is defined by taking this to be exact:

$$y_{n+1} = y_n + hf(x_n, y_n), \qquad 0 \le n \le N-1 \tag{8.23}$$

For the initial guess, use $y_0 = Y_0$ or some close approximation of Y_0. Sometimes Y_0 is obtained empirically and thus may be known only approximately. Formula (8.23) gives a rule for computing $y_1, y_2, \ldots, y_N$ in succession. This is typical of most numerical methods for solving ordinary differential equations.

Some geometric insight into Euler's method is given in Figure 8.3. The tangent line to the graph of $z = Y(x)$ at x_n has slope

$$Y'(x_n) = f(x_n, Y(x_n))$$

If we use this tangent line to approximate the curve near the point $(x_n, Y(x_n))$, the value of the tangent line at $x = x_{n+1}$ is given by the right side of (8.22).

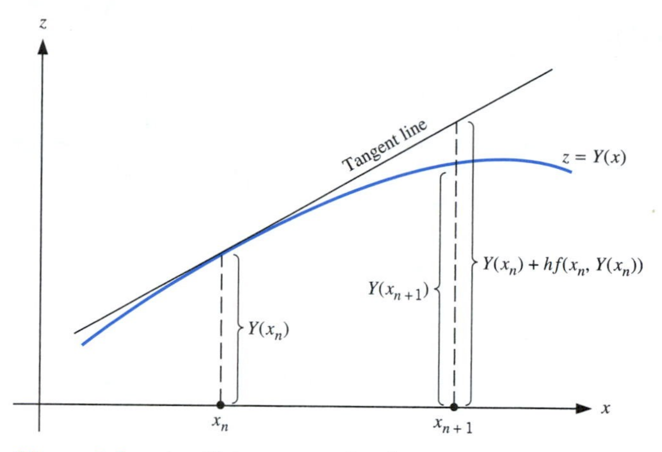

Figure 8.3. An illustration of Euler's method derivation

Example 8.2.1 (a) The true solution of the problem

$$Y'(x) = -Y(x), \qquad Y(0) = 1 \tag{8.24}$$

is $Y(x) = e^{-x}$. Euler's method is given by

$$y_{n+1} = y_n - hy_n, \qquad n \geq 0 \tag{8.25}$$

with $y_0 = 1$ and $x_n = nh$. The solution $y(x)$ is given in Table 8.1, for three values of h and selected values of x. To illustrate the procedure, we compute y_1 and y_2 when $h = 0.1$. From (8.25)

$$\begin{aligned} y_1 &= y_0 - hy_0 = 1 - (0.1)(1) = 0.9, \qquad x_1 = 0.1 \\ y_2 &= y_1 - hy_1 = 0.9 - (0.1)(0.9) = 0.81, \qquad x_2 = 0.2 \end{aligned}$$

For the error in these values,

$$\begin{aligned} Y(x_1) - y_1 &= e^{-0.1} - y_1 \doteq 0.004837 \\ Y(x_2) - y_2 &= e^{-0.2} - y_2 \doteq 0.001873 \end{aligned}$$

(b) Solve

$$Y'(x) = \frac{Y(x) + x^2 - 2}{x + 1}, \qquad Y(0) = 2 \tag{8.26}$$

whose true solution is

$$Y(x) = x^2 + 2x + 2 - 2(x + 1)\log(x + 1)$$

Table 8.1. Euler's Method for (8.24)

h	x	$y_h(x)$	Error	Relative Error
0.2	1.0	3.2768E − 1	4.02E − 2	0.109
	2.0	1.0738E − 1	2.80E − 2	0.207
	3.0	3.5184E − 2	1.46E − 2	0.293
	4.0	1.1529E − 2	6.79E − 3	0.371
	5.0	3.7779E − 3	2.96E − 3	0.439
0.1	1.0	3.4867E − 1	1.92E − 2	0.0522
	2.0	1.2158E − 1	1.38E − 2	0.102
	3.0	4.2391E − 2	7.40E − 3	0.149
	4.0	1.4781E − 2	3.53E − 3	0.193
	5.0	5.1538E − 3	1.58E − 3	0.234
0.05	1.0	3.5849E − 1	9.39E − 3	0.0255
	2.0	1.2851E − 1	6.82E − 3	0.0504
	3.0	4.6070E − 2	3.72E − 3	0.0747
	4.0	1.6515E − 2	1.80E − 3	0.0983
	5.0	5.9205E − 3	8.17E − 4	0.121

Euler's method for this differential equation is

$$y_{n+1} = y_n + \frac{h(y_n + x_n^2 - 2)}{x_n + 1}, \qquad n \geq 0$$

with $y_0 = 2$ and $x_n = nh$. The solution $y(x)$ is given in Table 8.2 for three values of h and selected values of x. A graph of the solution $y_h(x)$ for $h = 0.2$ is given in Figure 8.4. The node values $y_h(x_n)$ have been connected by straight-line segments in the graph. Note that the horizontal and vertical scales are different. ■

In both examples, observe the behavior of the error as h decreases. For each fixed value of x, note that the errors decrease by a factor of about 2 when h is halved. As an illustration, take Example 8.2.1(a) with $x = 5.0$. The errors for $h = 0.2$, 0.1, and 0.05, respectively, are

$$2.96\text{E} - 3, \qquad 1.58\text{E} - 3, \qquad 8.17\text{E} - 4$$

and these decrease by successive factors of 1.87 and 1.93. The reader should do the same calculation for other values of x, in both Example 8.2.1(a) and Example 8.2.1(b). Also, note that the behavior of the error as x increases may be quite different from the behavior of the relative error. In the second example, the relative errors increase initially, and then they decrease with increasing x.

Table 8.2. Euler's Method for (8.26)

h	x	$y_h(x)$	Error	Relative Error
0.2	1.0	2.1592	6.82E–2	0.0306
	2.0	3.1697	2.39E–1	0.0701
	3.0	5.4332	4.76E–1	0.0805
	4.0	9.1411	7.65E–1	0.129
	5.0	14.406	1.09	0.0703
	6.0	21.303	1.45	0.0637
0.1	1.0	2.1912	3.63E–2	0.0163
	2.0	3.2841	1.24E–1	0.0364
	3.0	5.6636	2.46E–1	0.0416
	4.0	9.5125	3.93E–1	0.0665
	5.0	14.939	5.60E–1	0.0361
	6.0	22.013	7.44E–1	0.0327
0.05	1.0	2.2087	1.87E–2	0.00840
	2.0	3.3449	6.34E–2	0.0186
	3.0	5.7845	1.25E–1	0.0212
	4.0	9.7061	1.99E–1	0.0337
	5.0	15.214	2.84E–1	0.0183
	6.0	22.381	3.76E–1	0.0165

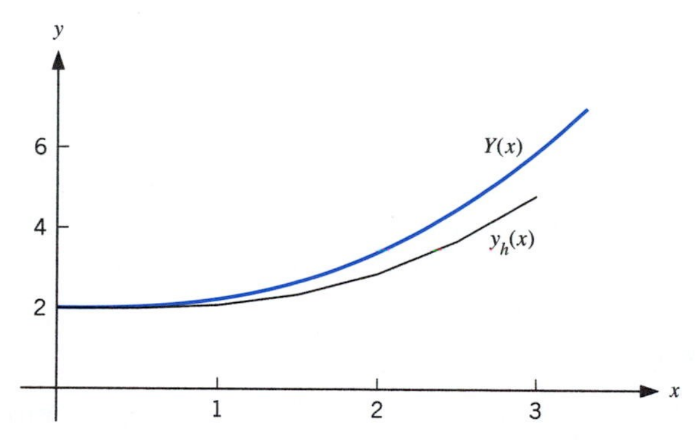

Figure 8.4. Euler's method for problem (8.26), $h = 0.2$

MATLAB PROGRAM. The following MATLAB program implements Euler's method. The Euler method is also called the *forward Euler method*. The backward Euler method is discussed in the next section.

```
function [x,y] = euler_for(x0,y0,x_end,h,fcn)
%
% function [x,y]=euler_for(x0,y0,x_end,h,fcn)
%
% Solve the initial value problem
%    y' = f(x,y),  x0 <= x <= b,  y(x0)=y0
% Use Euler's method with a stepsize of h.  The user must
% supply a program with some name, say, deriv, and a first
% line of the form
%   function ans=deriv(x,y)
% A sample call would be
%   [t,z]=euler_for(t0,z0,b,delta,'deriv')
%
% Output:
% The routine eulercls will return two vectors, x and y.
% The vector x will contain the node points
%   x(1)=x0, x(j)=x0+(j-1)*h, j=1,2,...,N
% with
%   x(N) <= x_end-h,  x(N)+h > x_end-h
% The vector y will contain the estimates of the solution Y
% at the node points in x.
%
n = fix((x_end-x0)/h)+1;
x = linspace(x0,x_end,n)';
y = zeros(n,1);
y(1) = y0;
for i = 2:n
  y(i) = y(i-1)+h*feval(fcn,x(i-1),y(i-1));
end
```

PROBLEMS

1. Solve the following problems using Euler's method with stepsizes of $h = 0.2, 0.1, 0.05$. Compute the error and relative error using the true answer $Y(x)$. For selected values of x, observe the ratio by which the error decreases when h is halved.

   (a) $Y'(x) = [\cos(Y(x))]^2, \quad 0 \le x \le 10, \quad Y(0) = 0;$
   $Y(x) = \tan^{-1}(x)$

   (b) $Y'(x) = \dfrac{1}{1+x^2} - 2[Y(x)]^2, \quad 0 \le x \le 10, \quad Y(0) = 0;$

   $$Y(x) = \frac{x}{1+x^2}$$

   (c) $Y'(x) = \frac{1}{4}Y(x)\left[1 - \frac{1}{20}Y(x)\right], \quad 0 \le x \le 20, \quad Y(0) = 1;$

   $$Y(x) = \frac{20}{1 + 19e^{-x/4}}$$

(d) $Y'(x) = -[Y(x)]^2, \qquad 1 \le x \le 10, \qquad Y(1) = 1; \quad Y(x) = \dfrac{1}{x}$

(e) $Y'(x) = xe^{-x} - Y(x), \qquad 0 \le x \le 10, \qquad Y(0) = 1;$
$y(x) = \left(1 = \frac{1}{2}x^2\right) e^{-x}$

(f) $Y'(x) = \dfrac{x^3}{Y(x)}, \qquad 0 \le x \le 10, \qquad Y(0) = 1; \quad Y(x) = \sqrt{\frac{1}{2}x^4 + 1}$

(g) $Y'(x) = \left(3x^2 - 1\right) Y(x)^2, \qquad 0 \le x \le 10, \qquad Y(0) = 1;$
$y(x) = \sqrt{3\left(x^3 - x\right) + 1}$

2. Consider the linear equation

$$Y'(x) = \lambda Y(x) + (1 - \lambda)\cos(x) - (1 + \lambda)\sin(x), \qquad Y(0) = 1$$

from (8.3) of Section 8.1. The true solution is $Y(x) = \sin(x) + \cos(x)$. Solve this problem using Euler's method with several values of λ and h, for $0 \le x \le 10$. Comment on the results.

(a) $\lambda = -1; \qquad h = 0.5, 0.25, 0.125$

(b) $\lambda = 1; \qquad h = 0.5, 0.25, 0.125$

(c) $\lambda = -5; \qquad h = 0.5, 0.25, 0.125, 0.0625$

(d) $\lambda = 5; \qquad h = 0.0625$

3. As a special case in which the error of Euler's method can be analyzed directly, consider Euler's method applied to

$$Y'(x) = Y(x), \qquad Y(0) = 1$$

The true solution is e^x.

(a) Show that the solution of Euler's method can be written as

$$y_h(x_n) = (1 + h)^{x_n/h}, \qquad n \ge 0$$

(b) Using l'Hospital's rule from calculus, show that

$$\lim_{h \to 0} (1 + h)^{1/h} = e$$

This then proves that for fixed $x = x_n$,

$$\lim_{h \to 0} y_h(x) = e^x$$

(c) Let us do a more delicate convergence analysis. Use the property $a^b = e^{b \log a}$ to write

$$y_h(x_n) = e^{x_n \log(1+h)/h}$$

Then use the formula

$$\log(1+h) = h - \frac{h^2}{2} + O(h^3)$$

and Taylor expansion of the natural exponential function to show that

$$Y(x_n) - y_h(x_n) = -\frac{x_n e^{x_n}}{2} h + O(h^2)$$

This shows that for small h, the error is almost proportional to h, a phenomenon already observed from the numerical results given in Tables 8.1 and 8.2.

8.3. CONVERGENCE ANALYSIS OF EULER'S METHOD

The purpose of analyzing Euler's method is to understand how it works, to be able to predict the error when using it, and to perhaps accelerate its convergence. Being able to do this for Euler's method will also make it easier to answer similar questions for other, more efficient numerical methods.

We begin by considering the error in approximation (8.22) that led directly to Euler's method. Using Taylor's theorem, write

$$Y(x_{n+1}) = Y(x_n) + hY'(x_n) + \frac{h^2}{2} Y''(\xi_n)$$

for some $x_n \le \xi_n \le x_{n+1}$. When we use the fact that $Y(x)$ satisfies the differential equation, this becomes

$$Y(x_{n+1}) = Y(x_n) + hf(x_n, Y(x_n)) + \frac{h^2}{2} Y''(\xi_n) \tag{8.27}$$

The term

$$T_{n+1} = \frac{h^2}{2} Y''(\xi_n) \tag{8.28}$$

is called the *truncation error* for Euler's method, and it is the error in the approximation

$$Y(x_{n+1}) \approx Y(x_n) + hf(x_n, Y(x_n))$$

To analyze the error in Euler's method, subtract

$$y_{n+1} = y_n + hf(x_n, y_n)$$

from (8.27) to obtain

$$Y(x_{n+1}) - y_{n+1} = Y(x_n) - y_n + h[f(x_n, Y(x_n)) - f(x_n, y_n)] + \frac{h^2}{2} Y''(\xi_n) \qquad (8.29)$$

The error in y_{n+1} consists of two parts: (1) the truncation error T_{n+1}, newly introduced at step x_{n+1}, and (2) the propagated error

$$Y(x_n) - y_n + h[f(x_n, Y(x_n)) - f(x_n, y_n)]$$

The propagated error can be simplified by applying the mean value theorem to $f(x, z)$, considering it as a function of z:

$$f(x_n, Y(x_n)) - f(x_n, y_n) = \frac{\partial f(x_n, z_n)}{\partial z} [Y(x_n) - y_n]$$

for some z_n between $Y(x_n)$ and y_n. Let $e_k \equiv Y(x_k) - y_k$, $k \geq 0$, and then use the above to rewrite (8.29) as

$$e_{n+1} = \left[1 + h \frac{\partial f(x_n, z_n)}{\partial z} \right] e_n + \frac{h^2}{2} Y''(\xi_n) \qquad (8.30)$$

These results can be used to give a general error analysis of Euler's method for the initial value problem. Let us consider a special case that will yield some intuitive understanding of the error in Euler's method.

Consider using Euler's method to solve the problem

$$Y'(x) = 2x, \qquad Y(0) = 0$$

whose true solution is $Y(x) = x^2$. The error formula (8.30) becomes

$$e_{n+1} = e_n + h^2, \qquad e_0 = 0$$

This leads, by induction, to

$$e_n = nh^2, \qquad n \geq 0$$

Since $nh = x_n$,

$$e_n = hx_n$$

For each fixed x_n, the error at x_n is proportional to h. The truncation error is $O(h^2)$, but the cumulative effect of these errors is a total error proportional to h.

Using (8.30) or (8.29), along with much algebraic manipulation, one can prove the following general result.

Theorem 8.3.1 Assume the true solution $Y(x)$ is twice continuously differentiable on $[x_0, b]$ and

$$K = \sup_{\substack{-\infty<z<\infty \\ x_0 \le x \le b}} \left| \frac{\partial f(x,z)}{\partial z} \right| < \infty \tag{8.31}$$

Then the Euler method solution $y_h(x)$ satisfies the error bound

$$\begin{aligned} |Y(x_n) - y_h(x_n)| &\le e^{(b-x_0)K} |Y_0 - y_0| \\ &\quad + h \left[\frac{e^{(b-x_0)K} - 1}{2K} \right] \max_{x_0 \le x \le b} |Y''(x)| \end{aligned} \tag{8.32}$$

for all $x_0 \le x_n \le b$.

The proof is omitted; it can be found in most higher-level numerical analysis textbooks. The hypothesis (8.31) makes the proof easier, but it can be weakened to be much the same as in Theorem 8.1.3.

When $y_0 = Y_0$, as is commonly the case, (8.32) can be written as

$$|Y(x_n) - y_h(x_n)| \le ch, \qquad x_0 \le x_n \le b$$

with c a constant. This is consistent with the behavior observed in Tables 8.1 and 8.2 in the preceding section, and it agrees with the special cases preceding the statement of the theorem. When h is halved, the bound ch is also halved, and that is the behavior in the error that was observed earlier. Euler's method is said to converge with order 1, because that is the power of h that occurs in the error bound. In general, if we have

$$|Y(x_n) - y_h(x_n)| \le ch^p, \qquad x_0 \le x_n \le b \tag{8.33}$$

then we say that the numerical method is *convergent with order* p. Naturally, the higher the order p, the faster the convergence we can expect.

We emphasize that for the error bound (8.32) to hold, the true solution must be assumed twice continuously differentiable. This assumption is not always valid. When the true solution does not have a continuous second derivative, the error bound (8.32) no longer holds. See Problem 6.

The error bound (8.32) is valid for a large family of the initial value problems. But, it usually produces a very pessimistic bound for the error, due to the presence of the exponential terms in the error bound. Under certain circumstances, we can improve the result. Assume

$$\frac{\partial f(x,z)}{\partial z} \le 0$$

and it is bounded in magnitude for all z and for all $x_0 \le x \le b$, the interval on which the differential equation is being solved. Note the relation of this to the stability condition

(8.19) in Section 8.1. Also assume that h has been chosen so small that

$$1 + h\frac{\partial f(x, z)}{\partial z} \geq -1, \qquad x_0 \leq x \leq b, \quad -\infty < z < \infty$$

in (8.30). Then (8.30) implies

$$|e_{n+1}| \leq |e_n| + ch^2, \qquad x_0 \leq x_n \leq b \tag{8.34}$$

where

$$c = \frac{1}{2} \cdot \max_{x_0 \leq x \leq b} \left|Y''(x)\right|$$

and $e_0 = 0$. Applying (8.34) inductively, we obtain

$$|e_n| \leq nch^2 = cx_n h \tag{8.35}$$

The error is bounded by a quantity proportional to h, and the coefficient of the h term increases linearly with respect to the point x_n, in contrast to the exponential growth given in the bound (8.32).

The error estimate in Theorem 8.3.1 is rigorous, and is useful in providing an insight into the convergence behavior of the numerical solution. However, it is usually not advisable to use (8.32) for an actual error bound as the next example shows.

Example 8.3.2 The problem

$$Y'(x) = -Y(x), \qquad Y(0) = 1 \tag{8.36}$$

was solved in Section 8.2, with the results given in Table 8.1. To apply (8.32), we have $\partial f(x, z)/\partial z = -1$, $K = 1$. The true solution is $Y(x) = e^{-x}$; thus,

$$\max_{0 \leq x \leq b} \left|Y''(x)\right| = 1$$

With $y_0 = Y_0 = 1$, the bound (8.32) becomes

$$\left|e^{-x_n} - y_h(x_n)\right| \leq \frac{h}{2}\left(e^b - 1\right), \qquad 0 \leq x_n \leq b \tag{8.37}$$

As $h \to 0$, this shows that $y_h(x)$ converges to e^{-x}. However, this bound is excessively conservative. As b increases, the bound increases exponentially. For $b = 5$, the above bound is

$$\left|e^{-x_n} - y_h(x_n)\right| \leq \frac{h}{2}\left(e^5 - 1\right) \approx 73.7h, \qquad 0 \leq x_n \leq 5$$

And this is far larger than the actual errors shown in Table 8.1, by several orders of magnitude. For the problem (8.36), the improved error bound (8.35) applies with $c = 1/2$ (cf. Problem 2). A more general approach for accurate error estimation is discussed in the following subsection. ■

8.3.1 Asymptotic Error Analysis

To obtain more accurate predictions of the error, we consider asymptotic error estimates, similar to what was done earlier in Chapter 5 with numerical integration. An asymptotic error formula is available for Euler's method, provided that certain conditions are satisfied. Assume Y is three times continuously differentiable and

$$\frac{\partial f(x,z)}{\partial z}, \qquad \frac{\partial^2 f(x,z)}{\partial z^2}$$

are both continuous for all values of (x, z) near $(x, Y(x))$, $x_0 \le x \le b$. Then one can prove that the error in Euler's method satisfies

$$Y(x_n) - y_h(x_n) = hD(x_n) + O(h^2), \qquad x_0 \le x_n \le b \tag{8.38}$$

If we assume $y_0 = Y_0$, the usual case, the function $D(x)$ satisfies an initial value problem for a linear differential equation

$$D'(x) = g(x)D(x) + \tfrac{1}{2}Y''(x), \qquad D(x_0) = 0 \tag{8.39}$$

where

$$g(x) = \left.\frac{\partial f(x,z)}{\partial z}\right|_{z=Y(x)}$$

When $D(x)$ can be obtained explicitly, the leading error term $hD(x_n)$ from the formula (8.38) usually provides a quite good estimate of the true error $Y(x_n) - y_h(x_n)$, and the quality of the estimation improves with decreasing stepsize h.

Example 8.3.3 Consider again the problem (8.36). Then $D(x)$ satisfies

$$D'(x) = -D(x) + \tfrac{1}{2}e^{-x}, \qquad D(0) = 0$$

The solution is

$$D(x) = \tfrac{1}{2}xe^{-x}$$

When we use (8.38), the error satisfies

$$Y(x_n) - y_h(x_n) \approx \frac{h}{2}x_n e^{-x_n} \tag{8.40}$$

We are neglecting the $O(h^2)$ term, since it should be smaller than the term $hD(x)$ in (8.38), for all sufficiently small values of h. To check the accuracy of (8.40), consider $x_n = 5.0$ with $h = 0.05$. Then

$$\frac{h}{2} x_n e^{-x_n} \doteq 0.000842$$

From Table 8.1, the actual error is 0.000818, which is quite close to our estimate of it. ■

8.3.2 Richardson Extrapolation

It is not practical to try to find the function $D(x)$ from the problem (8.39), since this requires knowledge of the true solution; and even if a good approximation of $Y''(x)$ is available, solving for $D(x)$ in (8.39) is costly. The real power of the formula (8.38) is that it describes precisely the error behavior. We can use (8.38) to estimate the solution error and to improve the quality of the numerical solution, without knowing the function $D(x)$. For this purpose, we will need two numerical solutions, say, y_h and y_{2h}. Assume x is a node point with the stepsize $2h$. Then by the formula (8.38), we have

$$Y(x) - y_h(x) = hD(x) + O(h^2)$$
$$Y(x) - y_{2h}(x) = 2hD(x) + O(h^2)$$

Eliminate $D(x)$ from the two relations to obtain

$$Y(x) - [2\,y_h(x) - y_{2h}(x)] = O(h^2)$$

which can also be written as

$$Y(x) - y_h(x) = y_h(x) - y_{2h}(x) + O(h^2)$$

By dropping the higher-order term $O(h^2)$, we obtain *Richardson's extrapolation formula*

$$Y(x) \approx \tilde{y}_h(x) \equiv 2y_h(x) - y_{2h}(x) \tag{8.41}$$

and *Richardson's error estimate*

$$Y(x) - y_h(x) \approx y_h(x) - y_{2h}(x) \tag{8.42}$$

With these formulas, we can estimate the error in Euler's method and can obtain a more rapidly convergent solution $\tilde{y}_h(x)$.

Table 8.3. Euler's Method with Richardson Extrapolation

x	$Y(x) - y_h(x)$	$y_h(x) - y_{2h}(x)$	$\tilde{y}_h(x)$	$Y(x) - \tilde{y}_h(x)$
1.0	9.39E − 3	9.80E − 3	3.6829346E − 1	−4.14E − 4
2.0	6.82E − 3	6.94E − 3	1.3544764E − 1	−1.12E − 4
3.0	3.72E − 3	3.68E − 3	4.9748443E − 2	3.86E − 5
4.0	1.80E − 3	1.73E − 3	1.8249877E − 2	6.58E − 5
5.0	8.17E − 4	7.67E − 4	6.6872853E − 3	5.07E − 5

Example 8.3.4 Consider (8.36) with stepsize $h = 0.05$, $2h = 0.1$. Then Table 8.3 contains Richardson's extrapolation results for selected values of h. Note that (8.42) is a fairly accurate estimator of the error, and that $\tilde{y}_h(x)$ is much more accurate than $y_h(x)$. ■

Using (8.38) with the procedure that led to (8.41), we can show that

$$Y(x_n) - \tilde{y}_h(x_n) = O(h^2) \tag{8.43}$$

an improvement on the speed of convergence of Euler's method. We will repeat this type of extrapolation for the methods introduced in later sections. However, the formulas will be different than (8.41) and (8.42), and they will depend on the order of the method.

PROBLEMS

1. Check the accuracy of the error bound (8.37) for $b = 1, 2, 3, 4, 5$ and $h = 0.2, 0.1, 0.05$. Compute the error bound and compare it with Table 8.1.

2. Consider again the problem (8.36) of Example 8.3.2. Let us derive a more accurate error bound than the one given in Theorem 8.3.1. From (8.30) we have

$$e_{n+1} = (1 - h)\, e_n + \frac{h^2}{2} e^{-\xi_n}$$

   Using this formula and recalling $e_0 = 0$, show the error bound

$$|e_n| \le \frac{x_n}{2} h$$

   Compare this error bound to the true errors in Table 8.1 of Section 8.2.

3. Compute the error bound (8.32), assuming $y_0 = Y_0$, for the problem (8.26) of Section 8.2. Compare the bound with the actual errors given in Table 8.2, for $b = 1, 2, 3, 4, 5$ and $h = 0.2, 0.1, 0.05$.

4. Repeat Problem 3 for the equation in Problem 1(a) of Section 8.2.

5. For Problems 1(b) to (d) of Section 8.2, the constant K in (8.31) will be infinite. To use the error bound (8.32) in such cases, let

$$K = 2 \cdot \max_{x_0 \le x \le b} \left| \frac{\partial f(x, Y(x))}{\partial z} \right|$$

This can be shown to be adequate for all sufficiently small values of h. Then repeat Problem 3 above for Problems 1(b) to (d) of Section 8.2.

6. Consider the initial value problem

$$Y'(x) = \alpha x^{\alpha-1}, \qquad Y(0) = 0$$

where $\alpha > 0$. The true solution is $Y(x) = x^\alpha$. When $\alpha \neq$ integer, the true solution is not infinitely differentiable. In particular, to have Y twice continuously differentiable, we need $\alpha > 2$. Use the Euler method to solve the initial value problem for $\alpha = 2.5, 1.5, 1.1$ with stepsize $h = 0.2, 0.1, 0.05$. Compute the solution errors at the nodes, and determine numerically the convergence orders of the Euler method for these problems.

7. The solution of

$$Y'(x) = \lambda Y(x) + \cos(x) - \lambda \sin(x), \qquad Y(0) = 0$$

is $Y(x) = \sin(x)$. Find the asymptotic error formula (8.38) in this case. Also compute the Euler solution for $0 \le x \le 6$, $h = 0.2, 0.1, 0.05$, and $\lambda = 1, -1$. Compare the true errors with those obtained from the asymptotic estimate

$$Y(x_n) - y_n \approx hD(x_n)$$

8. Repeat Problem 7 for Problem 1(d) of Section 8.2. Compare for $1 \le x \le 6$, $h = 0.2, 0.1, 0.05$.

9. For the example (8.26) of Section 8.2, with the numerical results in Table 8.2, use Richardson's extrapolation to estimate the error $Y(x_n) - y_h(x_n)$ when $h = 0.05$. Also, produce the Richardson extrapolate $\tilde{y}_h(x_n)$ and compute its error. Do this for $x_n = 1, 2, 3, 4, 5, 6$.

10. Repeat Problem 9 for Problems 1(a) to (d) of Section 9.2.

11. Derive (8.43) from (8.38).

8.4. NUMERICAL STABILITY, IMPLICIT METHODS

In Section 8.1, we discussed the stability property of the initial value problem (8.7), which states that a small perturbation in the initial value leads to a small change in the solution. With numerical methods for solving the initial value problem, it is desirable to have a similar stability property. This means that for any sufficiently small stepsize h, a small change in the initial value will lead to a small change in the numerical solution. Indeed, such a stability property is closely related to the convergence property. A well-known result is that under hypotheses similar to those of Theorem 8.1.3 on the initial value problem (8.7), a numerical method with truncation errors of order 2 or greater is convergent if and only if it is stable. For the Euler method, the truncation errors T_{n+1} are of order 2 (cf. (8.28)). Convergence of the method has been established in Theorem 8.3.1. Stability of the method can be shown by an argument similar to that leading to Theorem 8.3.1. For another example on the relations between convergence and stability, we refer to Problem 13 for a numerical method that is neither convergent nor stable.

The stability mentioned above states that a stable numerical method is well behaved, provided that the stepsize h is sufficiently small. In actual computations, however, the stepsize h cannot be too small since a very small stepsize decreases the efficiency of the numerical method. As is seen from the discussion of Section 5.4, the accuracy of difference approximations, such as $[Y(x+h) - Y(x)]/h$ to the derivative $Y'(x)$, deteriorates when h is too small. Hence, for actual computations, what matters is the performance of the numerical method when h is not assumed to be *very small*. We need to further analyze the stability of numerical methods when h is not assumed to be small.

Examining the stability question for the general problem

$$Y'(x) = f(x, Y(x)), \qquad Y(x_0) = Y_0 \tag{8.44}$$

is too complicated. Instead, we examine the stability of numerical methods for the *model equation*

$$Y'(x) = \lambda Y(x) + g(x), \qquad Y(0) = Y_0 \tag{8.45}$$

whose exact solution can be found from (8.4). Questions regarding stability and convergence are more easily answered for this equation; and the answers to these questions can be shown to usually be the answers to those same questions for the more general equation (8.44).

Let $Y(x)$ be the solution of (8.45), and let $Y_\epsilon(x)$ be the solution with the perturbed initial data $Y_0 + \epsilon$

$$Y_\epsilon'(x) = \lambda Y_\epsilon(x) + g(x), \qquad Y_\epsilon(0) = Y_0 + \epsilon$$

Let $Z_\epsilon(x)$ denote the change in the solution

$$Z_\epsilon(x) = Y_\epsilon(x) - Y(x)$$

Then, subtracting the equation for $Y_\epsilon(x)$ from (8.45), we get

$$Z_\epsilon'(x) = \lambda Z_\epsilon(x), \qquad Z_\epsilon(0) = \epsilon$$

The solution is

$$Z_\epsilon(x) = \epsilon e^{\lambda x}$$

Typically in applications, we are interested in the case that $\lambda < 0$ or that λ is complex with a negative real part. In such a case, $Z_\epsilon(x)$ will go to zero as $n \to \infty$ and, thus, the effect of the ϵ perturbation dies out for large values of x. We would like the same behavior to hold for the numerical method that is being applied to (8.45).

By considering the function $Z_\epsilon(x)/\epsilon$ instead of $Z_\epsilon(x)$, we obtain the following model problem that is commonly used to test the performance of various numerical methods:

$$\begin{cases} Y' = \lambda Y, & x > 0 \\ Y(0) = 1 \end{cases} \tag{8.46}$$

In the following, when we refer to the model problem (8.46), we always assume the constant $\lambda < 0$ or λ has a negative real part. The true solution of the problem (8.46) is

$$Y(x) = e^{\lambda x} \tag{8.47}$$

which decays exponentially in x since the parameter λ has a negative real part.

The kind of stability property we would like for a numerical method is that when it is applied to (8.46), the numerical solution satisfies

$$y_h(x_n) \to 0 \qquad \text{as } x_n \to \infty \tag{8.48}$$

for any choice of the stepsize h. This property is called *absolute stability*. The set of values $h\lambda$, considered as a subset of the complex plane, for which $y_n \to 0$ as $h \to 0$, is called the *region of absolute stability* of the numerical method.

Let us examine the performance of the Euler method on the model problem (8.46). We have

$$y_{n+1} = y_n + h\,\lambda\, y_n = (1 + h\,\lambda)\, y_n, \qquad n \ge 0, \quad y_0 = 1$$

By an inductive argument, it is not difficult to find

$$y_n = (1 + h\,\lambda)^n, \qquad n \ge 0 \tag{8.49}$$

Note that for a fixed grid point $x_n = n\,h \equiv \overline{x}$, as $n \to \infty$,

$$y_n = \left(1 + \frac{\lambda\,\overline{x}}{n}\right)^n \to e^{\lambda \overline{x}}$$

The limiting behavior is obtained using l'Hospital's rule from calculus. This confirms the convergence of the Euler method. We emphasize that this is an asymptotic property in the sense that it is valid in the limit $h \to 0$.

From the formula (8.49), we see that $y_n \to 0$ as $n \to \infty$ if and only if

$$|1 + h\lambda| < 1$$

For λ real and negative, the condition becomes

$$h\lambda > -2 \tag{8.50}$$

This sets a restriction on the range of h we can take to apply Euler's method: $0 < h < (-2)/\lambda$.

Example 8.4.1 Consider the model problem with $\lambda = -100$. Then the Euler method will perform well only when $h < 2 \cdot 100^{-1} = 0.02$. The true solution $Y(x) = e^{-100x}$ at $x = 0.2$ is 2.061×10^{-9}. Table 8.4 lists the Euler solution at $x = 0.2$ for several values of h. ■

8.4.1 The Backward Euler Method

Now we consider a numerical method that has the property (8.48) for any stepsize h when applied to the model problem (8.46). Such a method is said to be *absolutely stable*. In the derivation of the Euler method, we used the forward difference approximation

$$Y'(x) \approx \frac{1}{h}[Y(x+h) - Y(x)]$$

Let us use, instead, the backward difference approximation

$$Y'(x) \approx \frac{1}{h}[Y(x) - Y(x-h)]$$

Table 8.4. Euler's Solution at $x = 0.2$ for Several Values of h

h	$y_h(0.2)$
0.1	81
0.05	256
0.02	1
0.01	0
0.001	7.06E − 10

Then the differential equation $Y'(x) = f(x, Y(x))$ at $x = x_n$ is discretized as

$$y_n = y_{n-1} + h\, f(x_n, y_n)$$

Shifting the index by 1, we then obtain the backward Euler method

$$\begin{cases} y_{n+1} = y_n + h\, f(x_{n+1}, y_{n+1}), & 0 \le n \le N-1 \\ y_0 = Y_0 \end{cases} \tag{8.51}$$

Like the Euler method, the backward Euler method is of first-order accuracy, and a convergence result similar to Theorem 8.3.1 holds. Also, an asymptotic error expansion of the form (8.38) is valid.

Let us show that the backward Euler method has the desired property (8.48) on the model problem (8.46). We have

$$y_{n+1} = y_n + h\,\lambda\, y_{n+1}$$
$$y_{n+1} = (1 - h\,\lambda)^{-1} y_n, \qquad n \ge 0, \quad y_0 = 1$$

Then

$$y_n = (1 - h\,\lambda)^{-n} \tag{8.52}$$

For any stepsize $h > 0$, $|1 - h\,\lambda| > 1$ and so $y_n \to 0$ as $n \to \infty$. Continuing Example 8.4.1, we have the following table for numerical results from the backward Euler method.

A comparison between Table 8.4 and Table 8.5 reveals that the backward Euler method is substantially better than the Euler method on the model problem (8.46).

The major difference between the two methods is that for the backward Euler method, at each time step, we need to solve a nonlinear algebraic equation

$$y_{n+1} = y_n + h\, f(x_{n+1}, y_{n+1}) \tag{8.53}$$

for y_{n+1}. Methods in which y_{n+1} must be found by solving a rootfinding problem are called *implicit methods*, since y_{n+1} is defined implicitly. In contrast, methods that give y_{n+1} directly are called *explicit methods*. Euler's method is an explicit method, whereas

Table 8.5. Backward Euler's Solution at $x = 0.2$ for Several Values of h

h	$y_h(0.2)$
0.1	8.26E − 3
0.05	7.72E − 4
0.02	1.69E − 5
0.01	9.54E − 7
0.001	5.27E − 9

the backward Euler method is an implicit method. Under the Lipschitz continuity assumption of the function $f(x, y)$ with respect to y, it can be shown by applying Theorem 3.4.2 that if h is small enough, the equation (8.53) has a unique solution.

The rootfinding methods of Chapter 3 can be applied to (8.53) to find its root y_{n+1}; but usually that will be a very time-consuming process. Instead, (8.53) is usually solved by a simple iteration technique. Given an initial guess $y_{n+1}^{(0)} \approx y_{n+1}$, define $y_{n+1}^{(1)}, y_{n+1}^{(2)}$, etc., by

$$y_{n+1}^{(j+1)} = y_n + h\, f(x_{n+1}, y_{n+1}^{(j)}), \qquad j = 0, 1, 2, \ldots \tag{8.54}$$

It can be shown that if h is sufficiently small, then the iterates $y_{n+1}^{(j)}$ will converge to y_{n+1} as $j \to \infty$. Subtracting (8.54) from (8.53) gives us

$$y_{n+1} - y_{n+1}^{(j+1)} = h\,[f(x_{n+1}, y_{n+1}) - f(x_{n+1}, y_{n+1}^{(j)})]$$

$$y_{n+1} - y_{n+1}^{(j+1)} \approx h \cdot \frac{\partial f(x_{n+1}, y_{n+1})}{\partial z}[y_{n+1} - y_{n+1}^{(j)}]$$

The last formula is obtained by applying the mean value theorem to $f(x_{n+1}, z)$, considered a function of z. This formula gives a relation between the error in successive iterates. Therefore, if

$$\left| h \cdot \frac{\partial f(x_{n+1}, y_{n+1})}{\partial z} \right| < 1 \tag{8.55}$$

then the errors will converge to zero, as long as the initial guess $y_{n+1}^{(0)}$ is good enough.

The above iteration method (8.54) and its analysis are simply a special case of the theory of fixed point iteration from Section 3.4. In the notation of that earlier material, $\alpha = y_{n+1}$ is the fixed point, and

$$g(z) \equiv y_n + h\, f(x_{n+1}, z)$$

The convergence condition (8.55) is simply the earlier condition $|g'(\alpha)| < 1$ of (3.47). The remaining results of Section 3.4 can be applied to (8.53) and (8.54), but there is little benefit to doing so here.

In practice, one uses a good initial guess $y_{n+1}^{(0)}$, and one chooses an h that is so small that the quantity in (8.55) is much less than 1. Then the error $y_{n+1} - y_{n+1}^{(j)}$ decreases rapidly to a small quantity, and usually only one iterate needs to be computed. The usual choice of the initial guess $y_{n+1}^{(0)}$ for (8.54) is based on the Euler method

$$y_{n+1}^{(0)} = y_n + hf(x_n, y_n)$$

It is called a *predictor formula*, as it predicts the root of the implicit method.

For many equations, it is usually sufficient to do the iteration (8.54) once. Thus, a practical way to implement the backward Euler method is to do the following one-point iteration for solving (8.53) approximately:

$$\overline{y}_{n+1} = y_n + h\, f(x_{n+1}, y_n)$$
$$y_{n+1} = y_n + h\, f(x_{n+1}, \overline{y}_{n+1})$$

The resulting numerical method is then given by the formula

$$y_{n+1} = y_n + h\, f(x_{n+1}, y_n + h\, f(x_{n+1}, y_n)) \tag{8.56}$$

It can be shown that this method is still of first-order accurary. However, it is no longer absolutely stable (cf. Problem 1).

MATLAB PROGRAM. Now we turn to an implementation of the backward Euler method. At each step, with y_n available from the previous step, we use the Euler method to compute an estimate of y_{n+1}:

$$y_{n+1}^{(1)} = y_n + hf(x_n, y_n)$$

Then we use the trapezoidal formula to do the iteration

$$y_{n+1}^{(k+1)} = y_n + h\, f(x_{n+1}, y_{n+1}^{(k)})$$

until the difference between successive values of the iterates is sufficiently small, indicating a sufficiently accurate approximation of the solution y_{n+1}. To prevent an infinite loop of iteration, we require the iteration to stop if 10 iteration steps are taken without reaching a satisfactory solution; and in this latter case, an error message will be displayed.

```
function [x,y] = euler_back(x0,y0,x_end,h,fcn,tol)
%
% function [x,y] = euler_back(x0,y0,x_end,h,fcn,tol)
%
% Solve the initial value problem
%    y' = f(x,y),  x0 <= x <= b,  y(x0)=y0
% Use the backward Euler method with a stepsize of h.
% The user must supply an m-file to define the
% derivative f, with some name, say, 'deriv.m', and a
% first line of the form
%     function ans=deriv(x,y)
% tol is the user supplied bound on the difference
% between successive values of the trapezoidal
% iteration.  A sample call would be
%   [t,z]=euler_back(t0,z0,b,delta,'deriv',1.0e-3)
%
% Output:
```

```
% The routine euler_back will return two vectors,
% x and y. The vector x will contain the node points
%   x(1)=x0, x(j)=x0+(j-1)*h, j=1,2,...,N
% with
%   x(N) <= x_end,  x(N)+h > x_end
% The vector y will contain the estimates of the
% solution Y at the node points in x.
%

% Initialize.
n = fix((x_end-x0)/h)+1;
x = linspace(x0,x_end,n)';
y = zeros(n,1);
y(1) = y0;
i = 2;
% advancing
while i <= n
%
% forward Euler estimate
%
  yt1 = y(i-1)+h*feval(fcn,x(i-1),y(i-1));
% one-point iteration
  count = 0;
  diff = 1;
  while diff > tol & count < 10
    yt2 = y(i-1) + h*feval(fcn,x(i),yt1);
    diff = abs(yt2-yt1);
    yt1 = yt2;
    count = count +1;
  end
  if count >= 10
    disp('Not converging after 10 steps at x = ')
    fprintf('%5.2f\n', x(i))
  end
  y(i) = yt2;
  i = i+1;
end
```

8.4.2 The Trapezoidal Method

One main drawback of both the Euler method and the backward Euler method is the low convergence order. Next we present a method that has a higher convergence order, and at the same time, the stability property (8.48) is valid for any stepsize h in solving the model problem (8.46).

We integrate the differential equation

$$Y'(x) = f(x, Y(x))$$

from x_n to x_{n+1}:

$$Y(x_{n+1}) = Y(x_n) + \int_{x_n}^{x_{n+1}} f(x, Y(x))\, dx$$

and use the trapezoidal rule to approximate the integral

$$Y(x_{n+1}) \approx Y(x_n) + \frac{h}{2}\left[f(x_n, Y(x_n)) + f(x_{n+1}, Y(x_{n+1}))\right]$$

By equating both sides, we then obtain the trapezoidal method for solving the initial value problem (8.7):

$$\begin{cases} y_{n+1} = y_n + \dfrac{h}{2}\left[f(x_n, y_n) + f(x_{n+1}, y_{n+1})\right], & n \geq 0 \\ y_0 = Y_0 \end{cases} \tag{8.57}$$

It can be shown that the method is of second-order accuracy, and it is absolutely stable.

Notice that the trapezoidal method is an implicit method. In a general step, y_{n+1} is found from the equation

$$y_{n+1} = y_n + \frac{h}{2}\left[f(x_n, y_n) + f(x_{n+1}, y_{n+1})\right] \tag{8.58}$$

and it can be solved directly in only a small percentage of cases. The discussion for the solution of the backward Euler equation (8.53) applies to the solution of the equation (8.58), with a slight variation. The iteration formula (8.54) is now replaced by

$$y_{n+1}^{(j+1)} = y_n + \frac{h}{2}\left[f(x_n, y_n) + f(x_{n+1}, y_{n+1}^{(j)})\right], \qquad j = 0, 1, 2, \ldots \tag{8.59}$$

And if h is sufficiently small, then the iterates $y_{n+1}^{(j)}$ will converge to y_{n+1} as $j \to \infty$. The convergence condition (8.55) is replaced by

$$\left| \frac{h}{2} \cdot \frac{\partial f(x_{n+1}, y_{n+1})}{\partial z} \right| < 1 \tag{8.60}$$

Note that the condition (8.60) is easier to satisfy than (8.55), indicating the trapezoidal method is easier to use than the backward Euler method.

The usual choice of the initial guess $y_{n+1}^{(0)}$ for (8.59) is based on the Euler method

$$y_{n+1}^{(0)} = y_n + hf(x_n, y_n) \tag{8.61}$$

or an Adams–Bashforth method of order 2 (cf. Section 8.6)

$$y_{n+1}^{(0)} = y_n + \frac{h}{2}[3f(x_n, y_n) - f(x_{n-1}, y_{n-1})] \tag{8.62}$$

These are called *predictor formulas*. In either of the two above cases, compute $y_{n+1}^{(1)}$ from (8.59) and accept it as the root y_{n+1}. With both methods of choosing $y_{n+1}^{(0)}$, it can be shown that the global error in the resulting solution $\{y_h(x_n)\}$ is still $O(h^2)$. If the Euler predictor (8.61) is used to define $y_{n+1}^{(0)}$, and we accept $y_{n+1}^{(1)}$ as the value of y_{n+1}, then the resulting new scheme is

$$y_{n+1} = y_n + \frac{h}{2}[f(x_n, y_n) + f(x_{n+1}, y_n + h\,f(x_n, y_n))] \tag{8.63}$$

known as *Heun's method.* The Heun method is still of second-order accuracy. However, it is no longer absolutely stable.

MATLAB PROGRAM. In our implementation of the trapezoidal method, at each step, with y_n available from the previous step, we use the Euler method to compute an estimate of y_{n+1}:

$$y_{n+1}^{(1)} = y_n + hf(x_n, y_n)$$

Then we use the trapezoidal formula to do the iteration

$$y_{n+1}^{(k+1)} = y_n + \frac{h}{2}\left[f(x_n, y_n) + f(x_{n+1}, y_{n+1}^{(k)})\right]$$

until the difference between successive values of the iterates is sufficiently small, indicating a sufficiently accurate approximation of the solution y_{n+1}. To prevent an infinite loop of iteration, we require the iteration to stop if 10 iteration steps are taken without reaching a satisfactory solution; and in this latter case, an error message will be displayed.

```
function [x,y] = trapezoidal(x0,y0,x_end,h,fcn,tol)
%
% function [x,y] = trapezoidal(x0,y0,x_end,h,fcn,tol)
%
% Solve the initial value problem
%    y' = f(x,y),  x0 <= x <= b,  y(x0)=y0
% Use trapezoidal method with a stepsize of h.  The
% user must supply an m-file to define the derivative
% f, with some name, say, 'deriv.m', and a first line
% of the form
%   function ans=deriv(x,y)
```

```
% tol is the user supplied bound on the difference
% between successive values of the trapezoidal
% iteration. A sample call would be
%    [t,z]=trapezoidal(t0,z0,b,delta,'deriv',1e-3)
%
% Output:
% The routine trapezoidal will return two vectors,
% x and y. The vector x will contain the node points
%    x(1) = x0, x(j) = x0+(j-1)*h, j=1,2,...,N
% with
%    x(N) <= x_end,  x(N)+h > x_end
% The vector y will contain the estimates of the
% solution Y at the node points in x.
%

% Initialize.
n = fix((x_end-x0)/h)+1;
x = linspace(x0,x_end,n)';
y = zeros(n,1);
y(1) = y0;
i = 2;
% advancing
while i <= n
  fyt = feval(fcn,x(i-1),y(i-1));
%
% Euler estimate
%
  yt1 = y(i-1)+h*fyt;
% trapezoidal iteration
  count = 0;
  diff = 1;
  while diff > tol & count < 10
    yt2 = y(i-1) + h*(fyt+feval(fcn,x(i),yt1))/2;
    diff = abs(yt2-yt1);
    yt1 = yt2;
    count = count +1;
  end
  if count >= 10
    disp('Not converging after 10 steps at x = ')
    fprintf('%5.2f\n', x(i))
  end
  y(i) = yt2;
  i = i+1;
end
```

Table 8.6. Euler's Method for (8.64)

λ	x	Error: $h = 0.5$	Error: $h = 0.1$	Error: $h = 0.01$
-1	1	$-2.46\text{E}-1$	$-4.32\text{E}-2$	$-4.22\text{E}-3$
	2	$-2.55\text{E}-1$	$-4.64\text{E}-2$	$-4.55\text{E}-3$
	3	$-2.66\text{E}-2$	$-6.78\text{E}-3$	$-7.22\text{E}-4$
	4	$2.27\text{E}-1$	$3.91\text{E}-2$	$3.78\text{E}-3$
	5	$2.72\text{E}-1$	$4.91\text{E}-2$	$4.81\text{E}-3$
-10	1	$3.98\text{E}-1$	$-6.99\text{E}-3$	$-6.99\text{E}-4$
	2	$6.90\text{E}+0$	$-2.90\text{E}-3$	$-3.08\text{E}-4$
	3	$1.11\text{E}+2$	$3.86\text{E}-3$	$3.64\text{E}-4$
	4	$1.77\text{E}+3$	$7.07\text{E}-3$	$7.04\text{E}-4$
	5	$2.83\text{E}+4$	$3.78\text{E}-3$	$3.97\text{E}-4$
-50	1	$3.26\text{E}+0$	$1.06\text{E}+3$	$-1.39\text{E}-4$
	2	$1.88\text{E}+3$	$1.11\text{E}+9$	$-5.16\text{E}-5$
	3	$1.08\text{E}+6$	$1.17\text{E}+15$	$8.25\text{E}-5$
	4	$6.24\text{E}+8$	$1.23\text{E}+21$	$1.41\text{E}-4$
	5	$3.59\text{E}+11$	$1.28\text{E}+27$	$7.00\text{E}-5$

Example 8.4.2 Consider the problem

$$Y'(x) = \lambda Y(x) + (1 - \lambda)\cos(x) - (1 + \lambda)\sin(x), \qquad Y(0) = 1 \tag{8.64}$$

whose true solution is $Y(x) = \sin(x) + \cos(x)$. Euler's method is used for the numerical solution, and the results for several values of λ and h are given in Table 8.6. Note that according to the formula (8.28) for the truncation error,

$$T_{n+1} = \frac{h^2}{2} Y''(\xi_n)$$

The solution will not depend on λ, since $Y(x)$ does not depend on λ. But the actual global error depends strongly on λ, as illustrated in the table; and the behavior of the global error is directly linked to the size of λh and, thus, to the size of the stability region for Euler's method. The error is small, provided that λh is sufficiently small. The cases of an unstable and rapid growth in the error are exactly the cases in which λh is outside the range (8.50).

We then apply the backward Euler method and the trapezoidal method to the solution of the problem (8.64). The results are shown in Tables 8.7 and 8.8, with the stepsize $h = 0.5$. The error varies with λ, but there are no stability problems, in contrast to the Euler method. The solutions of the backward Euler method and the trapezoidal method for y_{n+1} were done exactly. This is possible because the differential equation is linear in Y. The fixed point iterations (8.54) and (8.59) do not converge when $|\lambda h|$ is large. ■

Table 8.7. Backward Euler Solution for (8.64); $h = 0.5$

x	Error: $\lambda = -1$	Error: $\lambda = -10$	Error: $\lambda = -50$
2	$2.08E-1$	$1.97E-2$	$3.60E-3$
4	$-1.63E-1$	$-3.35E-2$	$-6.94E-3$
6	$-7.04E-2$	$8.19E-3$	$2.18E-3$
8	$2.22E-1$	$2.67E-2$	$5.13E-3$
10	$-1.14E-1$	$-3.04E-2$	$-6.45E-3$

Table 8.8. Trapezoidal Solution for (8.64); $h = 0.5$

x	Error: $\lambda = -1$	Error: $\lambda = -10$	Error: $\lambda = -50$
2	$-1.13E-2$	$-2.78E-3$	$-7.91E-4$
4	$-1.43E-2$	$-8.91E-5$	$-8.91E-5$
6	$2.02E-2$	$2.77E-3$	$4.72E-4$
8	$-2.86E-3$	$-2.22E-3$	$-5.11E-4$
10	$-1.79E-2$	$-9.23E-4$	$-1.56E-4$

Equations of the form (8.45)

$$Y'(x) = \lambda Y(x) + g(x)$$

with λ negative but large in magnitude are so-called *stiff differential equations*. A stiff differential equation is characterized by the property that when it is solved by a numerical scheme without absolute stability, the stepsize h is usually forced to be excessively small in order for the scheme to generate reasonable approximate solution. See the discussions in Examples 8.4.1 and 8.4.2 for the restriction of the size of h in using the Euler method to solve the model equations (8.45) and (8.64). For stiff differential equations, one must use a numerical method that is absolutely stable or has a large region of absolute stability. Usually, implicit methods are preferred to the explicit ones in solving stiff differential equations. The MATLAB built-in function `ode23tb` is especially designed to solve stiff differential equations. Its use is similar to that of another MATLAB built-in function `ode45`, illustrated in Subsection 8.5.3.

PROBLEMS

1. Show that the method defined by the formula (8.56) is not absolutely stable.
2. Show that the trapezoidal method (8.57) is absolutely stable, but the scheme (8.63) is not.
3. Use the backward Euler's method to solve Problem 2 of Section 8.2.
4. Use the trapezoidal method to solve Problem 2 of Section 8.2.
5. Apply the backward Euler method to solve the initial value problem in Problem 6 of Section 8.3 for $\alpha = 2.5, 1.5, 1.1$, with $h = 0.2, 0.1, 0.05$. Compute the error

in the solution at the nodes, determine numerically the convergence orders, and compare the results with that of the Euler method.

6. Apply the trapezoidal method to solve the initial value problem in Problem 6 of Section 8.3 for $\alpha = 2.5, 1.5, 1.1$, with $h = 0.2, 0.1, 0.05$. Compute the error in the solution at the nodes, determine numerically the convergence orders, and compare the results with that of the Euler method and the backward Euler method.

7. Solve the equation

$$Y'(x) = \lambda Y(x) + \frac{1}{1+x^2} - \lambda \tan^{-1}(x), \qquad Y(0) = 0$$

$Y(x) = \tan^{-1}(x)$ is the true solution. Use Euler's method, the backward Euler method, and the trapezoidal method. Let $\lambda = -1, -10, -50$, and $h = 0.5, 0.1, 0.001$. Discuss the results.

8. Apply the backward Euler method to the numerical solution of $Y'(x) = \lambda Y(x) + g(x)$ with $\lambda < 0$ and large in magnitude. Investigate how small h must be chosen in order that the iteration

$$y_{n+1}^{(j+1)} = y_n + hf\left(x_{n+1}, y_{n+1}^{(j)}\right), \qquad j = 0, 1, 2, \ldots$$

will converge to y_{n+1}. Is this iteration practical for large values of $|\lambda|$?

9. Determine whether the midpoint method

$$y_{n+1} = y_n + h\, f\left(x_{n+1/2}, \frac{y_n + y_{n+1}}{2}\right)$$

where $x_{n+1/2} = (x_n + x_{n+1})/2$ is absolutely stable.

10. Let $\theta \in [0, 1]$ be a constant, and denote $x_{n+\theta} = (1-\theta)\, x_n + \theta\, x_{n+1}$. Consider the generalized midpoint method

$$y_{n+1} = y_n + h\, f\,(x_{n+\theta}, (1-\theta)\, y_n + \theta\, y_{n+1})$$

and its trapezoidal analogue

$$y_{n+1} = y_n + h\, [(1-\theta)\, f(x_n, y_n) + \theta\, f(x_{n+1}, y_{n+1})]$$

Show that the methods are absolutely stable when $\theta \in [1/2, 1]$. Determine the regions of absolute stability of the methods when $0 \le \theta < 1/2$.

11. As a special case in which the error of the backward Euler method can be analyzed directly, we consider the model problem (8.46) again, with λ an arbitrary real constant. The backward Euler solution of the problem is given by the formula (8.52). Following the procedure of doing Problem 3 (c) in Section 8.2, show that

$$Y(x_n) - y_h(x_n) = -\frac{\lambda^2 x_n e^{\lambda x_n}}{2} h + O(h^2)$$

12. As in the previous problems, consider the model problem (8.46) with a real constant λ. Show that the solution of the trapezoidal method is

$$y_h(x_n) = \left(\frac{1 + \lambda h/2}{1 - \lambda h/2}\right)^n, \qquad n \geq 0$$

Rewrite the solution formula as

$$y_h(x_n) = \exp \frac{x_n[\log(1 + \lambda h/2) - \log(1 - \lambda h/2)]}{h}$$

and use Taylor's expansions to show that

$$Y(x_n) - y_h(x_n) = -\frac{\lambda^3 x_n e^{\lambda x_n}}{12} h^2 + O(h^4)$$

So for small h, the error is almost proportional to h^2.

13. In this exercise, we consider a method with third-order truncation errors that is not convergent or stable.

(a) If $Y(x)$ is three times continuously differentiable, show that

$$\begin{aligned} Y(x_{n+1}) = 3Y(x_n) - 2Y(x_{n-1}) + \frac{h}{2}[Y'(x_n) - 3Y'(x_{n-1})] \\ + \frac{7}{12}h^3 Y'''(x_n) + O(h^4) \end{aligned} \tag{8.65}$$

Thus, a numerical method for solving the differential equation

$$Y'(x) = f(x, Y(x))$$

is

$$y_{n+1} = 3y_n - 2y_{n-1} + \frac{h}{2}[f(x_n, y_n) - 3f(x_{n-1}, y_{n-1})], \qquad n \geq 1$$

This is a numerical method whose truncation error is $O(h^3)$. It is an example of a multistep method (cf. Section 8.6). To use the method, we need a value for y_1, called an artificial initial value, in addition to the initial value $y_0 = Y_0$.

Hint: To prove (8.65), use Taylor expansions about the point x_n.

(b) Now apply the method to solve the very simple initial value problem

$$Y'(x) \equiv 0, \qquad Y(0) = 1$$

whose solution is $Y(x) \equiv 1$. Show that if the initial values are chosen to be $y_0 = 1$, $y_1 = 1 + h$, then the numerical solution is $y_n = 1 - h + h\,2^n$. Note that $|y_1 - Y(h)| = h \to 0$ as $h \to 0$. Let $x_n = 1$. Show that $|Y(1) - y_n| \to \infty$ as $h \to 0$. Thus, the method is not convergent.

(c) A slight variant of the arguments of (b) can be used to show the instability of the method. Show that with the initial values $y_0 = y_1 = 1$, the numerical solution is $y_n = 1$ for all n, whereas if the initial values are perturbed to $y_{\epsilon,0} = 1$, $y_{\epsilon,1} = 1 + \epsilon$, then the numerical solution becomes $y_{\epsilon,n} = 1 - \epsilon + \epsilon\, 2^n$. Show that at any fixed node point $x_n = \overline{x} > 0$, $|y_{\epsilon,n} - y_n| \to \infty$ as $h \to 0$. Hence, the method is unstable.

8.5. TAYLOR AND RUNGE–KUTTA METHODS

One way to improve on the speed of convergence of Euler's method is through looking for approximations to $Y(x_{n+1})$ that are more accurate than the approximation

$$Y(x_{n+1}) \approx Y(x_n) + hY'(x_n)$$

which led to Euler's method. Since this is a linear Taylor polynomial approximation, it is natural to consider higher-order Taylor approximations. Doing this will lead to a family of methods, depending on the order of the Taylor approximation being used.

To keep the initial explanations as intuitive as possible, we will develop a Taylor method for the problem

$$Y'(x) = -Y(x) + 2\cos(x), \qquad Y(0) = 1 \tag{8.66}$$

whose true solution is $Y(x) = \sin(x) + \cos(x)$. To approximate $Y(x_{n+1})$ by using information about Y at x_n, use the quadratic Taylor approximation

$$Y(x_{n+1}) \approx Y(x_n) + hY'(x_n) + \frac{h^2}{2}Y''(x_n) \tag{8.67}$$

Its truncation error is

$$T_{n+1} = \frac{h^3}{6}Y'''(\xi_n), \qquad \text{some } x_n \le \xi_n \le x_{n+1} \tag{8.68}$$

To evaluate the right side of (8.67), $Y'(x_n)$ can be obtained directly from (8.66). For $Y''(x)$, differentiate (8.66) to get

$$Y''(x) = -Y'(x) - 2\sin(x) = Y(x) - 2\cos(x) - 2\sin(x)$$

Then (8.67) becomes

$$\begin{aligned} Y(x_{n+1}) \approx{}& Y(x_n) + h[-Y(x_n) + 2\cos(x_n)] \\ &+ \frac{h^2}{2}[Y(x_n) - 2\cos(x_n) - 2\sin(x_n)] \end{aligned}$$

Table 8.9. Example of Second-Order Taylor Method (8.69)

h	x	$y_h(x)$	Error	Euler Error
0.1	2.0	0.492225829	9.25E − 4	−4.64E − 2
	4.0	−1.411659477	1.21E − 3	3.91E − 2
	6.0	0.682420081	−1.67E − 3	1.39E − 2
	8.0	0.843648978	2.09E − 4	−5.07E − 2
	10.0	−1.384588757	1.50E − 3	2.83E − 3
0.05	2.0	0.492919943	2.31E − 4	−2.30E − 2
	4.0	−1.410737402	2.91E − 4	1.92E − 2
	6.0	0.681162413	−4.08E − 4	6.97E − 3
	8.0	0.843801368	5.68E − 5	−2.50E − 2
	10.0	−1.383454154	3.62E − 4	1.39E − 2

By forcing equality, we are led to the numerical method

$$\begin{aligned} y_{n+1} &= y_n + h[-y_n + 2\cos(x_n)] \\ &\quad + \frac{h^2}{2}[y_n - 2\cos(x_n) - 2\sin(x_n)], \qquad n \geq 0 \end{aligned} \tag{8.69}$$

with $y_0 = 1$. This should approximate the solution of the problem (8.66). Because the truncation error (8.68) contains a higher power of h than was true for Euler's method [see (8.28)], it is hoped that the method (8.69) will converge more rapidly.

Table 8.9 contains numerical results for (8.69) and for Euler's method, and it is clear that (8.69) is superior. In addition, if the results for stepsizes $h = 0.1$ and 0.05 are compared, it can be seen that the errors are decreasing by a factor of about 4 when h is halved. This can be justified theoretically, as is discussed later.

In general, to solve the initial value problem

$$Y'(x) = f(x, Y(x)), \qquad x_0 \leq x \leq b, \qquad Y(x_0) = Y_0 \tag{8.70}$$

by the Taylor method, select an order of Taylor approximation and proceed as illustrated above. For order p, write

$$Y(x_{n+1}) \approx Y(x_n) + hY'(x_n) + \cdots + \frac{h^p}{p!}Y^{(p)}(x_n) \tag{8.71}$$

with the truncation error being

$$T_{n+1} = \frac{h^{p+1}}{(p+1)!}Y^{(p+1)}(\xi_n), \qquad x_n \leq \xi_n \leq x_{n+1} \tag{8.72}$$

Find $Y''(x), \ldots, Y^{(p)}(x)$ by differentiating the differential equation in (8.70) successively, obtaining formulas that implicitly involve only x_n and $Y(x_n)$. As an illustration, we have the following formulas:

$$Y''(x) = f_x + f_z f$$
$$Y^{(3)}(x) = f_{xx} + 2 f_{xz} f + f_{zz} f^2 + f_z(f_x + f_z f)$$

where

$$f_x = \frac{\partial f}{\partial x}, \qquad f_z = \frac{\partial f}{\partial z}, \qquad f_{xz} = \frac{\partial^2 f}{\partial x \partial z}$$

etc. are partial derivatives, and together with f, they are evaluated at $(x, Y(x))$. The formulas for the higher derivatives rapidly become very complicated as the differentiation order is increased.

Substitute these formulas into (8.71) and then obtain a numerical method of the form

$$y_{n+1} = y_n + hy_n' + \frac{h^2}{2} y_n'' + \cdots + \frac{h^p}{p!} y_n^{(p)} \tag{8.73}$$

by forcing (8.71) to be an equality.

If the solution $Y(x)$ and the function $f(x, z)$ are sufficiently differentiable, then it can be shown that the method (8.73) will satisfy

$$\max_{x_0 \le x_n \le b} |Y(x_n) - y_h(x_n)| \le ch^p \cdot \max_{x_0 \le x \le b} \left|Y^{(p+1)}(x)\right| \tag{8.74}$$

The constant c is similar to that appearing in the error formula (8.32) for Euler's method. In addition, there is an asymptotic error formula

$$Y(x_n) - y_h(x_n) = h^p D(x_n) + O(h^{p+1}) \tag{8.75}$$

with $D(x)$ satisfying a certain linear differential equation. The result (8.74) shows that higher-order Taylor approximations lead to an equal, higher order of convergence. The asymptotic result (8.75) justifies the use of Richardson's extrapolation to estimate the error and to accelerate the convergence.

Example 8.5.1 With $p = 2$, the formula (8.75) leads to

$$Y(x_n) - y_h(x_n) \approx \tfrac{1}{3}[y_h(x_n) - y_{2h}(x_n)] \tag{8.76}$$

Its derivation is left as Problem 3 for the reader. To illustrate the usefulness of the formula, use the entries from Table 8.9 with $x_n = 10$.

$$y_{0.1}(10) = -1.384588757$$
$$y_{0.05}(10) = -1.383454154$$

From (8.76),

$$Y(10) - y_{0.05}(10) \doteq \tfrac{1}{3}[0.001134603] \doteq 3.78\text{E} - 4$$

This is a good estimate of the true error 3.62E − 4, given in Table 8.9. ■

8.5.1 Runge–Kutta Methods

The Taylor method is conceptually easy to work with, but as we have seen, it is tedious and time-consuming to have to calculate the higher-order derivatives. To avoid the need for the higher-order derivatives, the Runge–Kutta methods evaluate $f(x, z)$ at more points, while attempting to equal the accuracy of the Taylor approximation. The methods obtained are fairly easy to program, and they are among the most popular methods for solving the initial value problem.

We begin with Runge–Kutta methods of order 2, and later we consider some higher-order methods. The Runge–Kutta methods have the general form

$$y_{n+1} = y_n + hF(x_n, y_n; h), \qquad n \geq 0, \qquad y_0 = Y_0 \tag{8.77}$$

The quantity $F(x_n, y_n; h)$ can be thought of as some kind of "average slope" of the solution on the interval $[x_n, x_{n+1}]$. But its construction is based on making (8.77) act like a Taylor method. For methods of order 2, we generally choose

$$F(x, y; h) = \gamma_1 f(x, y) + \gamma_2 f(x + \alpha h, \; y + \beta h f(x, y)) \tag{8.78}$$

and determine the constants $\{\alpha, \beta, \gamma_1, \gamma_2\}$ so that when the true solution $Y(x)$ is substituted into (8.77), the truncation error

$$T_{n+1} \equiv Y(x_{n+1}) - [Y(x_n) + hF(x_n, Y(x_n); h)] \tag{8.79}$$

will be $O(h^3)$, just as with the Taylor method of order 2.

To find the equations for the constants, we use Taylor expansions to compute the truncation error T_{n+1}. For the term $f(x + \alpha h, y + \beta h f(x, y))$, we first expand with respect to the second argument around y. Note that we need a remainder $O(h^2)$:

$$f(x + \alpha h, y + \beta h f(x, y)) = f(x + \alpha h, y) + f_z(x + \alpha h, y)\beta h f(x, y) + O(h^2)$$

We then expand the terms with respect to the x variable to obtain

$$f(x + \alpha h, y + \beta h f(x, y)) = f + f_x \alpha h + f_z \beta h f + O(h^2)$$

where the functions are all evaluated at (x, y). Also, recall that

$$Y'' = f_x + f_z f$$

Hence,

$$\begin{aligned} Y(x+h) &= Y + hY' + \frac{h^2}{2} Y'' + O(h^3) \\ &= Y + hf + \frac{h^2}{2}(f_x + f_z f) + O(h^3) \end{aligned}$$

Then

$$\begin{aligned} &Y(x+h) - [Y(x) + h\,F(x, Y(x); h)] \\ &\quad = Y + hf + \frac{h^2}{2}(f_x + f_z f) - [Y + h\gamma_1 f + \gamma_2 h\,(f + \alpha h f_x + \beta h f_z f)] + O(h^3) \\ &\quad = h\,(1 - \gamma_1 - \gamma_2)\,f + \frac{h^2}{2}[(1 - 2\gamma_2\alpha)\,f_x + (1 - 2\gamma_2\beta) f_z f] + O(h^3) \end{aligned}$$

So the coefficients must satisfy the system

$$\begin{cases} 1 - \gamma_1 - \gamma_2 = 0 \\ 1 - 2\gamma_2\alpha = 0 \\ 1 - 2\gamma_2\beta = 0 \end{cases}$$

Therefore,

$$\gamma_2 \neq 0, \qquad \gamma_1 = 1 - \gamma_2, \qquad \alpha = \beta = \frac{1}{2\gamma_2} \tag{8.80}$$

Thus, there is a family of Runge–Kutta methods of order 2, depending on the choice of γ_2. The three favorite choices are $\gamma_2 = \frac{1}{2}, \frac{3}{4}$, and 1. With $\gamma_2 = \frac{1}{2}$, we obtain the numerical method

$$y_{n+1} = y_n + \frac{h}{2}[f(x_n, y_n) + f(x_n + h, y_n + hf(x_n, y_n))], \qquad n \geq 0 \tag{8.81}$$

This is also Heun's method (8.63) discussed in Section 8.4. The number $y_n + hf(x_n, y_n)$ is the Euler solution at x_{n+1}. Using it, we obtain an approximation to the derivative at x_{n+1}, namely,

$$f(x_{n+1}, y_n + hf(x_n, y_n))$$

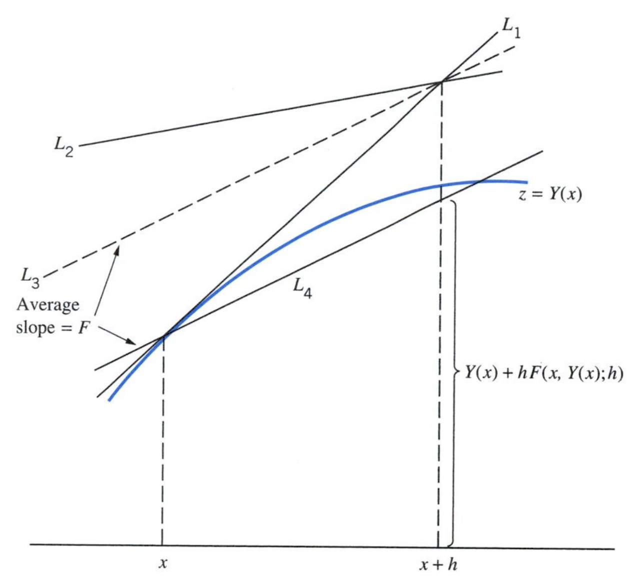

Figure 8.5. An illustration of the Runge–Kutta method (8.81); the slope of L_1 is $f(x, Y(x))$, that of L_2 is $f(x+h, Y(x)+hf(x, Y(x)))$, and those of L_3 and L_4 are the average $F(x, Y(x); h)$

This and the slope $f(x_n, y_n)$ are then averaged to give an "average" slope of the solution on the interval $[x_n, x_{n+1}]$, yielding

$$F(x_n, y_n; h) = \tfrac{1}{2}[f(x_n, y_n) + f(x_n + h, y_n + hf(x_n, y_n))]$$

This is then used to predict y_{n+1} from y_n, in (8.81). This definition is illustrated, in Figure 8.5, for $F(x, Y(x); h)$ as an average slope of Y' on $[x, x+h]$.

Example 8.5.2 Reconsider the problem (8.66)

$$Y'(x) = -Y(x) + 2\cos(x), \qquad Y(0) = 1$$

Here

$$f(x, y) = -y + 2\cos(x)$$

The numerical results from using (8.81) are given in Table 8.10. They show that the errors in this Runge–Kutta solution are comparable in accuracy to the results obtained with the Taylor method (8.69). In addition, the errors in Table 8.10 decrease by a factor of about 4 when h is halved, confirming the second-order convergence of the method. ■

Table 8.10. Example of Second Order Runge–Kutta Method

h	x	$y_h(x)$	Error
0.1	2.0	0.491215673	1.93E − 3
	4.0	−1.407898629	−2.55E − 3
	6.0	0.680696723	5.81E − 5
	8.0	0.841376339	2.48E − 3
	10.0	−1.380966579	−2.13E − 3
0.05	2.0	0.492682499	4.68E − 4
	4.0	−1.409821234	−6.25E − 4
	6.0	0.680734664	2.01E − 5
	8.0	0.843254396	6.04E − 4
	10.0	−1.382569379	−5.23E − 4

Runge–Kutta methods of higher order can also be developed. A popular classical method is the following fourth-order procedure:

$$\begin{aligned}
v_1 &= f(x_n, y_n) \\
v_2 &= f\left(x_n + \frac{h}{2}, y_n + \frac{h}{2}v_1\right) \\
v_3 &= f\left(x_n + \frac{h}{2}, y_n + \frac{h}{2}v_2\right) \\
v_4 &= f(x_n + h, y_n + hv_3) \\
y_{n+1} &= y_n + \frac{h}{6}[v_1 + 2v_2 + 2v_3 + v_4]
\end{aligned} \tag{8.82}$$

The truncation error in this method is $O(h^5)$. When the differential equation is simply

$$Y'(x) = f(x)$$

with no dependence of f on Y, this method reduces to Simpson's rule for numerical integration on $[x_n, x_{n+1}]$. The method (8.82) is easily programmed for a computer or hand calculator, and it is generally quite accurate. We leave its illustration as a problem for the reader.

If the solution $Y(x)$ of the initial value problem and the function $f(x, z)$ are sufficiently differentiable, and if the truncation error for the Runge–Kutta method is $O(h^{p+1})$, then it can be shown that the Runge–Kutta solution satisfies

$$\max_{x_0 \le x_n \le b} |Y(x_n) - y_h(x_n)| \le ch^p \tag{8.83}$$

The constant c depends on the derivatives of $Y(x)$ and the partial derivatives of $f(x, z)$. In addition, there is an asymptotic error formula of exactly the same general form (8.75) as was true for the Taylor method. Thus, Richardson's extrapolation can be justified for Runge–Kutta methods, and the error can be estimated. For the second-order method (8.81), we obtain the error estimate

$$Y(x_n) - y_h(x_n) \approx \tfrac{1}{3}[y_h(x_n) - y_{2h}(x_n)]$$

just as we obtained earlier for the second-order Taylor method; see Problem 3.

Example 8.5.3 Estimate the error for $h = 0.05$ and $x = 10$ in Table 8.10. then

$$Y(10) - y_{0.05}(10) \doteq \tfrac{1}{3}[-1.3825669379 - (-1.380966579)] \doteq -5.34\text{E} - 4$$

This compares closely with the actual error of $-5.23\text{E} - 4$. ■

8.5.2 Error Prediction and Control

The easiest way to predict the error in a numerical solution $y_h(x)$ is to use Richardson's extrapolation. Solve the initial value problem twice on the given interval $[x_0, b]$, with stepsizes $2h$ and h. Then use Richardson's extrapolation to estimate $Y(x) - y_h(x)$ in terms of $y_h(x) - y_{2h}(x)$, as was done in (8.76) for a second-order method. The cost of estimating the error in this way is an approximately 50% increase in the amount of computation, as compared with the cost of computing just $y_h(x)$. This may seem a large cost, but it is generally worth paying except for the most time-consuming of problems.

It would be desirable to have computer programs that would solve a differential equation on a given interval $[x_0, b]$ with an error less than a given error tolerance $\epsilon > 0$. Unfortunately, this is not possible with most types of numerical methods for the initial value problem. If at some point $\tilde{x}$, we discover that $Y(\tilde{x}) - y_h(\tilde{x})$ is too large, then the error cannot be made smaller by merely decreasing h from that point onward in the computation. The error $Y(\tilde{x}) - y_h(\tilde{x})$ depends on the cumulative effect of all preceding errors at points $x_n < \tilde{x}$. Thus, to decrease the error at $\tilde{x}$, it is necessary to repeat the solution of the equation from x_0, but with a smaller stepsize h. For this reason, most package programs for solving the initial value problem will not attempt to directly control the error, although they may try to monitor or bound it. Instead, they use indirect methods to influence the size of the error.

The error $Y(x_n) - y_h(x_n)$ is called the *global error* or total error at x_n. In contrast, the truncation error at x_n [see (8.72)] is called the *local error*, because it is the error introduced into the solution at step x_n. Most computer programs that contain error control are based on estimating the local error and then controlling it by varying h suitably; by doing so, they hope to keep the global error sufficiently small. If an error parameter $\epsilon > 0$ is given, the better programs choose the stepsize h to ensure that the local error T_{n+1} is much smaller, usually satisfying something like

$$|T_{n+1}| \leq \epsilon(x_{n+1} - x_n) \tag{8.84}$$

Then the global error is also kept small; for many differential equations, the global error will be less than $\epsilon(x_{n+1} - x_0)$.

To estimate the local error, various techniques can be used, including Richardson's extrapolation. Recently another technique has been devised, and it has led to the currently most popular Runge–Kutta methods. Rather than computing a method of fixed order, one simultaneously computes by using two methods of different orders. Then the higher-order formula is used to estimate the error in the lower-order formula. These methods are called *Fehlberg methods*; we will give one such pair of methods, of orders 4 and 5.

Define six intermediate slopes in $[x_n, x_{n+1}]$ by

$$\begin{aligned} v_0 &= f(x_n, y_n) \\ v_i &= f\left(x_n + a_i h,\ y_n + h\sum_{j=0}^{i-1} b_{ij} v_j\right), \qquad i = 1, 2, 3, 4, 5 \end{aligned} \tag{8.85}$$

Then the fourth- and fifth-order formulas are given by

$$y_{n+1} = y_n + h\sum_{i=0}^{4} c_i v_i \tag{8.86}$$

$$\hat{y}_{n+1} = y_n + h\sum_{i=0}^{5} d_i v_i \tag{8.87}$$

The coefficients a_i, b_{ij}, c_i, d_i are given in Tables 8.11 and 8.12.

The local error in the fourth-order formula (8.86) is estimated by

$$T_{n+1} \approx \hat{y}_{n+1} - y_{n+1} \tag{8.88}$$

Table 8.11. Fehlberg Coefficients a_i, b_{ij}

i	a_i	b_{i0}	b_{i1}	b_{i2}	b_{i3}	b_{i4}
1	1/4	1/4				
2	3/8	3/32	9/32			
3	12/13	1932/2197	−7200/2197	7296/2197		
4	1	439/216	−8	3680/513	−845/4104	
5	1/2	−8/27	2	−3544/2565	1859/4104	−11/40

Table 8.12. Fehlberg Coefficients c_i, d_i

i	0	1	2	3	4	5
c_i	25/216	0	1408/2565	2197/4104	−1/5	
d_i	16/135	0	6656/12,825	28,561/56,430	−9/50	2/55

Table 8.13. Example of Fourth-Order Fehlberg Formula (8.86)

h	x	$y_h(x)$	$Y(x) - y_h(x)$	$\hat{y}_h(x) - y_h(x)$
0.25	2.0	0.493156301	−5.71E − 6	−9.49E − 7
	4.0	−1.410449823	3.71E − 6	1.62E − 6
	6.0	0.680752304	2.48E − 6	−3.97E − 7
	8.0	0.843864007	−5.79E − 6	−1.29E − 6
	10.0	−1.383094975	2.34E − 6	1.47E − 6
0.125	2.0	0.493150889	−2.99E − 7	−2.35E − 8
	4.0	−1.410446334	2.17E − 7	4.94E − 8
	6.0	0.680754675	1.14E − 7	−1.76E − 8
	8.0	0.843858525	−3.12E − 7	−3.47E − 8
	10.0	−1.383092786	1.46E − 7	4.65E − 8

It can be shown that this is a correct asymptotic result as $h \to 0$. By using this estimate, if T_{n+1} is too small or too large, the stepsize can be varied so as to give a value of acceptable size for T_{n+1}. Note the two formulas (8.86) and (8.87) use the common intermediate slopes $v_0, \ldots, v_4$. At each step, we only need to evaluate six intermediate slopes.

Example 8.5.4 Solve

$$Y'(x) = -Y(x) + 2\cos(x), \qquad Y(0) = 1 \tag{8.89}$$

whose true solution is $Y(x) = \sin(x) + \cos(x)$. Table 8.13 contains numerical results for $h = 0.25$ and 0.125. Compare the global errors with those in Tables 8.9 and 8.10, where second-order methods are used. Also, it can be seen that the global errors in y_h decrease by factors of 17 to 21, which are fairly close to the theoretical value of 16 for a fourth-order method. The truncation errors, estimated from (8.88), are included to show that they are quite different from the global error. ■

The method (8.85) to (8.88) uses $\hat{y}_{n+1}$ only for estimating the truncation error in the fourth-order method. In practice, $\hat{y}_{n+1}$ is kept as the numerical solution rather than y_{n+1}; and thus $\hat{y}_n$ should replace y_n on the right sides of (8.85) to (8.87). The quantity in (8.88) will still be the truncation error in the fourth-order method. Programs based on this will be fifth-order, but they will vary their stepsize h to control the local error in the fourth-order method. This tends to make these programs very accurate with regard to global error.

Example 8.5.5 Repeat the last example, but use the fifth-order method described in the preceding paragraph. The results are given in Table 8.14. Note that the errors decrease by approximately 32 when h is halved, which is consistent with a fifth order method. ■

Table 8.14. Example of Fifth-Order Method (8.87)

h	x	$\hat{y}_n(x)$	$Y(x) - \hat{y}_n(x)$
0.25	2.0	0.493151148	−5.58E − 7
	4.0	−1.410446359	2.43E − 7
	6.0	0.680754463	3.26E − 7
	8.0	0.843858731	−5.18E − 7
	10.0	−1.383092745	1.05E − 7
0.125	2.0	0.493150606	−1.61E − 8
	4.0	−1.410446124	8.03E − 9
	6.0	0.680754780	8.65E − 9
	8.0	0.843858228	1.53E − 8
	10.0	−1.383092644	4.09E − 9

8.5.3 MATLAB Built-in Functions

An adaptive implementation of a method similar to the Fehlberg method based on the (4,5) pair (8.86) and (8.87) is given by the MATLAB built-in function `ode45`. The method implemented is the Dormand–Prince (4,5) pair [14]. Unlike the discussion so far in this chapter that the stepsize h is a constant, `ode45` adaptively adjusts the stepsize so that the estimated local error is within specified bounds `AbsTol` for absolute error and `RelTol` for relative error. The default values of the error tolerances are

$$\texttt{AbsTol} = 10^{-6}, \qquad \texttt{RelTol} = 10^{-3}$$

For the initial value problem (8.7), the `ode45` function can be called by

```
[x,y] = ode45('fcn',[x0 b],Y0)
```

where `fcn` is the name of an m-file that computes the right-hand-side function $f(x, z)$ of the differential equation. If $\texttt{x0} = 0$, then we can simply write

```
[x,y] = ode45('fcn',b,Y0)
```

Example 8.5.6 Consider the problem (8.89) again and use the MATLAB built-in function `ode45` to solve it. Define the right-hand-side function `z = fcn(x,y)` by

$$\texttt{z} = -\texttt{y} + 2 * \texttt{cos(x)};$$

and use

```
[x,y] = ode45('fcn',[0,10],1);
plot(x,y,'*-')
```

to generate a graph of the numerical solution. The graph is shown in Figure 8.6. To obtain solution values at $x = 2, 4, 6, 8, 10$ as in the previous example, we define

```
xvalues = 0:2:10;
```

Then

```
[x,y] = ode45('fcn',xvalues,1);
err = sin(x)+cos(x)-y;
```

will compute the numerical solution values and their errors at the specified points $x = 0, 2, 4, 6, 8, 10$. The numerical results are reported in Table 8.15.

To demand higher solution accuracy, we can assign smaller error tolerances through the use of a data structure `options` and the `odeset` function. As an example, we define

```
options = odeset('AbsTol',1e-9,'RelTol',1e-7);
```

Then use

```
[x,y] = ode45('fcn',xvalues,1,options);
err = sin(x)+cos(x)-y;
```

to compute the numerical solution values and their errors at the specified points $x = 0, 2, 4, 6, 8, 10$. The numerical results are reported in Table 8.16. ■

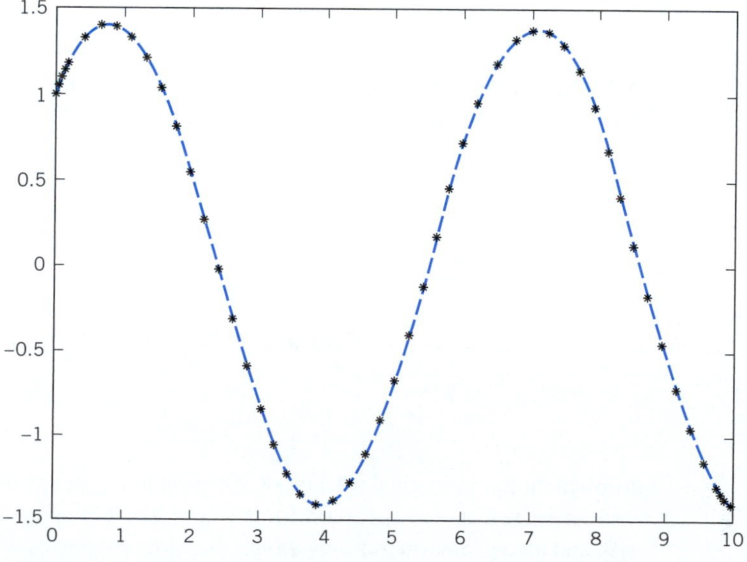

Figure 8.6. The numerical solution of (8.89) by `ode45`

Table 8.15. Example of `ode45` with Default Error Tolerances

x	$y_h(x)$	$Y(x) - y_h(x)$
2.0	0.492869823	2.81E − 4
4.0	−1.411610153	1.16E − 3
6.0	0.680895458	−1.41E − 4
8.0	0.843670835	1.87E − 4
10.0	−1.383138747	4.61E − 5

Table 8.16. Example of `ode45` with Smaller Error Tolerances

x	$y_h(x)$	$Y(x) - y_h(x)$
2.0	0.493150575	1.56E − 8
4.0	−1.410446139	2.28E − 8
6.0	0.680754810	−2.17E − 8
8.0	0.843858213	−2.8E − 10
10.0	−1.383092630	−1.01E − 8

Another MATLAB built-in function `ode23` uses the Bogacki–Shampine (2,3) pair introduced in [8]. The programs `ode45` and `ode23` are efficient for solving non-stiff ordinary differential equations. For stiff problems, the MATLAB built-in function `ode23tb` can be used, which implements an implicit Runge–Kutta-type algorithm. The use of the `ode23` and `ode23tb` functions is completely analogous to that of `ode45`, and the reader is encouraged to experiment with them.

PROBLEMS

1. A Taylor method of order 3 for problem (8.66) can be obtained using the same procedure that led to (8.69). Based on the third-order Taylor approximation

$$Y(x_{n+1}) \approx Y(x_n) + hY'(x_n) + \frac{h^2}{2}Y''(x_n) + \frac{h^3}{6}Y'''(x_n)$$

derive the numerical method

$$\begin{aligned} y_{n+1} = y_n + h[-y_n + 2\cos(x_n)] + \frac{h^2}{2}[y_n - 2\cos(x_n) - 2\sin(x_n)] \\ + \frac{h^3}{6}[-y_n + 2\sin(x_n)], \qquad n \geq 0 \end{aligned} \tag{8.90}$$

Implement the numerical method (8.90) for solving the problem (8.66). Compute with stepsizes of $h = 0.1, 0.05$ for $0 \leq x \leq 10$. Compare to the values in Table 8.9, and also check the ratio by which the error decreases when h is halved.

Hint: To make the programming easier, just modify the Euler program given in Section 8.2.

2. Compute solutions to the following problems with a second-order Taylor method. Use stepsizes $h = 0.2, 0.1, 0.05$.

(a) $Y'(x) = [\cos(Y(x))]^2, \quad 0 \le x \le 10, \quad Y(0) = 0;$
$Y(x) = \tan^{-1}(x)$

(b) $Y'(x) = \dfrac{1}{1+x^2} - 2[Y(x)]^2, \quad 0 \le x \le 10, \quad Y(0) = 0;$
$Y(x) = \dfrac{x}{1+x^2}$

(c) $Y'(x) = \frac{1}{4}Y(x)[1 - \frac{1}{20}Y(x)], \quad 0 \le x \le 20, \quad Y(0) = 1;$
$Y(x) = \dfrac{20}{1 + 19e^{-x/4}}$

(d) $Y'(x) = -[Y(x)]^2, \quad 1 \le x \le 10, \quad Y(1) = 1; \quad Y(x) = \dfrac{1}{x}$

These were solved previously in Problem 1 of Section 8.2. Compare your results with those earlier ones.

3. Recall the asymptotic error for Taylor methods, given in (8.75). For second-order methods, this yields

$$Y(x_n) - y_h(x_n) = h^2 D(x_n) + O(h^3)$$

From this, derive Richardson's extrapolation formula

$$\begin{aligned} Y(x_n) &= \tfrac{1}{3}[4y_h(x_n) - y_{2h}(x_n)] + O(h^3) \\ &\approx \tfrac{1}{3}[4y_h(x_n) - y_{2h}(x_n)] \equiv \tilde{y}_h(x_n) \end{aligned}$$

and the asymptotic error estimate

$$\begin{aligned} Y(x_n) - y_h(x_n) &= \tfrac{1}{3}[y_h(x_n) - y_{2h}(x_n)] + O(h^3) \\ &\approx \tfrac{1}{3}[y_h(x_n) - y_{2h}(x_n)] \end{aligned}$$

Hint: Consider the formula

$$Y(x_n) - y_{2h}(x_n) = 4h^2 D(x_n) + O(h^3)$$

and combine it suitably with the earlier formula for $Y(x_n) - y_h(x_n)$.

4. Repeat Problem 3 for methods of a general order $p \ge 1$. Derive the formulas

$$Y(x_n) \approx \frac{1}{2^p - 1}[2^p y_h(x_n) - y_{2h}(x_n)] \equiv \tilde{y}_h(x_n)$$

with an error proportional to h^{p+1}, and

$$Y(x_n) - y_h(x_n) \approx \frac{1}{2^p - 1}[y_h(x_n) - y_{2h}(x_n)]$$

5. Use Problem 3 to estimate the errors in the results of Table 8.9, for $h = 0.05$. Also produce the Richardson extrapolate $\tilde{y}_h(x_n)$ and calculate its error. Compare its accuracy to that of $y_h(x_n)$.

6. Derive the second-order Runge–Kutta methods (8.77) corresponding to $\gamma_2 = \frac{3}{4}$ and $\gamma_2 = 1$ in (8.78). For $\gamma_2 = 1$, draw an illustrative graph analogous to that of Figure 8.5 for $\gamma_2 = \frac{1}{2}$.

7. Solve the problem (8.66) with one of the formulas from Problem 6. Compare your results to those in Table 8.10 for the formula (8.81) with $\gamma_2 = \frac{1}{2}$.

8. Using (8.81), solve the equations in Problem 2. Estimate the error by using Problem 3, and compare it to the true error.

9. Implement the classical procedure (8.82), and apply it to the equation (8.66). Solve it with stepsizes of $h = 0.25$ and 0.125. Compare with the results in Table 8.13, the fourth-order Fehlberg example.

Hint: Modify the Euler program of Section 8.2.

10. Use the program of Problem 9 to solve the equations in Problem 2.

11. Solve the equations of Problem 2 with the built-in `ode45` function. Experiment with several choices of error tolerances.

12. Solve the equations of Problem 2 with the built-in `ode23` function. Experiment with several choices of error tolerances.

13. Consider the motion of a particle of mass in falling vertically under the earth's gravitational field, and suppose the downward motion is opposed by a frictional force $p(v)$ dependent on the velocity $v(t)$ of the particle. Then the velocity satisfies the equation

$$mv'(t) = -mg + p(v), \qquad t \geq 0, \qquad v(0) \text{ given}$$

Let $m = 1$ kg, $g = 9.8$ m/sec^2, and $v(0) = 0$. Solve the differential equation for $0 \leq t \leq 20$ and for the following choices of $p(v)$:

(a) $p(v) = -0.1v$, which is positive for a falling body.

(b) $p(v) = 0.1v^2$.

Find answers to at least three digits of accuracy. Graph the functions $v(t)$. Compare the solutions.

14. **(a)** Using the Runge–Kutta method (8.81), solve

$$Y'(x) = -Y(x) + x^{0.1}[1.1 + x], \qquad Y(0) = 0$$

whose solution is $Y(x) = x^{1.1}$. Solve the equation on $[0, 5]$, printing the solution and the errors at $x = 1, 2, 3, 4, 5$. Use stepsizes $h = 0.1, 0.05, 0.025, 0.0125, 0.00625$. Calculate the ratios by which the errors decrease when h is halved. How does this compare with the theoretical rate of convergence of $O(h^2)$. Explain your results as best as you can.

(b) What difficulty arises in attempting to use a Taylor method of order ≥ 2 to solve the equation of part (a)? What does it tell us about the solution?

8.6. MULTISTEP METHODS

The Taylor method and the Runge–Kutta methods are known as *single-step methods*, since at a typical step, y_{n+1} is solely determined from y_n. In this section, we consider multistep methods where to compute the numerical solution y_{n+1}, we use the solution values at several previous nodes. We present here the derivation of two families of multistep methods.

Reformulate the differential equation

$$Y'(x) = f(x, Y(x))$$

by integrating it over the interval $[x_n, x_{n+1}]$, obtaining

$$\int_{x_n}^{x_{n+1}} Y'(x)\,dx = \int_{x_n}^{x_{n+1}} f(x, Y(x))\,dx$$

$$Y(x_{n+1}) = Y(x_n) + \int_{x_n}^{x_{n+1}} f(x, Y(x))\,dx \tag{8.91}$$

We are going to develop numerical methods to find the solution $Y(x)$ by approximating the integral in (8.91). There are many such methods, and we will consider only the most popular of them: the Adams–Bashforth (AB) and Adams–Moulton (AM) methods. These methods are the basis of some of the most widely used computer codes for solving the initial value problem. They are generally more efficient than the Runge–Kutta methods, especially if one wishes to find the solution with a high degree of accuracy or if the derivative $f(x, z)$ is expensive to evaluate.

To evaluate the integral

$$\int_{x_n}^{x_{n+1}} g(x)\,dx, \qquad g(x) = f(x, Y(x)) \tag{8.92}$$

we approximate $g(x)$ by using polynomial interpolation and then integrate the interpolating polynomial. For a given nonnegative integer q, the AB methods use an interpolation polynomial of degree q on the set of points $\{x_n, x_{n-1}, \ldots, x_{n-q}\}$, and AM methods use an interpolation polynomial of degree q on the set of points $\{x_{n+1}, x_n, x_{n-1}, \ldots, x_{n-q+1}\}$. We first consider AB methods, beginning with the AB method based on linear interpolation.

The linear polynomial interpolating $g(x)$ at $\{x_n, x_{n-1}\}$ is

$$p_1(x) = \frac{1}{h}[(x_n - x)g(x_{n-1}) + (x - x_{n-1})g(x_n)] \tag{8.93}$$

Integrating over $[x_n, x_{n+1}]$, we obtain

$$\int_{x_n}^{x_{n+1}} g(x)\,dx \approx \int_{x_n}^{x_{n+1}} p_1(x)\,dx = \frac{3h}{2}g(x_n) - \frac{h}{2}g(x_{n-1})$$

By using ideas similar to those used in Sections 5.1 and 5.2 of Chapter 5, one can obtain the more complete result

$$\int_{x_n}^{x_{n+1}} g(x)\,dx = \frac{h}{2}[3g(x_n) - g(x_{n-1})] + \frac{5}{12}h^3 g''(\xi_n) \tag{8.94}$$

for some $x_{n-1} \le \xi_n \le x_{n+1}$; also see Problem 3. Applying this to the relation (8.91) gives us

$$Y(x_{n+1}) = Y(x_n) + \frac{h}{2}[3f(x_n, Y(x_n)) - f(x_{n-1}, Y(x_{n-1}))] + \frac{5}{12}h^3 Y'''(\xi_n)$$

Dropping the final term, the truncation error, we obtain the numerical method

$$y_{n+1} = y_n + \frac{h}{2}[3f(x_n, y_n) - f(x_{n-1}, y_{n-1})] \tag{8.95}$$

With this method, note that it is necessary to have $n \ge 1$. Both y_0 and y_1 are needed in finding y_2; and y_1 cannot be found from (8.95). The value of y_1 must be obtained by another method. The method (8.95) is an example of a two-step method, since values at x_{n-1} and x_n are needed in finding the value at x_{n+1}. If we can determine y_1 with an accuracy $O(h^2)$ (e.g., via Euler's method), the AB method (8.95) is of order 2, that is, its global error can be bounded by $O(h^2)$, since the truncation error is bounded by $O(h^3)$.

Example 8.6.1 Use (8.95) to solve

$$Y'(x) = -Y(x) + 2\cos(x), \qquad Y(0) = 1 \tag{8.96}$$

with the solution $Y(x) = \sin(x) + \cos(x)$. For illustrative purposes only, we take $y_1 = Y(x_1)$. The numerical results are given in Table 8.17, using $h = 0.05$. Note that the errors decrease by a factor of about 4 when h is halved, which is consistent with the numerical method being of order 2. The Richardson error estimate is also included in the

Table 8.17. An Example of the Second-Order Adams–Bashforth Method

x	$y_h(x)$	$Y(x) - y_{2h}(x)$	$Y(x) - y_h(x)$	Ratio	$\frac{1}{3}[y_h(x) - y_{2h}(x)]$
2	0.49259722	2.13E − 3	5.53E − 4	3.9	5.26E − 4
4	−1.41116963	2.98E − 3	7.24E − 4	4.1	7.52E − 4
6	−0.67500371	−3.91E − 3	−9.88E − 4	4.0	−9.73E − 4
8	0.84373678	3.68E − 4	1.21E − 4	3.0	8.21E − 5
10	−1.38398254	3.61E − 3	8.90E − 4	4.1	9.08E − 4

table, using the estimate in (8.76) for second-order methods. Where the error decreases like $O(h^2)$, the error estimate is quite accurate. ■

AB methods are often considered to be less "expensive" than Runge–Kutta methods, and the main reason can be seen by comparing (8.95) with the second-order Runge–Kutta method in (8.81). The main work of both methods lies in the evaluations of the function $f(x, z)$. With second-order Runge–Kutta methods, there are two evaluations of f for each step from x_n to x_{n+1}. In contrast, the AB formula (8.95) uses only one evaluation per step, provided that past values of f are reused. There are other factors that affect the choice of a numerical method, but the AB and AM methods are generally more efficient in the number of evaluations of f that are needed for a given amount of accuracy.

A problem with multistep methods is having to generate some of the initial values of the solution by using another method. For the second-order AB method in (8.95), we must obtain y_1. And since the global error in $y_h(x_n)$ is to be $O(h^2)$, we must ensure that $Y(x_1) - y_h(x_1)$ is also $O(h^2)$. There are two immediate possibilities, using methods from preceding sections.

Case (i) Use Euler's method:

$$y_1 = y_0 + hf(x_0, y_0) \tag{8.97}$$

Assuming $y_0 = Y_0$, we determine this has an error of

$$Y(x_1) - y_1 = \frac{h^2}{2} Y''(\xi_1)$$

based on (8.27). Thus, (8.97) meets our error criteria for y_1. Globally, Euler's method is only $O(h)$ accurate, but the error of a single step is $O(h^2)$.

Case (ii) Use a second-order Runge–Kutta method, such as (8.81). Since only one step in x is being used, $Y(x_1) - y_1$ will be $O(h^3)$, which is more than adequate.

Example 8.6.2 Use (8.97) with (8.95) to solve the problem (8.96) from the last example. For $h = 0.05$ and $x = 10$, the error in the numerical solution turns out to be

$$Y(10) - y_h(10) \doteq 8.90\text{E} - 4$$

almost the same as before for the results in Table 8.17. ■

Higher-order Adams–Bashforth methods are obtained by using higher-degree polynomial interpolation in the approximation of the integrand in (8.92). For example, let $p_2(x)$ denote the quadratic polynomial that interpolates $g(x)$ at x_n, x_{n-1}, x_{n-2}. Then use

$$\int_{x_n}^{x_{n+1}} g(x)\,dx \approx \int_{x_n}^{x_{n+1}} p_2(x)\,dx$$

Integrating this and including its error term (derived by other means) give us

$$\int_{x_n}^{x_{n+1}} g(x)\,dx = \frac{h}{12}[23g(x_n) - 16g(x_{n-1}) + 5g(x_{n-2})] + \frac{3}{8}h^4 g'''(\xi_n)$$

for some $x_{n-2} \le \xi_n \le x_{n+1}$. Applying this to (8.91), the integral formulation of the differential equation, we obtain

$$\begin{aligned} Y(x_{n+1}) = Y(x_n) &+ \frac{h}{12}[23f(x_n, Y(x_n)) - 16f(x_{n-1}, Y(x_{n-1})) \\ &+ 5f(x_{n-2}, Y(x_{n-2}))] + \frac{3}{8}h^4 Y^{(4)}(\xi_n) \end{aligned}$$

By dropping the last term, the truncation error, we obtain the third-order AB method

$$y_{n+1} = y_n + \frac{h}{12}[23y_n' - 16y_{n-1}' + 5y_{n-2}'], \qquad n \ge 2 \tag{8.98}$$

where $y_k' \equiv f(x_k, y_k)$, $k \ge 0$. This is a three-step method, requiring $n \ge 2$. Thus, y_1, y_2 must be obtained separately by other methods. We leave the illustration of (8.98) as Problem 2 for the reader.

In general, it can be shown that the AB method based on interpolation of degree q will be a $(q + 1)$-step method and that its truncation error will be of the form

$$T_{n+1} = c_q h^{q+2} Y^{(q+2)}(\xi_n)$$

for some $x_{n-q} \le \xi_n \le x_{n+1}$. The initial values $y_1, \dots, y_q$ will have to be generated by other methods. If the errors in these initial values satisfy

$$Y(x_n) - y_h(x_n) = O(h^{q+1}), \qquad n = 1, 2, \dots, q \tag{8.99}$$

Table 8.18. Adams–Bashforth Methods

q	Order	Method	Truncation Error
0	1	$y_{n+1} = y_n + hy_n'$	$\frac{1}{2}h^2 Y''(\xi_n)$
1	2	$y_{n+1} = y_n + \frac{h}{2}[3y_n' - y_{n-1}']$	$\frac{5}{12}h^3 Y'''(\xi_n)$
2	3	$y_{n+1} = y_n + \frac{h}{12}[23y_n' - 16y_{n-1}' + 5y_{n-2}']$	$\frac{3}{8}h^4 Y^{(4)}(\xi_n)$
3	4	$y_{n+1} = y_n + \frac{h}{24}[55y_n' - 59y_{n-1}' + 37y_{n-2}' - 9y_{n-3}']$	$\frac{251}{720}h^5 Y^{(5)}(\xi_n)$

then the global error in the $(q+1)$-step AB method will also be $O(h^{q+1})$. In addition, the global error will satisfy an asymptotic error formula

$$Y(x_n) - y_h(x_n) = D(x_n)h^{q+1} + O(h^{q+2})$$

much as was true earlier for the Taylor and Runge–Kutta methods described in Section 8.5. Thus, Richardson's extrapolation can be used to accelerate the convergence of the method and to estimate the error.

To generate the initial values $y_1, \ldots, y_q$ for the $(q+1)$-step AB method, and to have their errors satisfy the requirement (8.99), it is sufficient to use a Runge–Kutta method of order q. However, in many instances, people prefer to use a Runge–Kutta method of order $q+1$, the same order as that of the $(q+1)$-step AB method. There are other procedures used in the automatic computer programs for Adams methods, but we will not discuss them here.

The AB methods of orders 1 to 4 are given in Table 8.18. The order 1 formula is simply Euler's method. In the table, $y_k' \equiv f(x_k, y_k)$.

Example 8.6.3 Solve the problem (8.96) by using the fourth-order AB method. Generate the initial values y_1, y_2, y_3 by using the true solution

$$y_i = Y(x_i), \qquad i = 1, 2, 3$$

The results for $h = 0.125$ and $2h = 0.25$ are given in Table 8.19. Richardson's error estimate for a fourth-order method is given in the last column. For a fourth-order method, the error should decrease by a factor of about 16 when h is halved. In those cases where this is true, Richardson's error estimate is accurate. In no case is the error badly underestimated. ■

Comparing these results with those in Table 8.13 for the fourth-order Fehlberg method, we see that the present errors appear to be very large. But note that the Fehlberg formula uses five evaluations of $f(x, z)$ for each step of x_n to x_{n+1}; and the fourth-order AB method uses only one evaluation of f per step, assuming past evaluations are reused. If this AB method is used with an h that is only $1/5$ as large (in order to have a comparable

Table 8.19. Example of Fourth-Order Adams–Bashforth Method

x	$y_h(x)$	$Y(x) - y_{2h}(x)$	$Y(x) - y_h(x)$	Ratio	$\frac{1}{15}[y_h(x) - y_{2h}(x)]$
2	0.49318680	−3.96E − 4	−3.62E − 5	10.9	−2.25E − 5
4	−1.41037698	−1.25E − 3	−6.91E − 5	18.1	−7.37E − 5
6	0.68067962	1.05E − 3	7.52E − 5	14.0	6.12E − 5
8	0.84385416	3.26E − 4	4.06E − 6	80.0	2.01E − 5
10	−1.38301376	−1.33E − 3	−7.89E − 5	16.9	−7.82E − 5

number of evaluations of f), then the present errors will decrease by a factor of about $5^4 = 625$. Then the AB errors will be mostly smaller than those of the Fehlberg method in Table 8.13, and the work will be comparable (measured by the number of evaluations of f).

As with the AB methods, we begin our presentation of AM methods by considering the method based on linear interpolation. Let $p_1(x)$ be the linear polynomial that interpolates $g(x)$ at x_n and x_{n+1}

$$p_1(x) = \frac{1}{h}[(x_{n+1} - x)g(x_n) + (x - x_n)g(x_{n+1})]$$

Using it to approximate the integral in (8.92), we have

$$\int_{x_n}^{x_{n+1}} g(x)\,dx \approx \int_{x_n}^{x_{n+1}} p_1(x)\,dx = \frac{h}{2}[g(x_n) + g(x_{n+1})]$$

which is just the simple trapezoidal rule of Section 5.1 in Chapter 5. Applying this to the integral formulation (8.91), and including the error term (5.26) for the trapezoidal rule, we obtain

$$Y(x_{n+1}) = Y(x_n) + \frac{h}{2}[f(x_n, Y(x_n)) + f(x_{n+1}, Y(x_{n+1}))] - \frac{h^3}{12}Y'''(\xi_n) \tag{8.100}$$

Dropping the last term, the truncation error, we obtain the AM method:

$$y_{n+1} = y_n + \frac{h}{2}[f(x_n, y_n) + f(x_{n+1}, y_{n+1})], \qquad n \ge 0 \tag{8.101}$$

We recognize that this is the trapezoidal method studied in Section 8.4. It is a second-order method, that is, its global error is $O(h^2)$.

Example 8.6.4 Solve the earlier example (8.96) by using the AM method (8.101) (the trapezoidal method). The results are given in Table 8.20 for $h = 0.05$, $2h = 0.1$. Richardson's error estimate for second-order methods is given in the last column of the table. In this case, $O(h^2)$ error behavior is very apparent, and error estimation is very accurate. ■

Table 8.20. Example of Adams–Moulton Method of Order 2

x	$Y(x) - y_{2h}(x)$	$Y(x) - y_h(x)$	Ratio	$\frac{1}{3}[y_h(x) - y_{2h}(x)]$
2	−4.59E − 4	−1.15E − 4	4.0	−1.15E − 4
4	−5.61E − 4	−1.40E − 4	4.0	−1.40E − 4
6	7.98E − 4	2.00E − 4	4.0	2.00E − 4
8	−1.21E − 4	−3.04E − 5	4.0	−3.03E − 4
10	−7.00E − 4	−1.75E − 4	4.0	−1.25E − 4

Example 8.6.5 Repeat the last example, but using the procedure described in Section 8.4, with only one iterate being computed for each n. Then, the errors do not change by very much from those given in Table 8.17. For example, with $x = 10$ and $h = 0.05$, the error is

$$Y(10) - y_h(10) \doteq -2.02\text{E} - 4$$

This is not very different from the value of −1.75E − 4 given in Table 8.17. The use of the iterate $y_{n+1}^{(1)}$ as the root y_{n+1} will not affect significantly the solution of most differential equations. ■

By integrating the polynomial of degree q that interpolates at the nodes $\{x_{n+1}, x_n, \ldots, x_{n-q+1}\}$ to the function $g(x)$ of (8.92), we obtain the AM method of order $q + 1$. It will be an implicit method, but in other respects the theory is the same as for the AB methods described previously. The AM methods of orders 1 through 4 are given in Table 8.21. In the table, $y_k' \equiv f(x_k, y_k)$. Notice that the AM method of order 1 is the backward Euler method, and the AM method of order 2 is the trapezoidal method.

The effective cost of an AM method is two evaluations of the derivative $f(x, z)$ per step, assuming past function values are reused. This includes one evaluation of f for the predictor, and then one evaluation of f in the iteration formula for the AM method, as in (8.59). When this is taken into consideration, then, there is no significant gain in accuracy over the AB methods of the same order, for equal costs. Nonetheless, there are other properties of the AM methods that make them desirable to use for many

Table 8.21. Adams–Moulton Methods

q	Order	Method	Truncation Error
0	1	$y_{n+1} = y_n + hy_{n+1}'$	$-\frac{1}{2}h^2Y''(\xi_n)$
1	2	$y_{n+1} = y_n + \frac{h}{2}[y_{n+1}' + y_n']$	$-\frac{1}{12}h^3Y'''(\xi_n)$
2	3	$y_{n+1} = y_n + \frac{h}{12}[5y_{n+1}' + 8y_n' - y_{n-1}']$	$-\frac{1}{24}h^4Y^{(4)}(\xi_n)$
3	4	$y_{n+1} = y_n + \frac{h}{24}[9y_{n+1}' + 19y_n' - 5y_{n-1}' + y_{n-2}']$	$-\frac{19}{720}h^5Y^{(5)}(\xi_n)$

types of differential equations. The desirable features relate to stability characteristics of numerical methods. As we have seen before, the AM methods of orders 1 and 2 are absolutely stable, and thus they are particularly suitable for solving stiff differential equations.

The larger a region of absolute stability, the less restrictive the condition on h to have numerical stability of the type just discussed. Thus, a method with a large region of absolute stability is generally preferred over a method with a smaller region, provided that the accuracy of the two methods is similar. It can be shown that for AB and AM methods of equal order, the AM method will have the larger region of absolute stability. Consequently, Adams–Moulton methods are generally preferred over Adams–Bashforth methods.

Example 8.6.6 The AB method of order 2 has $-1 < h\lambda < 0$ as the real part of its region of stability; in contrast, the AM method of order 2 has $-\infty < h\lambda < 0$ as the real part of its region of stability. There is no stability restriction on h with this AM method. ■

Some of the most popular computer codes for solving the initial value problem are based on using AM and AB methods in combination, as suggested above. These codes control the truncation error by varying the stepsize h and by varying the order of the method. The possible order is allowed to be as large as 12 or more; and this results in a very efficient numerical method when the solution $Y(x)$ has several continuous derivatives and is slowly varying. A thorough discussion of Adams's methods and an example of one such computer code are given in Shampine (1994) [31].

Built-in MATLAB programs based on the multistep methods are `ode113` and `ode15s`. These programs implement explicit and implicit linear multistep methods of various orders, respectively. The program `ode113` is used to solve nonstiff ordinary differential equations, whereas `ode15s` is for stiff ordinary differential equations. The programs are used in precisely the same manner as the program `ode45` discussed in the previous section.

MATLAB PROGRAM. To aid in programming the methods of this section, we present a modification of the Euler program of Section 8.2. The program implements the Adams-Bashforth formula of order 2, given in (8.95); and it uses Euler's method to generate the first value y_1 as in (8.97).

```
function [x,y] = AB2(x0,y0,x_end,h,fcn)
%
% function [x,y]=AB2(x0,y0,x_end,h,fcn)
%
% Solve the initial value problem
%    y' = f(x,y),  x0 <= x <= b,  y(x0)=y0
% Use Adams-Bashforth formula of order 2 with
% a stepsize of h. Euler's method is used for
```

```
% the value y1.  The user must supply a program
% with some name, say, deriv, and a first  line of
% the form
%   function ans=deriv(x,y)
% A sample call would be
%   [t,z]=AB2(t0,z0,b,delta,'deriv')
%
% Output:
% The routine AB2 will return two vectors, x and y.
% The vector x will contain the node points
%   x(1)=x0, x(j)=x0+(j-1)*h, j=1,2,...,N
% with
%   x(N) <= x_end-h,  x(N)+h > x_end-h
% The vector y will contain the estimates of the
% solution Y at the node points in x.
%
n = fix((x_end-x0)/h)+1;
x = linspace(x0,x_end,n)';
y = zeros(n,1);
y(1) = y0;
ft1 = feval(fcn,x(1),y(1));
y(2) = y(1)+h*ft1;
for i = 3:n
  ft2 = feval(fcn,x(i-1),y(i-1));
  y(i) = y(i-1)+h*(3*ft2-ft1)/2;
  ft1 = ft2;
end
```

PROBLEMS

1. Use the MATLAB program for the AB method of order 2 to solve the equations in Problem 2 of Section 8.5. Include the Richardson error estimate for $y_h(x)$ when $h = 0.1$ and 0.05.

2. Modify the MATLAB program of this section to use the third-order AB method. For y_1 and y_2, use one of the second-order Runge–Kutta methods from Section 8.5. Then repeat Problem 1. Also solve the continuing example problem (8.96).

3. To make the error term in (8.94) a bit more believable, prove

$$\int_0^h g(x)\,dx - \frac{h}{2}[3g(0) - g(-h)] = \frac{5}{12}h^3 g''(0) + O(h^4)$$

Hint: Expand $g(x)$ as a quadratic Taylor polynomial about the origin, with an error term $R_3(x)$. Substitute that into the left side of the above equation, and obtain the right side.

4. Modify the MATLAB program of this section to use the AM method of order 2. For the predictor, use the AB method of order 2; for the first step x_1, use the Euler predictor. Iterate the formula (8.59) only once. Apply this to the solution of the equations considered in Problem 1, and produce Richardson's error estimate.

5. Check the differential equation programs available from your computer center, and see whether there is a multistep Adams's code. If so, use it to solve the equations in Problem 1 above. As error tolerances, use $\epsilon = 10^{-4}$ and $\epsilon = 10^{-8}$. Keep track of the number of evaluations of $f(x, z)$ that are used by the routine, and compare it to the number used in your own programs for Adams's methods.

6. **(a)** Using the program of Problem 1 for the AB method of order 2, solve

$$Y'(x) = -50Y(x) + 51\cos(x) + 49\sin(x), \qquad Y(0) = 1$$

for $0 \leq x \leq 10$. The solution is $Y(x) = \sin(x) + \cos(x)$. Use stepsizes of $h = 0.1, 0.02, 0.01$. In each case, print the errors as well as the answers.

(b) Using the program of Problem 4 for the AM method of order 2, repeat part (a). Check the condition of (8.60).

(c) When the AM method of order 2 is applied to the equation in (a), the value of y_{n+1} can be found directly. Doing so, repeat part (a). Compare your results.

8.7. SYSTEMS OF DIFFERENTIAL EQUATIONS

Although some applications of differential equations involve only a single first-order equation, most applications involve a system of several such equations or higher-order equations. In this section, we consider numerical solution of systems of first-order equations, showing how the methods of earlier sections apply to such systems. Numerical treatment of higher-order equations can be carried out by first converting them to equivalent systems of first-order equations.

To begin with a simple case, the general form of a system of two first-order differential equations is

$$\begin{cases} Y_1'(x) = f_1(x, Y_1(x), Y_2(x)) \\ Y_2'(x) = f_2(x, Y_1(x), Y_2(x)) \end{cases} \tag{8.102}$$

The functions $f_1(x, z_1, z_2)$ and $f_2(x, z_1, z_2)$ define the differential equations, and the unknown functions $Y_1(x)$ and $Y_2(x)$ are being sought. The initial value problem consists of solving (8.102), subject to the initial conditions

$$Y_1(x_0) = Y_{1,0}, \qquad Y_2(x_0) = Y_{2,0} \tag{8.103}$$

Example 8.7.1 (a) The initial value problem

$$\begin{aligned} Y_1' &= Y_1 - 2Y_2 + 4\cos(x) - 2\sin(x), \qquad & Y_1(0) &= 1 \\ Y_2' &= 3Y_1 - 4Y_2 + 5\cos(x) - 5\sin(x), \qquad & Y_2(0) &= 2 \end{aligned} \tag{8.104}$$

has the solution

$$Y_1(x) = \cos(x) + \sin(x), \qquad Y_2(x) = 2\cos(x)$$

This example will be used later in a numerical example of Euler's method for systems.

(b) Consider the system

$$\begin{aligned} Y_1' &= AY_1[1 - BY_2], \qquad & Y_1(0) &= Y_{1,0} \\ Y_2' &= CY_2[DY_1 - 1], \qquad & Y_2(0) &= Y_{2,0} \end{aligned} \tag{8.105}$$

with $A, B, C, D > 0$. This is called the *Lotka–Volterra predator-prey model*. The variable x denotes time, $Y_1(x)$ the number of prey (e.g., rabbits) at time x, and $Y_2(x)$ the number of predators (e.g., foxes). If there is only a single type of predator and a single type of prey, then this model is often a good approximation of reality. The behavior of the solutions Y_1 and Y_2 is illustrated in Problem 9. ■

The initial value problem for a system of m first-order differential equations has the general form

$$\begin{aligned} Y_1' &= f_1(x, Y_1, \dots, Y_m), \qquad & Y_1(x_0) &= Y_{1,0} \\ &\vdots \\ Y_m' &= f_m(x, Y_1, \dots, Y_m), \qquad & Y_m(x_0) &= Y_{m,0} \end{aligned} \tag{8.106}$$

We seek the functions $Y_1(x), \dots, Y_m(x)$ on some interval $x_0 \le x \le b$. An example of a system of three equations is given below in (8.120).

The general form (8.106) is clumsy to work with, and it is not a convenient way to specify the system when using a computer program for its solution. To simplify the form of (8.106), represent the solution and the differential equations by using column vectors. Denote

$$\mathbf{Y}(x) = \begin{bmatrix} Y_1(x) \\ \vdots \\ Y_m(x) \end{bmatrix}, \qquad \mathbf{Y}_0 = \begin{bmatrix} Y_{1,0} \\ \vdots \\ Y_{m,0} \end{bmatrix}, \qquad \mathbf{f}(x, \mathbf{z}) = \begin{bmatrix} f_1(x, z_1, \dots, z_m) \\ \vdots \\ f_m(x, z_1, \dots, z_m) \end{bmatrix} \tag{8.107}$$

with $\mathbf{z} = [z_1, z_2, \dots, z_m]^T$. Then (8.106) can be rewritten as

$$\mathbf{Y}'(x) = \mathbf{f}(x, \mathbf{Y}(x)), \qquad \mathbf{Y}(x_0) = \mathbf{Y}_0 \tag{8.108}$$

This looks like the earlier first-order single equation, but it is general as to the number of equations. Computer programs for solving systems will almost always refer to the system in this manner.

Example 8.7.2 System (8.104) can be rewritten as

$$\boldsymbol{Y}'(x) = A\boldsymbol{Y}(x) + \boldsymbol{G}(x), \qquad \boldsymbol{Y}(0) = \boldsymbol{Y}_0$$

with

$$\boldsymbol{Y} = \begin{bmatrix} Y_1 \\ Y_2 \end{bmatrix}, \qquad A = \begin{bmatrix} 1 & -2 \\ 3 & -4 \end{bmatrix},$$

$$\boldsymbol{G}(x) = \begin{bmatrix} 4\cos(x) - 2\sin(x) \\ 5\cos(x) - 5\sin(x) \end{bmatrix}, \qquad \boldsymbol{Y}_0 = \begin{bmatrix} 1 \\ 2 \end{bmatrix}$$

In the notation of (8.107),

$$\boldsymbol{f}(x, \boldsymbol{z}) = A\boldsymbol{z} + \boldsymbol{G}(x), \qquad \boldsymbol{z} = [z_1, z_2]^T \quad \blacksquare$$

8.7.1 Higher-Order Differential Equations

In physics and engineering, the use of *Newton's second law of motion* leads to systems of second-order differential equations, modeling some of the most important physical phenomena of nature. In addition, other applications lead to other higher-order equations. Higher-order equations can be either studied directly, or studied through equivalent systems of first-order equations.

As an example, consider the second-order equation

$$Y''(x) = f(x, Y(x), Y'(x)) \tag{8.109}$$

where $f(x, z_1, z_2)$ is given. The initial value problem consists of solving (8.109) subject to the initial conditions

$$Y(x_0) = Y_0, \qquad Y'(x_0) = Y_0' \tag{8.110}$$

To reformulate this as a system of first-order equations, denote

$$Y_1(x) = Y(x), \qquad Y_2(x) = Y'(x)$$

Then Y_1 and Y_2 satisfy

$$\begin{aligned} Y_1' &= Y_2, & Y_1(x_0) &= Y_0 \\ Y_2' &= f(x, Y_1, Y_2), & Y_2(x_0) &= Y_0' \end{aligned} \tag{8.111}$$

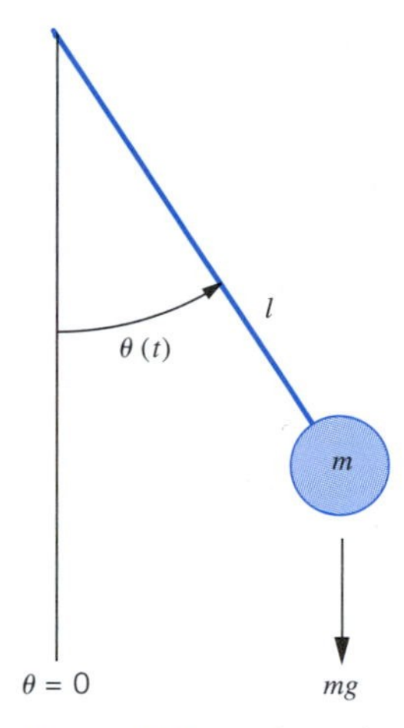

Figure 8.7. The schematic of pendulum

And if we start from this system, showing that the solution Y_1 of (8.111) will also have to satisfy (8.109) and (8.110) is straightforward, thus demonstrating the equivalence of the two formulations.

Example 8.7.3 Consider the pendulum shown in Figure 8.7, of mass m and length l. The motion of this pendulum about its center line $\theta = 0$ is modeled by a second-order differential equation derived from Newton's second law of motion. If the pendulum is assumed to move back and forth with negligible friction at its vertex, then the motion is modeled fairly accurately by the equation

$$ml\frac{d^2\theta}{dx^2} = -mg\sin(\theta(x)) \tag{8.112}$$

with x being time and $\theta(x)$ the angle between the vertical center line and the pendulum. The description of the motion is completed by specifying the initial position $\theta(0)$ and initial angular velocity $\theta'(0)$.

To convert this to a system of two first-order equations, write

$$Y_1(x) = \theta(x), \qquad Y_2(x) = \theta'(x)$$

Then (8.112) becomes

$$\begin{aligned} Y_1'(x) &= Y_2(x), & Y_1(0) &= \theta(0) \\ Y_2'(x) &= -\frac{g}{l}\sin(Y_1(x)), & Y_2(0) &= \theta'(0) \end{aligned} \tag{8.113}$$

This is equivalent to the original second-order equation. ■

A general differential equation of order m can be written as

$$\frac{d^m Y}{dx^m} = f\left(x, Y, \frac{dY}{dx}, \ldots, \frac{d^{m-1}Y}{dx^{m-1}}\right) \tag{8.114}$$

and the initial conditions needed in solving it are given by

$$Y(x_0) = y_0, \qquad Y'(x_0) = Y_0', \qquad \ldots, \qquad Y^{(m-1)}(x_0) = Y_0^{(m-1)} \tag{8.115}$$

It is reformulated as a system of m first-order equations by introducing

$$Y_1 = Y, \qquad Y_2 = Y', \qquad \ldots, \qquad Y_m = Y^{(m-1)}$$

Then the equivalent initial value problem for a system of first-order equations is

$$\begin{aligned} Y_1' &= Y_2 & Y_1(x_0) &= Y_0 \\ &\vdots & &\vdots \\ Y_{m-1}' &= Y_m & Y_{m-1}(x_0) &= Y_0^{(m-2)} \\ Y_m' &= f(x, Y_1, \ldots, Y_m) & Y_m(x_0) &= Y_0^{(m-1)} \end{aligned} \tag{8.116}$$

A special case of (8.114) is the order m linear differential equation

$$\frac{d^m Y}{dx^m} = a_0(x)Y + a_1(x)\frac{dY}{dx} + \cdots + a_{m-1}(x)\frac{d^{m-1}Y}{dx^{m-1}} + b(x) \tag{8.117}$$

This is reformulated as above, with

$$Y_m' = a_0(x)Y_1 + a_1(x)Y_2 + \cdots + a_{m-1}(x)Y_m + b(x) \tag{8.118}$$

replacing the last equation in (8.116).

Example 8.7.4 The initial value problem

$$\left\{\begin{array}{l} Y''' + 3Y'' + 3Y' + Y = -4\sin(x), \\ Y(0) = Y'(0) = 1, \quad Y''(0) = -1 \end{array}\right. \tag{8.119}$$

is reformulated as

$$\begin{aligned} Y_1' &= Y_2 & Y_1(0) &= 1 \\ Y_2' &= Y_3 & Y_2(0) &= 1 \\ Y_3' &= -Y_1 - 3Y_2 - 3Y_3 - 4\sin(x) & Y_3(0) &= -1 \end{aligned} \tag{8.120}$$

The solution of (8.119) is $Y(x) = \cos(x) + \sin(x)$, and the solution of (8.120) can be generated from it. This system will be solved numerically later in this section. ■

8.7.2 Numerical Methods for Systems

The numerical methods of earlier sections can be applied without change to the solution of systems of first-order differential equations. The numerical method should be applied to each equation in the system. Or more simply, apply the method in a straightforward way to the system written in the matrix form (8.108). The derivation of these numerical methods for the solution of systems is essentially the same as was done previously for a single equation. And the convergence and stability analyses are also done in the same manner.

To be more specific, we consider Euler's method for the general system of two first-order equations that is given in (8.102). By following the derivation given for Euler's method in obtaining (8.27), Taylor's theorem gives

$$Y_1(x_{n+1}) = Y_1(x_n) + hf_1(x_n, Y_1(x_n), Y_2(x_n)) + \frac{h^2}{2}Y_1''(\xi_n)$$
$$Y_2(x_{n+1}) = Y_2(x_n) + hf_2(x_n, Y_1(x_n), Y_2(x_n)) + \frac{h^2}{2}Y_2''(\zeta_n)$$

for some ξ_n, ζ_n in $[x_n, x_{n+1}]$. Dropping the error terms, we obtain Euler's method for a system of two equations for $n \geq 0$:

$$\begin{cases} y_{1,n+1} = y_{1,n} + hf_1(x_n, y_{1,n}, y_{2,n}) \\ y_{2,n+1} = y_{2,n} + hf_2(x_n, y_{1,n}, y_{2,n}) \end{cases} \tag{8.121}$$

If $Y_1(x)$, $Y_2(x)$ are twice continuously differentiable, then it can be shown that

$$\left|Y_1(x_n) - y_{1,n}\right| \leq ch, \qquad \left|Y_2(x_n) - y_{2,n}\right| \leq ch$$

for all $x_0 \leq x_n \leq b$, for some constant c. In addition, the earlier asymptotic error formula (8.38) will still be valid; for $j = 1, 2$

$$Y_j(x_n) - y_{j,n} = D_j(x_n)h + O(h^2), \qquad x_0 \leq x_n \leq b$$

Thus, Richardson's extrapolation and error estimation formulas will still be valid. The functions $D_1(x)$, $D_2(x)$ satisfy a particular linear system of differential equations, but we omit it here. Stability results for Euler's method generalize without any significant change, including the region of stability. Thus in summary, the earlier work for Euler's method generalizes without significant change to systems. The same is true of the other numerical methods given earlier, thus justifying our limitation to a single equation for introducing those methods.

MATLAB PROGRAM. The following is a MATLAB code `eulersys` implementing the Euler method to solve the initial value problem (8.108). One evident advantage of MATLAB over *Fortran* is that the program is valid for both a single equation and a system of *any* number of equations. Indeed, it can be seen that the code `eulersys` is just a slight modification of the code `euler_for` for solving a single equation in Section 8.2. The program can automatically determine the number of equations of the system.

```
function [x,y] = eulersys(x0,y0,x_end,h,fcn)
%
% function [x,y]=eulersys(x0,y0,x_end,h,fcn)
%
% Solve the initial value problem of a system
% of first-order equations
%    y' = f(x,y),  x0 <= x <= b,  y(x0)=y0
% Use Euler's method with a stepsize of h.
% The user must supply a program to compute the
% right-hand-side function with some name, say,
% deriv, and a first line of the form
%   function ans=deriv(x,y)
% A sample call would be
%   [t,z]=eulersys(t0,z0,b,delta,'deriv')
%
% The program automatically determines the
% number of equations from the dimension of
% the initial value vector y0.
%
% Output:
% The routine eulersys will return a vector x
% and matrix y. The vector x will contain the
% node points in [x0,x_end]:
%   x(1)=x0, x(j)=x0+(j-1)*h, j=1,2,...,N
% The matrix y is of size N by m, with m the
% number of equations.  The ith row y(i,:) will
% contain the estimates of the solution Y
% at the node points in x(i).
%
m = length(y0);
n = fix((x_end-x0)/h)+1;
x = linspace(x0,x_end,n)';
y = zeros(n,m);
y(1,:) = y0;
for i = 2:n
  y(i,:) = y(i-1,:) + h*feval(fcn,x(i-1),y(i-1,:));
end
```

Example 8.7.5 (a) Solve (8.104) using Euler's method. The numerical results are given in Table 8.22, along with Richardson's error estimate

$$Y_j(x_n) - y_{j,h}(x_n) \approx y_{j,h}(x_n) - y_{j,2h}(x_n), \qquad j = 1, 2$$

Table 8.22. Solution of (8.104) Using Euler's Method

j	x	$Y_j(x)$	$Y_j(x) - y_{j,2h}(x)$	$Y_j(x) - y_{j,h}(x)$	Ratio	$y_{j,h}(x) - y_{j,2h}(x)$
1	2	0.49315	−5.65E − 2	−2.82E − 2	2.0	−2.83E − 2
	4	−1.41045	−5.64E − 3	−2.72E − 3	2.1	−2.92E − 3
	6	0.68075	4.81E − 2	2.36E − 2	2.0	2.44E − 2
	8	0.84386	−3.60E − 2	−1.79E − 2	2.0	−1.83E − 2
	10	−1.38309	−1.81E − 2	−8.87E − 3	2.0	−9.40E − 2
2	2	−0.83229	−3.36E − 2	−1.70E − 2	2.0	−1.66E − 2
	4	−1.30729	5.94E − 3	3.19E − 3	1.9	2.75E − 3
	6	1.92034	5.19E − 2	2.49E − 2	2.1	2.70E − 2
	8	−0.29100	−2.08E − 2	−1.05E − 2	2.0	−1.03E − 2
	10	−1.67814	8.39E − 2	4.14E − 2	1.3	4.25E − 2

In the table, $h = 0.05$, $2h = 0.1$. It can be seen that this error estimate is quite accurate, except for the one case $j = 2$, $x = 10$. To get the numerical solution values and their errors at the specified node points $x = 2, 4, 6, 8, 10$, we used the following MATLAB commands, which can be included at the end of the program `eulersys` for this example.

```
n1 = (n-1)/5;
for i = n1+1:n1:n
  e(i,1) = cos(x(i))+sin(x(i))-y(i,1);
  e(i,2) = 2*cos(x(i))-y(i,2);
end
diary euler_sys1
fprintf(' h = 6.5f\n', h)
disp(' x y(1) e(1) y(2) e(2)')
for i = n1+1:n1:n
  fprintf('2.0f%10.2e%10.2e%10.2e%10.2e\n', ...
      x(i), y(i,1),e(i,1),y(i,2),e(i,2))
end
diary off
```

The right-hand-side function for this example is defined by the following:

```
function z = eulersys_fcn(x,y);
z = zeros(1,2);
z(1) = y(1)-2*y(2)+4*cos(x)-2*sin(x);
z(2) = 3*y(1)-4*y(2)+5*cos(x)-5*sin(x);
```

Table 8.23. Solution of (8.119) Using Euler's Method

x	$y(x)$	$y(x) - y_{2h}(x)$	$y(x) - y_h(x)$	Ratio	$y_h(x) - y_{2h}(x)$
2	0.49315	−8.78E − 2	−4.25E − 2	2.1	−4.53E − 2
4	−1.41045	1.39E − 1	6.86E − 2	2.0	7.05E − 2
6	0.68075	5.19E − 2	2.49E − 2	2.1	2.70E − 2
8	0.84386	−1.56E − 1	−7.56E − 2	2.1	−7.99E − 2
10	−1.38309	8.39E − 2	4.14E − 2	2.0	4.25E − 2

(b) Solve the third-order equation in (8.119), by using Euler's method to solve the reformulated problem (8.120). The results for $y(x) = Y_1(x) = \sin(x) + \cos(x)$ are given in Table 8.23, for stepsizes $2h = 0.1$ and $h = 0.05$. The Richardson error estimate is again quite accurate. ■

Other numerical methods apply to systems in the same straightforward manner. And by using the matrix form (8.108) for a system, there is no apparent change in the numerical method. For example, the Runge–Kutta method (8.81) is

$$y_{n+1} = y_n + \frac{h}{2}[f(x_n, y_n) + f(x_{n+1}, y_n + hf(x_n, y_n))], \qquad n \geq 0 \tag{8.122}$$

And if we interpret this for a system of two equations with

$$\mathbf{y}_n = \begin{bmatrix} y_{1,n} \\ y_{2,n} \end{bmatrix}, \qquad \mathbf{f}(x_n, \mathbf{y}_n) = \begin{bmatrix} f_1(x_n, y_{1,n}, y_{2,n}) \\ f_2(x_n, y_{1,n}, y_{2,n}) \end{bmatrix}$$

the method is

$$\begin{aligned} y_{j,n+1} = y_{j,n} + \frac{h}{2}[&f_j(x_n, y_{1,n}, y_{2,n}) \\ &+ f_j(x_{n+1}, y_{1,n} + hf_1(x_n, y_{1,n}, y_{2,n}), y_{2,n} + hf_2(x_n, y_{1,n}, y_{2,n}))] \end{aligned} \tag{8.123}$$

for $j = 1, 2$. It is easier to consider this in the form (8.122); for programming it on a computer, the matrix form is very convenient. We leave the illustration of (8.123) to the problems section below.

PROBLEMS

1. Let

$$A = \begin{bmatrix} 1 & -2 \\ 2 & -1 \end{bmatrix}, \qquad \mathbf{Y} = \begin{bmatrix} Y_1 \\ Y_2 \end{bmatrix}$$

$$\mathbf{G}(x) = \begin{bmatrix} -2e^{-x} + 2 \\ -2e^{-x} + 1 \end{bmatrix}, \qquad \mathbf{Y}_0 = \begin{bmatrix} 1 \\ 1 \end{bmatrix}$$

Write out the two equations that make up the system

$$Y' = AY + G(x), \qquad Y(x_0) = Y_0$$

The true solution is $Y = [e^{-x}, 1]^T$.

2. Convert the system (8.120) to the general form of Problem 1, giving the matrix A.

3. Convert the following higher-order equations to systems of first-order equations:

 (a) $Y''' + 4Y'' + 5Y' + 2Y = 2x^2 + 10x + 8$,
 $Y(0) = 1, \qquad Y'(0) = -1, \qquad Y''(0) = 3$
 The true solution is $Y(x) = e^{-x} + x^2$.

 (b) $Y'' + 4Y' + 13Y = 40\cos(x), \qquad Y(0) = 3, \qquad Y'(0) = 4$
 The true solution is $Y(x) = 3\cos(x) + \sin(x) + e^{-2x}\sin(3x)$.

4. Convert the following system of second-order equations to a larger system of first-order equations. It arises from studying the gravitational attraction of one mass by another.

$$x''(t) = \frac{-cx(t)}{r(t)^3}, \qquad y''(t) = \frac{-cy(t)}{r(t)^3}, \qquad z''(t) = \frac{-cz(t)}{r(t)^3}$$

with c positive constant and $r(t) = [x(t)^2 + y(t)^2 + z(t)^2]^{1/2}$, t = time.

5. Using Euler's method, solve the system in Problem 1. Use stepsizes of $h = 0.1, 0.05, 0.025$, and solve for $0 \le x \le 10$. Use Richardson's error formula to estimate the error for $h = 0.025$.

6. Repeat Problem 5 using the backward Euler method.

7. Repeat Problem 5 for the systems in Problem 3.

8. Modify the Euler program of this section to implement the Runge–Kutta method given in (8.123). With this program, repeat Problems 5 and 7.

9. Consider the predator-prey model of (8.105), with the particular constants $A = 4$, $B = 0.5$, $C = 3$, $D = \frac{1}{3}$.

 (a) Show that there is a solution $Y_1(x) = C_1$, $Y_2(x) = C_2$, with C_1 and C_2 constants. What would be the physical interpretation of such a solution $Y(x)$?

 (b) Solve this system (8.105) with $Y_1(0) = 3$, $Y_2(0) = 5$, for $0 \le x \le 4$, and use the Runge–Kutta method of Problem 8 with stepsizes of $h = 0.01$ and 0.005. Examine and plot the values of the output in steps of x of 0.1. In addition to these plots of $Y_1(x)$ versus x and $Y_2(x)$ versus x, also plot Y_1 versus Y_2.

 (c) Repeat (b) for the initial values $Y_1(0) = 3$, $Y_2(0) = 1, 1.5, 1.9$ in succession. Comment on the relation of these solutions to one another and to the solution of part (a).

10. Consider solving the pendulum equation (8.112) with $l = 1$ and $g = 32.2$ ft/sec^2. For the initial values, choose $0 < \theta(0) \leq \pi/2$, $\theta'(0) = 0$. Use Euler's method to solve (8.113), and experiment with various values of h so as to obtain a suitably small error in the computed solution. Graph t versus $\theta(t)$, t versus $\theta'(t)$, and $\theta(t)$ versus $\theta'(t)$. Does the motion appear to be periodic in time?

8.8. FINITE DIFFERENCE METHOD FOR TWO-POINT BOUNDARY VALUE PROBLEMS

In the previous section, we have seen that the initial value problem of the second-order equation

$$Y'' = f(x, Y, Y') \tag{8.124}$$

can be reformulated as an initial value problem of a system of first-order equations. In this section, we consider the solution of another type of problems for the second-order equation (8.124), where conditions on the solution Y are given at two distinct x values. Such problems are called *two-point boundary value problems*. For simplicity, our discussion in this section will be focused on the following boundary value problem of a second-order *linear* equation:

$$\begin{cases} Y''(x) = p(x)\,Y'(x) + q(x)\,Y(x) + r(x), & a \leq x \leq b \\ Y(a) = g_1, \quad Y(b) = g_2 \end{cases} \tag{8.125}$$

The conditions $Y(a) = g_1$ and $Y(b) = g_2$ are called the *boundary conditions*. Boundary conditions involving the derivative of the unknown function are also common in applications, and we will comment on the finite difference approximations of such boundary conditions later in the section.

We assume the given functions p, q, and r are continuous on $[a, b]$. A standard theoretical result states that if $q(x) > 0$ for $x \in [a, b]$, then the boundary value problem (8.125) has a unique solution. We will assume the problem has a unique smooth solution Y.

The main feature of the finite difference method is to obtain discrete equations by replacing derivatives with appropriate finite divided differences. We derive a finite difference system for the boundary value problem (8.125) in three steps.

In the first step, we discretize the domain of the problem: the interval $[a, b]$. Let N be a positive integer, and divide the interval $[a, b]$ into N equal parts:

$$[a, b] = [x_0, x_1] \cup [x_1, x_2] \cup \cdots \cup [x_{N-1}, x_N]$$

where $a = x_0 < x_1 < \cdots < x_{N-1} < x_N = b$ are the grid (or node) points. Denote $h = (b - a)/N$, called the *stepsize*. Then the node points are given by

$$x_i = a + i\,h, \qquad 0 \leq i \leq N$$

Nonuniform partition of the interval is also possible, and is in fact preferred if the solution of the boundary value problem (8.125) changes much more rapidly in some part of the interval than the remaining part. We restrict ourselves to the case of uniform partitions for simplicity of exposition. We use the notation $p_i = p(x_i)$, $q_i = q(x_i)$, $r_i = r(x_i)$, $0 \le i \le N$, and denote y_i, $0 \le i \le N$, to be numerical approximations of the true solution values $Y_i = Y(x_i)$, $0 \le i \le N$.

In the second step, we discretize the differential equation at the interior node points $x_1, \dots, x_{N-1}$. For this purpose, let us recall the following difference approximation formulas (cf. (5.85), (5.92)):

$$Y'(x_i) = \frac{Y_{i+1} - Y_{i-1}}{2h} + O(h^2) \tag{8.126}$$

$$Y''(x_i) = \frac{Y_{i+1} - 2Y_i + Y_{i-1}}{h^2} + O(h^2) \tag{8.127}$$

Then the differential equation at $x = x_i$ becomes

$$\frac{Y_{i+1} - 2Y_i + Y_{i-1}}{h^2} = p_i \frac{Y_{i+1} - Y_{i-1}}{2h} + q_i Y_i + r_i + O(h^2) \tag{8.128}$$

Dropping the remainder term $O(h^2)$ and replacing Y_i by y_i, we obtain the difference equations

$$\frac{y_{i+1} - 2y_i + y_{i-1}}{h^2} = p_i \frac{y_{i+1} - y_{i-1}}{2h} + q_i y_i + r_i, \qquad 1 \le i \le N-1$$

which can be rewritten as

$$-\left(1 + \frac{h}{2} p_i\right) y_{i-1} + (2 + h^2 q_i) y_i + \left(\frac{h}{2} p_i - 1\right) y_{i+1} = -h^2 r_i, \qquad 1 \le i \le N-1 \tag{8.129}$$

The third step is devoted to the treatment of the boundary conditions. The difference equations (8.129) consist of $(N-1)$ equations for $(N+1)$ unknowns $y_0, y_1, \dots, y_N$. We need two more equations and they come from discretization of the boundary conditions. For the model problem (8.125), the discretization of the boundary conditions is straightforward:

$$y_0 = g_1, \qquad y_N = g_2 \tag{8.130}$$

The equations (8.129) and (8.130) together form a linear system. Since the values of y_0 and y_N are explicitly given in (8.130), we can eliminate y_0 and y_N from the linear system. With $y_0 = g_1$, we can rewrite the equation in (8.129) with $i = 1$ as

$$(2 + h^2 q_1)\, y_1 + \left(\frac{h}{2} p_1 - 1\right) y_2 = -h^2 r_1 + \left(1 + \frac{h}{2} p_1\right) g_1 \tag{8.131}$$

Similarly, from the equation in (8.128) with $i = N - 1$, we obtain

$$-\left(1+\frac{h}{2}p_{N-1}\right)y_{N-2} + (2+h^2q_{N-1})\,y_{N-1} = -h^2r_{N-1} + \left(1-\frac{h}{2}p_{N-1}\right)g_2 \tag{8.132}$$

So finally, the finite difference system for the unknown numerical solution vector $\boldsymbol{y} = [y_1, \ldots, y_{N-1}]^T$ is

$$A\boldsymbol{y} = \boldsymbol{b} \tag{8.133}$$

where

$$A = \begin{bmatrix} 2+h^2q_1 & \frac{h}{2}p_1 - 1 & & & \\ -\left(1+\frac{h}{2}p_2\right) & 2+h^2q_2 & \frac{h}{2}p_2 - 1 & & \\ & \ddots & \ddots & \ddots & \\ & & -\left(1+\frac{h}{2}p_{N-2}\right) & 2+h^2q_{N-2} & \frac{h}{2}p_{N-2} - 1 \\ & & & -\left(1+\frac{h}{2}p_{N-1}\right) & 2+h^2q_{N-1} \end{bmatrix} \tag{8.134}$$

is the coefficient matrix and

$$\boldsymbol{b} = \left[-h^2r_1 + \left(1+\frac{h}{2}p_1\right)g_1, -h^2r_2, \ldots, -h^2r_{N-2}, -h^2r_{N-1} + \left(1-\frac{h}{2}p_{N-1}\right)g_2\right]^T \tag{8.135}$$

is the right-hand-side vector.

It can be shown that if the true solution $Y(x)$ is sufficiently smooth, say, it has continuous derivatives up to order 4, then the difference scheme (8.133) is a second-order method, that is,

$$\max_{0\le i\le N} |Y(x_i) - y_i| = O(h^2)$$

Moreover, the following asymptotic error expansion holds:

$$Y(x_i) - y_h(x_i) = h^2 D(x_i) + O(h^4), \qquad 0 \le i \le N \tag{8.136}$$

for some function $D(x)$ independent of h. The Richardson extrapolation formula for this case is

$$\tilde{y}_h(x_i) = \frac{4\,y_h(x_i) - y_{2h}(x_i)}{3} \tag{8.137}$$

and we have

$$Y(x_i) - \tilde{y}_h(x_i) = O(h^4) \tag{8.138}$$

The linear system (8.133) is tridiagonal. So it is natural to use the algorithm developed in Section 6.4 to solve the system (8.133).

MATLAB PROGRAM. The following MATLAB code implements the difference method (8.133) for solving the problem (8.125):

```
function z = ODEBVP(p,q,r,a,b,ga,gb,N)
% A program to solve the two-point boundary
% value problem
%   y"=p(x)y'+q(x)y+r(x),  a<x<b
%   y(a)=ga,  y(b)=gb
% Input
%   p, q, r: coefficient functions
%   a, b: the end-points of the interval
%   ga, gb: the prescribed function values
%           at the end-points
%   N: number of subintervals
% Output
%   z = [ xx yy ]: xx is an (N+1) column vector
%                   of the node points
%                 yy is an (N+1) column vector of
%                   the solution values
% A sample call would be
%   z=ODEBVP('p','q','r',a,b,ga,gb,100)
% The user must provide m-files to define the
% functions p, q and r.
%
% Other MATLAB program called: tridiag.m
%
% Initialization
N1 = N+1;
h = (b-a)/N;
h2 = h*h;
xx = linspace(a,b,N1)';
yy = zeros(N1,1);
yy(1) = ga;
yy(N1) = gb;
% Define the subdiagonal avec, main diagonal bvec,
% superdiagonal cvec
pp(2:N) = feval(p,xx(2:N));
avec(2:N-1) = -1-(h/2)*pp(3:N);
bvec(1:N-1) = 2+h2*feval(q,xx(2:N));
cvec(1:N-2) = -1+(h/2)*pp(2:N-1);
% Define the right-hand-side vector fvec
fvec(1:N-1) = -h2*feval(r,xx(2:N));
```

```
fvec(1) = fvec(1)+(1+h*pp(2)/2)*ga;
fvec(N-1) = fvec(N-1)+(1-h*pp(N)/2)*gb;
% Solve the tridiagonal system
yy(2:N) = tridiag(avec,bvec,cvec,fvec,N-1,0);
z = [xx'; yy']';
```

Example 8.8.1 Consider the boundary value problem

$$\begin{cases} Y'' = -\dfrac{2x}{1+x^2}Y' + Y + \dfrac{2}{1+x^2} - \log(1+x^2), & 0 \le x \le 1 \\ Y(0) = 0, \quad Y(1) = \log(2) \end{cases} \tag{8.139}$$

The true solution is $Y(x) = \log(1+x^2)$.

In Table 8.24, we report the finite difference solution errors $Y - y_h$ at selected node points for several values of h. In Table 8.25, we report the errors of the extrapolated solutions $Y - (4\,y_h - y_{2h})/3$ at the same node points and the associated ratios of the errors for different stepsizes. The column marked Ratio next to the column of the solution errors for a stepsize h consists of the ratios of the solution errors for the stepsize h with those for the stepsize $2h$. We clearly observe an error reduction of a factor of around 4 when the stepsize is halved, indicating a second-order convergence of the method. There is a dramatic improvement in the solution accuracy through extrapolation. The extrapolated solution $\tilde{y}_h$ with $h = 1/40$ is much more accurate than the solution y_h with $h = 1/160$. Note that the cost of obtaining $\tilde{y}_h$ with $h = 1/40$ is substantially smaller than that for y_h with $h = 1/160$.

Also observe that for the extrapolated solution $\tilde{y}_h$, the error decreases by approximately a factor of 16 when h is halved. Indeed, it can be shown that if the true solution $Y(x)$ is six times continuously differentiable, then we can improve the asymptotic error

Table 8.24. Numerical Errors $Y(x) - y_h(x)$ for Solving the Problem (8.139)

x	$h = 1/20$	$h = 1/40$	Ratio	$h = 1/80$	Ratio	$h = 1/160$	Ratio
0.1	5.10E − 5	1.27E − 5	4.00	3.18E − 6	4.00	7.96E − 7	4.00
0.2	7.84E − 5	1.96E − 5	4.00	4.90E − 6	4.00	1.22E − 6	4.00
0.3	8.64E − 5	2.16E − 5	4.00	5.40E − 6	4.00	1.35E − 6	4.00
0.4	8.08E − 5	2.02E − 5	4.00	5.05E − 6	4.00	1.26E − 6	4.00
0.5	6.73E − 5	1.68E − 5	4.00	4.21E − 6	4.00	1.05E − 6	4.00
0.6	5.08E − 5	1.27E − 5	4.00	3.17E − 6	4.00	7.94E − 7	4.00
0.7	3.44E − 5	8.60E − 6	4.00	2.15E − 6	4.00	5.38E − 7	4.00
0.8	2.00E − 5	5.01E − 6	4.00	1.25E − 6	4.00	3.13E − 7	4.00
0.9	8.50E − 6	2.13E − 6	4.00	5.32E − 7	4.00	1.33E − 7	4.00

Table 8.25. Extrapolation Errors for Solving the Problem (8.139)

x	$h = 1/40$	$h = 1/80$	Ratio	$h = 1/160$	Ratio
0.1	−9.23E − 09	−5.76E − 10	16.01	−3.60E − 11	16.00
0.2	−1.04E − 08	−6.53E − 10	15.99	−4.08E − 11	15.99
0.3	−6.60E − 09	−4.14E − 10	15.96	−2.59E − 11	15.98
0.4	−1.18E − 09	−7.57E − 11	15.64	−4.78E − 12	15.85
0.5	3.31E − 09	2.05E − 10	16.14	1.28E − 11	16.06
0.6	5.76E − 09	3.59E − 10	16.07	2.24E − 11	16.04
0.7	6.12E − 09	3.81E − 10	16.04	2.38E − 11	16.03
0.8	4.88E − 09	3.04E − 10	16.03	1.90E − 11	16.03
0.9	2.67E − 09	1.67E − 10	16.02	1.04E − 11	16.03

expansion (8.136) to

$$Y(x_i) - y_h(x_i) = h^2 D_1(x_i) + h^4 D_2(x_i) + O(h^6)$$

Then (8.137) is replaced by

$$Y(x_i) - \tilde{y}_h(x_i) = -4\,h^4 D_2(x_i) + O(h^6)$$

Therefore, we can perform an extrapolation procedure also on $\tilde{y}_h$ to get an even more accurate numerical solution through the formula

$$Y(x_i) - \frac{16\,\tilde{y}_h(x_i) - \tilde{y}_{2h}(x_i)}{15} = O(h^6)$$

As an example, at $x_i = 0.5$, with $h = 1/80$, the further extrapolated solution has an error approximately equal to $-1.88\text{E}{-12}$. ■

So far, our discussions have focused on the solution of the particular boundary value problem (8.125). Let us make some remarks on the various extensions.

In principle, difference schemes for solving the more general equation (8.124) can be derived similarly. For example, if we use the formulas (8.126) and (8.127), the equation (8.124) at an interior node point x_i can be approximated by the difference equation

$$\frac{y_{i+1} - 2\,y_i + y_{i-1}}{h^2} = f\left(x_i, y_i, \frac{y_{i+1} - y_{i-1}}{2\,h}\right)$$

The treatment of boundary conditions involving the derivative of the unknown is a little bit more involved. Assume the boundary condition at $x = b$ is replaced by

$$Y'(b) + k\,Y(b) = g_2 \tag{8.140}$$

One obvious possibility is to approximate $Y'(b)$ by $(Y_N - Y_{N-1})/h$. However,

$$Y'(b) - \frac{Y_N - Y_{N-1}}{h} = O(h)$$

and the accuracy of this approximation is one order lower than the remainder term $O(h^2)$ in (8.128). As a result, the corresponding difference solution with the following discrete boundary condition:

$$\frac{y_N - y_{N-1}}{h} + k\, y_N = g_2 \tag{8.141}$$

will have an accuracy of $O(h)$ only. To retain the second-order convergence of the difference solution, we need to approximate the boundary condition (8.140) more accurately. One such treatment is based on the formula

$$Y'(b) = \frac{3\,Y_N - 4\,Y_{N-1} + Y_{N-2}}{2\,h} + O(h^2) \tag{8.142}$$

Then the boundary condition (8.140) is approximated by

$$\frac{3\,y_N - 4\,y_{N-1} + y_{N-2}}{2\,h} + k\, y_N = g_2 \tag{8.143}$$

It can be shown that the resulting difference scheme is again second-order accurate. There are other possibilities to approximate the boundary condition (8.140) so that second-order accuracy of the overall difference scheme is maintained, e.g. the one employed in Section 9.3 to numerically treat the derivative initial condition through the use of artificial variables (cf. (9.32) and the discussion there).

PROBLEMS

1. In general, the study of the existence and uniqueness of a solution for boundary value problems is more complicated. Consider the boundary value problem

$$\begin{cases} Y'' = 0, & 0 < x < 1 \\ Y'(0) = g_1, & Y'(1) = g_2 \end{cases}$$

Show that the problem has no solution if $g_1 \neq g_2$, and infinite many solutions when $g_1 = g_2$.

Hint: For the case $g_1 \neq g_2$, integrate the differential equation over [0, 1].

2. As another example of solution nonuniqueness, verify that for any constant c, $Y(x) = c\,\sin(x)$ solves the boundary value problem

$$\begin{cases} Y''(x) + Y(x) = 0, & 0 < x < \pi \\ Y(0) = Y(\pi) = 0 \end{cases}$$

3. Verify that any function of the form $Y(x) = c_1 e^x + c_2 e^{-x}$ satisfies the equation

$$Y''(x) - Y(x) = 0$$

Determine c_1 and c_2 for the function $Y(x)$ to satisfy the following boundary conditions:

(a) $Y(0) = 1, \qquad Y(1) = 0;$

(b) $Y(0) = 1, \qquad Y'(1) = 0;$

(c) $Y'(0) = 1, \qquad Y(1) = 0.$

4. Assume Y is three times continuously differentiable. Use Taylor's theorem to prove the formula (8.142).

5. Prove the formula (8.138) by using the asymptotic expansion (8.136).

6. Verify that any function of the form $Y(x) = c_1 x + c_2 x^2$ satisfies the equation

$$x^2 Y''(x) - 2\,x\,Y'(x) + 2\,Y(x) = 0$$

Determine the solution of the equation with the boundary conditions

$$Y(1) = 0, \qquad Y(2) = 1$$

Use the MATLAB program ODEBVP to solve the boundary value problem for $h = 0.1, 0.05, 0.025$, print the errors of the numerical solutions at $x = 1.2, 1.4, 1.6$, and 1.8. Comment on how errors decrease when h is halved. Do the same for the extrapolated solutions.

7. The general solution of the equation

$$x^2 Y'' - x\,(x+2)\,Y' + (x+2)\,Y = 0$$

is $Y(x) = c_1 x + c_2 x e^x$. Determine the solution of the equation with the boundary conditions

$$Y(1) = e, \qquad Y(2) = 2\,e^2$$

Use the MATLAB program ODEBVP to solve the boundary value problem for $h = 0.1, 0.05, 0.025$, print the errors of the numerical solutions at $x = 1.2, 1.4, 1.6$, and 1.8. Comment on how errors decrease when h is halved. Do the same for the extrapolated solutions.

8. The general solution of the equation

$$x\,Y'' - (2\,x+1)\,Y' + (x+1)\,Y = 0$$

is $Y(x) = c_1 e^x + c_2 x^2 e^x$. Find the solution of the equation with the boundary conditions

$$Y'(1) = 0, \qquad Y(2) = e^2$$

Write down a formula for a discrete approximation of the boundary condition $Y'(1) = 0$ similar to (8.143) that has an accuracy $O(h^2)$. Implement the method by modifying the program ODEBVP, and solve the problem with $h = 0.1, 0.05, 0.025$. Print the errors of the numerical solutions at $x = 1, 1.2, 1.4, 1.6, 1.8$, and comment on how errors decrease when h is halved. Do the same for the extrapolated solutions.

9. Consider the boundary value problem (8.125) with p, q, and r constant. Modify the MATLAB program so that the command `feval` does not appear. Use the modified program to solve the following boundary value problem:

(a) $$\begin{cases} Y'' = -Y, & 0 < x < \dfrac{\pi}{2} \\ Y(0) = Y\left(\dfrac{\pi}{2}\right) = 1 \end{cases}$$

The true solution is $Y(x) = \sin x + \cos x$.

(b) $$\begin{cases} Y'' + Y = \sin x, & 0 < x < \dfrac{\pi}{2} \\ Y(0) = Y\left(\dfrac{\pi}{2}\right) = 0 \end{cases}$$

The true solution is $Y(x) = -\dfrac{x}{2}\cos x$.

10. Give a second-order scheme for the following boundary value problem:

$$\begin{cases} Y'' = \sin\left(xY'\right) + 1, & 0 < x < 1 \\ Y(0) = 0, \quad Y(1) = 1 \end{cases}$$

CHAPTER 9

FINITE DIFFERENCE METHOD FOR PARTIAL DIFFERENTIAL EQUATIONS

Many phenomena in sciences and engineering depend on more than one independent variable. For example, an unknown function of a real-world problem usually depends on both the time t and the location of the point (x, y, z). The differential equation for the unknown function then involves partial derivatives of the function with respect to these independent variables. As a result, we get partial differential equations. Some examples of PDEs of two independent variables are

$$\frac{\partial^2 u}{\partial x^2} + \frac{\partial^2 u}{\partial y^2} = f(x, y), \qquad (x, y) \in \Omega \subset \mathbb{R}^2 \tag{9.1}$$

$$\frac{\partial u}{\partial t} = a\frac{\partial^2 u}{\partial x^2} + f(x, t), \qquad x \in (0, L),\ t > 0 \tag{9.2}$$

$$\frac{\partial^2 u}{\partial t^2} = a\frac{\partial^2 u}{\partial x^2} + f(x, t), \qquad x \in (0, L),\ t > 0 \tag{9.3}$$

where Ω is a domain in $\mathbb{R}^2$, the constant $L > 0$, the coefficient $a > 0$, and the function $f(x, t)$ are given. These equations are called the *Poisson equation*, *heat equation*, and *wave equation*, respectively. The homogeneous Poisson equation, that is, the equation (9.1) with $f(x, y) = 0$, is called the *Laplace equation*.

The Poisson equation can be used to model the steady-state distribution of temperature in the domain Ω, where u is interpreted as the temperature and $-f$ is interpreted as a (scaled) heat source distribution. It also models the electrostatic potential with $f(x, y)$ representing a charge density. The Laplace equation models some forms of two-dimensional fluid flows. The heat equation can be used to describe the heat transfer process along a thin rod, the coefficient a being determined by the thermal conductivity, the specific heat, and the density. This equation can also model the diffusion process and is therefore also called the *diffusion equation*. The wave equation can be used to model the vibration of a string, with the coefficient a depending on the string tension and density. The quantity $\sqrt{a}$ can be interpreted as the speed of wave propagation. The two-dimensional version of the equation, where the term $\frac{\partial^2 u}{\partial x^2}$ on the right-hand-side is replaced by $\frac{\partial^2 u}{\partial x^2} + \frac{\partial^2 u}{\partial y^2}$, can model the vibration of an elastic membrane.

The Poisson equation is a representative model of *elliptic equations*, whereas the heat equation and the wave equation belong to the family of *parabolic equations* and that of *hyperbolic equations*, respectively. In general, consider a second-order partial differential equation of two independent variables that is of the form

$$A\,u_{xx} + B\,u_{xy} + C\,u_{yy} = F(x, y, u, u_x, u_y) \tag{9.4}$$

for some constants A, B, and C, and a function $F(x, y, u, p, q)$. Denote the discriminant $\Delta = B^2 - 4\,A\,C$. Then the equation is *elliptic* if $\Delta < 0$, *parabolic* if $\Delta = 0$, and *hyperbolic* if $\Delta > 0$. It is useful to classify the differential equations, as different types of differential equations have different solution properties, and these properties have major impacts on the choice and performance of numerical methods. For parabolic and hyperbolic equations from various applications, one of the two independent variables can be interpreted as the time. For this reason, we will use x and t for the independent variables in parabolic and hyperbolic equations.

The Poisson equation (9.1) is of the form (9.4) with $A = C = 1$ and $B = 0$. Then the discriminant is $\Delta = 0^2 - 4 \cdot 1 \cdot 1 = -4 < 0$. So the Poisson equation is elliptic. The heat equation (9.2) can be rewritten as

$$a\,\frac{\partial^2 u}{\partial x^2} = \frac{\partial u}{\partial t} - f(x, t)$$

and we may take $A = a$, $B = C = 0$. Then the discriminant is 0 and the equation is parabolic. Finally, write the wave equation (9.3) in the form of (9.4):

$$a\,\frac{\partial^2 u}{\partial x^2} - \frac{\partial^2 u}{\partial t^2} = -f(x, t)$$

Then $A = a$, $B = 0$ and $C = -1$. The discriminant $\Delta = 4\,a > 0$ and the equation is hyperbolic. It is suggested that the reader determine the types of the following equations:

$$4\,\frac{\partial^2 u}{\partial x^2} + \frac{\partial^2 u}{\partial y^2} = f(x, y)$$

$$4\,\frac{\partial^2 u}{\partial x^2} + 4\,\frac{\partial^2 u}{\partial x \partial y} + \frac{\partial^2 u}{\partial y^2} = f(x, y)$$

$$4\,\frac{\partial^2 u}{\partial x^2} - 8\,\frac{\partial^2 u}{\partial x \partial y} + \frac{\partial^2 u}{\partial y^2} = f(x, y)$$

Partial differential equations are more difficult to solve than ordinary differential equations. By a solution of a partial differential equation, we mean a function that satisfies the differential equation. A basic fact is that in general it is not possible to find closed-form formulas for solutions of partial differential equations. To uniquely determine a solution, we often need to specify some supplementary conditions, called boundary value conditions or initial value conditions. This forms a partial differential equation problem: an initial value problem, boundary value problem, or initial boundary value problem. Partial differential equation problems arising in applications are usually solved by numerical methods. Among the various numerical methods, the finite difference method is a general one that is easy to derive and to implement. In this chapter, we will study the finite difference method to numerically solve the model PDEs (9.1–9.3).

9.1. THE POISSON EQUATION

Let $\Omega \subset \mathbb{R}^2$ be a planar domain, and denote its boundary by $\partial\Omega$. The boundary value problem we consider in this section is

$$\begin{aligned} \frac{\partial^2 u}{\partial x^2} + \frac{\partial^2 u}{\partial y^2} &= f(x, y), \qquad (x, y) \in \Omega \\ u &= g(x, y), \qquad (x, y) \in \partial\Omega \end{aligned} \tag{9.5}$$

The value of the unknown solution is prescribed on the boundary. Such a condition is called a *Dirichlet boundary condition*. For second-order differential equations, boundary conditions involving the first partial derivatives can be specified as well, leading to *Neumann* or *Robin boundary conditions*. In the following, we will consider the Dirichlet boundary condition only.

For demonstration purposes, we assume Ω is a square: $\Omega = (0, 1) \times (0, 1)$. Then the boundary $\partial\Omega$ consists of four line segments, which are the four sides of the square. Let us develop a finite difference scheme for the problem. We divide the x interval $[0, 1]$ into n_x equal parts and denote $h_x = 1/n_x$ the x stepsize. Similarly, we divide the y interval $[0, 1]$ into n_y equal parts and denote $h_y = 1/n_y$ the y stepsize. Then the grid points are

$$(x_i, y_j), \qquad 1 \le i \le n_x + 1,\ 1 \le j \le n_y + 1$$

where $x_i = (i - 1)\,h_x$, $y_j = (j - 1)\,h_y$. See Figure 9.1 for such a finite difference grid. Note that traditionally, the grid points are numbered as

$$(x_i, y_j), \qquad 0 \le i \le n_x, \quad 0 \le j \le n_y$$

Our choice here makes it easier to implement the numerical schemes in MATLAB.

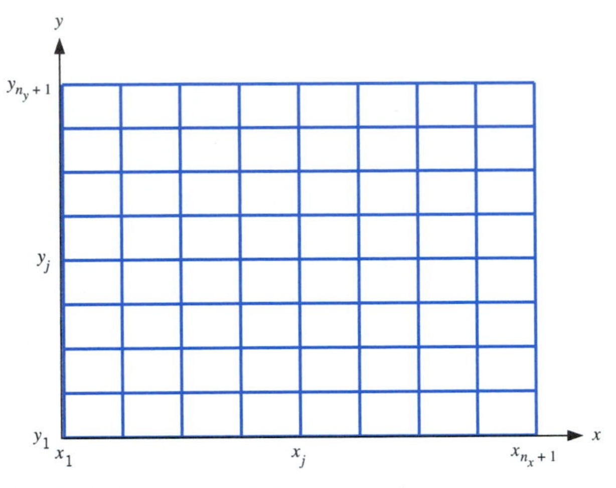

Figure 9.1. A finite difference grid

Consider the differential equation at an interior grid point (x_i, y_j), $2 \le i \le n_x$, $2 \le j \le n_y$. We use the three-point central difference to approximate the second derivative:

$$\frac{\partial^2 u}{\partial x^2}(x_i, y_j) = \frac{u(x_{i+1}, y_j) - 2\,u(x_i, y_j) + u(x_{i-1}, y_j)}{h_x^2} + O(h_x^2)$$

$$\frac{\partial^2 u}{\partial x^2}(x_i, y_j) = \frac{u(x_i, y_{j+1}) - 2\,u(x_i, y_j) + u(x_i, y_{j-1})}{h_y^2} + O(h_x^2)$$

Then the differential equation at (x_i, y_j) is approximately written as

$$\begin{aligned} &\frac{u(x_{i+1}, y_j) - 2\,u(x_i, y_j) + u(x_{i-1}, y_j)}{h_x^2} \\ &\quad + \frac{u(x_i, y_{j+1}) - 2\,u(x_i, y_j) + u(x_i, y_{j-1})}{h_y^2} \approx f_{ij} \end{aligned} \tag{9.6}$$

where $f_{ij} = f(x_i, y_j)$. Denote by u_{ij} the finite difference approximation of $u(x_i, y_j)$. Then we define the following difference scheme by replacing the true solution u by its approximate values at the grid points and equating both sides of the relation (9.6):

$$\frac{u_{i+1,j} - 2\,u_{i,j} + u_{i-1,j}}{h_x^2} + \frac{u_{i,j+1} - 2\,u_{i,j} + u_{i,j-1}}{h_y^2} = f_{ij}, \qquad 2 \le i \le n_x,\ 2 \le j \le n_y \tag{9.7}$$

These equations are supplemented by the discrete Dirichlet boundary condition:

$$u_{ij} = g_{ij}, \qquad i = 1 \text{ or } n_x + 1, \text{ or } j = 1 \text{ or } n_y + 1 \tag{9.8}$$

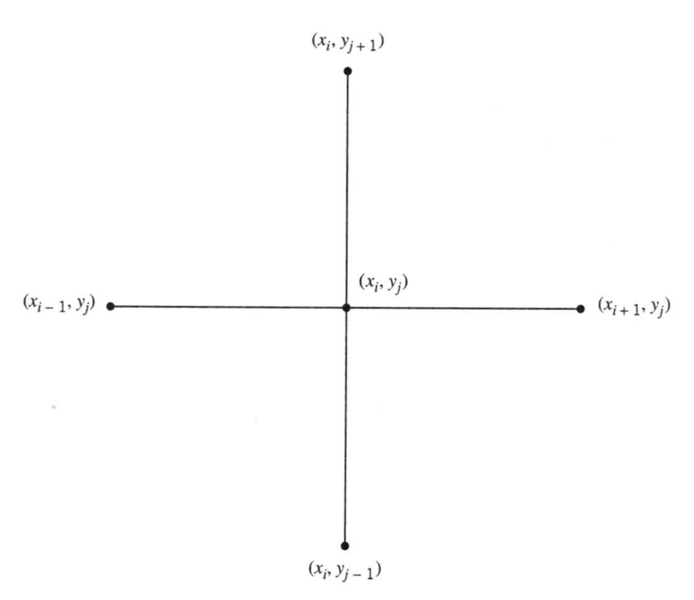

Figure 9.2. An interior grid point and its four neighboring grid points

The resulting method is known as the five-point scheme, since each difference equation in (9.7) is a relation of function values at an interior grid point and its four neighboring grid points; see Figure 9.2. Since the approximate solution at the boundary grid points is given from (9.8), we see that the unknowns are the approximate solution values u_{ij}, $2 \le i \le n_x$, $2 \le j \le n_y$, at the interior grid points, and there are $(n_x - 1)(n_y - 1)$ of them. This number matches the number of equations in (9.7). Indeed, it can be proved that the scheme (9.7–9.8) has a unique solution.

The truncation error of the scheme (9.7) is defined as

$$\frac{u(x_{i+1}, y_j) - 2u(x_i, y_j) + u(x_{i-1}, y_j)}{h_x^2} + \frac{u(x_i, y_{j+1}) - 2u(x_i, y_j) + u(x_i, y_{j-1})}{h_y^2} - f_{ij}$$

and it is $O(h_x^2 + h_y^2)$, assuming the solution u has continuous derivatives of order 4. This can be verified by using Taylor's theorem. It can be shown that the accuracy of the solution is of second order:

$$\max_{\substack{1 \le i \le n_x+1 \\ 1 \le j \le n_y+1}} |u(x_i, y_j) - u_{ij}| = O(h_x^2 + h_y^2) \tag{9.9}$$

Note that when we discuss the numerical solution of initial value problems of ordinary differential equations, the truncation error is defined to be the local error for the solution of a scheme at a grid point when true solution values are used at previous grid points. In the analysis of numerical methods for solving boundary value problems, the truncation error is defined to be the discrepancy between the difference equation and the differential equation. With Dirichlet boundary conditions, the convergence order of the method (9.7) for solving (9.5) is the same as that of the truncation error.

In the following, we focus on the particular case $n_x = n_y = n$, $h_x = h_y = h$. Then the difference equation (9.7) can be rewritten as

$$u_{i+1,j} + u_{i-1,j} + u_{i,j+1} + u_{i,j-1} - 4\,u_{i,j} = h^2 f_{ij}, \qquad 2 \le i, j \le n \tag{9.10}$$

The form of the discrete equation (9.10) naturally suggests iterative methods for the solution. We write the scheme as

$$\begin{aligned} u_{i,j} &= \frac{1}{4}\left(u_{i+1,j} + u_{i-1,j} + u_{i,j+1} + u_{i,j-1}\right) - \frac{h^2}{4} f_{ij}, \qquad 2 \le i, j \le n \\ u_{i,j} &= g_{i,j}, \qquad i = 1 \text{ or } n+1,\ 2 \le j \le n, \text{ or } j = 1 \text{ or } n+1,\ 2 \le i \le n \end{aligned} \tag{9.11}$$

Then an implementation of the Gauss–Seidel iteration method of Section 6.6 is as follows: Given an initial guess $\{u_{i,j}^{(1)}\}_{2\le i,j\le n}$, for $k = 1, 2, \ldots$, determine $\{u_{i,j}^{(k+1)}\}_{2\le i,j\le n}$ recursively by

for $i = 2, \ldots, n$
  for $j = 2, \ldots, n$
  $$u_{i,j}^{(k+1)} = \frac{1}{4}\left(u_{i+1,j}^{(k)} + u_{i-1,j}^{(k+1)} + u_{i,j+1}^{(k)} + u_{i,j-1}^{(k+1)}\right) - \frac{h^2}{4} f_{ij}$$
  end
end

When i or j is 1 or $n+1$, the boundary condition is used: $u_{i,j}^{(k+1)} = g_{i,j}$. In this implementation, the inner iteration is done with respect to the second index, and the outer iteration occurs on the first index. We can interchange the order of the inner and outer iterations and obtain another Gauss–Seidel implementation.

MATLAB PROGRAM. The following is a sample MATLAB program for solving the Dirichlet boundary value problem of the Poisson equation on the unit square $[0, 1]^2 \equiv [0, 1] \times [0, 1]$:

```
function U = Poisson(f,g,n,tol,max_it)
%
% function U = Poisson(f,g,n,tol,max_it)
%
% The five-point scheme for solving the Dirichlet BVP of the
% Poisson equation on the unit square.
% Input
%       f: the right-hand-side function
```

```
%       g: the Dirichlet boundary value function
%       n: the number of subintervals of [0,1]
%       tol: relative error tolerance of the iterative solution;
%            default value: 10^(-5)
%       max_it: maximal number of iterations allowed;
%            default value: 10,000
% Output
%        U: the solution u_{ij}, i,j=1,...,n+1
%
% To use the program, the user must supply two m-files, say,
% 'f.m' and 'g.m', to define the right-hand-side function f of
% the differential equation and the boundary value function g.
% The user should also choose a positive integer n for the
% number of subintervals of [0,1].
% A sample call would be
%       U = Poisson('f','g',n,1e-8,1000)
% with 10^(-8) as the tolerance for the relative errors of the
% iterative solution, and a maximal number of 1,000 iterations
% is allowed for solving the finite difference system.
% It is also possible to use
%      U = Poisson('f','g',n,1e-8)
% then the maximal number of iterations is the default
% value 10^4. If the default values 10^(-5) and 10^4 are
% to be used for the iteration relative error tolerance and
% maximal number of iterations, then one can simply use
%      U = Poisson('f','g',n)
% if max_it, or max_it and tol, are not provided, use the
% default values
if nargin < 5
  max_it = 10000
end
if nargin < 4
  tol = 1e-5
end
% compute some parameters
n1 = n+1;
h = 1/n;
h2 = h*h/4;
Fr = zeros(n,n);
Fr(2:n,2:n) = h2*feval(f,(1:n-1)*h,(1:n-1)*h);
% initialization
U = zeros(n1,n1);
% specify boundary conditions
U(1,1:n1) = feval(g,0,(0:n)*h);
U(n1,1:n1) = feval(g,1,(0:n)*h);
```

```
U(1:n1,1) = feval(g,(0:n)*h,0);
U(1:n1,n1) = feval(g,(0:n)*h,1);
% iteration
rel_err = 1;
itnum = 0;
while ((rel_err>tol) & (itnum<=max_it))
err = 0;
umax = 0;
  for j = 2:n
    for i = 2:n
      temp = (U(i+1,j)+U(i-1,j)+U(i,j+1)+U(i,j-1))/4-Fr(i,j);
      dif = abs(temp-U(i,j));
      if (err <= dif)
        err = dif;
      end
      U(i,j) = temp;
      temp = abs(temp);
      if(umax <= temp)
        umax = temp;
      end
    end
  end
itnum = itnum+1;
rel_err = err/umax;
end
% plot the numerical solution
X=(0:h:n*h)';
Y=X;
surf(X,Y,U')
xlabel('x-axis')
ylabel('y-axis')
zlabel('the numerical solution')
title('Plot of the numerical solution')
```

In the program, we use the MATLAB command `surf` to draw the three-dimensional graph of the numerical solution. Note that with the statement `surf(X,Y,Z)`, the vertices of the surface patches are the triples $(X(j), Y(i), Z(i, j))$, that is, the first vector X corresponds to the columns of the matrix Z and the second vector Y corresponds to the rows. In the above program, we use $u(i, j)$ for the numerical solution at (x_i, y_j). That is why the transpose `U'` is used in the statement `surf(X,Y,U')`. The same comment applies to other MATLAB three-dimensional commands: `mesh`, `waterfall`, `surfl`, and their color counterparts.

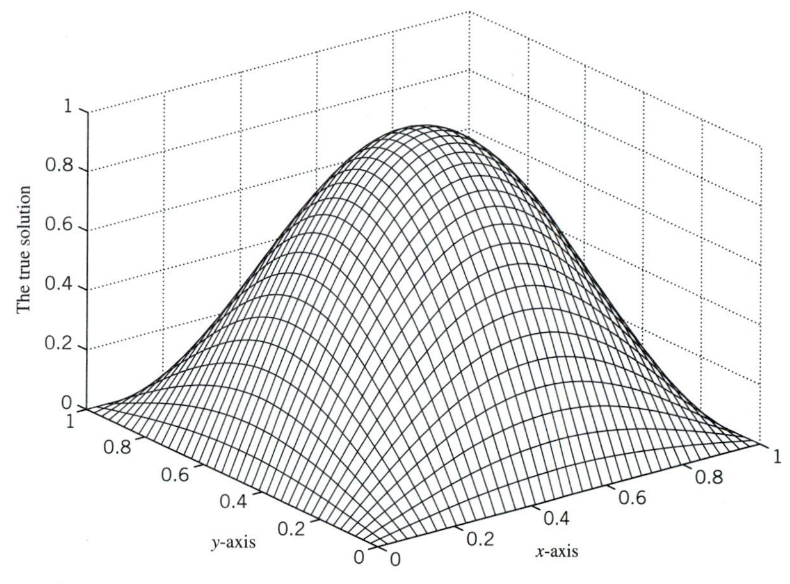

Figure 9.3. The true solution of Example 9.1.1

Example 9.1.1 We consider the boundary value problem

$$\begin{cases} \dfrac{\partial^2 u}{\partial x^2} + \dfrac{\partial^2 u}{\partial y^2} = -2\pi^2 \sin(\pi x)\, \sin(\pi y), & 0 < x, y < 1 \\ u(x, y) = 0, \quad x = 0, 1 \text{ or } y = 0, 1 \end{cases}$$

The true solution is

$$u(x, y) = \sin(\pi x)\, \sin(\pi y)$$

and it is shown in Figure 9.3.

We use the five-point scheme to solve the problem. In this example, we know the true solution. So it is possible to compute the errors of the numerical solutions. Table 9.1 lists the maximum error $\max_{1 \le i, j \le n+1} |u(x_i, y_j) - u_{ij}|$ for various n. ■

Table 9.1. Maximum Numerical Solution Errors for Example 9.1.1

n	Max. Error	Ratio
4	5.3029E − 2	
8	1.2951E − 2	4.09
16	3.2190E − 3	4.02
32	8.0347E − 4	4.01

The numerical results with $n = 16$ are given in Figure 9.4. This figure is generated when the graphing part of the program Poisson is replaced by the following:

```
% Plot the numerical solution
X = (0:h:n*h)'; Y = X;
subplot(1,2,1)
surf(X,Y,U')
xlabel('x-axis')
ylabel('y-axis')
zlabel('The numerical solution')
s1 = sprintf('h=6.4f',h)
title(s1)
hold on
%
% Plot the numerical error
Err = sin(pi*X)*sin(pi*Y')-U;
subplot(1,2,2)
surf(X,Y,Err')
xlabel('x-axis')
ylabel('y-axis')
zlabel('The error')
title(s1)
```

We observe from Table 9.1 that as the value n is doubled, that is, the grid size h is halved, the maximum error is reduced by a factor of approximately 4. This agrees with the theoretical error bound (9.9). To see more precisely the error behavior, we calculate in Table 9.2 the solution errors at six selective node points, and the corresponding ratios. The ratios are all close to 4. Notice that the boundary value problem has various symmetry properties. For instance, it is symmetric with respect to the diagonal lines and the midlines of the problem domain, defined by the equations $y = x$, $x + y = 1$, $x = 1/2$, and $y = 1/2$, respectively. This explains why the errors at several points are identical. To find out if a given code is wrong, one easy thing to check is to see whether the code maintains the symmetries of the problem being solved.

The theoretical error bound (9.9) can be improved. Under certain smoothness assumptions on the solution, it can be shown that for the numerical solution of the five-point scheme, there is an asymptotic error expansion

$$u(x_i, y_j) - u_{ij} = h^2 D(x_i, y_j) + O(h^4) \tag{9.12}$$

for any grid point (x_i, y_j). Here $D(x, y)$ denotes some function determined from a boundary value problem involving u in the data. This then suggests the possibility of doing the Richardson extrapolation. Denoting $u_h(x, y)$ for the numerical solution at a grid point (x, y) with the grid size h, we rewrite (9.12) as

$$u(x_i, y_j) - u_h(x_i, y_j) = h^2 D(x_i, y_j) + O(h^4) \tag{9.13}$$

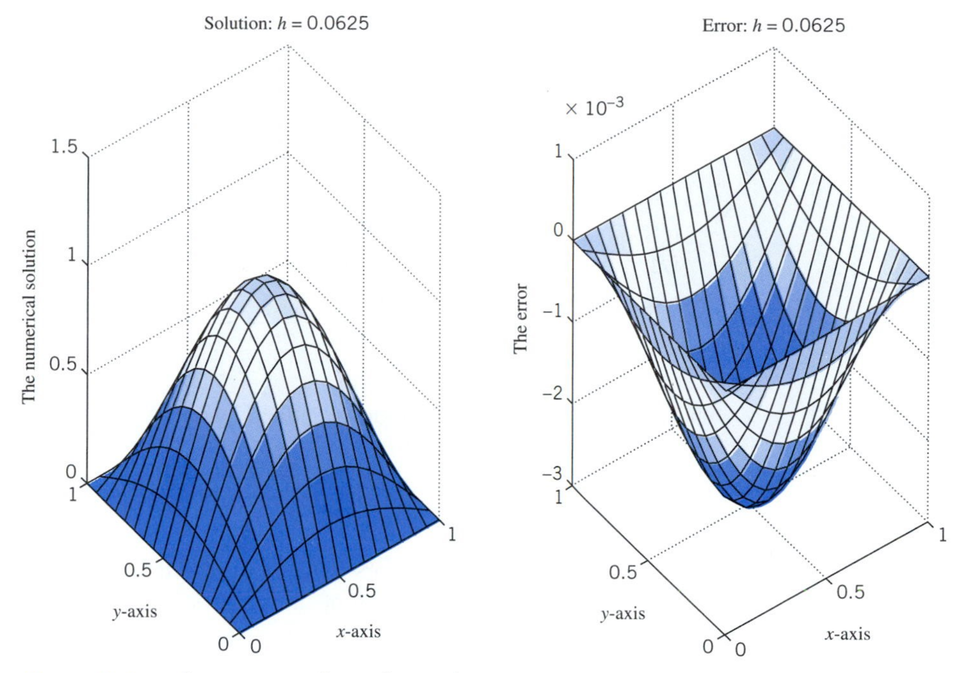

Figure 9.4. The numerical results with $n = 16$

Table 9.2. Numerical Solution Errors for Example 9.1.1

(x, y)	$n = 4$	$n = 8$	Ratio	$n = 16$	Ratio
(1/4, 1/4)	−2.65E − 2	−6.48E − 3	4.09	−1.61E − 3	4.02
(1/2, 1/4)	−3.75E − 2	−9.16E − 3	4.09	−2.28E − 3	4.02
(3/4, 1/4)	−2.65E − 2	−6.48E − 3	4.09	−1.61E − 3	4.02
(1/4, 1/2)	−3.75E − 2	−9.16E − 3	4.09	−2.28E − 3	4.02
(1/2, 1/2)	−5.30E − 2	−1.30E − 2	4.09	−3.22E − 3	4.02
(1/4, 3/4)	−2.65E − 2	−6.48E − 3	4.09	−1.61E − 3	4.02

If (x, y) is a grid point corresponding to the grid size $2h$, then it is a grid point also with the grid size h. From (9.13) we have

$$u(x, y) - u_h(x, y) = h^2 D(x, y) + O(h^4)$$
$$u(x, y) - u_{2h}(x, y) = (2h)^2 D(x, y) + O(h^4)$$

Eliminating the term $D(x, y)$ from the two relations, we obtain

$$u(x, y) - \tilde{u}_h(x, y) = O(h^4)$$

Table 9.3. Extrapolated Solution Errors for Example 9.1.1

(x, y)	$n = 8$	$n = 16$	Ratio
(1/4, 1/4)	2.04E − 4	1.25E − 5	16.35
(1/3, 1/4)	2.89E − 4	1.77E − 5	16.35
(3/4, 1/4)	2.04E − 4	1.25E − 5	16.35
(1/4, 1/2)	2.89E − 4	1.77E − 5	16.35
(1/2, 1/2)	4.09E − 4	2.50E − 5	16.35
(1/4, 3/4)	2.04E − 4	1.25E − 5	16.35

where $\tilde{u}_h$ is the extrapolated solution defined by

$$\tilde{u}_h(x, y) = \frac{4\,u_h(x, y) - u_{2h}(x, y)}{3}$$

In Table 9.3, we report the errors of the extrapolated solutions. We observe a good size improvement in the accuracy of the numerical solution. Since the errors of the extrapolated solutions are rather small, we have to solve the finite difference systems more accurately. This is achieved with the use of a small error tolerance in running the program `Poisson`. For the results reported in the table here, we used `tol` $= 10^{-9}$.

The numerical values reported in Tables 9.2 and 9.3 are obtained by the following MATLAB code. We define the right-hand-side function in `fP1` and the boundary value function in `gP1`.

```
% Define x and y coordinates of sampling points
xx = linspace(0.25,0.75,3)';
yy = xx;
% True solution at 9 points
tr = sin(pi*xx)*sin(pi*yy)'
% True solution values at 6 points (1/4,1/4),(1/2,1/4),...
Tr = [tr(1,1);tr(2,1);tr(3,1);tr(1,2);tr(2,2);tr(1,3)];
% Compute solution errors
N = 2.^(2:4)
err = zeros(6,3);
for i = 1:3
   n = N(i);
   n4 = n/4;
   zz = Poisson('fP1','gP1',n,1e-9);
   zv = zz(n4+1:n4:n,n4+1:n4:n);
   Zv = [zv(1,1);zv(2,1);zv(3,1);zv(1,2);zv(2,2);zv(1,3)];
   err(:,i) = Tr-Zv;
end
% Compute solution error ratios
```

```
R1 = zeros(6,3);
R1(:,1) = err(:,1)./err(:,2);
R1(:,2) = err(:,2)./err(:,3);
% Extrapolation
exte = zeros(6,2);
for i = 1:2
   exte(:,i) = (4.*err(:,i+1)-err(:,i))/3;
end
% Compute ratios of the extrapolation errors
R2 = zeros(6,1);
R2(:)  = exte(:,1)./exte(:,2);
% Print results
for i = 1:6
   fprintf(' 10.2e%10.2e%6.2f%10.2e%6.2f\n',...
   err(i,1),err(i,2),R1(i,1),err(i,3),R1(i,2))
end
for i = 1:6
   fprintf(' 10.2e%10.2e%6.2f\n',...
             exte(i,1),exte(i,2),R2(i))
end
```

PROBLEMS

1. Recall the definition of hyperbolic functions

$$\sinh(x) = \frac{e^x - e^{-x}}{2}, \qquad \cosh(x) = \frac{e^x + e^{-x}}{2}$$

Show that the function

$$u(x, y) = \frac{1}{\sinh(\pi)} [\sin(x)\sinh(\pi - y) + \sin(y)\sinh(\pi - x)]$$

solves the boundary value problem

$$\begin{cases} \dfrac{\partial^2 u}{\partial x^2} + \dfrac{\partial^2 u}{\partial y^2} = 0, & 0 < x, y < \pi \\ u(x, 0) = \sin(x), \quad u(x, \pi) = 0, & 0 \le x \le \pi \\ u(0, y) = \sin(y), \quad u(\pi, y) = 0, & 0 \le y \le \pi \end{cases}$$

2. It is well known that if $f(z)$, $z = x + iy$, is a differentiable complex-valued complex-variable function, then the real part $u(x, y)$ and the imaginary part $v(x, y)$ of the function $f(x + i\,y)$ satisfy the Cauchy–Riemann equations

$$u_x = v_y, \qquad u_y = -v_x$$

Moreover, u and v are infinitely differentiable. Show that both u and v satisfy the Laplace equation.

3. Show that $u(x, y) = (x - x^3)(y - y^2)$ is the solution of the boundary value problem

$$\begin{cases} \dfrac{\partial^2 u}{\partial x^2} + \dfrac{\partial^2 u}{\partial y^2} = 2x\,(x^2 + 3\,y^2 - 3\,y - 1), & 0 < x, y < 1 \\ u(x, y) = 0, \quad x = 0 \text{ or } 1, \text{ or } y = 0 \text{ or } 1 \end{cases}$$

Solve the problem by using the MATLAB program `Poisson`. Report numerical results similar to Example 9.1.1.

4. Modify the program `Poisson` so that it takes boundary value functions with different expressions on the four sides of the boundary of the unit square, that is, write a program of the form

```
function U = Poisson1(f, gr, gt, gl, gb, n, tol, max_it)
```

where `gr`, `gt`, `gl`, and `gb` are the function names of the boundary value conditions on the right part ($x = 1$), the top part ($y = 1$), the left part ($x = 0$), and the bottom part ($y = 0$) of the boundary. Use the code to solve the following problem:

$$\begin{cases} \dfrac{\partial^2 u}{\partial x^2} + \dfrac{\partial^2 u}{\partial y^2} = -2\,(x^2 + y^2), \quad 0 < x, y < 1 \\ u(x, 0) = 1 - x^2, \qquad u(x, 1) = 2\,(1 - x^2), \quad 0 \le x \le 1 \\ u(0, y) = 1 + y^2, \qquad u(1, y) = 0, \quad 0 \le y \le 1 \end{cases}$$

The true solution is $u(x, y) = (1 - x^2)(1 + y^2)$. The five point scheme reproduces the true solution at the grid points for this particular problem. Confirm this property by reporting the maximum numerical solution errors as in Table 9.2 for several values of the relative error tolerance (`max_it = 10000`): 10^{-5}, 10^{-6}, 10^{-7}, 10^{-8}.

5. Use the program from Problem 4 to solve the boundary value problem

$$\begin{cases} \dfrac{\partial^2 u}{\partial x^2} + \dfrac{\partial^2 u}{\partial y^2} = 0, \quad 0 < x, y < 1 \\ u(x, 0) = 1, \qquad u(x, 1) = 2, \quad 0 \le x \le 1 \\ u(0, y) = 1 + y^2, \qquad u(1, y) = 1 + y, \quad 0 \le y \le 1 \end{cases}$$

Use $n_x = n_y = 5, 10, 20$. Graph the corresponding numerical solutions. Produce a table of numerical values at the points $(i/5, i/5)$, $1 \le i \le 5$.

6. Modify the program `Poisson` so that as in the previous problem, it takes boundary value functions with different expressions on the four sides of the boundary of the unit square. Furthermore, the differential equation that can be solved is of the more general form

$$a_1 \frac{\partial^2 u}{\partial x^2} + a_2 \frac{\partial^2 u}{\partial y^2} = f(x, y)$$

where a_1 and a_2 are positive constants. Use the new program to solve the following problem:

$$\begin{cases} \pi^2 \dfrac{\partial^2 u}{\partial x^2} + \dfrac{\partial^2 u}{\partial y^2} = 0, \quad 0 < x, y < 1 \\ u(x, 0) = u(x, 1) = 0, \quad 0 \le x \le 1 \\ u(0, y) = \sin \pi y, \qquad u(1, y) = e \sin \pi y, \quad 0 \le y \le 1 \end{cases}$$

The true solution is $u(x, y) = e^x \sin \pi y$. Report numerical results in a style similar to that of Example 9.1.1.

7. Modify the program `Poisson` so that it can solve the Dirichlet boundary value problem of the Poisson equation on a square $[0, a]^2$ of any side $a > 0$. Use your new program to solve the following problem:

$$\begin{cases} \dfrac{\partial^2 u}{\partial x^2} + \dfrac{\partial^2 u}{\partial y^2} = (6y - y^3) \sin x, \quad 0 < x, y < \pi \\ u(x, 0) = 0, \qquad u(x, \pi) = \pi^3 \sin x, \quad 0 \le x \le \pi \\ u(0, y) = u(\pi, y) = 0, \quad 0 \le y \le \pi \end{cases}$$

The true solution is $u(x, y) = y^3 \sin x$. Report your numerical results in a style similar to that of Example 9.1.1.

8. The partial differential equation

$$\frac{\partial^2 u}{\partial x^2} + \frac{\partial^2 u}{\partial y^2} + a(x, y)\, u = f(x, y)$$

is called a Helmholtz equation. When the coefficient function $a(x, y) = 0$, it is reduced to the Poisson equation. Consider the special case of a nonpositive constant coefficient a, and the corresponding boundary value problem

$$\begin{cases} \dfrac{\partial^2 u}{\partial x^2} + \dfrac{\partial^2 u}{\partial y^2} + a\, u = f(x, y), \quad 0 < x, y < 1 \\ u(x, y) = g(x, y), \quad x = 0 \text{ or } 1, \text{ or } y = 0 \text{ or } 1 \end{cases}$$

Derive a difference scheme, similar to (9.11), to solve this boundary value problem. Implement the scheme in MATLAB and use it to solve the problem with $a = -2$, $g(x, y) = 0$, and

$$f(x, y) = x\, y \left[(x^2 - 7)(1 - y^2) + (1 - x^2)(y^2 - 7) \right]$$

The true solution is

$$u(x, y) = (x - x^3)\,(y - y^3)$$

Report numerical results similar to that of Example 9.1.1.

9. Consider the boundary value problem

$$\begin{aligned} \frac{\partial^2 u}{\partial x^2} + \frac{\partial^2 u}{\partial y^2} &= f(x, y), \qquad 0 < x, y < 1 \\ u_x\,(0, y) = g_\ell(y), &\qquad u(1, y) = g_r(y), \qquad 0 \le y \le 1 \\ u(x, 0) = g_b(x), &\qquad u(x, 1) = g_t(x), \qquad 0 \le x \le 1 \end{aligned}$$

with given functions f, g_ℓ, g_r, b_b, g_t. Show how to use the idea of the formula (8.143) to discretize the derivative boundary condition $u_x\,(0, y) = g_\ell(y)$.

9.2. ONE-DIMENSIONAL HEAT EQUATION

The model parabolic equation we consider in this section is

$$\frac{\partial u}{\partial t} = a\,\frac{\partial^2 u}{\partial x^2} + f \tag{9.14}$$

Here t is the time variable, and $x \in (0, L)$ is the spatial variable. Such an equation can be used to model a variety of physical phenomena, including heat conduction problems. To have a well-defined problem, the differential equation must be supplemented by an initial condition

$$u(x, 0) = u_0(x), \qquad x \in [0, L] \tag{9.15}$$

and boundary conditions

$$u(0, t) = g_1(t), \qquad u(L, t) = g_2(t) \tag{9.16}$$

We solve the initial boundary value problem for $t \le T$. The differential equation (9.14) and the conditions (9.15), (9.16) together form an *initial boundary value problem.* See Figure 9.5.

Two approaches can be adopted in the design of numerical methods for solving the initial boundary value problem (9.14–9.16). In the first approach, the derivation of numerical methods consists of two steps: Step 1 is a discretization of the spatial derivatives, leading to a semidiscrete system, a system of ordinary differential equations in the time variable; in step 2, an ODE solver is employed to solve the ODE system. In the second approach, we discretize with respect to both variable x and t simultaneously, and obtain fully discrete schemes that are ready to be used and solved.

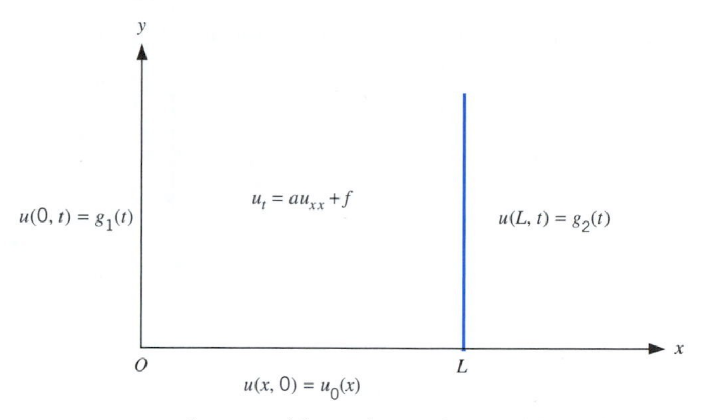

Figure 9.5. The initial boundary value problem of the parabolic equation

9.2.1 Semidiscretization

We first discuss the approach via semidiscretization. For this purpose, we divide the spatial interval $[0, L]$ into n_x equal parts. Denote $h_x = L/n_x$ for the grid size and $x_i = (i-1)\,h_x$, $1 \le i \le n_x + 1$, for the grid points. Let $u_i(t)$ be an approximation of $u(x_i, t)$, $1 \le i \le n_x + 1$. For $2 \le i \le n_x$, we consider the differential equation (9.14) and use the three-point central difference formula

$$\frac{\partial^2 u}{\partial x^2}(x_i, t) \approx \frac{u(x_{i+1}, t) - 2\,u(x_i, t) + u(x_{i-1}, t)}{h_x^2}$$

to do the approximation. As a result, we obtain

$$\dot{u}_i(t) = a\,\frac{u_{i+1}(t) - 2\,u_i(t) + u_{i-1}(t)}{h_x^2} + f_i(t), \qquad 2 \le i \le n_x$$

where $f_i(t) \equiv f(x_i, t)$, and a dot above a quantity denotes the time derivative of the quantity. These equations are to be supplemented by values of $u_1(t)$ and $u_{n_x+1}(t)$ from (9.16):

$$u_1(t) = g_1(t), \qquad u_{n_x+1}(t) = g_2(t)$$

as well as the initial values from (9.15):

$$u_i(0) = u_0(x_i), \qquad 1 \le i \le n_x + 1$$

In matrix/vector notation, the ODE system can be written as

$$\begin{aligned} \dot{\boldsymbol{u}}(t) &= A\boldsymbol{u}(t) + \boldsymbol{F}(t) \\ \boldsymbol{u}(0) &= \boldsymbol{u}_0 \end{aligned} \tag{9.17}$$

where $\boldsymbol{u}(t) = [u_2(t), \ldots, u_{n_x}(t)]^T$ is the vector of unknowns

$$A = \frac{a}{h_x^2} \begin{bmatrix} -2 & 1 & & & \\ 1 & -2 & 1 & & \\ & \ddots & \ddots & \ddots & \\ & & 1 & -2 & 1 \\ & & & 1 & -2 \end{bmatrix}_{(n_x-1)\times(n_x-1)}$$

$$\boldsymbol{F}(t) = \left[f_2(t) + \frac{a}{h_x^2} g_1(t),\, f_3(t), \ldots, f_{n_x-1}(t),\, f_{n_x}(t) + \frac{a}{h_x^2} g_2(t) \right]^T$$

$$\boldsymbol{u}_0 = [u_0(x_2), \ldots, u_0(x_{n_x})]^T$$

Generally, for a typical value n_x used in actual computation, the ODE system (9.17) is rather stiff. So it should be solved by an ODE solver that is effective for stiff equations. The main advantage of this approach is that state-of-the-art ODE solvers can be employed.

9.2.2 Explicit Full Discretization

We now turn to the discussion of fully discrete schemes. In addition to the partition of the spatial interval $[0, L]$, we need choose a stepsize for the time variable. Denote $h_t = T/n_t$ for some positive integer n_t and the grid points $t_k = (k-1)\,h_t$, $1 \le k \le n_t + 1$. We use u_i^k to denote the finite difference approximation value of $u(x_i, t_k)$. Let us discretize the differential equation at (x_i, t_k), $2 \le i \le n_x$, $2 \le k \le n_t$. Again use the three-point central difference approximation

$$\frac{\partial^2 u}{\partial x^2}(x_i, t_k) \approx \frac{u_{i+1}^k - 2\,u_i^k + u_{i-1}^k}{h_x^2}$$

for the second-order spatial derivative. There are several approaches to approximate the time derivative term $\frac{\partial u}{\partial t}(x_i, t_k)$. With

$$\frac{\partial u}{\partial t}(x_i, t_k) \approx \frac{u_i^{k+1} - u_i^k}{h_t}$$

we obtain the forward Euler formula

$$\frac{u_i^{k+1} - u_i^k}{h_t} = a\,\frac{u_{i+1}^k - 2\,u_i^k + u_{i-1}^k}{h_x^2} + f_i^k \tag{9.18}$$

The truncation error of the formula (9.18) is defined as

$$T_i^k \equiv \frac{u(x_i, t_{k+1}) - u(x_i, t_k)}{h_t} - a\,\frac{u(x_{i+1}, t_k) - 2\,u(x_i, t_k) + u(x_{i-1}, t_k)}{h_x^2} - f(x_i, t_k)$$

Since $(u(x_i, t_{k+1}) - u(x_i, t_k))/h_t$ is an $O(h_t)$ approximation to the derivative $u_t(x_i, t_k)$, and $(u(x_{i+1}, t_k) - 2\,u(x_i, t_k) + u(x_{i-1}, t_k))/h_x^2$ is an $O(h_x^2)$ approximation of $u_{xx}(x_i, t_k)$, we see that

$$\begin{aligned} T_i^k &= u_t(x_i, t_k) + O(h_t) - a\,u_{xx}(x_i, t_k) + O(h_x^2) - f(x_i, t_k) \\ &= O(h_t + h_x^2) \end{aligned}$$

Denote the ratio

$$\gamma = \frac{a\,h_t}{h_x^2}$$

The complete description of the method, incorporating the initial and boundary conditions, is

$$u_i^1 = u_0(x_i), \qquad 1 \le i \le n_x + 1 \tag{9.19}$$

and for $k \ge 1$,

$$\begin{cases} u_i^{k+1} = \gamma\,u_{i-1}^k + (1 - 2\,\gamma)\,u_i^k + \gamma\,u_{i+1}^k + h_t f_i^k, & 2 \le i \le n_x \\ u_1^{k+1} = g_1(t_{k+1}), \qquad u_{n_x+1}^{k+1} = g_2(t_{k+1}) \end{cases} \tag{9.20}$$

For this method, once the approximate solution at $t = t_k$ is known, we can compute the solution at the next time level $t = t_{k+1}$ directly. So the method is called an *explicit method*.

The main advantage of an explicit method is that there is no need to solve linear systems, as the approximate solution at time level $t = t_{k+1}$ is computed directly from that at the previous time steps. However, a useful numerical method must be stable, that is, error propagation through time advancing should be controllable. Suppose that the computed solution at level $t = t_k$ is $\tilde{u}_i^k = u_i^k + \epsilon_i^k$, $|\epsilon_i^k| \le \epsilon^k$, $1 \le i \le n_x + 1$. Then, easily, the computed solution from (9.20) (assuming no further error is introduced) is

$$\tilde{u}_i^{k+1} \equiv u_i^{k+1} + \epsilon_i^{k+1} = \gamma \tilde{u}_{i+1}^k + (1 - 2\,\gamma)\,\tilde{u}_i^k + \gamma \tilde{u}_{i-1}^k + h_t f_i^k$$

Hence,

$$\epsilon_i^{k+1} = \gamma \epsilon_{i+1}^k + (1 - 2\,\gamma)\,\epsilon_i^k + \gamma \epsilon_{i-1}^k, \qquad 2 \le i \le n_x$$

are the errors associated with the solution ϵ_i^{k+1}, $2 \le i \le n_x$. We observe that if

$$\gamma \equiv \frac{a\,h_t}{h_x^2} \le \frac{1}{2} \tag{9.21}$$

then

$$|\epsilon_i^{k+1}| \le \gamma |\epsilon_{i+1}^k| + (1 - 2\,\gamma)\,|\epsilon_i^k| + \gamma |\epsilon_{i-1}^k| \le \epsilon^k, \qquad 2 \le i \le n_x$$

that is, the error is not amplified in time advancing and the numerical method is stable. It can be shown that (9.21) is actually both a necessary and sufficient condition for the scheme (9.19–9.20) to be stable and hence useful.

Assume the stability condition (9.21) holds. It can be proved that if the true solution u is sufficiently smooth, the following error bound holds:

$$\max_{\substack{1\le i\le n_x+1\\ 1\le k\le n_t+1}} |u(x_i, t_k) - u_i^k| = O(h_t + h_x^2)$$

that is, the scheme is of second-order accuracy in h_x and of first-order accuracy in h_t. We notice that the convergence order is the same as that of the truncation error.

MATLAB PROGRAM. The following is a sample MATLAB code for the explicit method:

```
function U = Heat1(a,f,u0,g1,g2,L,T,nx,nt)
%
% function U = Heat1(a,f,u0,g1,g2,L,T,nx,nt)
%
% The forward difference scheme for solving the initial boundary
% value problem of the heat equation u_t = a u_{xx} + f for x
% in [0,L], and t in [0,T].
% Input
%      a: the coefficient of the u_{xx} term
%      f=f(x,t): the right-hand-side function
%      u0=u0(x): the initial value function
%      g1=g1(t): the boundary value function at x=0
%      g2=g2(t): the boundary value function at x=L
%      L:  right end-point of the spatial interval
%      T:  right end-point of the time interval
%      nx: the number of subintervals of [0,L]
%      nt: the number of subintervals of [0,T]
% Output
%       U: the solution (u_{k,i}), k the time grid point index,
%          i the space grid point index
% compute some parameters
nx1 = nx+1;
hx = L/nx;
nt1 = nt+1;
ht = T/nt;
r = a*ht/(hx*hx)
r1 = 1-2*r;
% stability test
if r > 0.5
  disp('The ratio is too big; the scheme is unstable!')
end
```

```
% generate grid point vectors
xvec = hx*(0:nx);
tvec = ht*(0:nt);
% initialization
U = zeros(nt1,nx1);
U(1,:) = feval(u0,xvec(:)');
% time advancing
U(2:nt1,1) = feval(g1,(1:nt)*ht);
U(2:nt1,nx1) = feval(g2,(1:nt)*ht);
for k = 1:nt
  U(k+1,2:nx) = r*(U(k,1:nx-1)+U(k,3:nx+1))+r1*U(k,2:nx) ...
                +ht*feval(f,xvec(2:nx),tvec(k));
end
% plot the numerical solution
surf(xvec,tvec,U)
xlabel('x-axis')
ylabel('t-axis')
zlabel('the numerical solution')
s1 = sprintf('h_t=%6.4f  h_x=%6.4f',ht,hx)
title(s1)
```

Example 9.2.1 Let us solve the following initial boundary value problem for $t \leq 0.2$:

$$\begin{cases} u_t = u_{xx}, & x \in (0,1),\ t > 0 \\ u(x,0) = \sin(\pi x), & x \in [0,1] \\ u(0,t) = u(1,t) = 0, & t > 0 \end{cases}$$

The true solution is $u(x,t) = e^{-\pi^2 t}\sin(\pi x)$, which is shown in Figure 9.6.

In this example, $a = 1$, $L = 1$, $T = 0.2$. Then $h_t = 0.2/n_t$, $h_x = 1/n_x$. So the stability condition (9.21) is

$$\gamma = \frac{0.2\,n_x^2}{n_t} \leq \frac{1}{2} \tag{9.22}$$

We choose $n_x = 3, 6, 12$ and correspondingly $n_t = 6, 24, 96$ so that $\gamma = 0.3$ and the stability condition (9.22) is satisfied. In Table 9.4, we list the maximum errors $\max_{1\leq i\leq n_x+1} |u(x_i, t_k) - u_i^k|$ for $t_k = 0.2$.

The numerical solution and the corresponding error for $n_x = 12$ and $n_t = 96$ are shown in Figure 9.7. This is done by replacing the last part of the program `Heat1` with the following:

```
% Plot the numerical solution
subplot(1,2,1)
surf(xvec,tvec,U)
```

```
xlabel('x-axis')
ylabel('t-axis')
zlabel('the numerical solution')
s1 = sprintf('h_t=%6.4f h_x=%6.4f',ht,hx)
title(s1)
hold on
% Plot the solution error
Err = exp(-pi*pi*tvec)'*sin(pi*xvec)-U;
subplot(1,2,2)
surf(xvec,tvec,Err)
xlabel('x-axis')
ylabel('t-axis')
zlabel('the error')
title(s1)
```
■

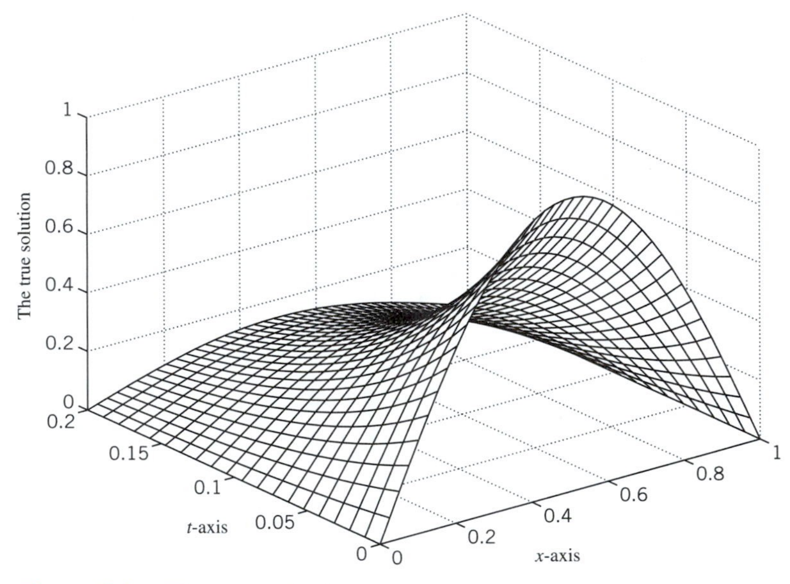

Figure 9.6. The true solution

Table 9.4. Maximum Solution Errors at $t = 0.2$

n_x	n_t	Max. Error
3	6	1.8414E − 2
6	24	5.0829E − 3
12	96	1.2573E − 3

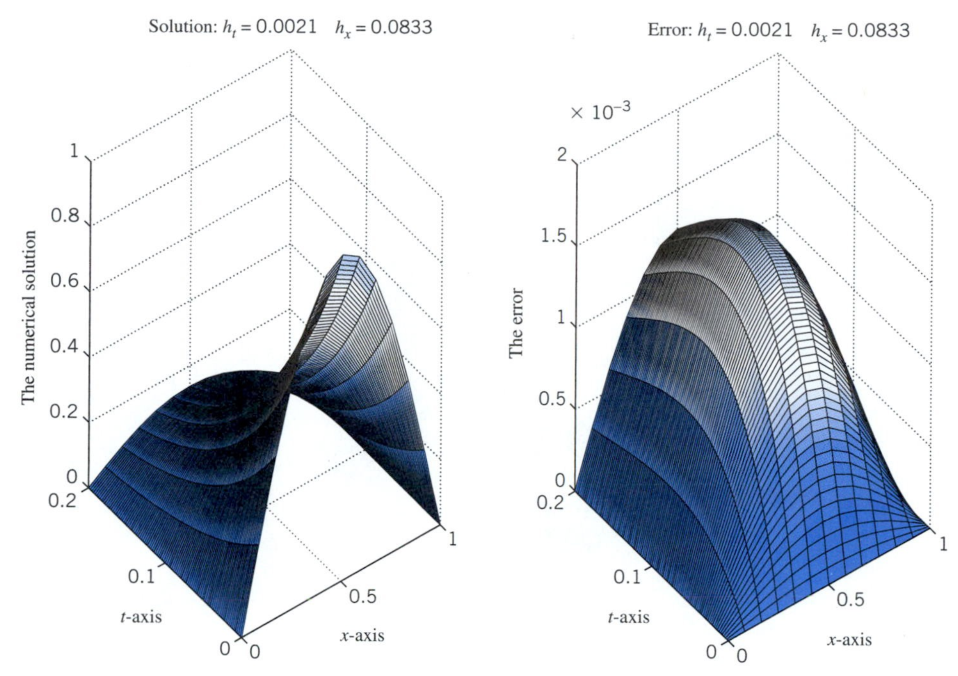

Figure 9.7. Numerical results of the explicit method: $n_x = 12$, $n_t = 96$

Example 9.2.2 To see the role played by the stability condition (9.21), we continue the numerical solution of the initial boundary value problem of Example 9.2.1. We first solve the problem by taking $n_x = 10$ and $n_t = 10$. The results are shown in Figure 9.8. In this case, the ratio $\gamma = 2$, and the scheme is not stable. However, since the difference scheme is applied only 10 times ($n_t = 10$), the accumulation of the round-off errors is not apparent.

To have a more dramatic illustration, we let $n_x = 20$ and $n_t = 20$. The ratio is $\gamma = 4$. The results are shown in Figure 9.9. We observe that the effect of round-off error accumulation becomes rather evident for t close to its upper bound 0.2.

This example shows the importance of maintaining the stability condition (9.21) when we use the explicit scheme (9.19–9.20). ■

9.2.3 Implicit Full Discretization

We have seen that for the explicit fully discrete scheme, h_t and h_x should be chosen so that the stability condition (9.21) is satisfied. The condition (9.21) imposes a restriction on the relative size of h_t with respect to h_x. Suppose the numerical solution corresponding to the current values of h_t and h_x is not accurate enough. It is natural to try smaller values for h_t and h_x. Now assume we use $\bar{h}_x = h_x/2$. To maintain the same value of

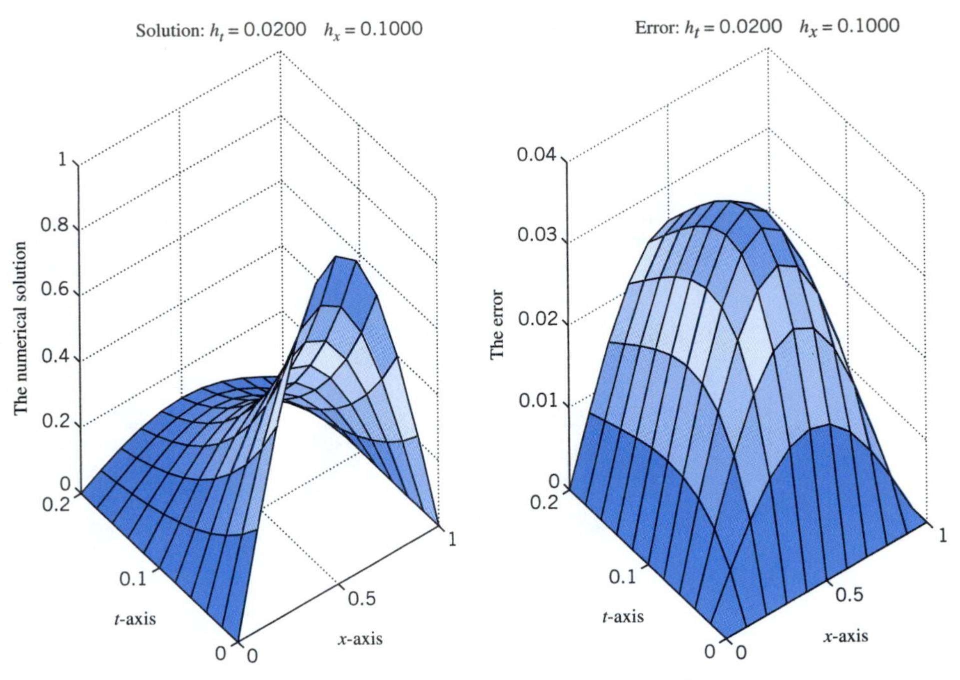

Figure 9.8. Numerical results of the explicit method, $n_x = 20$, $n_t = 10$

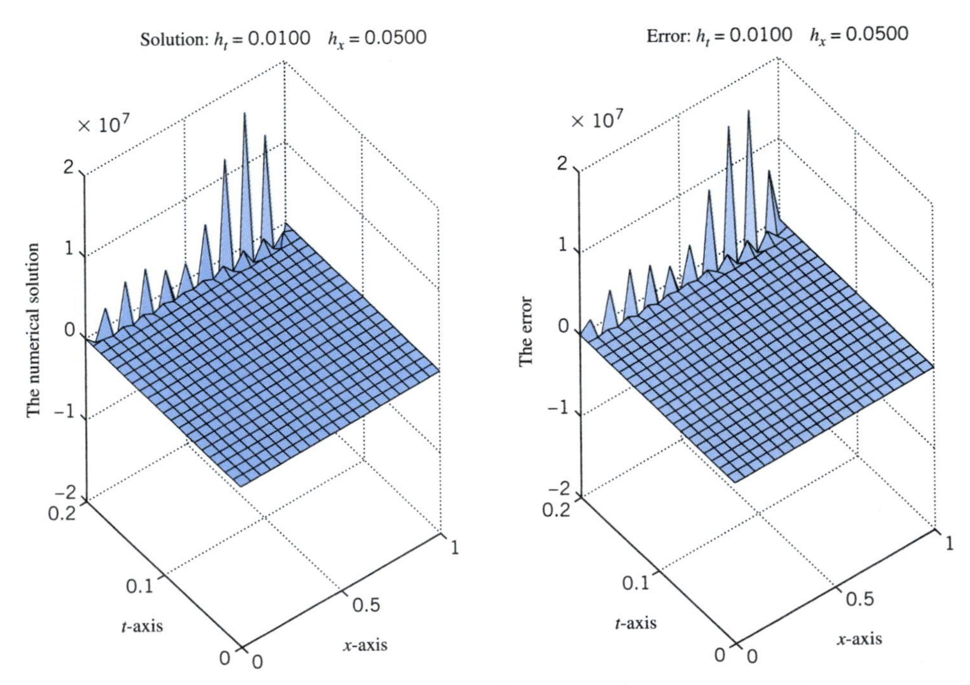

Figure 9.9. Numerical results of the explicit method, $n_x = 20$, $n_t = 20$

the ratio γ, we need to choose $\bar{h}_t$ such that

$$\frac{a\,\bar{h}_t}{\bar{h}_x^2} = \frac{a\,h_t}{h_x^2}$$

that is,

$$\bar{h}_t = \tfrac{1}{4} h_t$$

In other words, whenever h_x is halved, h_t must be quartered. This may lead to the use of prohibitively small time stepsize.

Let us explore the possibility of a numerical method that is always stable. With

$$\frac{\partial u}{\partial t}(x_i, t_k) \approx \frac{u_i^k - u_i^{k-1}}{h_t}$$

we obtain the backward Euler method

$$\frac{u_i^k - u_i^{k-1}}{h_t} = a\,\frac{u_{i+1}^k - 2\,u_i^k + u_{i-1}^k}{h_x^2} + f_i^k, \qquad 2 \le i \le n_x$$

Again, denote the ratio $\gamma = a\,h_t/h_x^2$. Then the numerical scheme is

$$u_i^1 = u_0(x_i), \qquad 1 \le i \le n_x + 1 \tag{9.23}$$

and for $k \ge 2$,

$$\begin{cases} -\gamma\, u_{i-1}^k + (1 + 2\gamma)\, u_i^k - \gamma\, u_{i+1}^k = u_i^{k-1} + h_t f_i^k, & 2 \le i \le n_x \\ u_1^k = g_1(t_k), \qquad u_{n_x+1}^k = g_2(t_k) \end{cases} \tag{9.24}$$

Here for given approximate solution at $t = t_{k-1}$, we need to solve a linear system to find the approximate solution at the next time level $t = t_k$. Such a method is called an *implicit method*.

Using an argument more delicate than that for the stability of the explicit scheme (9.19–9.20), we can show that for any value of the ratio, error propagation with the time advancing of the implicit scheme (9.23–9.24) is well controlled. We say that the implicit scheme is unconditionally stable.

Similar to the forward Euler scheme (9.19–9.20), it can be proved that if the true solution u is sufficiently smooth, the following error bound holds for the backward Euler scheme (9.23–9.24):

$$\max_{\substack{1 \le i \le n_x+1 \\ 1 \le k \le n_t+1}} |u(x_i, t_k) - u_i^k| = O(h_t + h_x^2) \tag{9.25}$$

that is, the scheme is of second-order accuracy in h_x and of first-order accuracy in h_t.

To implement the backward Euler scheme, at each time level $t = t_k$, $k \geq 2$, we need to solve a linear system:

$$A\boldsymbol{u}^k = \boldsymbol{F}^k$$

Here, the coefficient matrix is

$$A = \begin{bmatrix} 1+2\gamma & -\gamma & & & \\ -\gamma & 1+2\gamma & -\gamma & & \\ & \ddots & \ddots & \ddots & \\ & & -\gamma & 1+2\gamma & -\gamma \\ & & & -\gamma & 1+2\gamma \end{bmatrix}_{(n_x-1)\times(n_x-1)}$$

the unknown vector $\boldsymbol{u}^k = [u_2^k, \ldots, u_{n_x}^k]^T$, and the right-hand-side vector

$$\boldsymbol{F}^k = [u_2^{k-1} + h_t f_2^k + \gamma\, g_1(t_k), u_3^{k-1} + h_t f_3^k, \ldots, u_{n_x-1}^{k-1} + h_t f_{n_x-1}^k, \\ u_{n_x}^{k-1} + h_t f_{n_x}^k + \gamma\, g_2(t_{n_x+1})]^T$$

Note that since the coefficient matrix A is the same for any k, we only need to compute its LU factorization once, which can then be used to solve the linear systems for all k. Since A is tridiagonal, we can use the program tridiag from Subsection 6.4.2 for its LU factorization.

MATLAB PROGRAM. The following is a sample MATLAB program for the implicit method:

```
function U = Heat2(a,f,u0,g1,g2,L,T,nx,nt)
%
% function U = Heat2(a,f,u0,g1,g2,L,T,nx,nt)
%
% The backward difference scheme for solving the initial boundary
% value problem of the heat equation u_t=a u_{xx}+f for x in
% [0,L], and t in [0,T].
% Input
%       a: the coefficient of the u_{xx} term
%       f=f(x,t): the right-hand-side function
%       u0=u0(x): the initial value function
%       g1=g1(t): the boundary value function at x=0
%       g2=g2(t): the boundary value function at x=L
%       L:  right end-point of the spatial interval
%       T:  right end-point of the time interval
%       nx: the number of subintervals of [0,L]
%       nt: the number of subintervals of [0,T]
% Output
```

```
%       U: the solution (u_{k,i}), k the time grid point index,
%           i the space grid point index
% compute some parameters
nx1 = nx+1;
hx = L/nx;
nt1 = nt+1;
ht = T/nt;
r = a*ht/(hx*hx)
r1 = 1+2*r;
%  generate the grid points for x and t variables
xvec = hx*(0:nx);
tvec = ht*(0:nt);
%  generate the entries of the coefficient matrix for the linear
%  tridiagonal systems to be solved at each time level
avec(2:nx-1) = -r;
bvec(1:nx-1) = r1;
cvec(1:nx-2) = -r;
%  define the dimensions of the unknown
U = zeros(nt1,nx1);
%  the initial value
U(1,:) = feval(u0,xvec(:)');
%  compute the solution at t=ht; use the program tridiag to
%  factor and store A=LU and solve the system at t=ht
U(2,1) = feval(g1,ht);
U(2,nx1) = feval(g2,ht);
fvec(1) = U(1,2)+ht*feval(f,xvec(2),tvec(2))+r*U(2,1);
fvec(2:nx-2) = U(1,3:nx-1)+ht*feval(f,xvec(3:nx-1),tvec(2));
fvec(nx-1) = U(1,nx)+ht*feval(f,xvec(nx),tvec(2))+r*U(2,nx1);
[U(2,2:nx),avec,bvec,ier] = tridiag(avec,bvec,cvec,fvec,nx-1,0);
%  compute the solution at the other time levels
for k = 2:nt1
  U(k,1) = feval(g1,(k-1)*ht);
  U(k,nx1) = feval(g2,(k-1)*ht);
  fvec(1) = U(k-1,2)+ht*feval(f,xvec(2),tvec(k))+r*U(k,1);
  fvec(2:nx-2) = U(k-1,3:nx-1)+ht*feval(f,xvec(3:nx-1),tvec(k));
  fvec(nx-1) = U(k-1,nx)+ht*feval(f,xvec(nx),tvec(k))+r*U(k,nx1);
  U(k,2:nx) = tridiag(avec,bvec,cvec,fvec,nx-1,1);
end
%  plot the numerical solution
surf(xvec,tvec,U)
xlabel('x-axis')
ylabel('t-axis')
zlabel('the numerical solution')
s1 = sprintf('h_t=%6.4f  h_x=%6.4f',ht,hx)
title(s1)
```

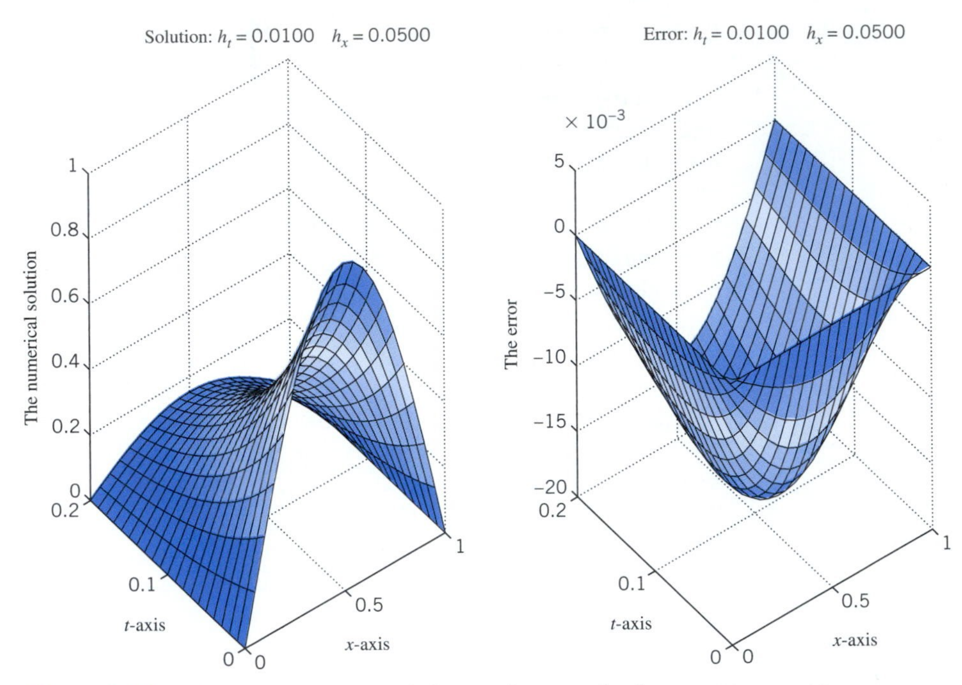

Figure 9.10. Numerical results of the implicit method: $n_x = 20$, $n_t = 20$

Example 9.2.3 Let us use the backward Euler method (9.23–9.24) to solve the problem of Example 9.2.1. Numerical results with $(n_x, n_t) = (20, 20)$ are shown in Figure 9.10. We see that there is no stability problem with the numerical solutions.

To take a closer look at the error behavior, we compute the numerical solution errors and related ratios for several pairs of n_x and n_t. The results are given in Table 9.5. First we use $n_t = n_x = 5$. Then we increase n_t and n_x separately, and compute the ratios of the numerical solution errors with $n_t = n_x = 5$ and those of larger values of n_t and n_x. It is evident that as we double both values of n_t and n_x, the errors are reduced by factors of approximately 2, indicating a linear convergence behavior. When we double the value of n_x and quadruple the value of n_t, the numerical solution errors are reduced

Table 9.5. Numerical Solution Errors at $t = 0.2$ for Example 9.2.3

x	$n_t = 5$ $n_x = 5$	$n_t = 10$ $n_x = 10$	Ratio	$n_t = 20$ $n_x = 10$	Ratio	$n_t = 20$ $n_x = 20$	Ratio
0.2	$-3.50E-2$	$-1.67E-2$	2.09	$-9.15E-3$	3.82	$-8.15E-3$	4.29
0.4	$-5.66E-2$	$-2.70E-2$	2.09	$-1.48E-2$	3.82	$-1.32E-2$	4.29
0.6	$-5.66E-2$	$-2.70E-2$	2.09	$-1.48E-2$	3.82	$-1.32E-2$	4.29
0.8	$-3.50E-2$	$-1.67E-2$	2.09	$-9.15E-3$	3.82	$-8.15E-3$	4.29

by factors of nearly 4 (in the table, 3.82). If we quadruple both n_t and n_x, the error reduction factors are again close to 4 (in the table, 4.29). When the starting values of n_t and n_x are larger, the ratios 3.82 and 4.29 given in Table 9.5 will be closer to 4. This phenomenon is consistent with the theoretical error bound (9.25) that states the method is of first-order in h_t and second-order in h_x. For such a method, to be more efficient, when we double the value n_t, we should quadruple the value n_x. ■

In this section, we have studied two difference methods for solving the initial boundary value problem for the one-dimensional parabolic equations. Both methods have the error bound $O(h_t + h_x^2)$. It is possible to derive schemes with higher convergence order. One example is the Crank–Nicholson scheme with an error bound $O(h_t^2 + h_x^2)$. We refer the reader to Morton and Mayers [25] for discussions of this and other higher-order difference schemes.

PROBLEMS

1. For any constants k, ω, c_1, and c_2, verify that the function

$$u(x,t) = e^{-k\omega^2 t}(c_1 \cos \omega x + c_2 \sin \omega x)$$

satisfies the differential equation

$$u_t = k\, u_{xx}$$

2. Show that the explicit fully discrete scheme (9.20) can be derived by discretizing the ODE system (9.17) with the forward Euler discretization for the time derivative.

3. Show that the implicit fully discrete scheme (9.24) can be derived by discretizing the ODE system (9.17) with the backward Euler discretization for the time derivative.

4. Verify that the function $u(x,t) = e^{-t} \sin x$ solves the initial boundary value problem

$$\begin{cases} u_t = u_{xx}, & 0 < x < \pi,\ t > 0 \\ u(0,t) = u(\pi,t) = 0, & t \ge 0 \\ u(x,0) = \sin x, & 0 \le x \le \pi \end{cases}$$

Use the programs `Heat1` and `Heat2` with several choices of n_x and n_t to solve the problem. Report maximum numerical solution errors at $t = 0.2$, draw graphs of the numerical solutions and the corresponding errors.

5. Verify that the function $u(x,t) = e^{-t} \cos x$ solves the initial boundary value problem

$$\begin{cases} u_t = u_{xx}, & 0 < x < \pi,\ t > 0 \\ u_x(0,t) = u_x(\pi,t) = 0, & t \ge 0 \\ u(x,0) = \cos x, & 0 \le x \le \pi \end{cases}$$

Note that here the boundary conditions are specified on the values of the spatial derivative. Such boundary conditions are examples of Neumann ones.

6. Show that

$$u(x,t) = x^2 - 1 + e^{-t}\cos(\pi x/2)$$

solves the initial boundary value problem

$$\begin{cases} u_t = \dfrac{4}{\pi^2}u_{xx} - \dfrac{8}{\pi^2}, & 0 < x < 1,\ t > 0 \\ u_x(0,t) = 0, \quad u(1,t) = 0, & t \ge 0 \\ u(x,0) = x^2 - 1 + \cos(\pi x/2), & 0 \le x \le 1 \end{cases}$$

7. Solve the equation $u_t = u_{xx}$ for $0 < x < 1$, $0 < t \le 1$ with the initial condition $u(x,0) = 0$ and boundary conditions $u(0,t) = t^2$, $u(1,t) = 0$. Use the explicit scheme (9.19–9.20) with $n_x = 4, 6, 8$ and correspondingly $n_t = 16, 36, 64$. Produce a table of numerical solutions at $x = 1/4, 1/2, 3/4$ and $t = 1$. Draw graphs of the numerical solutions.

8. Repeat Problem 7 but use the implicit scheme (9.23–9.24) with $n_x = n_t = 4$, 6, 8, 16.

9. Consider a heat conduction problem for a homogeneous rod of unit length. Assume the rod is sufficiently thin and its surface is thermally insulated. Then the temperature of the rod, $u = u(x,t)$, satisfies the equation

$$u_t = a\,u_{xx} + f(x,t)$$

where $a = k/(c\rho)$ with c the specific heat, ρ the density, k a thermal conductivity coefficient, and $f(x,t)$ a scaled heat source function. Suppose the temperatures at the two ends of the rod are fixed at 0, the initial rod temperature is 0, $a = 1$, and $f(x,t) = x\,(1-x)\,\sin\pi t$. Compute and plot the temperature at $t = 1$ for several values of n_t and n_x until the numerical solution at $x = 1/2$, $t = 1$ has three stabilized digits.

10. Give a difference scheme for solving the initial boundary value problem in Problem 6. Discretize the boundary condition $u_x(0,t) = 0$ by using the idea of formula (8.143).

11. The idea of the derivation of the difference schemes for the one-dimensional parabolic equation can be extended straightforward to higher-dimensional parabolic equations. As an example, consider the two-dimensional parabolic equation

$$\frac{\partial u}{\partial t} = a\left(\frac{\partial^2 u}{\partial x^2} + \frac{\partial^2 u}{\partial y^2}\right)$$

on the unit square $(0, 1)^2$, together with the initial condition

$$u(x, y, 0) = u_0(x, y)$$

and the boundary conditions

$$\begin{aligned} &u(0, y, t) = g_l(y, t), \quad &u(1, y, t) = g_r(y, t), \quad &0 \le y \le 1,\ t \ge 0 \\ &u(x, 0, t) = g_b(x, t), \quad &u(x, 1, t) = g_t(x, t), \quad &0 \le x \le 1,\ t \ge 0 \end{aligned}$$

Suppose we solve the problem for $t \le T$. Divide the unit interval into $n_x = n_y = n$ equal parts, and the interval $[0, T]$ into n_t parts. Denote $h = 1/n$, $h_t = T/n_t$, $x_i = (i-1)h$, $y_j = (j-1)h$, $t_k = (k-1)h_t$, $1 \le i, j \le n+1$, $1 \le k \le n_t + 1$. An approximation of the true solution value $u(x_i, y_j, t_k)$ is denoted by u_{ij}^k. Set up the finite difference systems for $\{u_{ij}^k\}$, similar to the explicit method (9.19–9.20) and the implicit method (9.23–9.24).

9.3. ONE-DIMENSIONAL WAVE EQUATION

The simplest second-order hyperbolic equation is

$$\frac{\partial^2 u}{\partial t^2} = a\,\frac{\partial^2 u}{\partial x^2} + f \tag{9.26}$$

This equation can describe one-dimensional wave propagations, as well as a variety of other physical processes. Assume the equation is specified for $x \in (0, L)$. To have a unique solution, we need to specify boundary conditions at $x = 0$ and L, and initial value conditions at $t = 0$. Suppose these conditions are chosen as follows:

$$u(0, t) = g_1(t), \qquad u(L, t) = g_2(t), \qquad t \ge 0 \tag{9.27}$$

$$u(x, 0) = u_0(x), \qquad u_t(x, 0) = v_0(x), \qquad 0 \le x \le L \tag{9.28}$$

We need two initial conditions since the equation is second-order in time. The differential equation (9.26), the boundary conditions (9.27), and the initial conditions (9.28) together form an initial boundary value problem. Compatibility between the boundary conditions and initial conditions requires

$$u_0(0) = g_1(0), \qquad u_0(L) = g_2(0)$$

Let us solve the initial boundary value problem for $0 \le t \le T$. As in the case of solving a parabolic equation problem, we can present two approaches to the design of numerical methods. In the first approach, we introduce a semidiscretization converting the problem to a system of ODEs that is then solved by an ODE solver. In the second approach, we discretize both the spatial and time variables simultaneously to obtain a fully discrete scheme. In the remaining part of this section, we consider a standard fully discrete scheme for the initial value problem.

Divide $[0, T]$ into n_t equal parts and denote $h_t = T/n_t$ the time stepsize. Divide $[0, L]$ into n_x equal parts and denote $h_x = L/n_x$ the spatial grid size. Let $x_i = (i-1)\,h_x$, $1 \le i \le n_x + 1$, $t_k = (k-1)\,h_t$, $1 \le k \le n_t + 1$. Denote u_i^k as an approximation of the solution $u(x_i, t_k)$. We approximate both $u_{tt}(x_i, t_k)$ and $u_{xx}(x_i, t_k)$ by second-order central differences. Then an explicit central difference equation at an interior grid point (x_i, t_k) is

$$\frac{u_i^{k+1} - 2\,u_i^k + u_i^{k-1}}{h_t^2} = a\,\frac{u_{i+1}^k - 2\,u_i^k + u_{i-1}^k}{h_x^2} + f_i^k \tag{9.29}$$

The truncation error of the scheme is defined to be the left-hand side minus the right-hand side of the difference equation (9.29) when the numerical solution values are replaced by the true solution values at the corresponding grid points. Since

$$\frac{u(x_i, t_{k+1}) - 2\,u(x_i, t_k) + u(x_i, t_{k-1})}{h_t^2}$$

is an $O(h_t^2)$ approximation of $u_{tt}(x_i, t_k)$, and

$$\frac{u(x_{i+1}, t_k) - 2\,u(x_i, t_k) + u(x_{i-1}, t_k)}{h_x^2}$$

is an $O(h_x^2)$ approximation of $u_{xx}(x_i, t_k)$, the truncation error of the scheme is $O(h_x^2 + h_t^2)$.

The formulas (9.29) at the interior grid points are to be supplemented by numerical boundary value and initial value conditions. The boundary conditions from (9.27) are

$$u_1^k = g_1(t_k),\ u_{n_x+1}^k = g_2(t_k), \qquad 1 \le k \le n_t + 1 \tag{9.30}$$

For initial conditions, we use

$$u_i^1 = u_0(x_i), \qquad 1 \le i \le n_x + 1 \tag{9.31}$$

$$\frac{u_i^2 - u_i^0}{2\,h_t} = v_0(x_i), \qquad 1 \le i \le n_x + 1 \tag{9.32}$$

Here u_i^0, $1 \le i \le n_x + 1$, are artificial variables, intended as approximations of $u(x_i, -h_t)$ when the true solution u is suitably extended for negative t—in particular, the extended solution solves the same equation (9.26) extended for negative t. The artificial variables are introduced so that we can use the divided difference

$$\frac{u(x_i, h_t) - u(x_i, -h_t)}{2\,h_t}$$

as an $O(h_t^2)$ approximation of the derivative $u_t(x_i, 0)$. Note that a simpler way to approximate the initial velocity condition $u_t(x, 0) = v_0(x)$ would be

$$\frac{u_i^2 - u_i^1}{h_t} = v_0(x_i), \qquad 1 \le i \le n_x + 1$$

However, this leads to a lower convergence order of the numerical method and will not be used here.

With the use of the artificial variables u_i^0, $1 \le i \le n_x + 1$, we need difference equations at $(x_i, 0)$ for $2 \le i \le n_x$. We require the difference equation (9.29) to be valid also for $k = 1$ (i.e., $t = 0$). Denote the ratio $\gamma = a\, h_t^2 / h_x^2$. Then we have the following equations:

$$u_i^2 = \gamma\, u_{i-1}^1 + 2\,(1-\gamma)\, u_i^1 + \gamma\, u_{i+1}^1 - u_i^0 + h_t^2 f_i^1$$

and

$$u_i^2 = u_i^0 + 2\, h_t v_0(x_i)$$

from (9.29) and (9.32), respectively. Adding the two equations, we can eliminate u_i^0 to get

$$2\, u_i^2 = \gamma\, u_{i-1}^1 + 2\,(1-\gamma)\, u_i^1 + \gamma\, u_{i+1}^1 + 2\, h_t v_0(x_i) + h_t^2 f_i^1$$

We then have

$$\left\{\begin{array}{l} u_1^2 = g_1(h_t), \qquad u_{n_x+1}^2 = g_2(h_t) \\ u_i^2 = \dfrac{\gamma}{2}\, u_{i-1}^1 + (1-\gamma)\, u_i^1 + \dfrac{\gamma}{2}\, u_{i+1}^1 + h_t v_0(x_i) + \dfrac{h_t^2}{2}\, f_i^1, \\ \qquad\qquad\qquad\qquad\qquad\qquad 2 \le i \le n_x \end{array}\right. \tag{9.33}$$

Summarizing, we use the following steps to determine the numerical solution: First, use (9.31) to compute $\{u_i^1\}_{i=1}^{n_x+1}$, and use (9.33) to compute $\{u_i^2\}_{i=1}^{n_x+1}$. Then, for $k = 2, \ldots, n_t$, compute

$$\left\{\begin{array}{l} u_1^{k+1} = g_1(k\, h_t), \qquad u_{n_x+1}^{k+1} = g_2(k\, h_t) \\ u_i^{k+1} = \gamma u_{i-1}^k + 2\,(1-\gamma)\, u_i^k + \gamma u_{i+1}^k - u_i^{k-1} + h_t^2 f_i^k, \\ \qquad\qquad\qquad\qquad\qquad\qquad 2 \le i \le n_x \end{array}\right. \tag{9.34}$$

It can be shown that the stability condition is $\gamma \le 1$, that is,

$$\sqrt{a}\, h_t \le h_x$$

This condition is not as restrictive as (9.21) for the case of solving a parabolic equation.

When the true solution u is sufficiently smooth, a theoretical result on the error bound is

$$\max_{\substack{1 \le i \le n_x+1 \\ 1 \le k \le n_t+1}} |u(x_i, t_k) - u_i^k| = O(h_t^2 + h_x^2) \tag{9.35}$$

that is, the order of the method is the same as that of the truncation error, when the approximations of the boundary conditions are at least of the same order of accuracy. So the scheme is of second-order in both x and t stepsizes.

MATLAB PROGRAM. A sample program implementing the above scheme is the following:

```
function U = Wave(a,f,u0,v0,g1,g2,L,T,nx,nt)
%
% function U=Wave(a,f,u0,v0,g1,g2,L,T,nx,nt)
%
% The centered difference scheme for solving the initial boundary
% value problem of the wave equation u_{tt}=a u_{xx}+f for
% x in [0,L], and t in [0,T].
% Input
%       a: the coefficient of the u_{xx} term
%       f=f(x,t): the right-hand-side function
%       u0=u0(x): the initial value function
%       v0=v0(x): the initial derivative value function
%       g1=g1(t): the boundary value function at x=0
%       g2=g2(t): the boundary value function at x=L
%       L:  right end-point of the spatial interval
%       T:  right end-point of the time interval
%       nx: the number of subintervals of [0,L]
%       nt: the number of subintervals of [0,T]
% Output
%        U: the solution (u_{k,i}), k the time grid point index,
%           i the space grid point index
% compute some parameters
nx1 = nx+1;
hx = L/nx;
nt1 = nt+1;
ht = T/nt;
ht2 = ht*ht;
r = a*ht*ht/(hx*hx)
r1 = 2*(1-r);
if r > 1
  disp('The ratio is too big; the scheme is unstable!')
end
```

```
% define the grid points
xvec = hx*(0:nx);
tvec = ht*(0:nt);
% compute the solution at t=0 and ht using the initial values
U = zeros(nt1,nx1);
U(1,:) = feval(u0,xvec);
U(2,1) = feval(g1,ht);
U(2,2:nx) = 0.5*(r*(U(1,1:nx-1)+U(1,3:nx+1))+r1*U(1,2:nx) ...
           +ht2*feval(f,xvec(2:nx),tvec(1)) ...
           +ht*feval(v0,xvec(2:nx));
U(2,nx1) = feval(g2,ht);
% compute the solution at the other times
for k = 2:nt
  U(k+1,1) = feval(g1,k*ht);
  U(k+1,nx1) = feval(g2,k*ht);
  U(k+1,2:nx) = r*(U(k,1:nx-1)+U(k,3:nx+1))+r1*U(k,2:nx) ...
               -U(k-1,2:nx)+ht2*feval(f,xvec(2:nx),tvec(k));
end
% plot the numerical solution
surf(xvec,tvec,U)
xlabel('x-axis')
ylabel('t-axis')
zlabel('the numerical solution')
s1=sprintf('h_t=%6.4f  h_x=%6.4f',ht,hx)
title(s1)
```

Example 9.3.1 Let us solve the following initial boundary value problem:

$$\begin{cases} \dfrac{\partial^2 u}{\partial t^2} = \dfrac{\partial^2 u}{\partial x^2} + 2\,e^{-t}\sin x, \quad x \in (0,\pi),\ t \in (0,1) \\ u(x,0) = \sin x, \qquad u_t(x,0) = -\sin x, \quad x \in [0,\pi] \\ u(0,t) = u(\pi,t) = 0, \quad t \in [0,1] \end{cases}$$

The true solution is $u(x,t) = e^{-t}\sin x$, and is shown in Figure 9.11.

We solve the problem with $(n_x, n_t) = (5,5)$, $(10,10)$, and $(20,20)$. Since the true solution is known in this example, we also compute the numerical solution errors. The numerical results with $(n_x, n_t) = (20,20)$ are shown in Figure 9.12.

To see more clearly the error behavior, we report in Table 9.6 the maximum numerical solution errors $\max_{1\le i\le n_x+1} |u(x_i,t_k) - u_i^k|$ at $t_k = 0.2, 0.4, 0.6, 0.8, 1$. We observe that the ratios are all close to 4, indicating a convergence order of 2 for the method.

In Table 9.7 we provide the numerical solutions errors and the corresponding ratios at $t = 1$ and $x = \pi/5, 2\pi/5, 3\pi/5$, and $4\pi/5$. We observe that the ratios are all close to 4, again agreeing with the theoretical result that the method is of second-order. ■

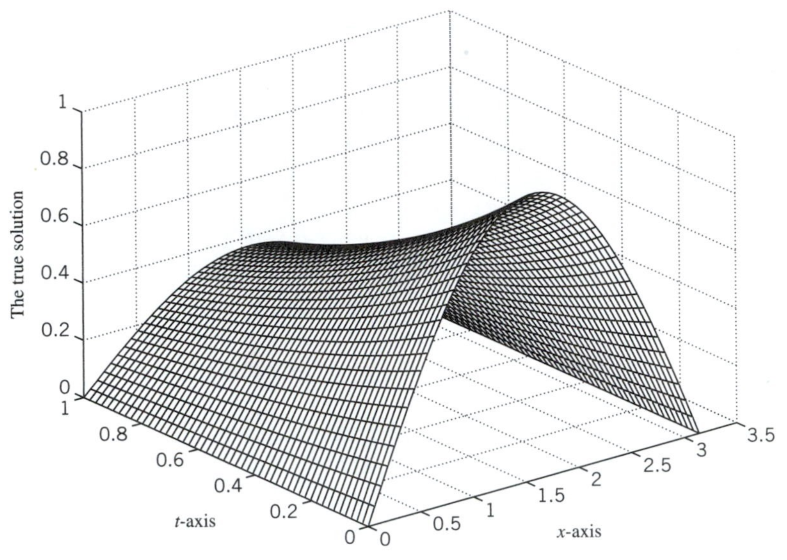

Figure 9.11. The true solution of Example 9.3.1

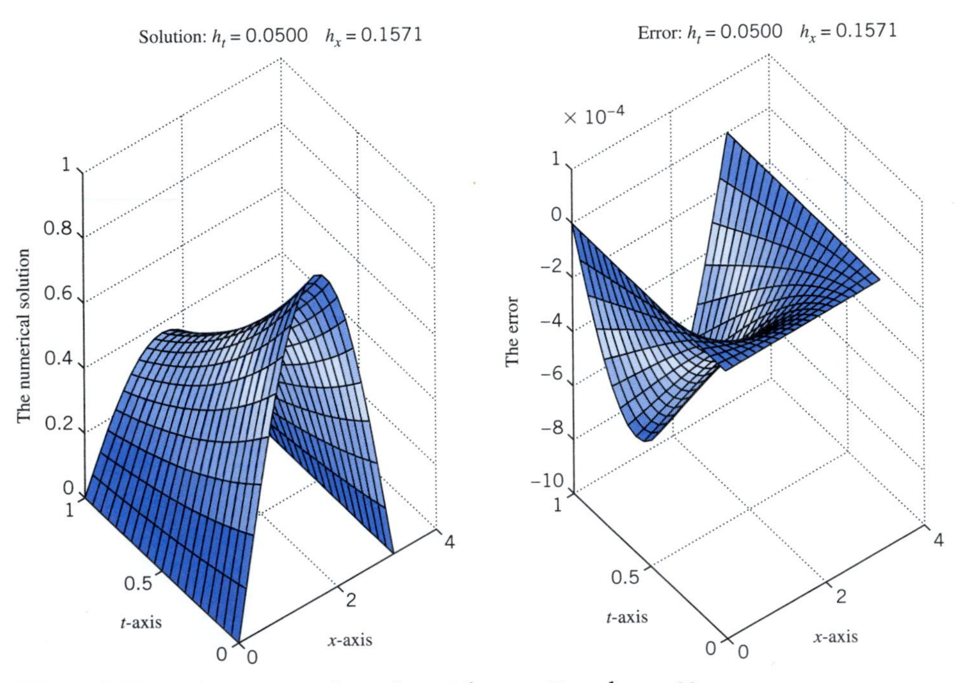

Figure 9.12. The numerical results with $n_x = 20$ and $n_t = 20$

Table 9.6. Maximum Numerical Solution Errors for Example 9.3.1

t	$n = 5$	$n = 10$	Ratio	$n = 20$	Ratio
0.2	1.8247E − 3	4.7175E − 4	3.87	1.1744E − 4	4.02
0.4	4.4861E − 3	1.1660E − 3	3.85	2.9066E − 4	4.01
0.6	7.7168E − 3	2.0116E − 3	3.84	5.0182E − 4	4.01
0.8	1.1257E − 2	2.9392E − 3	3.83	7.3354E − 4	4.01
1.0	1.4860E − 3	3.8829E − 3	3.83	9.6925E − 4	4.01

Table 9.7. Numerical Solution Errors at $t = 1$ for Example 9.3.1

x	$n = 5$	$n = 10$	Ratio	$n = 20$	Ratio
$\pi/5$	−9.18E − 3	−2.28E − 3	4.02	−5.70E − 4	4.01
$2\pi/5$	−1.49E − 2	−3.69E − 3	4.02	−9.22E − 4	4.01
$3\pi/5$	−1.49E − 2	−3.69E − 3	4.02	−9.22E − 4	4.01
$4\pi/5$	−9.18E − 3	−2.28E − 3	4.02	−5.70E − 4	4.01

PROBLEMS

1. Let F and G be twice differentiable functions. Show that the function

$$u(x, t) = F(x + \sqrt{a}\, t) + G(x - \sqrt{a}\, t)$$

satisfies the equation

$$u_{tt} = a\, u_{xx}$$

2. Consider the homogeneous wave equation

$$u_{tt} = a\, u_{xx}$$

on the whole real line $x \in \mathbb{R}$. Since the spatial domain is unbounded in both directions, we only need to specify the initial conditions

$$u(x, 0) = u_0(x), \qquad u_t(x, 0) = v_0(x), \qquad x \in \mathbb{R}$$

Verify the solution formula

$$\begin{aligned} u(x, t) &= \frac{1}{2}\left[u_0(x + \sqrt{a}\, t) + u_0(x - \sqrt{a}\, t)\right] + \frac{1}{2\sqrt{a}} \int_{x-\sqrt{a}\, t}^{x+\sqrt{a}\, t} v_0(s)\, ds \\ &\equiv u_1(x, t) + u_2(x, t) \end{aligned}$$

This is called D'Alembert's formula. The part $u_1(x, t)$ represents the propagation of the initial displacement for zero initial velocity, whereas the other part $u_2(x, t)$ describes the vibrations caused by the initial velocity with zero initial displacement. The quantity $\sqrt{a}$ can be interpreted as the wave propagation speed.

3. Consider a system of two first-order equations

$$\begin{aligned} u_t(x, t) &= v_x(x, t) + f_1(x, t) \\ v_t(x, t) &= u_x(x, t) + f_2(x, t) \end{aligned}$$

where $f_1(x, t)$ and $f_2(x, t)$ are differentiable functions. Show that u and v each satisfy a wave equation of the form (9.26).

4. The motion of a vibrating string is described by the wave equation

$$u_{tt} = \tfrac{1}{2} u_{xx}, \qquad 1 < x < 1,\ t > 0$$

Suppose the two ends of the string are fixed, the string is initially at rest, and the initial position is

$$u(x, 0) = e^x \sin(\pi x), \qquad 0 < x < 1$$

Compute and graph the finite difference approximations of the displacement $u(x, t)$ at $t = 1$ with $n_x = n_t = 5$, 10, and 20.

5. Repeat Problem 4 when the initial position condition is replaced by the following:

$$u(x, 0) = \frac{1}{4}\min(x, 1 - x) = \begin{cases} \dfrac{x}{4}, & 0 \le x \le \dfrac{1}{2} \\ \dfrac{1 - x}{4}, & \dfrac{1}{2} < x \le 1 \end{cases}$$

6. Repeat Problem 4 with the following changes in the data: The initial displacement and initial velocity are both assumed to be zero, the left end of the string is fixed, and for the right end, $u(1, t) = 0.1 \sin(\pi t)$.

7. Consider the planar problem of small transverse vibrations of an elastic string of length L, whose two ends are fixed at $x = 0$ and $x = L$ on the x-axis. Let the string be subject to the action of a vertical force of density $f(x, t)$ per unit mass. Assume the displacement vector is perpendicular to the x-axis and is small in size. Then the equation for the vertical displacement function, $u(x, t)$, is

$$u_{tt} = a\, u_{xx} + f(x, t)$$

where $a = T_0/\rho$, with T_0, the tension of the string, and ρ, the mass density of the string, both assumed constant. Suppose $L = 1$, $a = 1$. Compute the vertical displacement for the following cases:

(a) $f(x,t) = \sin \pi x$, $u(x,0) = 0$ and $u_t(x,0) = 0$.

(b) $f(x,t) = 0$, $u(x,0) = 0$ and $u_t(x,0) = 1$.

8. Use the idea of the difference formula (9.32) for the approximation of $u_t(x_i, 0)$ to derive similar difference formulas for approximations of the quantities $u_x(0, t_k)$ and $u_x(L, t_k)$. Consider the initial boundary value problem (9.26), (9.28), with (9.27) replaced by

$$u_x(0,t) = g_1(t), \qquad u(L,t) = g_2(t)$$

Derive the corresponding finite difference system.

9. Modify the MATLAB program `Wave` to solve the initial boundary value problem

$$\begin{cases} \dfrac{\partial^2 u}{\partial t^2} = \dfrac{\partial^2 u}{\partial x^2} + \left(1 + \dfrac{\pi^2}{4}\right) e^{-t} \cos \dfrac{\pi x}{2}, & x \in (0,1),\ t \in (0,1) \\ u(x,0) = \cos \dfrac{\pi x}{2}, \qquad u_t(x,0) = -\cos \dfrac{\pi x}{2}, & x \in [0,1] \\ u_x(0,t) = u(1,t) = 0, \quad t \in [0,1] \end{cases}$$

The true solution if $u(x,t) = e^{-t} \cos(\pi x/2)$. Compute the numerical solutions with $n_x = n_t = 5$, 10, and 20, and plot the error functions. Give the solution errors at $t = 1$ and $x = i/5$, $0 \le i \le 4$, and comment on the numerical results.

Appendix A

Mean Value Theorems

This appendix discusses mean value theorems. These are often viewed as unnecessary for someone interested in applications; but, in fact, they are quite useful in deriving numerical methods, obtaining error estimates, and simplifying complicated expressions.

The mean value theorems are about functions $f(x)$ on an interval $[a, b]$. The theorems assert the existence of special points $x = c$ in $[a, b]$ for which some property of $f(x)$ is true. Since these theorems are discussed thoroughly in most calculus texts, we will just state the theorems and illustrate their use.

Theorem A.1 (Intermediate Value Theorem) Let $f(x)$ be a continuous function on the interval $a \leq x \leq b$. Let

$$M = \max_{a \leq x \leq b} f(x), \qquad m = \min_{a \leq x \leq b} f(x)$$

Then for every value v satisfying $m \leq v \leq M$, there is at least one point c in $[a, b]$ for which $f(c) = v$.

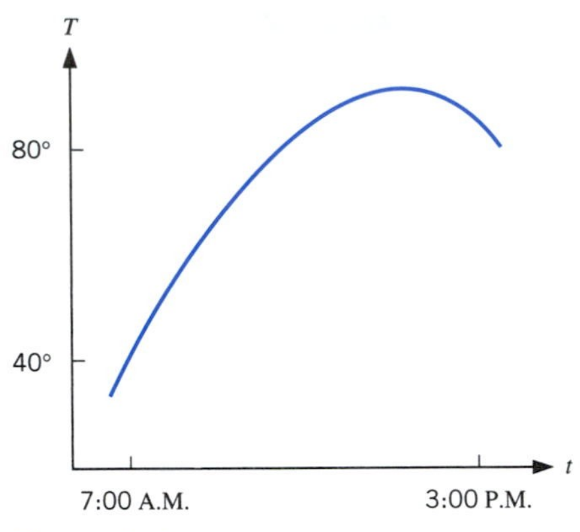

Figure A.1. Temperature versus time

This is an intuitive result. For example, consider the variation of temperature $T(t)$ throughout the day, where $T(t)$ denotes the temperature at time t. Generally, $T(t)$ is considered to be a continuous function of t. Thus, if T is 45°F at 7:30 a.m. and is 83°F at 3:00 p.m., then every temperature between 45°F and 83°F occurs at least once between 7:30 a.m. and 3:00 p.m. A possible graph of $T(t)$ is given in Figure A.1.

Example A.2 Find the values of c in $0 \le x \le \pi$ for which

$$\sin(c) = \tfrac{1}{2}$$

The graph of $\sin(x)$ is given in Figure A.2. Since the minimum of $\sin(x)$ on $[0, \pi]$ is $m = 0$ and the maximum is $M = 1$, we are guaranteed that there is a solution c to the equation. From a knowledge of the function $\sin(x)$, the critical values are easily shown to be $c = \pi/6$ and $5\pi/6$. ■

An important use of the intermediate value theorem is in simplifying complicated expressions. As a special case of a more general expression, consider

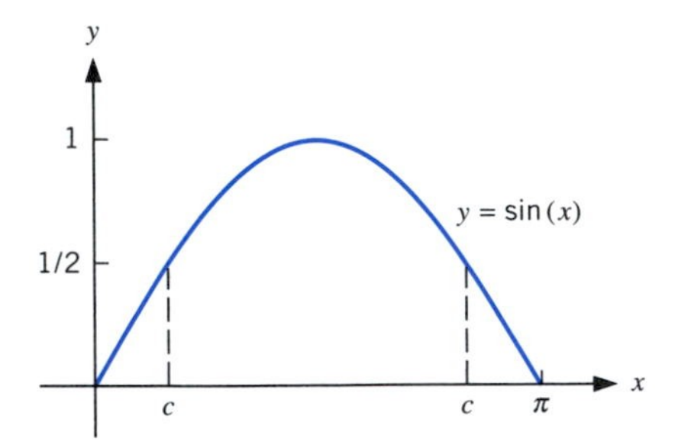

Figure A.2. Illustration of intermediate value theorem

$$S = \tfrac{1}{3}[f(x_1) + f(x_2) + f(x_3)]$$

where $f(x)$ is continuous on an interval $[a, b]$ and x_1, x_2, and x_3 are points in that interval. Using the notation of Theorem A.1, we can show

$$S \leq \tfrac{1}{3}[M + M + M] = M$$

and

$$S \geq \tfrac{1}{3}[m + m + m] = m$$

Thus, S is a value in the interval $[m, M]$; and by the theorem, there is a point c in the interval for which $S = f(c)$, or

$$\tfrac{1}{3}[f(x_1) + f(x_2) + f(x_3)] = f(c) \tag{A.1}$$

for some c in $[a, b]$.

Let $f(x)$ be continuous on $[a, b]$ and let $x_1, x_2, \ldots, x_n$ be points in $[a, b]$. Let $w_1, w_2, \ldots, w_n$ be positive numbers for which

$$\sum_{j=1}^{n} w_j = 1$$

Define

$$S = \sum_{j=1}^{n} w_j f(x_j)$$

Then by an argument analogous to that of the last paragraph, there is a point c in $[a, b]$ for which $S = f(c)$; or

$$\sum_{j=1}^{n} w_j f(x_j) = f(c) \tag{A.2}$$

for some c in $[a, b]$.

Example A.3 Let $f(x) = \sqrt{x}$, and consider

$$S = \tfrac{1}{5}[f(0) + f(0.25) + f(0.5) + f(0.75) + f(1)]$$

Here there are $n = 5$ points, and all $w_j = 1/5$. Let us find the value c of (A.2). First, $S \doteq 0.61463$. Thus, we want

$$\sqrt{c} \doteq 0.61463$$
$$c \doteq (0.61463)^2 \doteq 0.3777$$

S is called an "average value" or "mean value" of the function $\sqrt{x}$ on $[0, 1]$, and it is attained at the above value of c. ■

Theorem A.4 (Mean Value Theorem) Let $f(x)$ be continuous for $a \le x \le b$ and differentiable for $a < x < b$. Then there is at least one point c in (a, b) for which

$$f(b) - f(a) = f'(c)(b - a) \tag{A.3}$$

Example A.5 Find the values of c for which (A.3) is true for $f(x) = \sqrt{x}$ on $[0, 2]$. Figure A.3 is an illustration of the theorem in this case. The dash line connects the points $(0, 0)$ and $(2, \sqrt{2})$. By (A.3),

$$\sqrt{2} - \sqrt{0} = \frac{1}{2\sqrt{c}}(2 - 0)$$

This equation has the solution $c = 1/2$. The graph shows the tangent line at $\left(\frac{1}{2}, f\left(\frac{1}{2}\right)\right)$. ■

Note that by rewriting (A.3) as

$$\frac{f(b) - f(a)}{b - a} = f'(c)$$

we have a simple geometric interpretation of the theorem. It says that the straight line connecting the endpoints of the graph has the same slope as that of some tangent line to the curve at an intermediate point $(c, f(c))$. Check this with Figure A.3.

One of the uses of (A.3) is to estimate or bound the effects of an error in the argument of a function. Let x_T denote the true value of an argument, often unknown, and let x_A denote a known approximate value. Then

$$f(x_T) - f(x_A) = f'(c)(x_T - x_A)$$

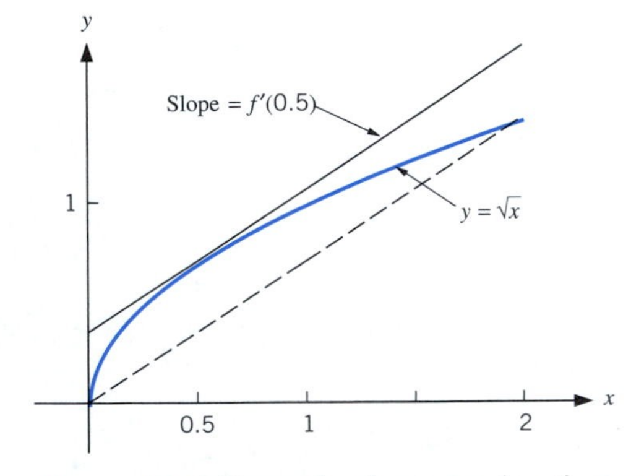

Figure A.3. Example of mean value theorem

with c some number between x_T and x_A. Since x_T and x_A are generally very close together, the value of $f'(x)$ will usually not vary a great deal for x between x_T and x_A and, thus,

$$f(x_T) - f(x_A) \approx \begin{cases} f'(x_T)(x_T - x_A) \\ \text{or} \\ f'(x_A)(x_T - x_A) \end{cases} \tag{A.4}$$

Use the more convenient formula for your particular problem.

Example A.6 Consider evaluating $\cos(\pi/6)$ using $\pi/6 \doteq 0.524$. To use (A.4), we let $x_T = \pi/6$, $x_A = 0.524$, and $f(x) = \cos(x)$. In general, $x_T - x_A$ will not be known and a bound for it will be used. But for illustrative purposes, we have $x_T - x_A \doteq -0.0004$. Using (A.4) gives us

$$\begin{aligned} \cos\left(\frac{\pi}{6}\right) - \cos(0.524) &\doteq -\sin\left(\frac{\pi}{6}\right)\left(\frac{\pi}{6} - 0.524\right) \\ &\doteq -\frac{1}{2}(-0.0004) = 0.0002 \end{aligned}$$

The actual error is 0.0002007. ■

The mean value theorem is also useful in obtaining bounds for the variation of a function based on the variation of the argument. If $|f'(x)|$ is bounded by a constant M everywhere on an interval $a \le x \le b$, then

$$|f(x) - f(z)| \le M\,|x - z| \tag{A.5}$$

for all x, z in $[a, b]$.

Example A.7 Take $f(x) = e^x$ for $0 \le x \le 1$. Then $f'(x)$ is bounded by e^1, and thus

$$\left|e^x - e^z\right| \le e\,|x - z|, \qquad 0 \le x, z \le 1 \quad ■$$

The mean value theorem is used in a number of ways other than those described here, and these occur in several places in the text. Also, the mean value theorem is a special case of Taylor's theorem, described in Sections 1.1 and 1.2 of Chapter 1.

Theorem A.8 (Integral Mean Value Theorem) Let $w(x)$ be a nonnegative integrable function on $[a, b]$, and let $f(x)$ be continuous on $[a, b]$. Then there is at least one point c in $[a, b]$ for which

$$\int_a^b f(x)w(x)\,dx = f(c)\int_a^b w(x)\,dx \tag{A.6}$$

In particular, if we take $w(x) \equiv 1$, then

$$\int_a^b f(x)\,dx = f(c)(b-a) \tag{A.7}$$

for some c in $[a, b]$.

A geometric interpretation of (A.7) is given in Figure A.4. For $f(x)$ nonnegative as well as continuous, the left side of (A.7) represents the area under the curve of $y = f(x)$ between $x = a$ and $x = b$. The right side of (A.7) represents the area of a rectangle with base $[a, b]$ and height $f(c)$. The theorem asserts that these two areas are equal for some choice of $f(c)$.

The result (A.6) is used in simplifying, estimating, and bounding integrals that occur in a number of situations. Many of these occur in connection with Taylor's theorem, in Section 1.2 of Chapter 1. As an example of the use of the integral mean value theorem, consider the integral

$$I = \int_a^b (b-x)(x-a)f(x)\,dx$$

Let $w(x) = (b-x)(x-a)$, and apply (A.6) to get

$$I = f(c)\int_a^b (b-x)(x-a)\,dx = \frac{(b-a)^3}{6} f(c) \tag{A.8}$$

for some c in $[a, b]$. This integral occurs in the analysis of the trapezoidal numerical integration rule, in Section 5.2 of Chapter 7.

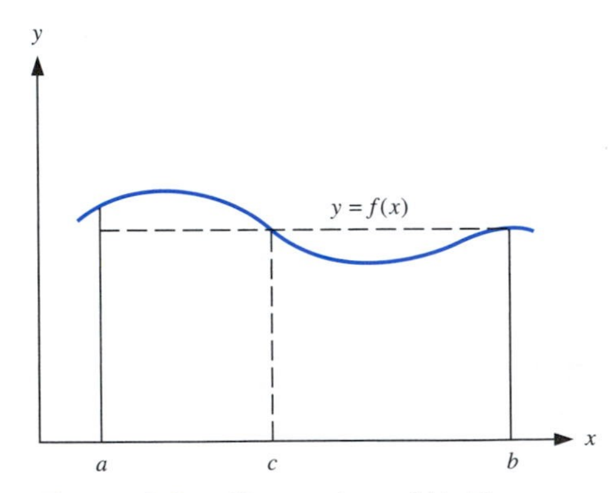

Figure A.4. Illustration of (A.7)

Example A.9 As a more concrete example, consider evaluating

$$g(t) = \int_0^t \frac{x^n}{1+x^2}\,dx, \qquad n \geq 0$$

Take $w(x) = x^n$ and $f(x) = 1/(1+x^2)$. Then from (A.6),

$$g(t) = \frac{1}{1+c^2}\int_0^t x^n\,dx = \left(\frac{1}{1+c^2}\right)\left(\frac{t^{n+1}}{n+1}\right)$$

for some c between 0 and t. Thus for $t \geq 0$,

$$\left(\frac{1}{1+t^2}\right)\frac{t^{n+1}}{n+1} \leq g(t) \leq \frac{t^{n+1}}{n+1} \qquad \blacksquare$$

PROBLEMS

1. Illustrate the intermediate value theorem by finding the values of c for which $f(c) = v$ for the given value of v.

   (a) $f(x) = 2x^2 - 5x + 2,\ v = -1,\ -5 \leq x \leq 5$

   (b) $f(x) = \cos(x),\ v = 1/2,\ \pi \leq x \leq 3\pi$

   (c) $f(x) = \sin(x) + \cos(x),\ v = \sqrt{2},\ 0 \leq x \leq \pi$

2. If $f'(x) > 0$ for $a \leq x \leq b$, what can be said about the number of solutions of $f(x) = v$?

3. Find the average value on $[a, b]$ of $f(x)$, based on (A.2) and the procedure given in the example following it. Also, find the point c in $[a, b]$ at which this average value is actually attained.

   (a) $f(x) = x^2,\ 0 \leq x \leq 1,\ n = 3$, and $n = 6$

   (b) $f(x) = \sin(x),\ 0 \leq x \leq \pi,\ n = 3$, and $n = 6$

4. Illustrate the mean value theorem by finding the value of c for which (A.3) is satisfied with the following cases of $f(x)$ and $[a, b]$:

   (a) $f(x) = \cos(x),\ 0 \leq x \leq \pi/2$

   (b) $f(x) = x^3,\ 0 \leq x \leq 1$

   (c) $f(x) = x^3 - x,\ -2 \leq x \leq 2$

5. Estimate the error in evaluating $f(x)$ with x_A rather than x_T.

   (a) $f(x) = \sqrt{x},\ x_T = \pi,\ x_A = 22/7$

   (b) $f(x) = \log(x),\ x_T = e,\ x_A = 2.72$

   (c) $f(x) = \cos(x),\ x_T = \sqrt{2},\ x_A = 1.414$

6. Derive the following inequalities:

(a) $|\cos(x) - \cos(z)| \le |x - z|$, for all x, z

(b) $|e^x - e^z| \le |x - z|$, for all $x, z \le 0$

(c) $|\tan(x) - \tan(z)| \ge |x - z|$, $-\pi/2 < x, z < \pi/2$

(d) $py^{p-1}(x - y) \le x^p - y^p \le px^{p-1}(x - y)$, $0 \le y \le x$, $p \ge 1$

7. Illustrate the integral mean value theorem by finding the points c for which the following results are true:

(a) $\displaystyle\int_0^1 x^3\,dx = c^3$, c in $[0, 1]$

(b) $\displaystyle\int_0^1 \frac{x - (1/2)}{\sqrt{x}}\,dx = \left(c - \frac{1}{2}\right)\int_0^1 \frac{dx}{\sqrt{x}} = 2\left(c - \frac{1}{2}\right)$, c in $[0, 1]$

(c) $\displaystyle\int_0^{\pi/2} x\sin(x)\,dx = \sin(c)\int_0^{\pi/2} x\,dx = \frac{\pi^2}{8}\sin(c)$, c in $[0, \pi/2]$

8. Use the integral mean value theorem to obtain the following results:

(a) $\displaystyle\int_0^h x^2(h - x)^2 g(x)\,dx = \frac{h^5}{30}g(c)$, c in $[0, h]$

(b) $\displaystyle\int_a^x (t - a)^n f(t)\,dt = \frac{(x - a)^{n+1}}{n + 1}f(c)$, c in $[a, x]$

9. An average value of $f(x)$ on $[a, b]$ can be found using (A.2), with

$$\text{Average value} = \frac{1}{n}\sum_{j=1}^{n} f(x_j)$$

Here

$$x_1 = a,\ x_2 = a + h,\ x_3 = a + 2h, \ldots, x_n = b, \text{ and } h = (b - a)/(n - 1)$$

If we write this as

$$\frac{1}{b - a}\sum_{j=1}^{n} f(x_j)\left(\frac{b - a}{n}\right)$$

the sum is recognizable as a Riemann sum or approximate integral. Letting $n \to \infty$, we have an increasingly accurate average value of $f(x)$ on $[a, b]$; in the limit, it is equal to

$$A = \frac{1}{b - a}\int_a^b f(x)\,dx$$

If we use (A.7), this too can be written as $A = f(c)$ for some c in $[a, b]$. Use the integral formula for A to compute the average value of the functions in Problem 3 and compare the outcome with the earlier results of that problem.

APPENDIX B

MATHEMATICAL FORMULAS

B.1. ALGEBRA

(a) The quadratic formula. The solutions of $ax^2 + bx + c = 0$, $a \neq 0$, are given by

$$x = \frac{-b \pm \sqrt{b^2 - 4ac}}{2a} = \frac{2c}{-b \mp \sqrt{b^2 - 4ac}}$$

(b) Logarithms and exponentials. For $a > 0$,

$$a^x = e^{x \log a}, \qquad \log_a x = \frac{\log x}{\log a}$$

with $\log(x) = \log_e x = \ln x$, the logarithm of x to the base e.

(c) Factorials. $0! = 1$

$$\begin{aligned} n! &= n \cdot [(n-1)!] \\ &= n(n-1)(n-2)\cdots(2)(1), \qquad n \geqq 1 \end{aligned}$$

(d) Sums.

$$\sum_{j=1}^{n} j = 1 + 2 + 3 + \cdots + n = \frac{n(n+1)}{2}$$

$$\sum_{j=1}^{n} j^2 = 1 + 4 + 9 + \cdots + n^2 = \frac{n(n+1)(2n+1)}{6}$$

$$\sum_{j=0}^{n} r^j = 1 + r + r^2 + \cdots + r^n = \frac{r^{n+1}-1}{r-1}, \qquad r \neq 1$$

$$(a+b)^n = \sum_{j=0}^{n} \binom{n}{j} a^{n-j} b^j = \binom{n}{0} a^n + \binom{n}{1} a^{n-1} b + \cdots + \binom{n}{n} b^n$$

$$\binom{n}{j} = \frac{n!}{j!(n-j)!}$$

(e) The fundamental theorem of algebra. If $p(x)$ is a polynomial of degree n, there is a set of numbers $r_1, r_2, \ldots, r_n$ (possibly complex), for which

$$p(x) = c(x - r_1)(x - r_2)\cdots(x - r_n)$$

where c is the coefficient of x^n in $p(x)$. The numbers $\{r_j\}$ are unique, except for their order.

B.2. GEOMETRY

(a) Areas.

(i) Area of a triangle

$$A = \tfrac{1}{2}bh \qquad (b = \text{base},\ h = \text{height})$$

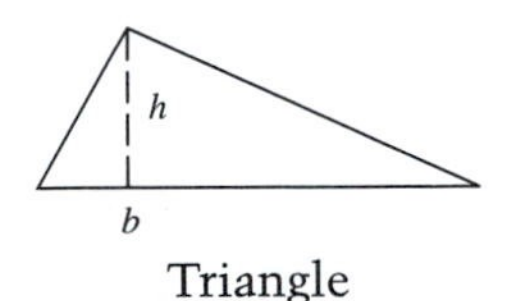

Triangle

(ii) Area of a trapezoid

$$A = h\left(\frac{b_1 + b_2}{2}\right)$$

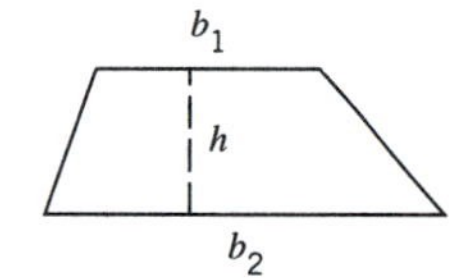

Trapezoid

with b_1 and b_2 the lengths of the parallel sides.

(iii) Area of a circle

$$A = \pi r^2 \qquad (r = \text{radius})$$

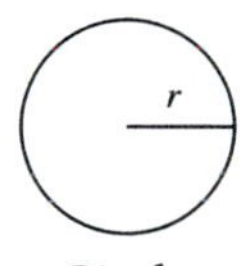

Circle

(iv) Area of a sector of a circle with central angle θ radians

$$A = \tfrac{1}{2}r^2\theta$$

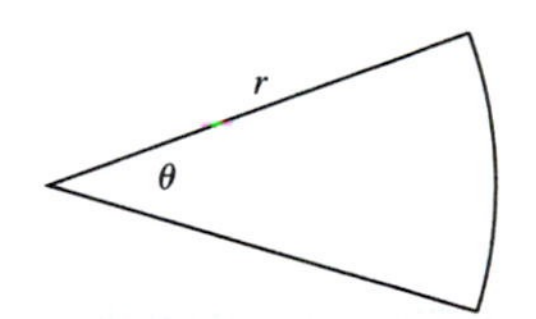

Circular sector

(v) Area of the ellipse with equation

$$\frac{x^2}{a^2} + \frac{y^2}{b^2} = 1 \qquad A = \pi ab$$

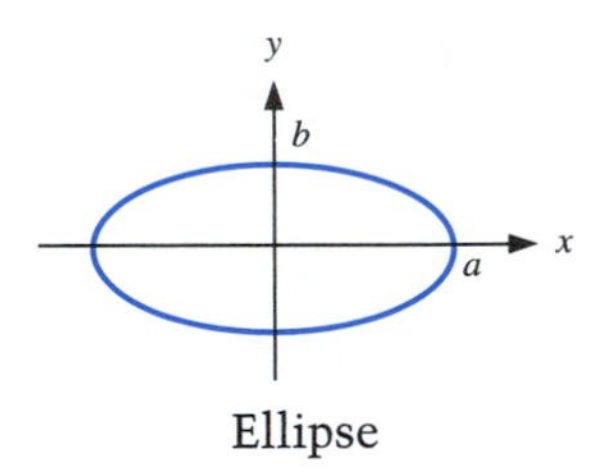

Ellipse

(vi) Length of a circular arc, with central angle θ radians

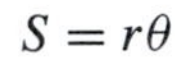

$$S = r\theta$$

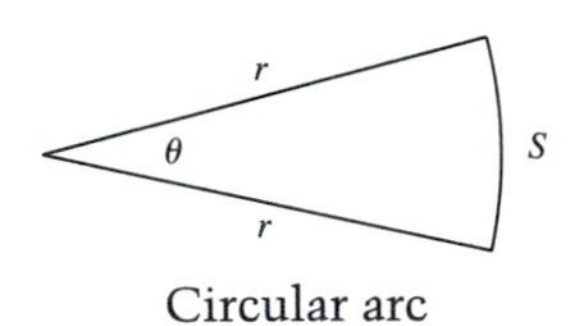

Circular arc

(b) Equations.

(i) Equation of a straight line in the xy-plane

$$ax + by = c$$

(ii) Slope of a straight line

$$m = \frac{y_2 - y_1}{x_2 - x_1}$$

with (x_1, y_1) and (x_2, y_2) points on the line

(iii) Equation of a plane in three-dimensional space

$$ax + by + cz = d$$

(iv) Equation of a circle in the xy-plane

$$(x - x_0)^2 + (y - y_0)^2 = r^2$$

with (x_0, y_0) the center and r the radius.

(v) Equation of a parabola, with its axis parallel to the y-axis

$$y = ax^2 + bx + c, \qquad a \neq 0$$

B.3. TRIGONOMETRY

(a) Conversion between angles in degrees (θ_D) and angles in radians (θ_R).

$$\theta_R = \frac{\pi}{180}\theta_D$$

(b) Graphs.

(i) $\sin(x)$ and $\cos(x)$

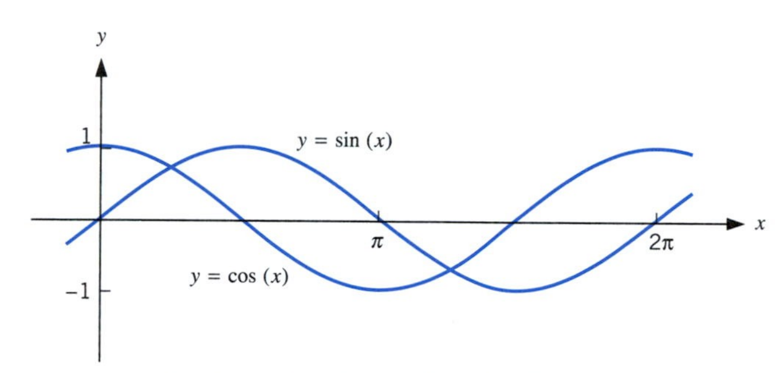

Graphs of the sine and cosine functions

(ii) $\tan(x)$

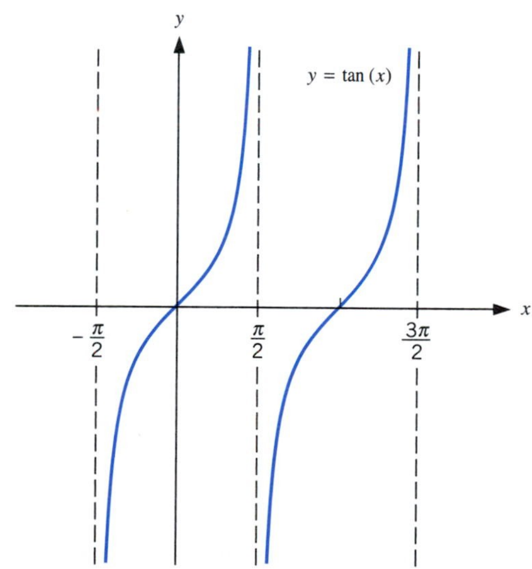

Graph of the tangent function

(iii) $\sin^{-1}(x) = \arcsin(x)$ and $\cos^{-1}(x) = \arccos(x)$

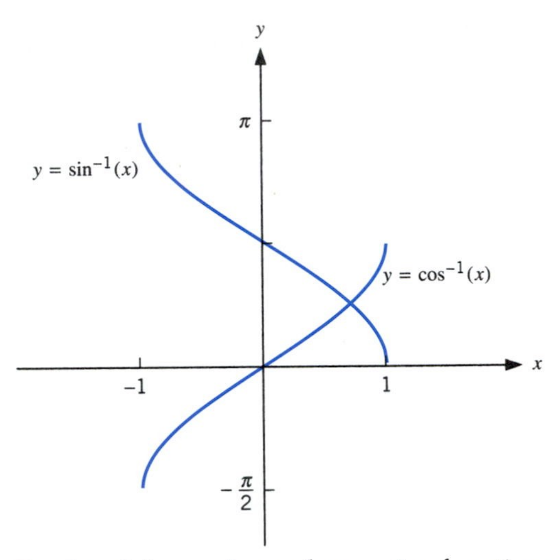

Graphs of the arcsine and arccosine functions

(iv) $\tan^{-1}(x) = \arctan(x)$

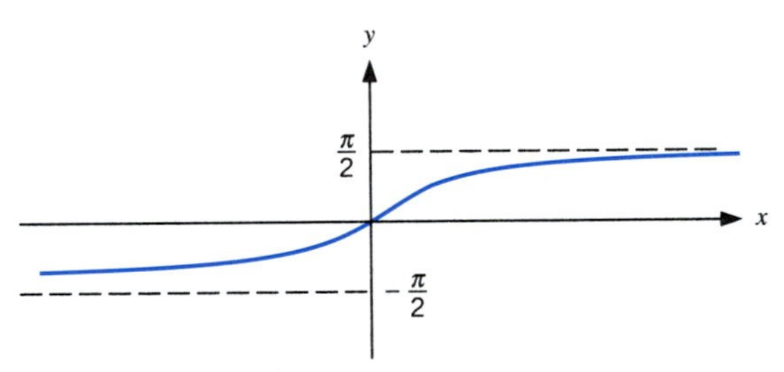

Graph of the arctangent function

(c) Definitions.

$$\tan(x) = \frac{\sin(x)}{\cos(x)}, \qquad \cot(x) = \frac{1}{\tan(x)} = \frac{\cos(x)}{\sin(x)}$$

$$\sec(x) = \frac{1}{\cos(x)}, \qquad \csc(x) = \frac{1}{\sin(x)}$$

(d) Trigonometric identities.

$\sin(-\alpha) = -\sin(\alpha), \qquad \cos(-\alpha) = \cos(\alpha)$

$\cos^2(\alpha) + \sin^2(\alpha) = 1, \qquad 1 + \tan^2(\alpha) = \sec^2(\alpha)$

$\sin(\alpha \pm \beta) = \sin(\alpha)\cos(\beta) \pm \cos(\alpha)\sin(\beta)$

$\cos(\alpha \pm \beta) = \cos(\alpha)\cos(\beta) \mp \sin(\alpha)\sin(\beta)$

$$\tan(\alpha \pm \beta) = \frac{\tan(\alpha) \pm \tan(\beta)}{1 \mp \tan(\alpha)\tan(\beta)}$$

$$\sin(2\alpha) = 2\sin(\alpha)\cos(\alpha)$$

$$\cos(2\alpha) = \cos^2(\alpha) - \sin^2(\alpha) = 2\cos^2(\alpha) - 1 = 1 - 2\sin^2(\alpha)$$

$$\sin^2(\alpha) = \frac{1 - \cos(2\alpha)}{2}, \qquad \cos^2(\alpha) = \frac{1 + \cos(2\alpha)}{2}$$

$$\sin(\alpha) \pm \sin(\beta) = 2\sin\left(\frac{\alpha \pm \beta}{2}\right)\cos\left(\frac{\alpha \mp \beta}{2}\right)$$

$$\cos(\alpha) + \cos(\beta) = 2\cos\left(\frac{\alpha + \beta}{2}\right)\cos\left(\frac{\alpha - \beta}{2}\right)$$

$$\cos(\alpha) - \cos(\beta) = -2\sin\left(\frac{\alpha + \beta}{2}\right)\sin\left(\frac{\alpha - \beta}{2}\right)$$

$$\sin^{-1}(x) + \cos^{-1}(x) = \frac{\pi}{2}$$

$$\tan^{-1}(x) = \frac{\pi}{2} - \tan^{-1}\left(\frac{1}{x}\right), \qquad x > 0$$

(e) Law of cosines

$$c^2 = a^2 + b^2 - 2ab\cos(\gamma)$$

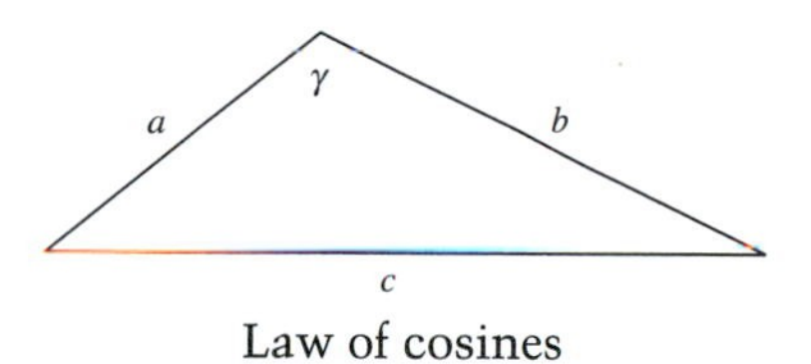

Law of cosines

The angle γ is opposite to the side with length c.

B.4. CALCULUS

(a) Definition of derivative.

$$f'(x) = \lim_{h \to 0} \frac{f(x+h) - f(x)}{h}$$

Other notations: $f'(x) = \dfrac{df(x)}{dx} = Df(x)$

(b) Properties of the derivative.

$$[f(x) + g(x)]' = f'(x) + g'(x)$$

$$[f(x)g(x)]' = f(x)g'(x) + f'(x)g(x)$$

$$[cf(x)]' = cf'(x), \quad c = \text{constant}$$

$$[f(x)/g(x)]' = \frac{f'(x)g(x) - f(x)g'(x)}{[g(x)]^2}$$

$$[f(g(x))]' = f'(g(x)) \cdot g'(x) \qquad \text{(chain rule)}$$

(c) $$\frac{d}{dx}(x^n) = nx^{n-1}, \qquad n \neq 0$$

$$\frac{d}{dx}[\cos(x)] = -\sin(x), \qquad \frac{d}{dx}[\sin(x)] = \cos(x)$$

$$\frac{d}{dx}[\tan(x)] = \sec^2(x), \qquad \frac{d}{dx}[\sec(x)] = \sec(x)\tan(x)$$

$$\frac{d}{dx}[\sin^{-1}(x)] = \frac{1}{\sqrt{1-x^2}}, \qquad \frac{d}{dx}[\cos^{-1}(x)] = \frac{-1}{\sqrt{1-x^2}}$$

$$\frac{d}{dx}[\tan^{-1}(x)] = \frac{1}{1+x^2}$$

$$\frac{d}{dx}[e^x] = e^x, \qquad \frac{d}{dx}[a^x] = [\log a]a^x$$

$$\frac{d}{dx}[\log(x)] = \frac{1}{x}, \qquad \frac{d}{dx}[\log_a x] = \frac{1}{x \log a}$$

(d) Definition of the definite integral.

$$\int_a^b f(x)\,dx = \lim_{n\to\infty} \sum_{j=1}^{n} f(p_j)(x_j - x_{j-1})$$

with $a = x_0 < x_1 < x_2 < \cdots < x_n = b$, p_j in $[x_{j-1}, x_j]$, and $\max(x_j - x_{j-1}) \to 0$ as $n \to \infty$. The above sums are called Riemann sums.

(e) Properties of the definite integral.

$$\int_a^b [f(x) + g(x)]dx = \int_a^b f(x)\,dx + \int_a^b g(x)\,dx$$

$$\int_a^b cf(x)\,dx = c\int_a^b f(x)\,dx, \qquad c = \text{constant}$$

$$\int_a^b f(x)\,dx = \int_a^c f(x)\,dx + \int_c^b f(x)\,dx$$

$$\int_a^b f(x)\,dx = -\int_b^a f(x)\,dx$$

$$\int_a^b f(x)g'(x)\,dx = [f(x)g(x)]_a^b - \int_a^b f'(x)g(x)\,dx$$

$$\int_a^b f(g(x))g'(x)\,dx = \int_{g(a)}^{g(b)} f(u)\,du$$

(f) Fundamental theorem of the calculus.

Form 1 Let $F(x)$ be an antiderivative of $f(x)$ for all x in $[a, b]$. Then

$$\int_a^b f(x)\,dx = [F(x)]_{x=a}^{x=b} = F(b) - F(a)$$

Form 2

$$\frac{d}{dx}\int_a^x f(t)\,dt = f(x)$$

(g) Table of antiderivatives.

$$\int x^n\,dx = \frac{x^{n+1}}{n+1} + c, \qquad n \neq -1$$

$$\int \frac{dx}{x} = \log|x| + c$$

$$\int e^x\,dx = e^x + c, \qquad \int a^x\,dx = \frac{a^x}{\log(a)} + c$$

$$\int x^\alpha \log(x)\,dx = \frac{x^{\alpha+1}}{\alpha+1}\cdot\log(x) - \frac{x^{\alpha+1}}{(\alpha+1)^2} + c, \qquad \alpha \neq -1$$

$$\int \frac{dx}{x^2+a^2} = \frac{1}{a}\tan^{-1}\left(\frac{x}{a}\right) + c$$

$$\int \frac{dx}{\sqrt{a^2-x^2}} = \sin^{-1}\left(\frac{x}{a}\right) + c$$

$$\int \sin(x)\,dx = -\cos(x) + c, \qquad \int \cos(x)\,dx = \sin(x) + c$$

$$\int \sec^2(x)\,dx = \tan(x) + c, \qquad \int \sec(x)\tan(x)\,dx = \sec(x) + c$$

$$\int \tan(x)\,dx = \log|\sec(x)| + c$$

$$\int \sec(x)\,dx = \log|\sec(x) + \tan(x)| + c$$

$$\int \sin^2(x)\,dx = \frac{x}{2} - \frac{\sin(2x)}{4} + c$$

$$\int \cos^2(x)\,dx = \frac{x}{2} + \frac{\sin(2x)}{4} + c$$

Appendix C

Numerical Analysis Software Packages

The programs included in this book are designed to give insight into the numerical methods being studied. They are not intended to be used in a production setting that requires highly accurate, efficient, robust, and convenient codes. For such computer codes, one should take advantage of the many high-quality numerical analysis program packages that have been written in the past three decades. In this appendix we list some of these packages, along with information on obtaining them.

Fortran is the traditional programming language for numerical computations. It is not surprising that most available packages are written in Fortran. Recently, however, C, C++, and Java have been gaining more and more popularity in computations. Many of the widely used packages are available also in C. In addition, on can use the `f2c` converter developed at Bell Laboratories to convert Fortran source codes to C. The `f2c` converter is available freely online at `http://netlib.bell-labs.com/netlib/f2c/`.

C.1. COMMERCIAL PACKAGES

There are two widely used general libraries in numerical analysis, each containing programs for solving most standard numerical analysis problems. Both libraries are commercially available on most lines of computers and minicomputers, and both have subsets available on some microcomputers. In addition, both companies have additional numerical analysis packages that are not discussed in this appendix.

1. *IMSL (International Mathematics and Statistics Library)*: product of Visual Numerics Inc., Houston, Texas, `www.vni.com/products/imsl`. It consists of three IMSL libraries: IMSL Fortran 90 MP Library, IMSL C Numerical Library, and JMSL for Java application development. The IMSL libraries are a large collection of mathematical and statistical functions that can be used as reliable building blocks for programming in scientific computing.
2. *NAG (Numerical Algorithms Group)*: product of NAG Inc., Downers Grove, Illinois, `www.nag.com`. The NAG libraries contain numerical, symbolic, statistical, visualization, and simulation programs for solving a variety of problems in applications. The programs are mostly in Fortran or C.

C.2. PUBLIC DOMAIN PACKAGES

Although some software packages are commercialized, others are available freely over the Internet.

One good source of free software is Netlib. This is a large collection of mathematical software. The Netlib repository is maintained by AT&T Bell Laboratories, the University of Tennessee, and the Oak Ridge National Laboratory, and all programs in the collection are available at no charge. The Netlib software or documents can be retrieved through the World Wide Web at the URL

`www.netlib.org`

or email (send to `netlib@netlib.org` the message `send index` to receive a contents summary and instructions). A good starting point to get acquainted with Netlib is its FAQ (Frequently Asked Questions) at

`http://www.netlib.org/misc/faq.html#2.1`

Listed below are some of the packages available from Netlib:

1. *ARPACK*. ARPACK is a collection of Fortran77 subroutines designed to compute a few eigenvalues and corresponding eigenvectors of general matrices of large size. A parallel version, PARPACK, is also available. A main reference is

   R. B. Lehoucq, D. C. Sorensen, and C. Yang, *ARPACK User's Guide: Solution of Large-Scale Eigenvalue Problems with Implicitly Restarted Arnoldi Methods*, SIAM, Philadelphia, 1998.

2. *BLAS* (*Basic Linear Algebra Subprograms*). The BLAS are efficient, portable, high-quality "building block" routines for matrix and vectors operations. There are three levels of BLAS, doing vector-vector, matrix-vector, and matrix-matrix operations, respectively. They are commonly used in the development of high-quality numerical linear algebra software such as LINPACK and LAPACK. In addition to Fortran implementation, C interface and Java BLAS are also available. Related references are

C. L. Lawson, R. J. Hanson, D. Kincaid, and F. T. Krogh, Basic Linear Algebra Subprograms for FORTRAN Usage, *ACM Trans. Math. Soft.* **5** (1979), 308–323.

J. J. Dongarra, J. Du Croz, S. Hammarling, and R. J. Hanson, An Extended Set of FORTRAN Basic Linear Algebra Subprograms, *ACM Trans. Math. Soft.* **14** (1988), 1–17.

J. J. Dongarra, J. Du Croz, S. Hammarling, and R. J. Hanson, Algorithm 656: An Extended Set of FORTRAN Basic Linear Algebra Subprograms, *ACM Trans. Math. Soft.* **14** (1988), 18–32.

J. J. Dongarra, J. Du Croz, I. S. Duff, and S. Hammarling, A Set of Level 3 Basic Linear Algebra Subprograms, *ACM Trans. Math. Soft.* **16** (1990), 1–17.

J. J. Dongarra, J. Du Croz, I. S. Duff, and S. Hammarling, Algorithm 679: A Set of Level 3 Basic Linear Algebra Subprograms, *ACM Trans. Math. Soft.* **16** (1990), 18–28.

3. *EISPACK*. This package solves eigenvalue-eigenvector problems for a wide variety of types of matrices, in double precision Fortran. The single precision version routines are in SEISPACK. Like several of the other packages, it was developed jointly by researchers at Argonne National Laboratory and several universities, with financial support from the National Science Foundation. For a detailed description of the package and a listing of all of the programs in it, see the following two monographs:

B. T. Smith, J. M. Boyle, J. J. Dongarra, B. S. Garbow, Y. Ikebe, V. C. Klema, and C. B. Moler, *Matrix Eigensystem Routines–EISPACK Guide*, 2nd ed., *Lecture Notes in Computer Science*, Volume 6, Springer-Verlag, New York, 1976.

B. S. Garbow, J. M. Boyle, J. J. Dongarra, and C. B. Moler, *Matrix Eigensystem Routines–EISPACK Guide Extension, Lecture Notes in Computer Science,* Volume 51, Springer-Verlag, New York, 1977.

4. *FFTPACK*. It is a package of Fortran subprograms for the fast Fourier transform of periodic and other symmetric sequences. A main reference is

P. N. Swarztrauber, Vectorizing the FFTs, in *Parallel Computations*, G. Rodrigue, ed., Academic Press, New York, 1982, pp. 51–83.

5. *FISHPACK*. This package contains Fortran programs for the finite difference methods for solving elliptic boundary value problems.

6. *LAPACK*. This package combines, updates, and extends the programs in EISPACK and LINPACK, for solving systems of linear algebraic equations, linear least-squares solutions, eigenvalue problems, and singular value problems. Versions of the programs are available for four kinds of computer arithmetic: single and double precision real, and single and double precision complex. The programs are especially well-adapted to the new generation of vector and parallel computers. A complete description of the package is given in the book

E. Anderson, Z. Bai, C. Bischof, J. Demmel, J. Dongarra, J. Du Croz, A. Greenbaum, S. Hammarling, A. McKenney, S. Ostrouchov, and D. Sorensent, *LAPACK Users' Guide*, third edition, SIAM, Philadelphia, 1999.

The LAPACK programs are written in Fortran77. Its C version is called CLAPACK, C++ version is LAPACK++, and Fortran90 interface is LAPACK90.

7. *LINPACK.* This package is for the solution of simultaneous systems of linear algebraic equations. Special subprograms are included for many common types of coefficient matrices for these systems. Error monitoring is available, along with checks on the conditioning of the matrix of coefficients for the linear system. LINPACK has been largely superceded by LAPACK. For documentation and a listing of all programs in LINPACK, see

J. J. Dongarra, J. R. Bunch, C. B. Moler, and G. W. Stewart, *LINPACK Users' Guide*, SIAM Publications, 1979.

8. *MINPACK.* This contains programs for solving systems of nonlinear algebraic equations and nonlinear least-squares problems. It is developed at Argonne National Laboratory.

9. *ODEPACK.* It is a collection of Fortran77 solvers for the initial value problem of ordinary differential equation systems. It can be used for solving both stiff and nonstiff systems. More information is found in

A. C. Hindmarsh, ODEPACK, A Systematized Collection of ODE Solvers, in *Scientific Computing*, R. S. Stepleman et al., eds., North-Holland, Amsterdam, 1983 (Vol. 1 of *IMACS Transactions on Scientific Computation*), pp. 55–64.

10. *PLTMG.* This package contains Fortran programs for solving elliptic partial differential equations in general planar regions. It features adaptive local mesh refinement, multigrid iteration, and a pseudo-arclength continuation option for parameter dependencies. The main reference is

R. Bank, *PLTMG: A Software Package for Solving Elliptic Partial Differential Equations—Users' Guide 8.0*, SIAM, Philadelphia, 1998.

11. *QUADPACK.* It is a Fortran subroutine package for the numerical computation of definite one-dimensional integrals. Most programs are automatic and attempt to perform the integration to within the requested absolute or relative error. For more information, see

R. Piessens, E. deDoncker-Kapenga, C. Uberhuber, and D. Kahaner, *Quadpack: A Subroutine Package for Automatic Integration*, Springer Verlag, New York, 1983.

Another good source of online mathematical and statistical software is the Guide to Available Mathematical Software from NIST (National Institute of Standards and

Technology), `gams.nist.gov`. In addition to the packages found in Netlib, this site contains many other packages.

C.3. INTERACTIVE NUMERICAL COMPUTATION ENVIRONMENTS

In both scientific research and education, interactive computing environments are becoming increasingly popular. These interactive environments are easy to learn, easy to use, with powerful built-in mathematical capabilities. With such interactive environments, one can focus more on problem modeling and solving, results interpretation, rather than on programming details.

A very widely used interactive computing environment is MATLAB. It is a commercial product of The Mathworks, Inc., Natick, MA, `www.mathworks.com`. This is the interactive computing environment we chose to work with in this textbook. A brief introduction of MATLAB is given in Appendix D; and a more detailed description of MATLAB can be found in the references listed there.

In the market, there are many competing MATLAB-like computing environments. They all offer an interpreted mathematical programming language with matrices as the fundamental data type and simple syntax. Some examples follow:

- *GAUSS* of APTECH Systems, Inc., `www.aptech.com`
- *MathViews* of MathWizards, `www.mathwizards.com`
- *Mathcad* of Mathsoft Engineering & Education Inc., `www.mathsoft.com`
- $\mathcal{O}$-MATRIX, `www.omatrix.com`
- *PV-WAVE* of Visual Numerics, Inc., `www.vni.com/products/wave`

There also exist freely distributed MATLAB-like computing environments. Some examples follow:

- *Octave* by John W. Eaton and others, `www.octave.org`
- *RLaB*, which is no longer under active development, `rlab.sourceforge.net`
- *Scilab* by the Scilab Group at INRIA of France, `www-rocq.inria.fr/scilab`
- *Tela* (TEnsor LAnguage), `www.geo.fmi.fi/prog/tela.html`

C.4. SYMBOLIC COMPUTATION ENVIRONMENTS

During the 1980s, computer programs to do symbolic mathematics, rather than numerical mathematics, came of age. The increasing use of symbolic mathematics programs is altering both the teaching and research of mathematics, including numerical analysis.

It is unclear as to where the use of these new computer tools will take us; but they are likely to have a significant effect on the practice of all areas of mathematics. Here we list some of the major computer algebra systems:

- *Derive* of MathWare Ltd., `www.mathware.com`
- *Macsyma*, `www.scientek.com/macsyma/mxmain.htm`
- *Maple* of the Symbolic Computation Group, University of Waterloo, `www.scg.uwaterloo.ca`
- *Mathematica* of Wolfram Research Inc., `www.wolfram.com`
- *MuPAD*, `www.mupad.de`
- *Reduce*, `www.uni-koeln.de/REDUCE`

MATLAB has a Symbolic Math Toolbox that performs a variety of symbolic computations based on linking to a *Maple* engine.

C.5. LITERATURE ON MATHEMATICAL SOFTWARE

There are several periodicals in which new programs or information about them are published. Several of these are published by the Association for Computing Machinery (ACM), `www.acm.org`, and we will limit our references to those periodicals.

1. *ACM Transactions on Mathematical Software*. This journal contains articles on the development and testing of numerical analysis software. It also contains new programs for a wide variety of numerical analysis problems. These programs are usually written very carefully, and they are in a portable form for use on different computers.
2. *Collected Algorithms of the ACM*. This is a collection of all programs that have been published in ACM journals, and this collection is available as a separate subscription from the ACM. All new mathematics programs now appear in the *ACM Transactions Software*, listed above. Machine readable copies of the algorithms are available from NETLIB.

The discussion of this appendix is far from inclusive of all that is available in quality mathematical software. In addition, the choice of program packages is heavily skewed toward the United States, since the authors are better acquainted with these programs.

APPENDIX D

MATLAB: AN INTRODUCTION

MATLAB is an interactive computer language that has been designed especially to carry out the types of numerical computations needed in numerical analysis. In MATLAB we can give direct commands, as on a hand calculator, and we can write programs. The best way to learn MATLAB is to spend time using it. Many MATLAB expressions and commands have a simple-to-use default format; and these can be extended easily to allow a larger number of options. In this appendix we give some guidance to help get you started with MATLAB. There are a number of books and online guides that have been written to help you learn MATLAB, and we reference some of them at the end of this appendix.

Standard mathematical expressions can be evaluated as with most computer languages. For example, the expression

$$y = 6 - 4x + 7x^2 - 3x^5 + \frac{3}{x+2} \tag{D.1}$$

is evaluated using

```
y = 6 - 4*x + 7*x*x - 3*x^5 + 3/(x+2);
```
(D.2)

based on a previously defined value of x. Note that the statement ends with a semicolon. This is standard in MATLAB; omitting the semicolon will cause the value of y to be printed on the computer screen. MATLAB has many built-in functions, including standard elementary functions such as e^x, $\log x$, trigonometric and inverse trigonometric functions, $\sqrt{x}$, and others. More sophisticated functions are also available. For example, gamma(x) is the standard gamma function evaluated at x; for a nonnegative integer n, gamma$(n+1) = n!$, the factorial function. The important constant π is available by using the name `pi`.

The default arithmetic used in MATLAB is double precision and real; however, where appropriate it defaults to complex arithmetic. For example, `sqrt(-4)` results in an answer of `2i`. The default output to the screen is to have four digits to the right of the decimal point. To control the formatting of output to the screen, use the command `format`. The default formatting is obtained using

```
format short
```

To obtain the full accuracy available in a number, you can use

```
format long
```

The commands

```
format short e
format long e
```

will use "scientific notation" for the output. Other `format` options are also available.

An important aspect of MATLAB is that it works very efficiently with arrays, and many tasks are best done with arrays. We will illustrate this with a simple code to plot $\sin x$ and $\cos x$ on the interval $0 \le x \le 10$.

```
t = 0:.1:10;
x = cos(t); y = sin(t);
plot(t,x,t,y)
```
(D.3)

The first line sets up a row vector of length 101:

$$t = [0, 0.1, 0.2, \dots, 1.0, 1.1, \dots, 9.9, 10]$$

In general, the statement

```
t = a:h:b;
```

creates a row vector of the form

$$t = [a, a + h, a + 2h, \ldots]$$

giving all values $a + jh$ that are $\leq b$. When h is omitted, it is assumed to be 1. Thus,

```
n = 1:5
```

creates the row vector

$$n = [1, 2, 3, 4, 5]$$

The second line of our code in (D.3) above defines two new row vectors. The command `x=cos(t)` creates a row vector of the same size as that of `t`, and `x` consists of the value of the cosine function at all of the components of `t`. The command `x=sin(t)` is similar in its operation. The third line in (D.3) creates a two-dimensional plot of both the cosine function and sine function on [0, 10] based on the values stored in `t`, `x`, and `y`. The reader should implement this simple example (and the other examples given in this appendix). This example also illustrates the simplicity of creating plots in MATLAB.

Arrays can be created in a number of ways in MATLAB. For example,

```
b = [1, 2, 3]
```

creates a row vector of length 3; and

```
A = [1 2 3; 4 5 6; 7 8 9]
```

creates the square matrix

$$A = \begin{bmatrix} 1 & 2 & 3 \\ 4 & 5 & 6 \\ 7 & 8 & 9 \end{bmatrix}$$

Spaces or commas can be used as delimiters in giving the components of an array; and a semicolon will separate the various rows of a matrix, as in the above example. There are also other options available when directly defining an array. When wanting to define a column vector, we can use the transpose operation. For example,

```
b = [1 3 -6]'
```

results in the column vector

$$\begin{bmatrix} 1 \\ 3 \\ -6 \end{bmatrix}$$

When two arrays have the same size, we add them in the standard way matrices are added, namely component-wise. For example, adding the 2×3 matrices

```
A = [1, 2; 3, -2; -6, 1];
B = [2, 3; -3, 2; 2, -2];
C = A + B;
```

results in the answer

$$C = \begin{bmatrix} 3 & 5 \\ 0 & 0 \\ -4 & -1 \end{bmatrix}$$

When multiplying an array by a constant, we multiply the constant with each component of the array. For example, using the above matrix A and the command

```
D = 2*A;
```

results in the answer

$$D = \begin{bmatrix} 2 & 4 \\ 6 & -4 \\ -12 & 2 \end{bmatrix}$$

Matrix multiplication has the standard meaning as well. For example, multiplying as follows:

```
E = [1, -2; 2, -1; -3, 2];
F = [2, -1, 3; -1, 2, 3];
G = E*F;
```

results in the answer

$$G = \begin{bmatrix} 1 & -2 \\ 2 & -1 \\ -3 & 2 \end{bmatrix} \begin{bmatrix} 2 & -1 & 3 \\ -1 & 2 & 3 \end{bmatrix} = \begin{bmatrix} 4 & -5 & -3 \\ 5 & -4 & 3 \\ -8 & 7 & -3 \end{bmatrix}$$

A nonstandard notation is the addition of a constant to an array. This is implemented by adding the constant to each component of the array. For example, using the above matrix F, the command

```
H = 3 + F;
```

results in the computation

$$H = 3\begin{bmatrix} 1 & 1 & 1 \\ 1 & 1 & 1 \end{bmatrix} + \begin{bmatrix} 2 & -1 & 3 \\ -1 & 2 & 3 \end{bmatrix} = \begin{bmatrix} 5 & 2 & 6 \\ 2 & 5 & 6 \end{bmatrix}$$

These operations and their implementation are discussed at greater length in Chapter 6, Sections 6.1, 6.2.

MATLAB also has component-wise operations for multiplication, division, and exponentiation, operations you will not have seen used in most linear algebra classes. These three operations are denoted by using a period to precede the usual symbol for the operation. As examples, we have the following. For

```
a = [1  2  3];     b = [2  -1  4];
```

we have

```
a.*b = [2  -2  12]
a./b = [0.5  -2.0  0.75]
a.^3 = [1  8  27]
2.^a = [2  4  8]
b.^a = [ 2  1  64]
```

As a further example of component-wise operations in MATLAB, consider that the algebraic expression in (D.1) is to be evaluated at all of the components of an array `x`. Then the MATLAB command (D.2) should be rewritten as

```
y = 6 - 4*x + 7*x.*x - 3*x.^5 + 3./(x+2);
```

Note the use of the `.*`, `.^`, and `./` component-wise operations. The remaining operations have the standard meanings discussed earlier from linear algebra.

As was seen in the program segment in (D.3), we can also apply standard functions to arrays, and the functions are interpreted as component-wise function evaluations. For example,

```
sqrt([1  4  9]) = [1  2  3]
```

There are a number of MATLAB operations that deal with "housekeeping" details. For example, to see the currently active variables, use the command `who`. To see the size and type of these variables, use the command `whos`. To remove the current variables from use, use the command `clear`. To clear the output screen, use the command `cls`. To obtain a description of a MATLAB command, use the format

```
help command_name
```

For example,

```
help sqrt
```

results in the output

```
SQRT Square root.
SQRT(X) is the square root of the elements of X. Complex
results are produced if X is not positive.
```

There are also commands to produce special matrices. The most commonly used ones in this text are illustrated as follows:

```
A = zeros(2,3)
```

produces an array with two rows and three columns, with all components set to zero,

$$\begin{bmatrix} 0 & 0 & 0 \\ 0 & 0 & 0 \end{bmatrix}$$

This functions like the DIMENSION statement in *Fortran*, with initialization of the array to zero. The command

```
B = ones(2,3)
```

produces an array with two rows and three columns, with all components set to 1,

$$\begin{bmatrix} 1 & 1 & 1 \\ 1 & 1 & 1 \end{bmatrix}$$

This is different from the identity matrix, which can be produced using the command `eye`. For example, the command `eye(3)` results in the 3×3 identity matrix

$$\begin{bmatrix} 1 & 0 & 0 \\ 0 & 1 & 0 \\ 0 & 0 & 1 \end{bmatrix}$$

There are many MATLAB commands that operate on arrays; we include only a very few here. For a vector `x`, row or column, of length n, we have the following functions:

$$\begin{aligned} \texttt{max(x)} &= \text{maximum component of } \texttt{x} \\ \texttt{min(x)} &= \text{minimum component of } \texttt{x} \\ \texttt{abs(x)} &= \text{vector of absolute values of components of } \texttt{x} \\ \texttt{sum(x)} &= \text{sum of the components of } \texttt{x} \\ \texttt{norm(x)} &= \sqrt{|x_1|^2 + \cdots + |x_n|^2} \end{aligned}$$

For more information on these functions, use the `help` command.

Script files and functions Although MATLAB is an interactive language and the commands can be typed and executed successively, it is often useful to create a file of the commands that you want MATLAB to execute. For example, the code in (D.3) could have been stored as a file and then executed. Such files must have a name that ends in .m (a period followed by the letter m). For example, we could save (D.3) under the name `demo_plot.m`, and then it could have been executed by simply typing `demo_plot` followed by a carriage return. A file such as this is called a *script* file. Such files have many advantages. Most immediately, it is all too easy to make typing errors when giving interactive commands, and having a script file makes it easier to correct such errors and then to rerun the computation. We can also experiment with writing a program, making changes as we find them necessary. For example, we could add commands to `demo_plot.m` to add labels and a title to the graph produced by the script. More important, we can write complex programs that operate on data in MATLAB's current workspace. An example of such a program, named `exp_taylor`, is given at the end of Section 1.1 of Chapter 1, producing a table of approximation values for various Taylor polynomial approximations of e^x; and other examples appear elsewhere in this textbook.

You can also create functions within MATLAB, allowing you to augment the many functions already available. A function has input and output parameters; and all other variables used in the function are unknown outside of that function. As an example, consider a function for evaluating the polynomial

$$p(x) = a_1 + a_2 x + a_3 x^2 + \cdots + a_n x^{n-1}$$

We have begun the indexing with 1 rather than 0 because MATLAB does not allow zero subscripts for arrays. The following function would be stored under the name `polyeval.m`. The coefficients $\{a_j\}$ are given to the function in the array named `coeff`, and the polynomial is to be evaluated at all of the components of the array `x`.

```
function value = polyeval(x,coeff);
%
% function value = polyeval(x,coeff)
%
% Evaluate a polynomial at the points given in x.  The
% coefficients are to be given in coeff.  The constant term
% in the polynomial is coeff(1).

n = length(coeff)
value = coeff(n)*ones(size(x));
for i = n-1:-1:1
  value = coeff(i) + x.*value;
end
```

A more general version of this is given at the end of Section 1.1 of Chapter 1. The polynomial is being evaluated in *nested form*, a topic considered in Section 1.3 of Chapter 1. The output from this function is given in the variable named `value`. In this code, the use of the character % signifies that everything which follows it on that line is considered a comment. As an example of `polyeval`, let

```
cf = [1  2  3];
 t = [3 4];
```

Then the command

```
v = polyeval(t,cf);
```

will set `v` to the array [34, 57], which is the polynomial $1 + 2x + 3x^2$ evaluated at the two components of `t`. There are many functions given in this text, and you can use them to learn more about MATLAB.

There are many online Web sites to aid you in learning about MATLAB, and you can locate many of those by using a Web search engine (e.g., *Google*). Simply type in "Matlab tutorial", and a number of such online sites will be listed. In addition, there are a number of excellent texts on MATLAB, some general and some specific to some subarea of MATLAB. Our favorite general text is that of Higham and Higham [18]. For a very complete introduction to MATLAB graphics, we have used Marchand [24].

APPENDIX E

THE BINARY NUMBER SYSTEM

For representing numbers and doing calculations with them, most computers use the binary number system, or some variant of it, rather than using the decimal system. In this appendix, we give an introduction to the binary number system and to the conversion between it and the decimal system.

In the decimal system, a number such as 342.105 means

$$3 \cdot 10^2 + 4 \cdot 10^1 + 2 \cdot 10^0 + 1 \cdot 10^{-1} + 0 \cdot 10^{-2} + 5 \cdot 10^{-3}$$

Numbers written in the decimal system are interpreted as a sum of multiples of integer powers of 10. There are 10 digits, denoted by $0, 1, \dots, 9$. We say 10 is the base of the decimal system.

The binary system represents all numbers as a sum of multiples of integer powers of 2. There are two digits, 0 and 1; and 2 is the base of the binary system. The digits 0 and 1 are called *bits*, which is short for *binary digits*. For example, the number 1101.11

in the binary system has the value

$$1 \cdot 2^3 + 1 \cdot 2^2 + 0 \cdot 2^1 + 1 \cdot 2^0 + 1 \cdot 2^{-1} + 1 \cdot 2^{-2} \tag{E.1}$$

in the decimal system. For clarity when discussing a number with respect to different bases, we will enclose the number in parentheses and give the base as a subscript. For example,

$$(1101.11)_2 = (13.75)_{10}$$

By default, a number without the parentheses is a decimal 1.

To convert a general binary number to its decimal equivalent, we proceed as in (E.1). For this purpose, sometimes we need the formula for the geometric series

$$\sum_{i=0}^{n} r^i = \frac{1 - r^{n+1}}{1 - r}, \qquad r \neq 1 \tag{E.2}$$

or its variant

$$\sum_{i=1}^{n} r^i = \frac{r - r^{n+1}}{1 - r}, \qquad r \neq 1 \tag{E.3}$$

Letting $n \to \infty$ in (E.2), we obtain the formula

$$\sum_{i=0}^{\infty} r^i = \frac{1}{1 - r}, \qquad |r| < 1 \tag{E.4}$$

which will also be needed later.

Example E.1 Consider $(111\ldots1)_2$ with n consecutive 1's to the left of the binary point. This has the decimal equivalent

$$2^{n-1} + 2^{n-2} + \cdots + 2^1 + 2^0 = 2^n - 1$$

Thus,

$$(\underbrace{11\ldots11}_{n})_2 = (2^n - 1)_{10} \tag{E.5}$$

Similarly,

$$(0.\underbrace{11\cdots1}_{n})_2 = 2^{-1} + 2^{-2} + \cdots + 2^{-n} = \frac{2^{-1} - 2^{-(n+1)}}{1 - 2^{-1}} = 1 - 2^{-n}$$

where we have applied the formula (E.3) with $r = 2^{-1}$. ■

Example E.2 Convert the infinite repeating binary fraction

$$x = (0.1010101010101\ldots)_2 \tag{E.6}$$

to its decimal equivalent. Converting to powers of 2 gives us

$$\begin{aligned} x &= 2^{-1} + 2^{-3} + 2^{-5} + 2^{-7} + \cdots \\ &= \tfrac{1}{2}\left[1 + \tfrac{1}{4} + \tfrac{1}{16} + \tfrac{1}{64} + \cdots\right] \end{aligned}$$

The quantity in brackets is an infinite geometric series. Using (E.4), we obtain

$$x = \frac{1}{2} \cdot \frac{1}{1 - \frac{1}{4}} = \frac{1}{2} \cdot \frac{4}{3} = \frac{2}{3}$$

Written as a decimal fraction,

$$x = (0.66666\ldots)_{10} \quad ■$$

The principles behind the arithmetic operations are the same in the binary system and the decimal system, with the major difference being that in the binary system fewer digits are allowed. As examples, consider the following addition and multiplication calculations:

$$\begin{array}{r} 11110 \\ +\quad 1101 \\ \hline 101011 \end{array} \qquad\qquad \begin{array}{r} 111 \\ \times\ 110 \\ \hline 000 \\ 111 \\ 111 \\ \hline 101010 \end{array}$$

Tables E.1 and E.2 contain addition and multiplication tables for the binary numbers corresponding to the decimal digits 1, 2, 3, 4, and 5.

Table E.1. Binary Addition

+	1	10	11	100	101
1	10	11	100	101	110
10	11	100	101	110	111
11	100	101	110	111	1000
100	101	110	111	1000	1001
101	110	111	1000	1001	1010

Table E.2. Binary Multiplication

×	1	10	11	100	101
1	1	10	11	100	101
10	10	100	110	1000	1010
11	11	110	1001	1100	1111
100	100	1000	1100	10000	10100
101	101	1010	1111	10100	11001

E.1. CONVERSION FROM DECIMAL TO BINARY

We will give methods for converting decimal integers and decimal fractions to binary integers and binary fractions. These methods will be in a form convenient for hand computation. Different algorithms are required when doing such conversions within a binary computer, but we will not discuss them here.

Suppose that x is an integer written in decimal. We want to find coefficients $a_0, a_1, \ldots, a_n$, all 0 or 1, $a_n = 1$, for which

$$a_n \cdot 2^n + a_{n-1} \cdot 2^{n-2} + \cdots + a_1 \cdot 2^1 + a_0 \cdot 2^0 = x \tag{E.7}$$

The binary integer will be

$$(a_n a_{n-1} \cdots a_0)_2 = (x)_{10} \tag{E.8}$$

To find the coefficients, begin by dividing x by 2, and denote the quotient by x_1. The remainder is a_0. Next divide x_1 by 2, and denote the quotient by x_2. The remainder is a_1. Continue this process, finding $a_2, a_3, a_4, \ldots, a_n$ in succession.

Example E.3 The following shortened form of the above method is convenient for hand computation. Convert $(19)_{10}$ to binary.

$$\begin{array}{llllllll}
\lfloor 2\overline{)19}\rfloor & = 9 & = x_1 & & & & & a_0 = 1 \\
 & \lfloor 2\overline{)9}\rfloor & = 4 & = x_2 & & & & a_1 = 1 \\
 & & \lfloor 2\overline{)4}\rfloor & = 2 & = x_3 & & & a_2 = 0 \\
 & & & \lfloor 2\overline{)2}\rfloor & = 1 & = x_4 & & a_3 = 0 \\
 & & & & \lfloor 2\overline{)1}\rfloor & = 0 & = x_5 & a_4 = 1
\end{array}$$

In this, the notation $\lfloor b \rfloor$ denotes the largest integer $\leq b$, and the notation $2\overline{)n}$ denotes the quotient resulting from dividing 2 into n. From the above calculation, $(19)_{10} = (10011)_2$. ■

Suppose now that x is a decimal fraction and that it is positive and less than 1.0. Then we want to find coefficients $a_1, a_2, a_3, \ldots$, all 0 or 1, for which

$$a_1 \cdot 2^{-1} + a_2 \cdot 2^{-2} + a_3 \cdot 2^{-3} + \cdots = x \tag{E.9}$$

The binary fraction will be

$$(.a_1a_2a_3\ldots)_2 = (x)_{10} \tag{E.10}$$

To find the coefficients, begin by denoting $x = x_1$. Multiply x_1 by 2, and denote $x_2 = \text{Frac}(2x_1)$, the fractional part of $2x_1$. The integer part $\text{Int}(2x_1)$ equals a_1. Repeat the process. Multiply x_2 by 2, letting $x_3 = \text{Frac}(2x_2)$ and $a_2 = \text{Int}(2x_2)$. Continue in the same manner, obtaining $a_3, a_4, \ldots$ in succession.

Example E.4 Find the binary form of 5.578125. We break the number into an integer and a fraction part. As in the previous example, we find that

$$(5)_{10} = (101)_2$$

For the fractional part $x_1 = x = 0.578125$, use the above algorithm.

$$
\begin{array}{lll}
2x_1 = 1.15625 & x_2 = 0.15625 & a_1 = 1 \\
2x_2 = 0.3125 & x_3 = 0.3125 & a_2 = 0 \\
2x_3 = 0.625 & x_4 = 0.625 & a_3 = 0 \\
2x_4 = 1.25 & x_5 = 0.25 & a_4 = 1 \\
2x_5 = 0.5 & x_6 = 0.5 & a_5 = 0 \\
2x_6 = 1.0 & x_0 = 0 & a_6 = 1
\end{array}
$$

Thus, the binary equivalent is 0.100101; combining it with the earlier result, we get

$$(5.578125)_{10} = (101.100101)_2 \quad \blacksquare$$

Example E.5 Convert the decimal fraction 0.1 to its binary equivalent. By using the above procedure, we obtain

$$(0.1)_{10} = (0.00011001100110\ldots)_2 \tag{E.11}$$

an infinite repeating binary fraction. Numbers have a finite binary fractional form if and only if they are expressible as a sum of a finite number of negative powers of 2. The decimal number 0.1 is not expressible as such a finite sum. In other words, the seemingly simple number 0.1 cannot be represented exactly in a computer. This has the implication that arithmetic operations involving simple numbers such as 0.1 (not to mention more complicated numbers) cannot be done exactly with the computer! ■

E.2. HEXADECIMAL NUMBERS

The base of the hexadecimal system is 16, and the digits are usually denoted by

$$0, 1, \ldots, 9, A, B, C, D, E, F$$

Thus, $(F)_{16} = (15)_{10}$. As before, a number written in the hexadecimal system is a sum of multiples of integer powers of 16. For example,

$$\begin{aligned}(AC.17)_{16} &= (10 \cdot 16^1 + 12 \cdot 16^0 + 1 \cdot 16^{-1} + 7 \cdot 16^{-2})_{10} \\ &= (172.08984375)_{10}\end{aligned}$$

Algorithms can be given to convert from decimal to hexadecimal, in analogy with those given above for binary numbers; see Problem 7.

There is a close connection between the binary and hexadecimal systems, and it is easy to convert from one to the other. To convert from hexadecimal to binary, replace each hex digit by its equivalent binary representation that uses four binary digits.

Example E.6 Convert $(AC.17)_{16}$ to binary. It is given by

$$(10101100.00010111)_2$$

since

$$(A)_{16} = (1010)_2, \quad \ldots, \quad (7)_{16} = (0111)_2 \quad \blacksquare$$

To convert x from binary to hexadecimal, break x into packets of four bits, going both left and right from the binary point. Then convert those to their hex equivalent.

Example E.7 Convert $x = (110101.11001)_2$ to hexadecimal. We write this as

$$(00110101.1100\ 1000)_2 = (35.C8)_{16} \quad \blacksquare$$

Example E.8 Convert the binary fraction in (E.6) to hexadecimal.

$$\begin{aligned}x &= (0.1010\ 1010\ 1010\ 1010\ldots)_2 \\ &= (.AAA\ldots)_{16}\end{aligned}$$

and by what was shown following (E.6), this is equivalent to the fraction $\frac{2}{3}$. $\blacksquare$

The hexadecimal system is used on many large mainframe IBM computers, and the binary system is used on most other kinds of computers.

PROBLEMS

1. Convert the following binary numbers to decimal form:
   (a) 110011 (b) 101.1101
   (c) 0.1000001 (d) 1010101010101

2. Expand Tables E.1 and E.2 to include one more column and row.

3. Convert the following decimal numbers to binary format:
   (a) 366 (b) 4.25 (c) 1/6
   (d) π (e) $\sqrt{2}$

4. Find the binary expressions for the decimal expressions 0.2, 0.4, 0.8, and 1.6.
   Hint: Use (E.11).

5. Find the rational number or fraction corresponding to the infinite repeating binary fraction

$$x = (0.110110110\ldots)_2$$

Hint: Write

$$\begin{aligned} x &= 2^{-1} + 2^{-2} + 2^{-4} + 2^{-5} + 2^{-7} + 2^{-8} + \cdots \\ &= [2^{-1} + 2^{-4} + 2^{-7} + 2^{-10} + \cdots] + [2^{-2} + 2^{-5} + 2^{-8} + \cdots] \end{aligned}$$

Treat each series in brackets in a way similar to that used for (E.6).

6. Convert the following hexadecimal numbers to both decimal and binary:
   (a) $1F.C$ (b) $FF\ldots F$ repeated n times
   (c) 11.1 (d) $.CCCC\ldots$

7. Generalize the algorithms for decimal → binary conversion to decimal → hexadecimal conversion. For conversion of integers, divide by 16 rather than 2; and for decimal fractions less than 1, multiply by 16 rather than 2. Convert the following decimal numbers to hexadecimal:
   (a) 161 (b) 0.3359375
   (c) 0.1 (d) 1.9

8. Let x be written in binary form $0 \le x \le 1$

$$x = (.a_1a_2a_3\ldots)_2$$

Give a geometric meaning to the coefficients $a_1, a_2, a_3, \ldots$ in terms of where x is located in the interval [0, 1].

9. Which number is bigger, $(1.000\ldots)_2$ or $(.1111\ldots)_2$? Explain.

Answers to Selected Problems

CHAPTER 1

Section 1.1

2(a). $p_1(x) = 1 + \frac{1}{2}(x-1)$

$p_2(x) = 1 + \frac{1}{2}(x-1) - \frac{1}{8}(x-1)^2$

3(c). $p_n(x) = 1 + \binom{1/2}{1}x + \binom{1/2}{2}x^2 + \cdots + \binom{1/2}{n}x^n$

9(a). $p_4(x) = 1 - x + \frac{1}{2}x^2 - \frac{1}{6}x^3 + \frac{1}{24}x^4$

11. $g(x) \approx 1 + \frac{1}{2}x + \frac{1}{6}x^2$, for small values of x

13(a). $q(x) = f(a) + [(x-a)/(b-a)][f(b) - f(a)]$

14. $b_0 = f(0), \quad b_1 = f'(0)$; solve for b_2, b_3 from the linear system

$b_2 + b_3 = f(1) - b_0 - b_1$

$2b_2 + 3b_3 = f'(1) - b_1$

Section 1.2

1. $e^x - p_3(x) = \dfrac{x^4}{4!}e^c$, c between 0 and x

$$\max_{-1\le x\le 1} |e^x - p_3(x)| \le (1)^4 \frac{e^1}{24} \doteq 0.1133$$

From Table 1.1, $e^1 - p_3(1) \doteq 0.05161$.

4(b). $\left|\dfrac{\sin(x) - x}{x}\right| \le 0.01 \qquad \text{for } |x| \le 0.245$

5. Use degree $= 7$.

9(a). $\frac{1}{2}$

14(b). $\tan^{-1}(x) = x - \dfrac{x^3}{3} + \cdots + \dfrac{(-1)^n x^{2n+1}}{2n+1} + R_{2n+1}(x)$

$$R_{2n+1}(x) = (-1)^{n+1} \int_0^x \frac{t^{2n+2}\,dt}{1+t^2}$$

19. With n odd, $f(x) = 2\left[x + \dfrac{x^3}{3} + \dfrac{x^5}{5} + \cdots + \dfrac{x^n}{n} + \dfrac{1}{n+2}\displaystyle\int_0^x \frac{t^{n+2}\,dt}{1-t^2}\right]$.

Section 1.3

4. Use degree $= 8$.

6. Using (1.35)–(1.38),
$q(x) = 4x^6 + x^5 - x^4 + x^2 + x + 1, \qquad b_0 = 0$
$p(x) = (x-1)q(x)$

8. Form $p'(x)$ by direct differentiation in (1.38), and then let $x = z$.

9. Seven multiplications are needed, if we assume that all coefficients have been calculated already.

10. $f(x) = 1 + z\left(5 + z^2(-1 + 2z)\right), \quad z = e^x$.

CHAPTER 2

Section 2.1

1(b). $(12)_{10}$ has the IEEE double precision hexadecimal representation 4028000000000000. The significand (in binary) is 1.100…0. The biased exponent (in binary) is 10000000010.

2. $n = 31$ bits in significand, $M = 2^{31} \doteq 2.15 \times 10^9$, $-2^{-31} \le \epsilon \le 2^{-31}$.

Section 2.2

1(c). $x_T = 2.7182818\ldots$ (not repeating)

$x_A = 2.71428571\ldots$ (repeating)

$x_T - x_A = 0.003996\ldots$

$\text{Re}(x_A) \doteq 0.00147$

x_A has three significant digits.

3. $\lim_{t\to\infty} N(t) = N_c = 2N(0)$

5(a). $\dfrac{1-\cos(x)}{x^2} = \dfrac{2\sin^2(x/2)}{x^2}$

5(d). $\sqrt[3]{1+x} - 1 = \dfrac{x}{(\sqrt[3]{x+1})^2 + \sqrt[3]{x+1} + 1}$

6(c). Use degree 5 Taylor approximations to e^x and e^{-x}; then $\dfrac{e^x - e^{-x}}{2x} \approx 1 + \dfrac{x^2}{3!} + \dfrac{x^4}{5!}$
Error bounds can be calculated using the error terms for e^x and e^{-x}.

7. Using a Taylor polynomial of degree 4 for $\cos u$ about $u = 0$, obtain $\dfrac{1-\cos(x^2)}{x} = \dfrac{1}{2}x^3 - \dfrac{1}{24}x^7 + \dfrac{1}{720}x^{11}\cos(c_x), \qquad 0 \le c_x \le x^2$

10. The roots are $x_1 = 20 + \sqrt{399}$, $x_2 = 20 - \sqrt{399}$. To calculate x_2, use $\sqrt{399} = 19.975$ and then form $x_2 = \dfrac{1}{20+\sqrt{399}} \doteq \dfrac{1}{39.975} \doteq 0.025016$

19. For IEEE normalized double precision floating-point representation, it underflows if $|x| < 2^{-102.2} \doteq 1.72 \times 10^{-31}$

Section 2.3

1(a). $2.05265 \le x_T + y_T \le 2.05375$

1(d). $\dfrac{8.4725}{0.0645} \le \dfrac{x_T}{y_T} \le \dfrac{8.4735}{0.0635}$

$$131.356 \le \frac{x_T}{y_T} \le 133.441$$

5(a). Use (2.42), with $x_A = 1.473$, $|x_T - x_A| \le 0.0005$. Then $\cos(x_T) - \cos(x_A) = -\sin(c)$ $(x_T - x_A)$, $|\cos(x_T) - \cos(x_A)| \le [\sin(1.4735)]\,(0.0005) < 0.0005$.

5(c). $x_A = 1.4712$, $|x_T - x_A| \le 0.00005$

$$|\log(x_T) - \log(x_A)| = \frac{1}{c}\,|x_T - x_A| \le \frac{0.00005}{1.47115} \doteq 0.000034$$

6. $\text{Rel}(\sqrt{x_A}) \approx \frac{1}{2}\,\text{Rel}(x_A)$ for x_A close to x_T.

11. $f(1+10^{-4}) \approx (-1)^{n-1}(n-1)! \cdot 10^{-4}$, $f(1) = 0$
For $n = 8$, $f(1+10^{-4}) \doteq -0.5040$.

Section 2.4

6. $p - p_n \approx (\epsilon_2 + \cdots + \epsilon_n)p$

CHAPTER 3

Section 3.1

1(a). 1.839287

1(c). 0.424031

3. $r = 0.10134$ or 10.13%

9. $n = 32$

15(a). *Hint:* Consider first the case with $z \geq 1$.

15(c). *Hint:* Use the floating-point representation of z.

Section 3.2

2(a). 1.839286755

2(c). 0.424031039

4. $x_{n+1} = \frac{1}{m}\left[(m-1)x_n + \frac{a}{x_n^{m-1}}\right]$

6.

B	Root
1	−0.5884017765
10	−0.3264020101
50	−0.1832913333

Section 3.3

8(b). $q_{n+1} = q_n + q_{n-1}, \qquad q_0 = q_1 = 1$

Section 3.4

2. Use $[a, b] = [0, 0.5]$. Then $|g'(x)| \leq 0.32$ for $a \leq x \leq b$.

3. Since $e^{-x} > 0$ for all x, the solution α must be positive. Thus considering $x > 0$, we have $0 < g(x) \leq 1$. Then $[a, b] = [0, 1]$ satisfies (3.37) in Lemma 3.4.1, but not (3.38) in Theorem 3.4.2. For the latter, consider

$$a = \min_{0 \leq x \leq 1} g(x) \qquad b = \max_{0 \leq x \leq 1} g(x)$$

5. Use $[a, b] = [c - |d|, c + |d|]$.

8(b). Converges; order of convergence = 2.

11. $c = \frac{1}{2}$ for quadratic convergence.

13(a). The Aitken error estimate is -9.175×10^{-4}.

18. 3

Section 3.5

4. For $\alpha(0) = 3$, $\alpha(\epsilon) \approx 3 - 15.2\epsilon = 3.03$.

7. $\alpha(\epsilon) - 5 \approx 3125\epsilon$, an ill-conditioned problem

8. The root has multiplicity 4.

CHAPTER 4

Section 4.1

1(a). $P_1(x) = 2 - x$

1(b). $f(x) = \left(\frac{1-2e}{1-e}\right) + \left(\frac{1}{1-e}\right)e^x$

2(a). $p(x) = 3 - \cos(\pi x) + 2\sin(\pi x)$

7. $-x^2 + 4x - 3$

8(a). $P_2(x) = P_1(x) = x + 1$

10. $L_0(x) = \dfrac{(x - x_1)(x - x_2)(x - x_3)}{(x_0 - x_1)(x_0 - x_2)(x_0 - x_3)}$

16. $q(x) = 4x^2 - 4x - 1$

19. $f[x_0, x_1] = 2.363200, \qquad f'\left(\dfrac{x_0 + x_1}{2}\right) = 2.363161$

$f[x_0, x_1, x_2] = 1.19349521$

21(a). $p(x) = f(x) - f(x_0)$ is a polynomial of degree m, and $p(x_0) = 0$. Thus, $(x - x_0)$ is a factor of $p(x)$, and $p(x)/(x - x_0)$ is a polynomial of degree $m - 1$.

23(a). $\dfrac{f'(x_1) - 2f[x_0, x_1] + f'(x_0)}{(x_1 - x_0)^2}$

24.

n	x_n	y_n	Dy_n	D^2y_n
0	0.1	0.2		
			0.4	
1	0.2	0.24		1
			0.6	
2	0.3	0.30		

with Dy_n and D^2y_n denoting first- and second-order divided differences. Then $P_1(0.15) = 0.22$, $P_2(0.15) = 0.2175$.

31. Using (4.24), with $m = n$, we get $f[x_0, x_1, \ldots, x_n] = \dfrac{n!}{n!} = 1$.

Section 4.2

1(a). 1.25×10^{-5}

1(b). 6.42×10^{-8}

3(a). 1.07×10^{-5}

3(b). 1.28×10^{-7}

5(a). $h \le 0.00632$, say, $h = 0.006$ or 0.005.

10. $h^4 e/24$

14. $|e^x - P_n(x)| \le \dfrac{e}{(n+1)n^{n+1}}, \qquad 0 \le x \le 1$

Section 4.3

1(c).
$$s(x) = \begin{cases} x^3 + 1 - x, & 0 \le x \le 1 \\ (2-x)^3 + (4x-3) - (2-x), & 1 \le x \le 2 \end{cases}$$

3(c). The equations for $M_i = s''(x_i)$ are

$\frac{1}{3}M_2 + \frac{1}{12}M_3 = 1$

$\frac{1}{12}M_2 + \frac{1}{2}M_3 + \frac{1}{6}M_4 = -\frac{7}{2}$

$\frac{1}{6}M_3 + \frac{2}{3}M_4 = 2$

The solution is

$M_2 = \frac{38}{7}$, $M_3 = -\frac{68}{7}$, $M_4 = \frac{38}{7}$.

7. The equations for $M_i = s''(x_i)$ are

$\frac{1}{3}M_1 + \frac{1}{6}M_2 = \frac{1}{2}$

$\frac{1}{6}M_1 + \frac{2}{3}M_2 + \frac{1}{6}M_3 = \frac{1}{3}$

$\frac{1}{6}M_2 + \frac{2}{3}M_3 + \frac{1}{6}M_4 = \frac{1}{12}$

$\frac{1}{6}M_3 + \frac{1}{3}M_4 = \frac{1}{48}$

The solution is

$M_1 = 173/120, \qquad M_2 = 7/60, \qquad M_3 = 11/120, \qquad M_4 = 1/60$

Using the resulting value of $s(x)$, we obtain

$s\left(\frac{3}{2}\right) \doteq 0.652604 \qquad$ True $f\left(\frac{3}{2}\right) = \frac{2}{3} \doteq 0.666667$

$s\left(\frac{5}{2}\right) \doteq 0.403646 \qquad$ True $f\left(\frac{5}{2}\right) = 0.4$

$s\left(\frac{7}{2}\right) \doteq 0.284896 \qquad$ True $f\left(\frac{7}{2}\right) = \frac{2}{7} \doteq 0.285714$

10. Yes.

15. $\beta = 20$; Not a natural cubic spline.

Section 4.4

2(a). $\max\limits_{-1 \le x \le 1} \left|\tan^{-1}(x) - t_1(x)\right| \doteq 0.215$

$t_3(x) = x - \frac{1}{3}x^3$

2(b). $\max\limits_{-1\le x\le 1} \left|\tan^{-1}(x) - m_3(x)\right| \doteq 0.00495$

4. Use trigonometric identities. For example, $\cos(-x) = \cos(x)$, $\cos(x + 2n\pi) = \cos(x)$, any integer n.

5(a). From (4.82), $\rho_3(f) \le 0.00198$.

5(b). $\max\limits_{0\le x\le \pi/2} |\cos(x) - m_3(x)| \doteq 0.00137$

8(a). Let $\theta = \tan^{-1}(x)$, $x > 0$. Then $0 < \theta < \pi/2$, and $\tan(\theta) = x$. From using a right triangle with angle θ, it is clear that $\tan[(\pi/2) - \theta] = 1/x$, or $(\pi/2) - \theta = \tan^{-1}(1/x)$, as desired.

Section 4.5

1. $T_5(x) = 16x^5 - 20x^3 + 5x$

3. $x_m = \cos(m\pi/n), \qquad m = 0, 1, 2, \dots, n$

6. $\left(\frac{1}{2}\right)^{n-1}$

8(c). $S_{n+1}(x) = 2xS_n(x) - S_{n-1}(x), \qquad n \ge 1$

Section 4.6

2. $x_1 = \cos(\pi/4) = 1/\sqrt{2}, \qquad x_2 = \cos(3\pi/4) = -1/\sqrt{2}$

$$P_1(x) = e^{x_1} + (x - x_1)\frac{e^{x_1} - e^{x_2}}{x_1 - x_2} \approx 1.2606 + 1.0854x$$

3(b). Using the zeros of $T_5(x)$, $P_3(x) \approx 0.9670576x - 0.1871296x^3$
$\max\limits_{-1\le x\le 1} \left|\tan^{-1}(x) - P_3(x)\right| \doteq 0.00590$

4. Use $t = \frac{1}{2}(1 + x)$, $-1 \le x \le 1$, and approximate $f(x) = e^{(1+x)/2}$ on $[-1, 1]$.

7.

n	max error
1	3.72E−2
2	4.37E−3
3	5.72E−4
4	7.94E−5
5	1.14E−5
6	1.69E−6

10. Evaluate $p(-x)$ and equate it to $p(x)$.

Section 4.7

1. $\ell_1(x) = (4e - 10) + x(-6e + 18)$
$\doteq 0.873127 + 1.690309x$

3(a). $\ell_3(x) = \ell_4(x) = x\left(-\frac{15}{2\pi} + \frac{315}{2\pi^3}\right) + x^3\left(\frac{35}{2\pi} - \frac{525}{2\pi^3}\right)$

$\doteq 2.6922925x - 2.8956048x^3$

6. $1 = P_0(x)$
$x = P_1(x)$
$x^2 = \frac{1}{3}(2P_2(x) + P_0(x))$
$x^3 = \frac{1}{5}(2P_3(x) + 3P_1(x))$

7. $p(x) = \left(a_0 + \frac{1}{3}a_2\right)P_0(x) + a_1P_1(x) + \frac{2}{3}a_2P_2(x)$

CHAPTER 5

Section 5.1

1. $T_4(f) \doteq 0.697024$ $\quad I - T_4(f) \doteq -0.00388$
$S_4(f) \doteq 0.693254$ $\quad I - S_4(f) \doteq -0.000107$

2(a).

n	T_n	Error	Ratio
2	26.516336	−25.2	
4	3.2490505	−1.95	13.0
8	1.6245252	−3.22E − 1	6.04
16	1.3757225	−7.33E − 2	4.39
32	1.3203119	−1.79E − 2	4.09
64	1.3068479	−4.45E − 3	4.02
128	1.3035057	−1.11E − 3	4.01
256	1.3026716	−2.78E − 4	4.00
512	1.3024632	−6.95E − 5	4.00

3(a).

n	S_n	Error	Ratio
2	22.715077371	−21.4	
4	−4.5067112930	5.81	−3.69
8	1.0830168315	2.19E − 1	26.5
16	1.2927882745	9.61E − 3	22.8
32	1.3018416653	5.52E − 4	17.4
64	1.3023598879	3.38E − 5	16.3
128	1.3023915828	2.10E − 6	16.1
256	1.3023935531	1.31E − 7	16.0
512	1.3023936761	8.20E − 9	16.0

10(b). $B_4(f) \doteq 0.69317460$, $I - B_4(f) \doteq -0.0000274$

13. 2

Section 5.2

1(a). $\pi h^2/24$

1(b). $h^2/6$

3(a). $\tilde{E}_n^T(f) = \dfrac{-h^2}{12}[e^\pi - 1]$

$\tilde{E}_{32}^T(f) \doteq -0.01778$. True $E_{32}^T(f) \doteq -0.0179$

6(a). $\tilde{E}_n^S(f) = \dfrac{13h^4}{1215}, \qquad h = 2/n$

Choose $h \le 0.031$, or $n \ge 65$. The number of nodes $= n + 1 \ge 66$.

6(d). $n \ge 396$ for Simpson's rule.

13(b). Use $\dfrac{I_{2n} - I_n}{I_{4n} - I_{2n}} = \dfrac{(I - I_n) - (I - I_{2n})}{(I - I_{2n}) - (I - I_{4n})}$.

Apply (5.39) and simplify.

14(a). $p = 1.5$

17(a). From Problem 2(a) in Section 5.1, with $n = 32$, $I - T_{32} \approx \frac{1}{3}(T_{32} - T_{16}) \doteq -0.01847$
True $I - T_{32} \doteq -0.0179$.

Section 5.3

1. $I_3 \doteq 2.3503369$, $I - I_3 \doteq 0.0000655$

2(a).

n	I_n	Error
2	−19.244871326	20.5
3	9.0897287084	−7.79
4	0.9220525215	3.80E − 1
5	1.1462063447	1.56E − 1
6	1.3298098510	−2.74E − 2
7	1.3003438200	2.05E − 3
8	1.3024781041	−8.44E − 5
9	1.3023918529	1.83E − 6
10	1.3023936884	−4.10E − 9

10(a). $\displaystyle\int_0^1 f(x)\log\left(\tfrac{1}{x}\right)dx \approx f\left(\tfrac{1}{4}\right) \equiv I_1$

10(b). $w_1 + w_2 = 1$

$w_1x_1 + w_2x_2 = \frac{1}{4}$

$w_1x_1^2 + w_2x_2^2 = \frac{1}{9}$

$w_1x_1^3 + w_2x_2^3 = \frac{1}{16}$

10(c). $I_1 \doteq 0.968912$, error $\doteq -0.0228$

Section 5.4

1(a).

h	$D_h f(0)$	Error Estimate	Error	Ratio
0.1	1.051709	$-5.00E-2$	$-5.17E-2$	
0.05	1.025422	$-2.50E-2$	$-2.54E-2$	2.03
0.025	1.012605	$-1.25E-2$	$-1.26E-2$	2.02
0.0125	1.006276	$-6.25E-3$	$-6.28E-3$	2.01
0.00625	1.003132	$-3.12E-3$	$-3.13E-3$	2.00

3(a).

h	$D_h f(0)$	Error Estimate	Error	Ratio
0.1	1.00166775	$-1.67E-3$	$-1.67E-3$	
0.05	1.00041672	$-4.17E-4$	$-4.17E-4$	4.00
0.025	1.00010417	$-1.04E-4$	$-1.04E-4$	4.00
0.0125	1.00002604	$-2.60E-5$	$-2.60E-5$	4.00
0.00625	1.00000651	$-6.51E-6$	$-6.51E-6$	4.00

5. $f'(x_2) \approx D_h f(x_2) \equiv (1/12h)\{f(x_0) - 8f(x_1) + 8f(x_3) - f(x_4)\}$

$f'(x_2) - D_h f(x_2) = (h^4/30) f^{(5)}(c),\ x_0 \le c \le x_4$

7. Denote $\hat{D}_h f(x) = 2D_h f(x) - D_{2h} f(x)$, with $D_h f(x)$ defined by (5.76).

h	$D_h f(x)$	$\hat{D}_h f(x)$	$f'(x) - \hat{D}_h f(x)$
0.1	-0.54243		
0.05	-0.52144	-0.50045	0.00045
0.025	-0.51077	-0.50010	0.00010
0.0125	-0.50540	-0.50003	0.00003

10(b). $f''(t) \approx \dfrac{1}{h^2}[-f(t+3h) + 4f(t+2h) - 5f(t+h) + 2f(t)]$

12. $|f'(x) - \tilde{D}f(x)| \le \dfrac{h^2}{6}\left|f'''(c)\right| + \dfrac{\delta}{h},\ x-h \le c \le x+h$, with $\tilde{D}f(x)$ the version of $D_h f(x)$ with rounded entries, and $\delta = \max\{|\epsilon_0|, |\epsilon_1|\}$.

14. With $h = 0.2$, $D_h^{(2)} f(0.5) \doteq 0.0775$. The part of the error due to rounding in table entries is $\dfrac{4\delta}{h^2} = \dfrac{4(0.00005)}{(0.2)^2} = 0.005$.

CHAPTER 6

Section 6.1

1(a). Write $p(x) = a_0 + a_1 x + a_2 x^2 + a_3 x^3$. Then a_0, a_1, a_2, a_3 satisfy $a_0 + a_1 x_i + a_2 x_i^2 + a_3 x_i^3 = y_i, \qquad i = 0, 1, 2, 3$

4.
$$3x_1 + x_2 = 4$$
$$x_{i-1} + 3x_i + x_{i+1} = 5, \qquad i = 2, 3, \ldots, n-1$$
$$x_{n-1} + 3x_n = 4$$

5.
```
sign = 1;
b = zeros(n,1)
A = zeros(n,n);
for i=1:n
   b(i) = sign;
   sign = -sign;
A(i,1:i) = (1/i)*[1:i];
A(1:i-1,i) = A(i,1:i-1)';
end
```

Section 6.2

1(a). $\begin{bmatrix} 5 & 6 \\ -1 & 6 \end{bmatrix}$

1(b). $\begin{bmatrix} a & b+2c \\ 0 & 3c \end{bmatrix}$

1(c). $A = \begin{bmatrix} \frac{7}{9} & -\frac{4}{9} & -\frac{4}{9} \\ -\frac{4}{9} & \frac{1}{9} & -\frac{8}{9} \\ -\frac{4}{9} & -\frac{8}{9} & \frac{1}{9} \end{bmatrix}, \qquad B = I$

2. Write $A = [A_{*1}, A_{*2}, A_{*3}]$, with A_{*j} denoting the jth column of A, a column matrix. Then $AD = [\lambda_1 A_{*1}, \lambda_2 A_{*2}, \lambda_3 A_{*3}]$. Column j of A is multiplied by λ_j.

5. $A = ww^T$,
$A^2 = (ww^T)(ww^T) = w(w^T w)w^T = w(1)w^T = ww^T = A$

8. mnp

17. $a = d, b = c$

28. $[1, 1, -2, -2]^T$

Section 6.3

1(a).
$$\left[\begin{array}{rrr|r} 2 & 1 & -1 & 6 \\ 4 & 0 & -1 & 6 \\ -8 & 2 & 2 & -8 \end{array}\right] \xrightarrow[m_{21}=2,\, m_{31}=-4]{} \left[\begin{array}{rrr|r} 2 & 1 & -1 & 6 \\ 0 & -2 & 1 & -6 \\ 0 & 6 & -2 & 16 \end{array}\right]$$
$$\xrightarrow[m_{32}=-3]{} \left[\begin{array}{rrr|r} 2 & 1 & -1 & 6 \\ 0 & -2 & 1 & -6 \\ 0 & 0 & 1 & -2 \end{array}\right]$$
$x_3 = -2, x_2 = 2, x_1 = 1$

4. If the third equation is a multiple of the second, then there are an infinite number of solutions. Solve for x_3 and then substitute into the first equation, to determine the relation between x_1 and x_2. If the third equation is not a multiple of the second, then the system is inconsistent and there is no solution. To better understand the possible behavior, try particular systems of the given special form.

6(a). $\begin{bmatrix} 0.4 & -0.1 & -0.1 \\ -0.1 & 0.4 & -0.1 \\ -0.1 & -0.1 & 0.4 \end{bmatrix}$

6(d). $\begin{bmatrix} -0.5 & -0.5 & -0.5 & -0.5 \\ -0.5 & -1.5 & -1.5 & -1.5 \\ -0.5 & -1.5 & -2.5 & -2.5 \\ -0.5 & -1.5 & -2.5 & -3.5 \end{bmatrix}$

7. $MD(A \to U) = \frac{1}{2}n(n+1)$
$MD(b \to g) = n - 1$
$MD(g \to x) = \frac{1}{2}n(n+1)$

Section 6.4

1. $L = \begin{bmatrix} 1 & 0 & 0 \\ 2 & 1 & 0 \\ -1 & \frac{1}{2} & 1 \end{bmatrix}, \qquad U = \begin{bmatrix} 1 & 2 & 1 \\ 0 & -2 & 1 \\ 0 & 0 & \frac{1}{2} \end{bmatrix}$

2(a). $L = \begin{bmatrix} 1 & 0 & 0 \\ 2 & 1 & 0 \\ -4 & -3 & 1 \end{bmatrix}, \qquad U = \begin{bmatrix} 2 & 1 & -1 \\ 0 & -2 & 1 \\ 0 & 0 & 1 \end{bmatrix}$

6(a). $\begin{bmatrix} 1 & 0 \\ -1 & 2 \end{bmatrix}$

This is not the only possible value for L.

8. $[\frac{5}{6}, -\frac{2}{3}, \frac{1}{2}, -\frac{1}{3}, \frac{1}{6}]^T$

11(a). $MD(A \to LU) = 2n - 2$

11(b). $MD(x) = 3n - 2$

Section 6.5

3(a). $\text{cond}(A) = 289$

5. $\text{cond}(A) = \dfrac{(1 + |c|)^2}{|1 - c^2|}$

6. Use $I = AA^{-1}$ and property (P3) in (6.88).

7(b). $\text{cond}(A_n) = n2^{n-1}$

Section 6.6

2. For Gauss–Jacobi iteration, the ratios approach 0.4069, and $\|M\| = 1/2$. For Gauss–Seidel, the ratios do not converge; but $\|M\| = 1/3$.

4(a). $\|M\| = 1/2$

4(b). $\|M\| = 1/2$

9. For Gauss–Jacobi, the ratios alternate between 0 and $1/2$.

14. $|\epsilon| \le \sqrt{2}$

CHAPTER 7

Section 7.1

1. $f^*(x) \approx 2.0810909x - 1.0837273$, $E \doteq 0.236$

3. $f^*(x) \approx 7.3301079x^2 - 2.2158961x - 0.98132527$

4.
$$\begin{bmatrix} 1 & \frac{1}{2} & \frac{1}{3} \\ \frac{1}{2} & \frac{1}{3} & \frac{1}{4} \\ \frac{1}{3} & \frac{1}{4} & \frac{1}{5} \end{bmatrix} \begin{bmatrix} a_1 \\ a_2 \\ a_3 \end{bmatrix} = \begin{bmatrix} w_1 \\ w_2 \\ w_3 \end{bmatrix}$$

$$w_i = \int_0^1 x^{i-1} f(x)\, dx, \qquad i = 1, 2, 3$$

10.
$$na + b \sum_{j=1}^{n} e^{-x_j} = \sum_{j=1}^{n} y_j$$

$$a \sum_{j=1}^{n} e^{-x_j} + b \sum_{j=1}^{n} e^{-2x_j} = \sum_{j=1}^{n} y_j e^{-x_j}$$

Section 7.2

1(a). $f(\lambda) = \lambda^2 - 5\lambda - 6$, $\lambda_1 = 6$, $\lambda_2 = -1$
$v^{(1)} = [2, 5]^T$, $v^{(2)} = [1, -1]^T$

2. $v^{(2)} = [1, 0, -1]^T$, $v^{(3)} = [1, 2, 1]^T$

12.
$$U = \begin{bmatrix} \frac{1}{\sqrt{3}} & \frac{1}{\sqrt{2}} & \frac{1}{\sqrt{6}} \\ -\frac{1}{\sqrt{3}} & 0 & \frac{2}{\sqrt{6}} \\ \frac{1}{\sqrt{3}} & -\frac{1}{\sqrt{2}} & \frac{1}{\sqrt{6}} \end{bmatrix}$$

14. In both cases, $f(\lambda) = \lambda^2 - 2\lambda + 1 = (\lambda - 1)^2$.

17(b). $\lambda_1 = -12$, $\lambda_2 = 9$, $\lambda_3 = 3$, $v^{(1)} = [1, -1, 1]^T$

17(d). $\lambda_1 = 3$, $\lambda_2 = 2$, $\lambda_3 = 1$, $v^{(1)} = [-4, 3, 1]^T$

17(f). $\lambda_1 = 15$, $\lambda_2 = 5$, $\lambda_3 = 5$, $\lambda_4 = -1$, $v^{(1)} = [1, 1, 1, 1]^T$

Section 7.3

1. Another root is $\alpha \doteq (2.1372167369636,\ 1.0526519628114)$.

2(a). The four exact roots are $(\pm\sqrt{2.5}, \pm\sqrt{1.5})$.

2(b). One of the roots is $(1.7701689217513,\ 0.4654304428338)$.

CHAPTER 8

Section 8.1

1(a). $y(x) = \sin(x) + e^{-x}$

1(b). $y(x) = x/(x - 0.5)$

3(a). $y(x) = ce^{\lambda x} + \dfrac{1}{\lambda}[e^{\lambda x} - 1]$, general solution.

$y(x) = [1 + (1/\lambda)]e^{\lambda x} - 1/\lambda$

4(a). $y = \sqrt{x^2 + 4}$

5(a). $\partial f/\partial z = -1$, equation well-conditioned.

Section 8.2

1(a).

x	$Y(x)$	$Y(x) - y_{0.2}(x)$	$Y(x) - y_{0.1}(x)$	$Y(x) - y_{0.05}(x)$
2	1.10715	−3.32E − 2	−1.63E − 2	−8.11E − 3
4	1.32582	−1.67E − 2	−8.34E − 3	−4.17E − 3
6	1.40565	−9.70E − 3	−4.86E − 3	−2.44E − 3
8	1.44644	−6.38E − 3	−3.20E − 3	−1.60E − 3
10	1.47113	−4.54E − 3	−2.28E − 3	−1.14E − 3

2(c).

x	$Y(x)$	$Y(x) - y_{0.5}(x)$	$Y(x) - y_{0.0625}(x)$
2	0.49315	2.23E − 2	−4.44E − 3
4	−1.41045	1.34E + 0	8.67E − 3
6	0.68075	6.39E + 0	−2.78E − 3
8	0.84386	3.25E + 1	−6.36E − 3
10	−1.38309	1.65E + 2	8.07E − 3

Section 8.3

3. $K = \sup_{0 \le x \le b} (1/(x+1)) = 1, \qquad$ for any $b \ge 0$

$\max_{0 \le x \le b} |Y''(x)| = \max_{0 \le x \le b} (2x/(x+1)) = 2b/(b+1)$

$|Y(b) - y_h(b)| \le h \left[\dfrac{e^b - 1}{2}\right]\left[\dfrac{2b}{b+1}\right]$

For $h = 0.05$,

b	$Y(b) - y_h(b)$	Error Bound
1	0.0187	0.0430
2	0.0634	0.213
3	0.125	0.716
4	0.199	2.14

5(b). Use $K = \max_{0 \le x \le b} |4Y(x)| = 2, \qquad$ for $b \ge 1$.

$\max_{0 \le x \le b} \left|Y''(x)\right| = \max_{0 \le x \le b} \left|\dfrac{-6x + 2x^3}{(1+x^2)^3}\right| \doteq 1.46, \qquad$ for $b \ge 0.41$

7. Solve $D'(x) = \lambda D(x) - \frac{1}{2}\sin(x), \qquad D(0) = 0.$
The solution $D(x) = \left\{[\lambda \sin(x) + \cos(x)] - e^{\lambda x}\right\}/[2(\lambda^2 + 1)]$ and
$Y(x_n) - y_h(x_n) \approx D(x_n)h.$
For $\lambda = -1$ and $h = 0.1$, the answers are as follows:

x	$D(x)h$	$Y(x) - y_n(x)$
1	−0.0167	−0.0169
2	−0.0365	−0.0372
3	−0.0295	−0.0299
4	0.00212	0.00258
5	0.0309	0.0319
6	0.0309	0.0316

10(a). For the case of $2h = 0.1$ $h = 0.05$,

x	$Y(x) - y_h(x)$	$y_h(x) - y_{2h}(x)$	$\tilde{y}_h(x)$	$Y(x) - \tilde{y}_h(x)$
2	−8.11E − 3	−8.22E − 3	1.1070296	1.19E − 4
4	−4.17E − 3	−4.16E − 3	1.3258165	1.12E − 6
6	2.44E − 3	−2.43E − 3	1.4056549	−7.28E − 6
8	−1.60E − 3	−1.60E − 3	1.4464472	−5.84E − 6
10	−1.14E − 3	−1.14E − 3	1.4711317	−4.06E − 6

Section 8.4

7. We give the values obtained with the backward Euler method when $\lambda = -50$. Compare these with the values for $\lambda = -1, -10$ for the same values of h. Note that the errors decrease by a factor of about 10 when h decreases from 0.1 to 0.01; this is consistent with the method being of order 1 in its convergence rate.

	Errors $Y(x) - y_h(x)$		
x	$h = 0.5$	$h = 0.1$	$h = 0.01$
2	1.01E − 3	1.70E − 4	1.64E − 5
4	1.59E − 4	2.88E − 5	2.81E − 6
6	4.82E − 5	9.00E − 6	8.87E − 7
8	2.03E − 5	3.86E − 6	3.82E − 7
10	1.04E − 5	1.99E − 6	1.97E − 7

Section 8.5

2(a). Following are some results for $h = 0.05$ and $2h = 0.1$. The Richardson error estimate (8.76) is also included.

x	$Y(x) - y_{2h}(x)$	$Y(x) - y_h(x)$	Ratio	$\frac{1}{3}[y_h(x) - y_{2h}(x)]$
2	2.05E − 4	5.12E − 5	4.0	5.12E − 5
4	1.75E − 4	4.28E − 5	4.1	4.40E − 5
6	1.01E − 4	2.48E − 5	4.1	2.70E − 5
8	6.40E − 5	1.57E − 5	4.1	1.61E − 5
10	4.37E − 5	1.07E − 5	4.1	1.10E − 5

5.

x	$Y(x) - y_h(x)$	$\frac{1}{3}[y_h(x) - y_{2h}(x)]$	$\tilde{y}_h(x)$	$Y(x) - \tilde{y}_h(x)$
2	2.31E − 4	2.31E − 4	0.4931513	−7.24E − 7
4	2.91E − 4	3.07E − 4	−1.4104300	−1.61E − 5
6	−4.08E − 4	−4.19E − 4	0.6807432	1.16E − 5
8	5.68E − 5	5.08E − 5	0.8438522	6.05E − 6
10	3.62E − 4	3.78E − 4	−1.3830760	−1.67E − 5

6. With $\gamma_2 = 1$, the Runge–Kutta method is $y_{n+1} = y_n + hf\left(x_n + \frac{h}{2}, y_n + \frac{h}{2}f(x_n, y_n)\right)$

7. Using the formula for $\gamma_2 = 1$ in Problem 6, we obtain the error results below for $h = 0.05,\ 2h = 0.1$. Richardson's error estimate is included.

x	$Y(x) - y_{2h}(x)$	$Y(x) - y_h(x)$	Ratio	$\frac{1}{3}[y_h(x) - y_{2h}(x)]$
2	1.46E − 3	3.53E − 4	4.1	3.69E − 4
4	−6.68E − 4	−1.67E − 4	4.0	−1.67E − 4
6	−8.31E − 4	−1.97E − 4	4.2	−2.11E − 4
8	1.37E − 3	3.33E − 4	4.1	3.46E − 4
10	−3.08E − 4	−8.01E − 5	3.8	−7.60E − 5

8(a). For $h = 0.05$ and $2h = 0.1$,

x	$Y(x) - y_{2h}(x)$	$Y(x) - y_h(x)$	Ratio	$\frac{1}{3}[y_h(x) - y_{2h}(x)]$
2	6.46E − 4	1.56E − 4	4.1	1.63E − 4
4	2.51E − 4	6.10E − 5	4.1	6.34E − 5
6	1.26E − 4	3.07E − 5	4.1	3.18E − 5
8	7.49E − 5	1.82E − 5	4.1	1.89E − 5
10	4.94E − 5	1.20E − 5	4.1	1.25E − 5

9. The results for $h = 0.125$ and $2h = 0.25$ are given below, along with Richardson's error estimate for $y_h(x)$, taken from Problem 4.

x	$Y(x) - y_{2h}(x)$	$Y(x) - y_h(x)$	Ratio	$\frac{1}{15}[y_h(x) - y_{2h}(x)]$
2	3.02E − 5	1.69E − 6	18	1.90E − 6
4	−3.64E − 5	−2.20E − 6	17	−2.28E − 6
6	−1.44E − 6	3.41E − 8	−42*	−9.85E − 8
8	3.74E − 5	2.16E − 6	17	2.35E − 6
10	−2.97E − 5	−1.83E − 6	16	−1.86E − 6

*The Richardson extrapolation is not justified, since Ratio < 0.

13(a). The true solution is $v(t) = -98[1 - e^{-0.1t}]$, $t \geq 0$.

Section 8.6

1(a). Let $h = 0.05$, $2h = 0.1$.

x	$Y(x) - y_{2h}(x)$	$Y(x) - y_h(x)$	Ratio	$\frac{1}{3}[y_h(x) - y_{2h}(x)]$
2	5.10E − 4	1.28E − 4	4.0	1.27E − 4
4	4.37E − 4	1.07E − 4	4.1	1.10E − 4
6	2.54E − 4	6.23E − 5	4.1	6.40E − 5
8	1.61E − 4	3.93E − 5	4.1	4.04E − 5
10	1.09E − 4	2.68E − 5	4.1	2.76E − 5

2(a). Let $h = 0.05$, $2h = 0.1$. Richardson's error formula is from Problem 4 of Section 8.5. The order of convergence is $p = 3$.

x	$Y(x) - y_{2h}(x)$	$Y(x) - y_h(x)$	Ratio	$\frac{1}{7}[y_h(x) - y_{2h}(x)]$
2	1.27E − 4	1.49E − 5	8.5	1.60E − 5
4	5.17E − 6	6.53E − 7	7.9	6.45E − 7
6	−1.55E − 6	−1.65E − 7	9.4	−1.98E − 7
8	−1.72E − 6	−1.94E − 7	8.9	−2.17E − 7
10	−1.36E − 6	−1.56E − 7	8.7	−1.72E − 7

4(a). Let $h = 0.05$, $2h = 0.1$.

x	$Y(x) - y_{2h}(x)$	$Y(x) - y_h(x)$	Ratio	$\frac{1}{3}[y_h(x) - y_{2h}(x)]$
2	−1.47E − 4	−3.14E − 5	4.7	−3.86E − 5
4	−1.08E − 4	−2.40E − 5	4.5	−2.80E − 5
6	−6.11E − 5	−1.37E − 5	4.5	−1.58E − 5
8	−3.81E − 5	−8.61E − 6	4.4	−9.84E − 6
10	−2.58E − 5	−5.85E − 6	4.4	−6.66E − 6

6(a). The errors for the three stepsizes are given below. Sharply differing values of the errors are due to different stability regions of the methods.

x	$Y(x) - y_{0.1}(x)$	$Y(x) - y_{0.02}(x)$	$Y(x) - y_{0.01}(x)$
2	3.82E + 13	1.37E − 4	1.09E − 6
4	2.06E + 30	1.33E − 4	−5.54E − 8
6	1.11E + 47	1.29E − 4	−1.05E − 6
8	5.98E + 63	1.37E − 4	9.27E − 7
10	3.23E + 80	1.34E − 4	2.76E − 7

Section 8.7

1.

$$Y_1' = Y_1 - 2Y_2 - 2e^{-x} + 2, \qquad Y_1(0) = 1$$
$$Y_2' = 2Y_1 - Y_2 - 2e^{-x} + 1, \qquad Y_2(0) = 1$$

3(b). Let $Y_1 = Y,\ Y_2 = Y'$. Then

$$\mathbf{Y}' = \begin{bmatrix} 0 & 1 \\ -13 & -4 \end{bmatrix}\mathbf{Y} + \begin{bmatrix} 0 \\ 40\cos(x) \end{bmatrix}, \qquad \mathbf{Y}(0) = \begin{bmatrix} 3 \\ 4 \end{bmatrix}$$

7(b). Use the notation of 3(b). The error estimate is

$$y(x) - y_h(x) \approx y_h(x) - y_{2h}(x)$$

Errors in $Y_1(x) = y(x)$

x	$h = 0.1$	$h = 0.05$	$h = 0.025$	Error estimate for $h = 0.025$
2	−2.13E − 2	−5.12E − 3	−1.65E − 3	−3.47E − 3
4	4.03E − 2	2.09E − 2	1.06E − 2	1.03E − 2
6	−5.14E − 2	−2.54E − 2	−1.26E − 2	−1.28E − 2
8	4.17E − 4	−3.74E − 4	−3.31E − 4	−4.35E − 5
10	5.10E − 2	2.57E − 2	1.29E − 2	1.28E − 2

8(b). Use the notation of 3(b). The error estimate is

$$y(x) - y_h(x) \approx \frac{1}{3}[y_h(x) - y_{2h}(x)]$$

Errors in $Y_1(x) = y(x)$				
x	$h = 0.1$	$h = 0.05$	$h = 0.025$	Error estimate for $h = 0.025$
2	8.20E − 4	1.74E − 4	4.06E − 5	4.44E − 5
4	−7.97E − 3	−2.02E − 3	−5.04E − 4	−5.06E − 4
6	4.77E − 3	1.32E − 3	3.44E − 4	3.27E − 4
8	3.95E − 3	9.06E − 4	2.15E − 4	2.30E − 4
10	−8.05E − 3	−2.08E − 3	−5.23E − 4	−5.19E − 4

9(a). $c_1 = 3, c_2 = 2$

Section 8.8

3(a). $Y(x) = \dfrac{e^{2-x} - e^x}{e^2 - 1}$

3(b). $Y(x) = \dfrac{e^{2-x} + e^x}{e^2 + 1}$

3(c). $Y(x) = \dfrac{e^x - e^{2-x}}{e^2 + 1}$

7. The true solution is $Y(x) = x\,e^x$. Numerical errors $Y(x) - y_h(x)$ at selected points are listed in the following table.

x	$h = 1/10$	$h = 1/20$	$h = 1/40$
1.2	3.40E − 3	8.43E − 4	2.11E − 4
1.4	6.30E − 3	1.57E − 3	3.91E − 4
1.6	7.71E − 3	1.92E − 3	4.78E − 4
1.8	6.24E − 3	1.55E − 3	3.87E − 4

CHAPTER 9

Section 9.1

5. With $n_x = n_y = 5$, the numerical values are

	$x = 1/5$	$x = 2/5$	$x = 3/5$	$x = 4/5$
$y = 1/5$	1.1161	1.1558	1.1776	1.1906
$y = 2/5$	1.2685	1.3294	1.3639	1.3848
$y = 3/5$	1.4685	1.5294	1.5639	1.5848
$y = 4/5$	1.7161	1.7558	1.7776	1.7906

7. The maximum numerical solution errors are

n	Max Error	Ratio
4	2.1761E − 1	
8	5.8340E − 2	3.73
16	1.4843E − 2	3.93
32	3.7087E − 3	4.00

8. For $2 \le i,\, j \le n$, the difference equations are

$$u_{i,j} = \frac{1}{4 - a\,h^2}\,(u_{i+1,j} + u_{i-1,j} + u_{i,j+1} + u_{i,j-1}) - \frac{h^2}{4 - a\,h^2} f_{ij}$$

Section 9.2

4. Maximum numerical solution errors of the explicit scheme are

	$n_t = 5$	$n_t = 10$	$n_t = 20$	$n_t = 40$
$n_x = 5$	2.0660E − 3	3.5877E − 3	4.3348E − 3	4.7049E − 3
$n_x = 10$	1.9646E − 3	2.8996E − 4	5.3178E − 4	9.3886E − 4
$n_x = 20$	3.0090E − 3	1.3153E − 3	4.8428E − 4	7.2630E − 5
$n_x = 40$	3.2707E − 3	1.5721E − 3	1.4839E − 2	2.3863E − 3
$n_x = 80$	3.3365E − 3	5.5905E + 0	2.1356E + 11	3.8206E + 26

Maximum numerical solution errors of the implicit scheme are

	$n_t = 5$	$n_t = 10$	$n_t = 20$	$n_t = 40$
$n_x = 5$	7.9392E − 3	6.5229E − 3	5.8022E − 3	5.4386E − 3
$n_x = 10$	4.4933E − 3	2.9373E − 3	2.1452E − 3	1.7455E − 3
$n_x = 20$	3.5212E − 3	1.9481E − 3	1.1472E − 3	7.4308E − 4
$n_x = 40$	3.2776E − 3	1.7002E − 3	8.9713E − 4	4.9189E − 4
$n_x = 80$	3.2167E − 3	1.6382E − 3	8.3457E − 4	4.2906E − 4

9. We use the implicit scheme and start with $n_t = n_x = 4$. Since the scheme is of first order in time and of second order in space, we consecutively double n_x and quadruple n_t. The sequence of the numerical values is ($\times 10^{-3}$):

$$6.1552,\ 7.2695,\ 7.4874,\ 7.5365,\ 7.5485,\ 7.5514,\ 7.5522$$

The three stabilized digits are 7, 5, 5, and the last number of the sequence corresponds to $n_x = 256$, $n_t = 16{,}384$.

Section 9.3

4. A few numerical solution values at $t = 1$:

	$x = 1/5$	$x = 2/5$	$x = 3/5$	$x = 4/5$
$n_x = n_t = 5$	-0.4420	-0.8776	-1.0394	-0.6661
$n_x = n_t = 10$	-0.4656	-0.9463	-1.0068	-0.6635
$n_x = n_t = 20$	-0.4721	-0.9624	-1.0051	-0.6608

7(a). With $n_x = n_t = 5$, numerical solution values ($t \leq 1$) are given in the following table, and the graph of the approximate vertical displacement is shown in the figure.

	$x = 1/5$	$x = 2/5$	$x = 3/5$	$x = 4/5$
$t = 0.2$	0.0118	0.0190	0.0190	0.0118
$t = 0.4$	0.0425	0.0688	0.0688	0.0425
$t = 0.6$	0.0806	0.1304	0.1304	0.0806
$t = 0.8$	0.1114	0.1802	0.1802	0.1114
$t = 1$	0.1231	0.1992	0.1992	0.1231

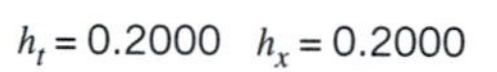

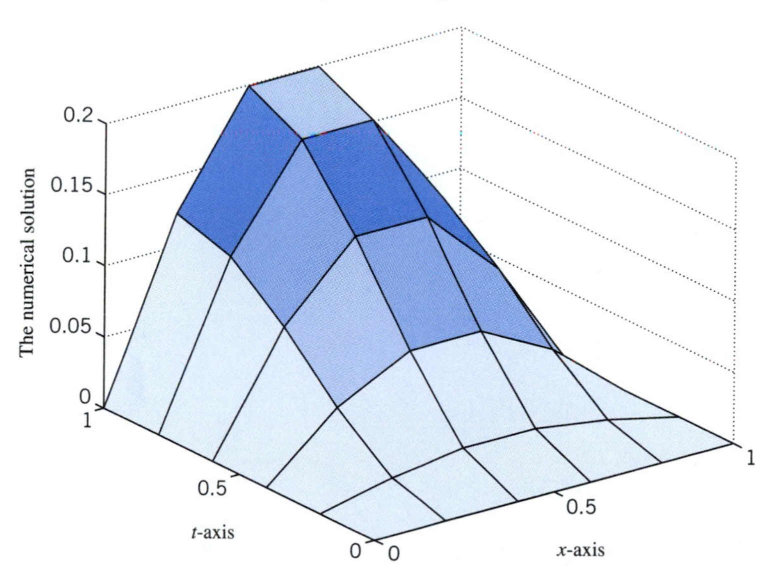

Approximate vertical displacement

APPENDIX E

1(b). $(5.8125)_{10}$

3(b). $(100.01)_2$

3(d). $\pi \doteq (11.001001)_2$

4. $(0.8)_{10} = (0.110011001100\ldots)_2$

5. 6/7

BIBLIOGRAPHY

[1] M. Abramowitz and I. Stegun. *Handbook of Mathematical Functions*, Dover Press, 1964, New York.

[2] M. Allen III and E. Isaacson. *Numerical Analysis for Applied Science*, John Wiley, 1998, New York.

[3] E. Anderson, Z. Bai, C. Bischof, J. Demmel, J. Dongarra, J. DuCroz, A. Greenbaum, S. Hammarling, A. McKenney, S. Ostrouchov, and D. Sorenson. *LAPACK Users' Guide*, SIAM, 1992, Philadelphia.

[4] U. Ascher, R. Mattheij, and R. Russell. *Numerical Solution of Boundary Value Problems for Ordinary Differential Equations*, Prentice-Hall, 1988, Englewood Cliffs, New Jersey.

[5] K. Atkinson. *An Introduction to Numerical Analysis*, 2nd ed., John Wiley, 1993, New York.

[6] O. Axelsson. *Iterative Solution Methods*, Cambridge University Press, 1994, Cambridge, U.K.

[7] A. Bjorck. *Numerical Methods for Least Squares Problems*, SIAM, 1996, Philadelphia.

[8] P. Bogacki and L. Shampine. A 3(2) Pair of Runge–Kutta Formulas, *Appl. Math. Lett.* **2** (1989), 321–325.

[9] C. de Boor. *A Practical Guide to Splines*, Springer-Verlag, 1978, New York.

[10] C. Chui. *An Introduction to Wavelets*, Academic Press, 1992, Burlington, Massachusetts.

[11] P. Davis and P. Rabinowitz. *Methods of Numerical Integration*, 2nd ed., Academic Press, New York.

[12] J. Demmel. *Applied Numerical Linear Algebra*, SIAM, 1997, Philadelphia.

[13] J. Dongarra, I. Duff, D. Sorensen, and H. van der Vorst. *Numerical Linear Algebra for High-Performance Computers*, SIAM, 1998, Philadelphia.

[14] J. Dormand and P. Prince. A Family of Embedded Runge–Kutta Formulae, *J. Comp. Appl. Math* **6** (1980), 19–26.

[15] W. Gander and J. Hrebicek. *Solving Problems in Scientific Computing Using Maple and Matlab*, 3rd ed., Springer, 1997, New York.

[16] W. Gautschi. *Numerical Analysis: An Introduction*, Birkhäuser, 1997, Boston.

[17] G. Golub and C. Van Loan. *Matrix Computations*, 3rd ed., John Hopkins University Press, 1996, Baltimore.

[18] D. Higham and N. Higham. *Matlab Guide*, SIAM, 2000, Philadelphia.

[19] A. Householder. *The Numerical Treatment of a Single Nonlinear Equation*, McGraw-Hill, 1970, New York.

[20] A. Iserles. *A First Course in the Numerical Analysis of Differential Equations*, Cambridge University Press, 1996, Cambridge, U.K.

[21] C. Kelley, *Iterative Methods for Linear and Nonlinear Equations*, SIAM, 1995, Philadelphia.

[22] R. Kress. *Numerical Analysis*, Springer-Verlag, 1998, New York.

[23] A. Krommer and C. Ueberhuber. *Computational Integration*, SIAM, 1998, Philadelphia.

[24] P. Marchand. *Graphics and GUIs with MATLAB*, 2nd ed., CRC Press, 1999, Boca Raton, Florida.

[25] K. W. Morton and D. F. Mayers. *Numerical Solution of Partial Differential Equations: An Introduction*, Cambridge University Press, 1994, Cambridge, U.K.

[26] J. Nocedal and S. Wright. *Numerical Optimization*, Springer-Verlag, 1999, New York.

[27] M. Overton. *Numerical Computing with IEEE Floating Point Arithmetic*, SIAM, 2001, Philadelphia.

[28] B. Parlett. *The Symmetric Eigenvalue Problem*, Classics in Applied Mathematics, SIAM, 1998, Philadelphia.

[29] M. Powell. *Approximation Theory and Methods*, Cambridge University Press, 1981, Cambridge, U.K.

[30] A. Quarteroni, R. Sacco, and F. Saleri. *Numerical Mathematics*, Springer-Verlag, 2000, New York.

[31] L. Shampine. *Numerical Solution of Ordinary Differential Equations*, Chapman & Hall, 1994, New York.

[32] A. Stroud and D. Secrest. *Gaussian Quadrature Formulas*, Prentice-Hall, 1966, Englewood Cliffs, New Jersey.

[33] L. Trefethen and D. Bau. *Numerical Linear Algebra*, SIAM, 1997, Philadelphia.

INDEX